Act 2000

M. S. Welling

Dictionary of
Plastics and Rubber Technology
German/English

Wörterbuch
Kunststoff- und Kautschuktechnologie
Deutsch/Englisch

Dedicated, with gratitude, to the memory of my father – and to my wife, as always.

To deliberate whenever I doubted, to enquire whenever I was ignorant, would have protracted the undertaking without end, and, perhaps, without much improvement. I saw that one enquiry only gave occasion to another, that book referred to book, that to search was not always to find, and to find was not always to be informed; and that thus to pursue perfection, was, like the first inhabitants of Arcadia, to chase the sun, which when they had reached the hill where he seemed to rest, was still beheld at the same distance from them.

Samuel Johnson: English Dictionary, Preface

© M. S. Welling, 1994

Vertrieb:
Bundesrepublik Deutschland, Österreich: VCH, Postfach 101161,
D-69451 Weinheim (Bundesrepublik Deutschland)
Schweiz: VCH, Postfach, CH-4020 Basel (Schweiz)
übrige Länder: Pentech Press Ltd., 3 Graham Lodge, Graham Road,
London NW4 3DG (England)

ISBN 3-527-28204-1 (VCH, Weinheim)   ISSN 0930-6862
ISBN 0-7273-0411-9 (Pentech Press)

M. S. Welling

parat

Dictionary of Plastics and Rubber Technology German/English

Wörterbuch Kunststoff- und Kautschuktechnologie Deutsch/Englisch

Weinheim · New York
Basel · Cambridge · Tokyo

Titel der Originalausgabe: German – English Dictionary of Plastics and
Rubber Technology, Volume 1
erschienen im Verlag Pentech Press Ltd., London, UK
© M. S. Welling

Dipl.-Chem. M. S. Welling
5 Campbell Croft
Edgeware, Middlesex HA8 8DS
United Kingdom

Herausgeber der Reihe „parat"
Dr.-Ing. H.-D. Junge
Cavaillonstraße 78/I
69469 Weinheim
Bundesrepublik Deutschland

Das vorliegende Werk wurde sorgfältig erarbeitet. Dennoch übernehmen Autoren, Herausgeber und Verlag für die Richtigkeit von Angaben, Hinweisen und Ratschlägen sowie für eventuelle Druckfehler keine Haftung.
This book was carefully produced. Nevertheless, authors, editors and publisher do not warrant the information contained therein to be free of errors. Readers are advised to keep in mind that statements, data, illustrations, procedural details or other items may inadvertently be inaccurate.

---

2. Auflage 1994

Lektorat: Roland Wengenmayr

Die Deutsche Bibliothek – CIP-Einheitsaufnahme
Dictionary of plastics and rubber technology =
Wörterbuch Kunststoff- und Kautschuktechnologie. – Weinheim ;
New York ; Basel ; Cambridge ; Tokyo : VCH.
(Parat)
NE: PT
Vol. 1. German/English = Deutsch/Englisch / M. S. Welling. – 2. Aufl. – 1994
 ISBN 3-527-28204-1
NE: Welling, Manfred S.

© M. S. Welling, 1994

Alle Rechte, insbesondere die der Übersetzung in andere Sprachen, vorbehalten. Kein Teil dieses Buches darf ohne schriftliche Genehmigung des Verlages in irgendeiner Form – durch Photokopie, Mikroverfilmung oder irgendein anderes Verfahren – reproduziert oder in eine von Maschinen, insbesondere von Datenverarbeitungsmaschinen, verwendbare Sprache übertragen oder übersetzt werden. Die Wiedergabe von Warenbezeichnungen, Handelsnamen oder sonstigen Kennzeichen in diesem Buch berechtigt nicht zu der Annahme, daß diese von jedermann frei benutzt werden dürfen. Vielmehr kann es sich auch dann um eingetragene Warenzeichen oder sonstige gesetzlich geschützte Kennzeichen handeln, wenn sie nicht eigens als solche markiert sind.
All rights reserved (including those of translation into other languages). No part of this book may be reproduced in any form – by photoprinting, microfilm, or any other means – nor transmitted or translated into a machine language without written permission from the publishers. Registered names, trademarks, etc. used in this book, even when not specifically marked as such, are not to be considered unprotected by law.

Preface

This dictionary is a greatly expanded, completely revised and updated version of my "German-English Dictionary of Plastics Technology", first published in 1985. It includes the latest terminology relating to plastics recycling and waste disposal, environmental technology, CAD/CAM/CAE and computer controls, quality assurance and engineering polymers. A great deal of material relating to rubber technology has been added - hence the slightly amended title of the book. The present volume contains some 28,000 entries, an increase of about 9,000 on the first edition.

Since the dictionary covers plastics and rubbers in the widest sense, it will enable translators to tackle with confidence any text directly or indirectly concerned with plastics and rubber technology. They should find 90-95% of the words they are looking for. Words not given can usually be derived from existing entries.

In the second edition, too, I have included a great many general, non-technical words that are frequently encountered, in the hope that this will save translators time looking through general dictionaries.

The growing tendency to use obscure abbreviations without any indication as to their meaning has become a minor nightmare for translators. I have therefore included as many of these as I could find - both English and German - since the former, too, are widely used in German texts.

When this dictionary was first published 8 years ago, it was thought that the inclusion of translation examples would be helpful, given the extreme complexity of technical German. Now, however, translators are faced with a new problem, namely that the quality of written technical German is not always what it was. Many sentences are obscure, repetitive and contain irrelevancies so that a great deal of ingenuity is needed to interpret them correctly and render them into good English. Speed has become more important than clarity, and authors rarely seem to check what they have written.

I therefore decided to take the idea of translation examples a step further by assembling, in an appendix, poorly written sentences and passages taken from published material, analysing these and then giving suggested translations. I hope that the study of these pages will prove as useful as the translation examples contained in the body of the text.

I have again been fortunate in having the help of friends and colleagues in the industry. My special thanks go to Ernst Kröner, Jim Paxton and Jon Relton of the Central Language Service of Bayer AG, Leverkusen, Germany, to Grant Edwards, Sally Laichena, Chris Thomlinson and Judith Wallace of the translation department of Hoechst UK Ltd., and to Dr. W.G. Barb, Derek Light and Ed van Renouard of RWS Translations Ltd. My thanks are also due to Dr. Hartmut Bangert, Dr. Heinz Gupta, Prof. Dr. Gerd Habenicht, Dr. John Haim, Messrs. Herbert Stern, Matthew Wiggans and Graham Wright. All of them helped to solve and clarify a great many technical and linguistic problems. Finally, my thanks are due to the British Plastics Federation, the Oil and Colour Chemists' Association, Süddeutsches Kunststoff-Zentrum, Verband Kunststofferzeugende Industrie and Die Sprachberatungsstelle der Dudenredaktion for their help and advice.

Suggestions for amendments and improvements to the dictionary are always welcome and should be sent to me at 5 Campbell Croft, Edgware, Middlesex HA8 8DS, England. Suggested additional German keywords should, if possible, be given in the original context.

M.S. Welling

Extract from the preface to the first edition

This dictionary has been compiled for professional translators and others who have to render German plastics texts into English, as well as for those who have to read such texts in the course of their work. It contains over 19,000 words and expressions covering every aspect of plastics technology, ranging from the physics and chemistry of high polymers, properties, testing, mixing and compounding, through melt rheology and filtration, to additives and the many different processes whereby polymers are converted into finished and semi-finished products. Also included are terms relating to microprocessor controls, end uses, business and finance.

The purpose of compiling this dictionary has been to clarify the ambiguities which puzzle the translator and reader of German plastics texts, and to provide a reliable tool to help translators produce technically correct work, free from the solecisms so often encountered. Wherever necessary, the English terms have been researched and verified by reference to technical articles, books and company literature published in Britain and the United States. They are therefore the words commonly used and understood in those countries. None of the English equivalents given have been taken from existing dictionaries or published translations of technical articles, official specifications, recommendations and similar documents because these have not always been found to be reliable.

The book contains numerous examples of how certain difficult as well as seemingly straightforward words should be handled in translation, showing how, by slightly or completely departing from the original text, or even by omitting a certain word or part of a word altogether, an elegant sentence can be produced without losing the meaning of the original. Technical explanations have been kept to an absolute minimum, having been included only where they were thought to be of real value to the translator.

General Information

Abbreviations used in this dictionary

abbr.	abbreviation
(bfe)	blown film extrusion
(bm)	blow moulding
(c)	calendering
(cm)	compression moulding
(e)	extrusion
e.g.	for example
(grp)	glass reinforced plastics processing
i.e.	that is
(im)	injection moulding
(me)	microelectronics
(mf)	melt filtration
(ptfe)	polytetrafluoroethylene processing
q.v.	which see
(sr)	size reduction
(t)	thermoforming
(tm)	transfer moulding
(txt)	textile industry
(w)	welding

Use of typefaces and symbols

Bold type has been used for German keywords and phrases, e.g. **Druckbeanspruchung** and ... **bei Raumtemperatur unter Druckbeanspruchung.**

Italics have been used

(1) for comments and explanations, e.g. **Klebneigung** tendency to stick (*e.g. compound to metal rolls*)
(2) for explanations of words which cannot be translated as such, e.g. **zwangsläufig** *denotes something being done forcibly*
(3) for abbreviations indicating the field to which the term is relevant, e.g. *(e), (mf), (txt),* etc.

Brackets have been used

(1) to enclose English words, or parts thereof, which can be omitted without altering the meaning, e.g. "(film) bubble stability", "ionomer (resin)" and "blow(ing) mould"
(2) to enclose explanatory English words printed in italics, e.g. bleeding (*of pigments or dyes*)
(3) to enclose abbreviations printed in italics, indicating the field to which the term is relevant, e.g. *(bm), (me), (w)* etc.

Oblique strokes have been used to indicate that either of the two words separated by the stroke can be used, e.g. "rigid/unplasticised PVC" and "non-stick/release effect" (hyphenated words here count as one word).

Note:

Where several English equivalents are given for a German word, the order in which these are printed does NOT indicate an order of preference. Only the context - and the translator's approach to the subject - can determine which of several alternatives is the most suitable.

Sources of reference

The German keywords and phrases which form the basis of this dictionary were extracted from the technical and sales literature published by the major German and Swiss plastics and machinery companies, German plastics and rubber journals, various German technical books, VDI Guidelines and DIN standard specifications. Their English equivalents were verified, where necessary, by reference to the following English and American books, technical journals and literature published by the companies listed.

BOOKS

Bristow, J.H., *Plastics Films*, Iliffe, London (1974)
Brown, R.P. (editor), *Handbook of Plastics Test Methods*, George Godwin, London (1981)
Brydson, J.A. *Flow Properties of Polymer Melts*, George Godwin, London (1981)
Brydson, J.A. and D.G. Peacock, *Principles of Plastics Extrusion*, Applied Science Publishers, London (1973)
Butler, J., *Compression and Transfer Moulding of Plastics*, Iliffe, London (1959)
Button, J., *A Dictionary of Green Ideas*, Routledge, London (1988)
Chandor, A., *The Penguin Dictionary of Computers*, Penguin Books (1981)
Cogswell, F.N., *Polymer Melt Rheology*, George Godwin, London (1971)
Collin, P.H., *Dictionary of Ecology and the Environment*, Peter Collin Publishing Ltd. (1989)
Dubois, J.H. and W.I. Pribble (editors), *Plastics Mold Engineering Handbook*, van Nostrand Reinhold, New York (1978)
Elden, R.A. and A.D. Swan, *Calendering of Plastics*, Iliffe, London (1971)
Elson, D., *Atmospheric Pollution*, Basil Blackwell Ltd., Oxford (1987)
Fenner, R.T., *Extruder Screw Design*, Iliffe, London (1970)
Fisher, E.G., *Blow Moulding of Plastics*, Iliffe, London (1971)
Fisher, E.G., *Extrusion of Plastics (2nd edition)*, Newnes-Butterworth, London (1976)
Frados, J. (editor), *Plastics Engineering Handbook of the Society of the Plastics Industry Inc. (4th edition)*, van Nostrand Reinhold, New York (1976)
Glanvill, A.B., *The Plastics Engineer's Data Book*, Machinery Publishing Co., Ltd., Brighton (1971)
Gordon, M., *The Structure and Physical Properties of High Polymers*, Plastics Institute (1957)
Groves, W.R., *Plastics Moulding Plant, Vol. 1: Hydraulics, Compression and Transfer Equipment*, Iliffe, London (1963)
Illingworth, V. (editor), *Dictionary of Computing*, Oxford University Press (1990)
Laurie, P., *The Micro Revolution*, Futura Publications (1980)
Manson, J.A. and L.H. Sperling, *Polymer Blends and Composites*, Plenum Press, New York (1976)

Miller, G.T., *Environmental Science - An Introduction*, Wadsworth Publishing Co., Belmont, California, (1988)
Morgan, E., *Microprocessors: a Short Introduction*, Dept. of Industry (1980)
Ogorkiewicz, R.M., *Thermoplastics: Effects of Processing*, Iliffe, London (1969)
Penn, W.S. (editor), *Injection Moulding of Elastomers*, Maclaren, London (1969)
Penn, W.S., *PVC Technology (3rd edition)*, Applied Science Publishers, London (1971)
Pye, R.G.W., *Injection Mould Design (2nd edition)*, Godwin, London (1978)
Ritchie, P.D., *Plasticisers, Stabilisers and Fillers*, Iliffe, London (1972)
Roff, W.J. and J.R. Scott, *Fibres, Films, Plastics and Rubbers*, Butterworth, London (1971)
Rubin, I.I., *Injection Molding - Theory and Practice*, Wiley, New York (1972)
Smith, P.I., *Plastics as Metal Replacements*, Scientific Publications (G.B.) Ltd., Brosely (1968)
Staudinger, J.J.P. (editor), *Plastics and the Environment*, British Plastics Federation, London (1973)
Walker, J.S. and E.R. Martin, *Injection Moulding of Plastics*, Iliffe, London (1966)
Willshaw, H., *Calenders for Rubber Processing*, Institution of the Rubber Industry, London (1956)
Wake, W.C., *Fillers for Plastics*, Iliffe, London (1973)
Chambers Dictionary of Science and Technology, Chambers, Edinburgh (1974)
Collins Dictionary of the English Language, Collins (1979)
The New Collins Thesaurus, Collins, London & Glasgow (1984)
The Concise Oxford Dictionary (6th Edition), Oxford University Press (1976)
Glossary of Terms relating to Rubber and Rubber-Like Materials, American Society for Testing Materials, Philadelphia (1956)
IBM Dictionary of Computing (1987)
Modern Plastics Encyclopedia (various editions)
British Standards ASTM Standards, ISO Standards and *Recommendations* and various papers, guidelines and reprints published by the Plastics and Rubber Institute and the British Plastics Federation
Department of the Environment Waste Management Paper No. 26: 'Landfilling Wastes', HMSO, London (1990)

TECHNICAL JOURNALS AND OTHER PUBLICATIONS

Advanced Composites Engineeering
British Plastics
British Plastics & Rubber
The Ecologist
European Adhesives & Sealants
European Plastics News
Europlastics Monthly
Financial Times
Green Magazine
High Performance Plastics
The Independent
The Industrial Robot
Internation Plastics Engineering
Journal of the Oil and Colour Chemists' Association
Modern Plastics
Modern Plastics International
New Scientist
Newsweek
The Observer
Packaging News
PC User
Personal Computer World
Plastics
Plastics Design Forum
Plastics Engineering
Plastics Machinery and Equipment
Plastics and Rubber Today
Plastics Technologyy
Plastics Today (ICI Ltd.)
Plastics and Polymers
Plastics and Rubber International
Plastics and Rubber Weekly
Plastics and Rubber Processing
Plastics and Rubber Processing and Applications
Reinforced Plastics
Shell Polymers (Shell Chemical Co. Ltd.)
The Sunday Times

LITERATURE PUBLISHED BY THE FOLLOWING FIRMS

ABB Instrumentation Ltd.
Baker Perkins Chemical Machinery Ltd.
Ball & Jewell Div. Sterling Inc.
Barber Colman Company
Beken Engineering Ltd.
Beringer Co. Inc.
Bone Craven Daniels Ltd.
B.P. Chemicals Ltd.
British Industrial Plastics Ltd.
Brown Boveri Kent Ltd.
Celanese Corporation Ltd.
Christy & Norris Ltd.
Churchill Fluid Heat Ltd.
Cresta Technology (UK) Ltd.
Desoutter Die Sets Ltd.
D.M.E. Company
Dow Chemical Company
Du Pont de Nemours International S.A.
Dynisco
Engineering Polymers Ltd.
Eurotherm Corporation
Farrel Bridge Ltd.
Fibreglass Ltd.
T.K. Fielder Ltd.
GE Plastics Ltd
GKN Windsor Ltd.
Gloucester Engineering Co. Inc.
Honeywell Corporation Inc.
Hunkar Laboratories Inc
IBM Corporation
Iddon Brothers Ltd.
Imperial Chemical Industries Ltd.
Incoe Corporation
Kent Process Control Ltd.
Killion Extruders Inc.
Leesona Plastics Machinery
Lin Pac Plastics Ltd.
J.J. Lloyd Instruments Ltd.
Mitts & Merrill Inc.
Modular Robotic Systems Ltd.
NDC Systems
Negretti Automation Ltd.
Nelmor Co. Inc.
The Packaging & Industrial Films Association Ltd.
Plasticisers Engineering Ltd.
Process Control Corporation
Ramco Industries
Rapra Technology Ltd.
Reprise Ltd.
Scott Bader Co. Ltd.
Francis Shaw & Co. Ltd.
Shell Chemicals UK Ltd.
Silverson Machines Ltd.
Sinclair Research Ltd.
Spectral Ltd.
Testometric Co. Ltd.
U.S. Industrial Chemicals Co.
Warren Spring Laboratory

A

Abarbeitung processing *(of data) (me)*
Abbau degradation
abbaubar degradable
abbaubar, biologisch biodegradable
abbaubar, photochemisch photodegradable
Abbaubarkeit degradability
abbauen 1. to relieve, to dissipate *(e.g. stresses)*. 2. to degrade, to disintegrate, to break down
Abbauprodukt degradation product
Abbauprozeß degradation process
Abbaureaktion degradation reaction
Abbauverhalten degradation behaviour
Abbildegenauigkeit accurate reproduction of detail
Abbildung, elektronenmikroskopische electron micrograph
Abbindegeschwindigkeit setting speed
abbinden to set, to harden *(adhesives, cement etc.)*
abbindend, hydraulisch hydraulic *(e.g. cement, mortar etc.)*
Abbindeverhalten setting characteristics
Abbindevermögen setting characteristics
Abbindezeit 1. setting time. 2. fibre time *(PU foam)*
Abblätterungen peeling
Abbrandgeschwindigkeit burning rate
Abbrandlänge burning distance, amount of material which has burnt away
Abbrandstrecke burning distance, amount of material which has burnt away
abbremsen to slow down
Abbruchreaktion termination reaction
Abdampfen evaporation
Abdeckfolie backing/protective film
Abdeckpapier backing paper
Abdeckplane tarpaulin
Abdeckung cover
Abdichtbacken distance pieces, check plates *(c)*
abdichten to seal, to waterproof
Abdichtung seal
Abdichtungshaut sealing membrane
Abdunsten cvaporation
Abfall 1. scrap *(when originating from the production of plastics mouldings, extrusions etc.* 2. waste, refuse, rubbish *(in a post-consumer context)* **gebrauchte Abfälle** post-consumer waste 3. decrease, decline

Abfallaufbereitung scrap reprocessing
Abfallaufbereitungsanlage scrap reprocessing plant
Abfallaufwicklung edge trim wind-up (unit)
Abfallbehandlung waste treatment, treatment of waste
Abfallbeseitigung waste/refuse disposal
Abfallbeseitigungsanlage waste disposal plant
Abfallbutzen flash, parison waste *(bm)*
Abfallentfernung 1. removal of waste/scrap. 2. flash removal device
Abfallentfernungssystem scrap removal system
Abfallentfernvorrichtung scrap removal unit
Abfallentsorgung waste/refuse disposal
Abfallentsorgungsprobleme waste disposal problems
Abfallentsorgungsunternehmen waste disposal firm
Abfallentstehung production/generation of waste
Abfallfolien film scrap
Abfallgesetz waste directive
Abfallgesetzgebung waste legislation
Abfallgranulator granulator *(sr)*
Abfallmaterial waste, scrap
Abfallminderung reducing the amount of waste produced
Abfallminimierung minimising of waste/scrap: **aus Gründen der Abfallminimierung** to reduce the amount of scrap produced to a minimum
Abfallmühle granulator, shredder *(sr)*
Abfallprodukt waste product
Abfallrecht waste legislation
Abfallrecycling waste recycling
Abfallschwefelsäure used/waste sulphuric acid
Abfallstoffe waste, refuse
Abfallteile reject mouldings
Abfalltonne dustbin, rubbish bin
Abfallverbrennungsanlage waste/refuse incinerator
Abfallvermeidung preventing the production of waste
Abfallverminderung reducing the amount of waste produced
Abfallverwertung waste/scrap recycling
Abfallwirtschaft waste industry
Abfallzerkleinerer granulator, shredder *(sr)*

Abfallzerkleinerungsmühle granulator, shredder *(sr)*
Abfedern to cushion
abfiltrieren to filter off
Abflammen flame treatment *(of a surface)*
Abflußrohr waste pipe
abformen 1. to cast *(i.e. to make an object from a mould by pouring in resin).* 2. to make a mould *(from an original pattern or model).* 3. to mould *(something from an original pattern or model)*
Abformgenauigkeit accurate reproduction of surface detail
Abformhäufigkeit the number of castings that can be taken *(from a mould)*
Abformmasse mould-making compound/material
Abformmaterial mould-making compound/material
Abformzahl number of castings: **...falls eine hohe Abformzahl gefordert wird** ...if many castings are to be produced
Abfrage interrogation, retrieval *(of data) (me)*
abfragen to interrogate, to retrieve *(me)*
Abfragesprache interrogation language *(me)*
abführen to take away
Abfüllautomat automatic filling machine
abfüllen 1. to bottle 2. to fill *(into sacks, containers etc.)*
Abgaberate emission rate
Abgas flue/waste gas, fumes
Abgasreinigungsanlage flue gas scrubber
Abgasschalldämpfer exhaust silencer
abgefahren retracted **bei abgefahrenem Zylinder** with the cylinder in the retracted position
abgegebene Leistung power output
abgenutzt worn
abgerundet rounded
abgesättigt saturated
abgeschirmt screened
abgesetzt offset
abgesetzte Überlappung recessed lap joint
abgesichert protected
abgesichert, elektrisch electrically interlocked *(e.g. safety guard)*
abgespeichert stored *(me)*
abgestuft gradated
abgewandelt modified
Abguß casting
Abhängigkeit dependence
abhäsiv non-stick
Abhäsivbeschichtung non-stick coating
Abhebegeschwindigkeit nozzle retraction speed *(im)*
Abhebehub nozzle retraction stroke *(im)*
abheben to detach *(paint film)*
abkalten to cool down
Abkantbank folding/bending press
abkanten 1. to bend, to fold. 2. to chamfer
Abkantwerkzeug chamfering tool
Abkömmling derivative

abkühlbedingt due to cooling
Abkühlbedingungen cooling conditions
Abkühldauer 1. cooling time. 2. setting time *(im)*
Abkühleigenspannungen internal cooling stresses
Abkühlgeschwindigkeit cooling rate, rate of cooling
Abkühlphase cooling phase *(im)*
Abkühlspannungen cooling stresses
Abkühlstrecke cooling section
Abkühlung cooling
Abkühlungsverhältnisse cooling conditions
Abkühlvorgang cooling process
Abkühlzeit 1. cooling time. 2. setting time *(im)*
abkuppeln to uncouple, to disconnect
Ablage 1. stacking. 2. stacking unit
ablagern to deposit
Ablagerungen deposits *(e.g. of charred polymer)*
Ablagetisch stacking table
ablängen to cut into lengths *(extruded pipe, profiles or sheets)*
Ablängvorrichtung flying knife
ablativ ablative
Ablauf, verfahrenstechnischer processing sequence
Ablaufdiagramm flow diagram
Ablaufneigung tendency to run *(paint)*
Ablaufsschema flow diagram/sheet
Ablaufsteuerung 1. sequence control. 2. sequence control unit. 3. process control. 4. process control unit
Ablegeeinrichtung stacker, stacking unit
Ableimen glueing
Ableitstrom leakage current
ablöschen to quench
ablösen to detach, to separate
Ablösung detachment
Abluft 1. waste air, outgoing air 2. fumes **Die beim Vernetzen von Elastomeren entstehende Abluft...** the fumes produced when elastomers are vulcanised...
ablüften to allow to evaporate *(e.g. solvent)*
Ablüftungszeit drying time, time required (for the solvent) to evaporate
Abluftventilator exhaust fan
Abluftwäscher waste/exhaust air scrubber
Abmaß dimensional deviation: BUT: **Schwindung ist eine Abmaßänderung in 3 Richtungen** shrinkage is a change in dimensions in three directions
Abmessungen dimensions, measurements
Abmessungsschwankungen dimensional variations
Abmessungstoleranz dimensional tolerance
Abminderung reduction, decrease
Abmischung 1. mixing, blending. 2. blend **Abmischungen mit Naturharzen sind ebenfalls möglich** blends with natural resins are also possible; **in Abmischung mit...** blended with

Abmischungsverhältnis mixing ratio
Abmusterung mould proving *(to check mould performance)*, carrying out mould trials
Abnahmebescheinigung acceptance certificate
Abnahmeprotokoll acceptance certificate
Abnahmeprüfung incoming goods control
Abnahmewalze take-off roll
abnehmbar detachable, removable
Abnehmerindustrie consumer industry
Abplatzen flaking *(e.g. of paint)*
***Abquetschgrat** flash *(bm)*
Abquetschkante pinch-off edge
Abquetschlinge neck and base flash *(bm)*
Abquetschmarkierungen pinch-off welds *(bm)*
Abquetschstelle pinch-off weld *(bm)*
Abquetschtemperatur pinch-off temperature *(bfe)*
Abquetschvorrichtung nip rolls, pinch rolls *(bfe)*
Abquetschwalzen nip rolls, pinch rolls *(bfe)*
Abquetschwalzenpaar nip rolls, pinch rolls *(bfe)*
Abquetschwerkzeug flash mould, semi-positive mould *(cm)*
Abrasionsbeständigkeit abrasion resistance
abrasionsfest abrasion resistant
Abrasionsfestigkeit abrasion resistance
Abrasionsverschleiß abrasive wear
Abrasionswiderstand abrasion resistance
Abrasionswirkung abrasive effect
abrasiv abrasive
Abrasivität abrasiveness
Abrasivkörner abrasive particles
Abrasivkörper abrasive substance
Abrasivstoff abrasive
Abrasivverschleiß abrasive wear
Abrasivwirkung abrasive effect
abrastern to scan
abreiben 1. to grind, to pass through the mill *(e.g. PVC paste, pigment paste etc.)*. 2. to wipe *(e.g. surface with solvent)*
Abreißanschnitt sprue puller gate *(im)*
Abreißbacken deflashing device, tear-off jaws *(bm)*
Abreißperforation tear-off perforation
Abreißplatte sprue puller plate *(im)*
Abreißpunktanschnitt sprue puller pin gate *(im)*
Abreißwalze take-off roll
Abrieb (surface) abrasion
abriebarm abrasion resistant
Abriebbeständigkeit abrasion resistance
abriebfest abrasion resistant
Abriebfestigkeit abrasion resistance
Abriebgeschwindigkeit abrasion/wear rate
Abriebprüfmaschine abrasion tester
Abriebprüfung abrasion test
Abriebspuren traces of abrasion
Abriebswert abrasion resistance
Abriebverhalten 1. abrasion characteristics. 2. abrasion resistance **PTFE verbessert das Abriebverhalten** PTFE improves abrasion resistance

Abriebwiderstand abrasion resistance
Abriebwirkung abrasive effect
Abrolleinrichtung unwinding unit
Abrollstation unwinding station
Abrollvorrichtung unwinding unit
Abruf retrieval, interrogation *(of data) (me)*
abrufbereit ready for retrieval *(me)*
abrufen to retrieve *(data) (me)*
Absackanlage bagging unit
Absättigung saturation
Absatz sales
Absatzausweitung increasing sales
Absatzmarkt market
Absatzrückgang drop in sales
Absatzstrategie sales policy
Absatzverbesserung improvement in sales
Absatzzahlen sales figures
Absauganlage extraction equipment
Absaughaube extraction hood, suction hood
Absaugleitung extraction line, exhaust line
Absaugöffnung vent *(e)*
Absaugung extractor fan
Absaugvorrichtung extractor fan
Abschälfestigkeit peel strength
Abschälmoment peel moment
Abschaltautomatik automatic switch-off mechanism
abschalten to switch/turn off
Abschälverfahren peel test
abschätzen to estimate
Abschätzmethode method of estimating
Abschätzung estimate
Abscheidegrad degree of separation
Abscheideleistung filtration efficiency/performance *(mf)*
Abscheidewirkung separation/filtration effect *(mf)*
Abscheidung separation
Abscheranschnitt tunnel gate, submarine gate *(im)*
Abscherkraft shear force at failure
Abscherspannung shear stress at failure
Abschiebekraft stripping force *(im)*
Abschiebeplatte stripper plate *(im)*
Abschirmeffekt screening/shielding effect
abschirmen to screen/shield
Abschirmmedium screening medium
Abschirmung, elektromagnetische electromagnetic screening/shielding
Abschirmungseigenschaften screening properties
Abschlußemulsion sealant emulsion
Abschnitte offcuts
abschotten to partition *(off)*
Abschreibung depreciation
Abschreibungsvolumen sum set aside for depreciation
Abschreibungszeitraum depreciation period
Abschwächungsbeiwert attenuation constant
abschwenkbar swing-back, swing-hinged, swivel-mounted, swivel-type, hinge-mounted, hinged *(machine unit)*

* *Words starting with* **Abquetsch-***, not found here, may be found under* **Quetsch-**

abschwinden to shrink away *(e.g. resin from reinforcing fibres)*
Absetzbehälter settling tank
Absetzen settling out, sedimentation
Absetzerscheinungen settling out, (signs of) sedimentation
Absetzneigung tendency to settle out
Absetzverhalten sedimentation behaviour
Absetzverhinderungsmittel anti-settling agent
Absicherung protection
absieben to sieve
Absolutdruck absolute pressure
Absolutmessung absolute determination
Absoluttemperatur absolute temperature
Absolutwert absolute value
Absorberturm absorption tower
Absorption absorption
Absorptionsband absorption band
Absorptionsgeschwindigkeit absorption rate
Absorptionskoeffizient absorption coefficient
Absorptionskurve absorption curve
Absorptionsspektroskopie absorption spectroscopy
Absorptionsspektrum absorption spectrum
Absorptionsturm absorption tower
absorptiv absorptive
Abspaltung separation *(e.g. of water or volatile compounds during a chemical reaction)*
Abspeicherung storage *(of data) (me)*
Absperrhahn stopcock
Absperrklappe throttle valve
Absperrmechanismus shut-off mechanism
Absperrschieber shut-off slide valve
Absprengungen spalling
Abspulgatter pay-off creel *(txt)*
Abstandhalter distance piece
Abstandstück distance piece
Abstellzeit downtime, shut-down period
Abstimmen tuning, balancing, harmonising
Abstoßung repulsion
Abstrahlung radiation
Abstreifdruck stripper pressure *(im)*
Abstreifer stripper device *(im)*
Abstreifhülse stripper bush *(im)*
Abstreifplatte stripper plate *(im)*
Abstreifrakel stripper knife
Abstreifring stripper ring *(im)*
Abstreifsystem stripper mechanism/device *(im)*
Abstreifvorrichtung stripper mechanism/device *(im)*
Abströmkanal outlet channel
Abstützfläche supporting surface
Abszisse abscissa
Abtastkopf scanning head/device
Abtastlaser(strahl) scanning laser beam
Abtastung scanning
Abteilung, anwendungstechnische application research (and technical service) department
Abteilungsleiter departmental manager
abtönen to tint
Abtönfarbe tinter
Abtönpaste tinting paste

Abtrag abrasion: **flächiger Abtrag** surface abrasion
Abtragtiefe depth of abrasion
Abtragung, funkenerosive electric discharge machining, EDM, spark erosion
Abtransport removal: **automatischer Abtransport der Hohlkörper** automatic removal of the blow mouldings
Abtrennung separation
abtropfen to drip
Abwärme waste heat
Abwärmenutzung waste heat recycling
Abwärtsbewegung downward movement
Abwärtsextrusion downward extrusion
Abwasser 1. effluent, waste water *(industrial)*. 2. sewage *(domestic)*
Abwasserkanal 1. effluent pipe *(industrial)*. 2. waste/sewage pipe *(domestic)*
Abwasserreinigungsanlage sewage purification plant
Abwasserrohr 1. effluent pipe *(industrial)*. 2. waste/sewage pipe *(domestic)*
Abweichung deviation
Abwickeleinrichtung unwind unit
Abwickelgerät unwind unit
Abwickelgeschwindigkeit unwinding speed
Abwickelvorrichtung unwind unit
Abwickelwalze unwinding roll
Abwickler unwind unit
Abwicklung 1. unwind unit. 2. production of a layflat *(the term is used in connection with flow pattern analysis and is the conversion of a three- dimensional object into a two- dimensional drawing)*. **Abwicklung in die Zeichenebene** production of a layflat. 3. layflat
Abziehkraft peel force
Abziehlack strip lacquer
Abziehwinkel peel-off angle
Abzug take-off, haul-off
Abzugsabquetschwalzenpaar take-off pinch rolls *(bfe)*
Abzugsaggregat take-off unit
Abzugseinrichtung take-off unit
Abzugsgeschwindigkeit take-off speed
Abzugshöhe distance between die and nip rolls *(bfe)*
Abzugskraft take-off tension
Abzugsraupe caterpillar take-off
Abzugsrichtung take-off direction
Abzugsrollen take-off rolls
Abzugsstation take-off station
Abzugsstrecke take-off section
Abzugsturm take-off tower *(bfe)*
Abzugstyp type of take-off
Abzugsverhältnis haul-off ratio *(i.e. haul-off speed/melt delivery rate from the die)*
Abzugsvorrichtung take-off unit
Abzugswalze take-off roll
Abzugswalzenpaar take-off rolls *(bfe)*
Abzugswalzenspalt take-off nip *(bfe)*
Abzugswalzenstation take-off rolls/unit *(bfe)*

Abzugswerk take-off unit
Abzugswerk-Wickler-Kombination combined take-off and wind-up unit
Acetal acetal
Acetalcopolymerisat acetal copolymer
Acetalharz acetal resin
Acetalisierung acetalisation
Acetalisierungsgrad degree of acetalisation
Acetat acetate
Acetatgruppe acetate group
Aceton acetone
Acetonitril acetonitrile
acetonlöslich acetone-soluble
Acetophenon acetophenone
Acetylacetonperoxid acetylacetone peroxide
Acetylcellulose acetyl cellulose
Acetylen acetylene
Acetylenalkohol acetylene alcohol
Acetylenruß acetylene black
Acetylgruppe acetyl group
acetyliert acetylated
Acetylierung acetylation
Achse axle
Achsmanschette bellows
achsparallel axially parallel
Achsrichtung axial direction
Achsschenkel axle journal
Achswickler centre-drive winder *(bfe)*
Achtfach-Spritzwerkzeug eight-cavity injection mould
Achtfachverteilerkanal eight-runner arrangement *(im)*
Achtfachwerkzeug eight-cavity mould
Acidität acidity
ACM *abbr. of* polyacrylate rubber
ACM-Mischung polyacrylate rubber mix
ACN-Anteil acrylonitrile content
ACN-Gehalt acrylonitrile content
AC-Regelung adaptive control
Acrylamid acrylamide
Acrylat acrylate
acrylatbasiert acrylate-based
Acrylatbasis, auf acrylate-based
Acrylatcopolymer acrylic copolymer
Acrylatcopolymerisat acrylic copolymer
Acrylatformmasse acrylic moulding compound
Acrylatharz acrylic resin
Acrylatharzlack acrylic paint/enamel
Acrylatkautschuk polyacrylate rubber
acrylatmodifiziert acrylate-modified
Acrylatmonomer acrylic monomer
Acrylatoligomer acrylic oligomer
Acrylatpolymer acrylic polymer, polyacrylate
Acrylatpolymerisat acrylic polymer, polyacrylate
Acrylcopolymer acrylic copolymer
Acrylcopolymerisat acrylic copolymer
Acryldecklack acrylic paint/enamel
Acrylester acrylate
Acrylestercopolymerisat acrylic copolymer
Acrylfaser acrylic fibre
Acrylglas acrylic sheet, acrylic glazing sheet/material
Acrylglasplatte acrylic sheet
Acrylharz acrylic resin
Acrylharzbindemittel acrylic binder
acrylharzmodifiziert acrylate-modified
Acrylkautschuk polyacrylate rubber
Acrylklarlack clear acrylic lacquer
Acryllack acrylic paint/enamel
Acryllackschicht acrylic coating/film
Acrylmonomer acrylic monomer
Acrylnitril acrylonitrile
Acrylnitril-Butadienkautschuk acrylonitrile-butadiene rubber, nitrile rubber
Acrylnitril-Butadien-Styrolkautschuk acrylonitrile-butadiene-styrene rubber, ABS rubber
Acrylnitrilcopolymerisat acrylonitrile copolymer
Acrylnitrilgehalt acrylonitrile content
Acrylnitrilgruppe acrylonitrile group
Acrylnitril-Methacrylatcopolymerisat acrylonitrile-methacrylate copolymer
Acrylnitril-Methylmethacrylatcopolymer acrylonitrile-methacrylate copolymer
Acrylnitrilrestgehalt residual acrylonitrile content
Acryloxygruppe acryloxy group
Acrylplatte acrylic sheet
Acrylpolymer acrylic polymer
Acrylrest acrylic radical/group
Acrylsäure acrylic acid
Acrylsäurealkylester alkyl acrylate
Acrylsäurebutylester butyl acrylate
Acrylsäureester acrylate
Acrylsäureethylester ethyl acrylate
Acrylsäureharz acrylic resin
Acrylsäuremethylester methyl acrylate
acylierend acylating
Acylierung acylation
Acylisocyanat acyl isocyanate
Acylurethan acyl urethane
Adapter adaptor, adapter
Adaptersystem adaptor/adapter system
adaptiv adaptive
Addition addition
additionsfähig capable of addition
Additionspolymer addition polymer
Additionsreaktion addition reaction
additionsvernetzend addition crosslinking/curing
Additionsvernetzung addition crosslinkage/cure
additiv additive
Additiv additive
Additiv, dispersionsstabilisierendes dispersing agent
Additiv, kratzfestmachendes scratchproofing additive
Additiv, thixotropierendes thixotropic agent
additiventhaltend containing additives
additivfrei additive-free, free from additives
Additivmischung additive blend
Additivpaket additive blend

Additivsystem additive blend/system
Addukt adduct
Ader wire
Adermischung wire covering compound
Aderummantelung wire covering
ADG-Verfahren *abbr. of* **automatisches Druck-Gelierverfahren**, automatic pressure gelation
Adhäsion adhesion
Adhäsionsbruch adhesive fracture/failure
Adhäsionseigenschaften adhesive properties
Adhäsionsfestigkeit adhesive strength
Adhäsionskleben adhesive bonding
Adhäsionskraft adhesive force
Adhäsionsriß adhesive failure
Adhäsionsversagen adhesive fracture/failure
adhäsiv adhesive
Adhäsivbruch adhesive fracture/failure
Adhäsivfestigkeit adhesive strength
adiabatisch adiabatic
Adipat adipate
Adipinsäure adipic acid
Adipinsäureester adipate
Adreßbus address bus *(me)*
Adresse address *(me)*
Adressenverwaltung address management *(me)*
Adressierung addressing *(me)*
Adsorbat adsorbate
Adsorbens adsorbent
Adsorption adsorption
Adsorptions-Desorptionsgleichgewicht adsorption-desorption equilibrium
Adsorptionsenthalpie adsorption enthalpy
Adsorptionsgleichgewicht adsorption equilibrium
Adsorptionspumpe adsorption pump
Adsorptionsverhalten adsorption behaviour
Adsorptionsvorgang adsorption process
Adsorptionswärme heat of adsorption
Adsorptionszentrum adsorption centre
A/D-Wandler analog-digital converter *(me)*
aerob aerobic
aerodynamisch aerodynamic
Aerosol 1. aerosol 2. mist
Aerosolbehälter aerosol (container)
Aerosoldosenlack aerosol paint
Aerosollack aerosol paint
Aerosolventil aerosol valve
AFA *abbr. of* **Absetzung für Abnutzung** depreciation for wear and tear
affin affinitive
Affinität affinity
AFK *abbr. of* **aramidfaserverstärkte Kunststoffe**, aramid fibre reinforced plastics
Agglomerat agglomerate
Agglomeration agglomeration
Agglomerationsgrad degree of agglomeration
agglomeriert agglomerated
Agglomerierung agglomeration
Aggregat 1. aggregate. 2. unit *(of machinery or equipment)*
Aggregatzustand state of aggregation

aggressiv aggressive, corrosive
Aggressivität aggressiveness
A-Glas high alkali glass
Agrarfolie agricultural film
AH *abbr. of* **Aluminiumhydroxid**, aluminium hydroxide
Airless-Gerät airless spraygun
Airless-Spritzen airless spraying
Airless-Spritzverfahren airless spraying
Akkukopf accumulator head *(bm)*
Akkumulator accumulator *(me)*
Akkumulierzylinder accumulator cylinder *(bm)*
Akku-Wirkung accumulator effect
Aktie share
Aktienkapital share capital
Aktivanteil active substance content
Aktivator activator
aktiviert activated
Aktivierung activation
Aktivierungsenergie activation energy
Aktivierungsmittel activating agent
Aktivkohle activated charcoal
Aktivkohlefilter activiated charcoal filter
Aktivruß active carbon black
Aktivsauerstoff active oxygen
Aktivsauerstoffgehalt active oxygen content
Aktivsubstanz active substance
aktualisierbar capable of being updated
aktualisiert updated, brought up-to-date
Aktualisierung updating
Aktualität topicality
akustisch 1. acoustic. 2. audible *(e.g. a signal)*
Akustomikroskop ultrasonic microscope
akut acute
Akzeptor acceptor
Alarmanzeige alarm signal, warning signal
Alarmeingang alarm input
Alarmgeber alarm signal, warning signal
Alarmgedächtnis alarm memory
Alarmlampe warning light
Alarmmeldung alarm signal
Aldehyd aldehyde
Aldehydaddukt aldehyde adduct
Aldehydgruppe aldehyle group
Aldehydharz aldehyde resin
Al-Folie aluminium foil
Algen algae
Algorithmus algorithm
alicyklisch alicyclic
Alimentierung filling *(e.g. of a mould)*: one would normally expect to encounter this word in a different context altogether, i.e. the feeding of foodstuffs. Its use in this particular context is most unusual but, since one author has used it, others may follow - which is why it has been included here
Aliphat 1. aliphatic compound. 2. aliphatic solvent
aliphatisch aliphatic
Alkali alkali
Alkalialkoholat alkali alcoholate
alkalibeständig alkali resistant

Alkalibeständigkeit alkali resistance
alkalifrei alkali-free, non-alkaline
Alkalihydroxid alkali hydroxide
alkalilöslich alkali soluble
Alkalimetall alkali metal
Alkaliresistenz alkali resistance
alkalisch alkaline
Alkalisulfit alkali sulphite
Alkalität alkalinity
Alkanolaminopolyglykolether alkanolaminopolyglycol ether
Alkansulfonsäure alkanesulphonic acid
Alkenylgruppe alkenyl group
Alkohol alcohol
Alkoholat alcoholate
alkoholfrei non-alcoholic
alkoholisch alcoholic
Alkoholkette alcohol chain
alkohollöslich alcohol soluble
Alkoholverträglichkeit compatibility with alcohol
Alkoholyse alcoholysis
Alkoxyalkylgruppe alkoxyalkyl group
alkoxyfunktionell alkoxy-functional
Alkoxygrupe alkoxy group
Alkoxymethylengruppe alkoxymethylene group
Alkoxymethylgruppe alkoxymethyl group
Alkoxyradikal alkoxy radical
Alkoxyrest alkoxy radical/group
Alkoxysilan alkoxysilane
Alkyd alkyd
Alkydeinbrennlack alkyd stoving paint/enamel
Alkydestergruppe alkyd ester group
Alkydharz alkyd resin
Alkydharz, fettes long-oil alkyd (resin)
Alkydharz, mittelfettes medium-oil alkyd (resin)
Alkydharzemulsion alkyd (resin) emulsion
Alkydharzlack alkyd paint
~~Alkydharzlack alkyd paint~~
Alkydharzlackierung alkyd finish
Alkylacetat alkyl acetate
Alkylalkohol alkyl alcohol
Alkylalkoxysilan alkylalkoxy silane
Alkylamin alkylamine
Alkylaminogruppe alkylamino group
Alkylbenzol alkyl benzene
Alkylcarbonsäure alkyl carboxylic acid
Alkylderivat alkyl derivative
Alkylenglykol alkylene glycol
Alkylenkette alkylene chain
Alkylester alkyl ester
Alkylgruppe alkyl group
Alkylhydroperoxid alkyl hydroperoxide
alkyliert alkylated
Alkylierung alkylation
Alkylkette alkyl chain
Alkylkomponente alkyl component
Alkylmethylpolysiloxan alkylmethyl polysiloxane
Alkylolverbindung alkylol compound
Alkylorganosilan alkyl organosilane
Alkylperester alkyl per-ester

Alkylperoxidradikal alkylperoxy radical
Alkylphenol alkyl phenol
Alkylphenoldisulfid alkylphenol disulphide
Alkylphenoxyrest alkylphenoxy group
Alkylphosphit alkyl phosphite
Alkylpolysulfid alkyl polysulphide
Alkylrest alkyl radical
Alkylsilan alkyl silane
Alkylsilikonharz alkyl silicone resin
Alkylsulfid alkyl sulphide
Alkylsulfonat alkyl sulphonate
Alkylsulfonsäureester alkyl sulphonate, alkyl sulphonic acid ester
Alkyltrichlorsilan alkyl trichlorosilane
Alkylvinylether alkyl vinyl ether
Alkylvinylpolysilan alkyl vinyl polysilane
Alkylzinncarboxylat alkyl tin carboxylate
Alkylzinnverbindung alkyl tin compound
Alleinbindemittel sole binder
Alleinweichmacher sole plasticiser
allergieneutral not producing an allergic reaction
Allgemeineigenschaften general properties
Allophanat allophanate
Allophanatbindung allophanate linkage
Allophanatbindungsanteil allophanate linkage content
Allophanatstruktur allophanate structure
Allroundspritzgußtyp general purpose injection moulding grade
allseitig on all sides
Alltagsbetrieb everyday use
Allylchlorid allyl chloride
Allylether allyl ether
Allylharz allyl resin
Allylidengruppe allylidene group
Allzweckextruder general purpose extruder
Allzweckkautschuk general purpose rubber
Allzweckschnecke general purpose screw
Allzweckvernetzungsmittel general purpose catalyst
Alphanumerikbildschirm alphanumeric display (me)
alphanumerisch alphanumeric (me)
Altanstrich old paint film
Altautos scrap cars
Altbausanierung restoration/refurbishment of old buildings
Altbauten old buildings
Altbeton old concrete
Alterung ageing
Alterung, thermische heat ageing
Alterungsanfälligkeit susceptibility to ageing
Alterungsauswirkungen effects of ageing
alterungsbedingt due to ageing
Alterungsbedingungen ageing conditions
alterungsbeständig ageing resistant
Alterungsbeständigkeit ageing resistance
Alterungeigenschaften ageing properties/characteristics
Alterungseinfluß effect of ageing
alterungsempfindlich susceptible to ageing

Alterungserscheinungen ageing phenomena
Alterungsgeschwindigkeit ageing speed
Alterungskennwerte ageing properties *(for translation example see under* **Alterungswerte***)*
Alterungsprozeß ageing process
Alterungsprüfung ageing test
Alterungsrißbildung cracking due to ageing
Alterungsschutzmittel antioxidant
Alterungstemperatur ageing temperature
Alterungsuntersuchung ageing test
Alterungsursachen causes of ageing
Alterungsverhalten ageing characteristics
Alterungsvorgang ageing process
Alterungswerte ageing properties: **verbesserte Alterungswerte** improved ageing resistance
Alterungszeit ageing period
Altfenster scrap windows
Altfolien film scrap
Altglas scrap/waste glass, cullet
Altkabel cable scrap
Altkatalysator spent catalyst
Altkunststoffe plastics waste
Altmetall scrap metal
Altöl used oil
Altpapier waste paper
Altputz old plaster
Alt-PVC scrap PVC, PVC scrap
Altreifen scrap tyres
Altstoff waste, refuse, scrap
Altstoßfänger scrap bumpers
Altware scrap material
Alufolie aluminium foil
Aluguß cast aluminium
Alugußgehäuse cast aluminium housing
Aluminium aluminium
Aluminiumacetylacetonat aluminium acetylacetonate
Aluminiumblech aluminium sheet
Aluminiumdruckguß diecast aluminium
Aluminiumfolien aluminium flakes
Aluminiumgrieß aluminium granules/pellets
Aluminiumguß cast aluminium
Aluminiumguß-Werkzeug cast aluminium mould
Aluminiumhydroxid aluminium hydroxide
Aluminiumklebung 1. bonding of aluminium. 2. bonded aluminium joint
Aluminiumkonstruktion aluminium structure/construction
Aluminiumlegierung aluminium alloy
aluminiumorganisch organo-aluminium
Aluminiumoxid 1. alumina *(when referring to the filler)*. 2. aluminium oxide *(chemical name)*
Aluminiumsilikat aluminium silicate
Aluminiumstearat aluminium stearate
Aluminiumverklebung 1. bonding of aluminium. 2. bonded aluminium joint
Ameisensäure formic acid
Amidwachs amide wax
Amin amine
Aminacrylat aminoacrylate

Aminaddukt amine adduct
Aminaldehydharz aminoaldehyde resin
Aminalkylsilan aminoalkyl silane
Aminanteil amine content
Aminbeschleuniger amine accelerator
Aminepoxidharz amino-epoxy resin
amingehärtet amine-cured
Amingruppe amine group
Aminhärter amine hardener
Aminhärtung amine cure
aminisch amine-type: **aminischer Härter** amine hardener
Aminkatalysator amine catalyst
Aminoalkohol aminoalcohol
Aminoamid aminoamide
Aminocapronsäure aminocaproic acid
Aminochinolin aminoquinoline
Aminodiphenylamin aminodiphenylamine
Aminoendgruppe terminal amino group
aminofunktionell amino-functional
Aminogruppe amino group
Aminoharz amino resin
Aminophenol aminophenol
Aminoplast aminoplastic
Aminoplastharz amino resin
Aminoplastpreßmasse amino moulding compound
Aminorest amino radical
Aminosäure aminoacid
Aminosilan aminosilane
Aminoundecansäure aminoundecanoic acid
Aminsalz amine salt
aminvernetzend amine-curing/-crosslinking
Aminzahl amine value
Ammoniak ammonia
Ammoniumpolyphosphat ammonium polyphosphate
amorph amorphous
Amortisation amortisation
Amortisationszeit amortisation period
Amortisationszeitraum amortisation period
Amperemeter ammeter
amphoionisch amphoionic
amphoter amphoteric
amphoterisch amphoteric
Amplitude amplitude
Amplitudendämpfung amplitude damping
Amplitudenverhältnis amplitude ratio
Amplitudenverteilung amplitude distribution
amtlich zugelassen officially approved
amtliche Prüfanstalt official test establishment
amtliche Zulassung official approval/authorisation
Amylopektin amylopectin
Amylose amylose
anaerob anaerobic
Analogausgabe analog output *(me)*
Analogausgang analong output *(me)*
Analogbaugruppe analog module *(me)*
Analog-Digital-Wandler analog-digital converter *(me)*
Analogeingabe analog input *(me)*

Analogeingang analog input *(me)*
Analogsignal analog signal *(me)*
Analogventil analog valve
Analogwerte analog data *(me)*
Analyse, dynamisch-mechanische dynamic-mechanical analysis
Analyse, thermische thermoanalysis
Analyse, thermogravimetrische thermogravimetric analysis, TGA
Analyse, thermomechanische thermomechanical analysis, TMA
Analysenergebnis analytical result
Analysenmethode method of analysis, analytical technique
Analysenverfahren method of analysis, analytical technique
Analysenwaage analytical balance
Analytiker analyst
Anatas anatase
anbacken to stick *(e.g. substance to walls)*
Anbaumischkopf add-on mixing head
Anbieter supplier
anbinden to gate (into) *(im)*
Anbinden, mehrfaches multi-point gating *(im)*
Anbindung gate *(im)*
Anblaswinkel cooling air impingement angle *(bfe)*
Änderungen, konstruktive design modifications
Andickung thickening
Andrückwalze pressure roll
Andwendbarkeit applicability, practical usefulness
Anfahrausschuß start-up waste/scrap
Anfahrautomatik automatic starting/start-up mechanism
Anfahrbrocken machine purgings
Anfahrdrehmoment starting torque
anfahren to start-up *(a machine)*
Anfahrfladen machine purgings
Anfahrgeschwindigkeit nozzle approach speed *(im)*
Anfahrphase starting-up phase
Anfahrschwierigkeiten starting-up problems/difficulties
Anfahrstellung start-up position
Anfahrverhalten start-up behaviour
Anfahrzeit starting-up period
Anfangsfestigkeit 1. initial strength. 2. green strength
Anfangsflexibilität initial flexibility
Anfangshaftung initial adhesion
Anfangsjahren, in den in the early years
Anfangskapazität initial capacity
Anfangskerbe original notch
Anfangsorientierung initial orientation
Anfangsphase initial stage
Anfangsposition initial/original position
Anfangsquerschnitt initial/original cross-section
Anfangsspannung initial stress
Anfangsspritzdruck initial injection pressure *(im)*
Anfangstack initial tack

Anfangstemperatur initial/starting temperature
Anfangsviskosität initial viscosity
Anfangsvolumen initial/original volume
anfärben to colour
anfasen to chamfer
Anforderungen, lebensmittelrechtliche food regulations
Anforderungen, verfahrenstechnische processing requirements
anforderungsgerecht geared to requirements
Anforderungskatalog list of requirements **Dies scheint einer der schwierigsten Punkte im Anforderungskatalog für Außenanstriche zu sein.** This would appear to be one of the most problematical requirements for exterior finishes.
Anforderungsprofil range/list of requirements
angeätzt etched
angebaut attached
Angebot 1. offer. 2. supply: **Angebot und Nachfrage** supply and demand
Angebotspalette range of products
angebrochen partly used *(contents of a drum etc.)*
angebunden, mehrfach multiple-gated *(im)*
angebunden, seitlich side-gated, edge-gated, side-fed *(im)*
angebunden, zentral centre-gated, centre-fed *(im)*
angeformt moulded-on
angegossen 1. gated, fed *(im)*. 2. moulded-on
angekerbt notched
angekoppelt coupled, linked, joined, connected
angelegt applied *(voltage, stress etc.)*
Angelieren 1. partial gelation (UP, EP resins). 2. gelation *(PVC paste)*
angeliert 1. partly gelled (UP, EP resins). 2. gelled *(PVC paste)*
Angelrute fishing rod
angeordnet, fliegend floating
angeordnet, stationär fixed
angerauht roughened
angereichert enriched
angeschmolzen partly fused
angeschnitten, seitlich side-gated, edge-gated *(im)*
angespeist, seitlich side-fed *(e,bm,bfe)*
angespeist, zentral centre-fed *(e,bm,bfe)*
angespritzt 1. gated, fed *(im)*. 2. moulded-on: **Niet mit angespritztem Kopf** rivet with a moulded-on head
angespritzt, mehrfach multiple-gated *(im)*
angespritzt, seitlich side-gated, edge-gated, side-fed *(im)*
angespritzt, zentral centre-gated, centre-fed *(im)*
Angestellte 1. employees *(general term)*. 2. staff, white-collar workers *(if contrasted with* **gewerbliche Mitarbeiter**, *(q.v.)*
angesteuert actuated
angeströmt, axial centre-fed *(e,bm,bfe)*
angeströmt, quer side-fed *(e,bm,bfe)*

angeströmt, radial side-fed *(e,bm,bfe)*
angeströmt, seitlich side-fed *(e,bm,bfe)*
angeströmt, zentral centre-fed *(e,bm,bfe)*
angetrieben powered *(e.g. a machine)*: **Die Maschine wird von einem 50 PS Motor angetrieben** The machine is powered by a 50 hp motor
angewandte Forschung applied research
Angieß- *see* **Anguß-**
Angleichdruck hot plate contact pressure *(w)*
Angleichen placing surfaces to be joined in contact with the hote plate OR with the heated tool *(w)*
Angriffsflüssigkeit attacking medium
Angriffsmittel attacking medium/agent
Anguß sprue *(im)*: **-anguß** *is often used in place of* **-anschnitt** *when naming the different types of gate, e.g.* **Punktanguß** *instead of* **Punktanschnitt***. The -anguß part of such words should always be translated as* gate*. The translator may also occasionally find that an author has mixed up the two nouns, as in the following example, taken from a paragraph discussing the best place inside a mould cavity for a pressure transducer:* **Da die Schmelze in der Nähe des Angußes zuletzt erstarrt ist es vorteilhaft den Druckfühler angußnah anzuordnen***. From the context it is obvious that* **Anguß** *must here be translated as* gate*, not as* sprue*. The same applies to* **angußnah***, which must here be translated as* near the gate*, since the discussion centres around the mould, not the moulded part. Similar considerations apply to the following sentence, where the author should have said* **Anschnitt: Am Beginn dieser Phase ist der Anguß noch offen***. The word to use here is obviously* gate*, not* sprue*.*
Angußabfall sprue wastes *(im)*
Angußabführung 1. disposal of sprues. 2. sprue disposal unit
Angußabtrennvorrichtung degating device *(im) see explanatory note under* **Angußentfernung (2.)**
Anguß-Anschnitt-System runners and gate, feed system *(im)*: **Die Schnecke drückt die Schmelze mit hoher Geschwindigkeit über das Anguß-Anschnitt-System in das Formnest** the screw forces the melt at high speed along the runners and through the gate into the cavity
Angußart type of gate/gating system *(im)*
Angußausdrückstift sprue ejector pin *(im)*
Angußausreißer sprue puller *(im)*
Angußausstoßzylinder sprue ejector cylinder *(im)*
Angußauswerfer sprue ejector *(im)*
Angußauswerfvorrichtung sprue ejector *(im)*
Angußauszieher sprue puller *(im)*
Angußbereich, im 1. near the sprue. 2. near the gate *(see explanatory note under***angußnah***)*
Angußbohrung gate *(im)*

Angußbuchse sprue bush *(im)*
Angußbutzen sprue *(im)*
Angußeinsatz sprue insert *(im)*
Angußentfernung 1. sprue removal *(im)*: ...besonders dann, wenn keine andere Möglichkeit zur Angußentfernung besteht ...especially if there is no other way of removing the sprue. 2. degating *(despite the English terms given for* **Anguß** *and* **Anschnitt***, and the comments made on their use, this is a very common expression)*
angußfern 1. away from the sprue *(if referring to the moulded part)*. 2. away from the gate *(if referring to the mould)*
Angußferne, in 1. away from the sprue. 2. away from the gate *(see explanatory note under* **angußfern***)*
angußfrei sprueless *(im)*
Angußgestaltung gate design *(im)*
Angußgröße gate size *(im)*
Angußhöhe gate length *(im)*: *This is more acceptable than a literal translation and corresponds more closely to what the authors of the source article had in mind, as is confirmed by an accompanying diagram (Kunststoffe 71(1981)5, p.274 fig.3)*
Angußkanal 1. runner *(if relating to a single-cavity mould)*. 2. main runner *(if relating to a multi-cavity mould)*. 3. feed channel *(if in a general context)*
Angußkegel sprue
Angußkonstruktion gate design *(im)*
Angußlage 1. sprue location, position of the sprue *(if referring to the moulded part)*. 2. gate location, position of the gate *(if referring to the mould)*
Angußloch gate *(im)*
angußlos sprueless *(im)*
Angußmarkierung gate mark *(im)*
Angußmühle sprue granulator *(sr)*
angußnah 1. near the sprue *(if referring to the moulded part)*. 2. near the gate *(if referring to the mould)*
Angußnähe, in 1. near the sprue. 2. near the gate *(see explanatory note under* **angußnah***)*
Angußöffnung gate *(im)*
Angußplatte feed plate *(im)*
Angußpositionierung positioning of the gate *(im)*
Angußpunkt 1. gate, feed point, injection point *(if referring to a machine)*. 2. gate mark *(if referring to a moulded part)*. For translation examples see under **Angußstelle**
Angußquerschnitt 1. sprue, sprue cross-section *(if referring to the moulded part)*: **bei noch plastischem Angußquerschnitt** as long as the sprue remains plastic. 2. gate, gate cross-section *(if referring to the mould)*: **Kleine Angußquerschnitte sind deswegen vorteilhaft** narrow gates are therefore advantageous
Angußseite fixed (mould) half *(im)*

angußseitig on the feed side *(im)*
Angußseparator sprue separator *(sr)*
Angußspinne 1. feed system, runner system *(im)*. 2. radial system of runners *(im)*. 3. sprues and runners *(sr) (when referring to thermoplastic scrap)*
Angußstange sprue *(im)*
Angußsteg gate land *(im)*
Angußstelle 1. gate, feed point, injection point *(im) (if referring to a machine)*: **Bis auf wenige Ausnahmen liegen bei den bisher bekannten Heißkanalsystem die Angußstellen in der Längsachse des Spritzgießwerkzeugs** With but a few exceptions, the gates in the hot runner systems known so far lie in the longitudinal axis of the mould. 2. gate mark *(im) (if referring to a moulded part)*: **Die Angußstelle eines derartig angespritzten Teils wirkt meist optisch nicht störend** The gate mark of a part which has been gated in this way does not normally affect its appearance
Angußstern 1. feed system, runner system *(im)*. 2. radial system of runners *(im)*. 3. sprues and runners *(sr) (when referring to thermoplastic scrap)*
Angußsystem 1. feed system. 2. gating system *(im)*
Angußtechnik gating (technique/method) *(im)*
Angußtunnel 1. runner *(if referring to a single-cavity mould)*. 2. main runner *(if referring to a multi-cavity mould)*. 3. feed channel *(if in a general context)*
Angußverlust sprue wastes
Angußversiegeln gate sealing *(im)*
Angußverteiler runner *(im)*
Angußverteilerplatte feed plate *(im)*
Angußverteilersystem runner system, feed system *(im)*
Angußvorkammer ante-chamber, hot well *(im)*
Angußweg runner *(im)*
Angußzäpfchen sprue *(im)*
Angußzapfen sprue *(im)*
Angußziehbuchse sprue puller bush *(im)*
Angußziehstift sprue puller pin *(im)*
Anhaltspunkt guide
Anhaltswert approximate value/figure
Anhärtung gelation
Anhäufungen (material) accumulations
Anhydrid anhydride
anhydridgehärtet anhydride-cured
Anhydridgruppe anhydride group
Anhydridhärter anhydride hardener
Anhydridhärtung anydride cure
anhydridvernetzend anhydride-curing/-crosslinking
Anhydridvernetzung anhydride cure
Anilin-Formaldehydharz aniline-formaldehyde resin
Anion anion
anionaktiv anionic
anionisch anionic

anionogen anionic
anisotrop anisotropic
Anisotropie anisotropy
Ankerrührer anchor-type stirrer/mixer
Ankerstrom armature curent
Ankopplung coupling, linkage
ankuppeln to couple, to connect
Anlage 1. plant, equipment 2. production line
Anlage, halbtechnische pilot plant
Anlagefläche contact surface
Anlagenabgänge disposal of fixed assets
Anlagenaggregat (machine) unit
Anlagenbau plant construction
Anlagenbetreiber plant operator
Anlagenelement (machine) unit
Anlagenkapazität plant capacity
Anlagenkonzept 1. plant design and lay-out. 2. production line, plant: **Anlagenkonzept zur Herstellung coextrudierter Blasfolie** plant for making coextruded blown film
Anlagenprogramm range of machines/equipment
Anlagenspektrum range of machines/equipment
Anlagerung addition
Anlagevermögen capital/fixed assets
Anlaufphase starting-up phase
anlenken to couple
Anlernling trainee
Anlieferungsviskosität viscosity as supplied
Anlieferungszustand, im as supplied
anlösen to partly dissolve, to attack
Anmachwasser mixing water
anmeldepflichtig officially notifiable
Anmischen mixing
Annäherungsschalter proximity switch
Anode anode
anodisch anodic
Anordnung arrangement
anorganisch inorganic
Anpaßbarkeit adaptability
Anpaßbaustein adaptor module *(me)*
Anpassungsfähigkeit adaptability
Anpasten (initial) mixing *(e.g. of PVC and plasticiser or of pigment and plasticiser, on a triple roller or in a mixer)*: **Diese Type läßt sich leicht zu einer Paste mit niedriger Viskosität anpasten** This grade is readily converted into a low viscosity paste
Anplastifizieren partial plasticisation/plastification
Anpreßdruck contact pressure
Anpreßgummiwalze rubber covered pressure roll
Anpreßkraft contact pressure
Anpreßwalze pressure roll
Anquellung 1. swelling. 2. solvation *(of PVC particles)*
Anregungsfrequenz excitation frequency
Anregungsspannung excitation voltage
Anreibeaggregat roll mill, rolls
Anreibebindemittel pigment paste binder

Anreichern enrichment
Anreißfestigkeit tear initiation resistance
Anreißkraft tear initiation force
Anrißschälwiderstand initial peel strength
Anrißzähigkeit crack initiation resistance
Ansatz 1. formula, equation, (mathematical) statement. 2. mix, mixture. 3. batch: **Verarbeitungszeit eines 250 g Ansatzes** pot life of a 250g batch
Ansatzgröße batch size *(e.g. of a two-pack adhesive after mixing)*
Ansatznummer batch number
ansäuern to acidify
Ansaugfilter suction filter
Ansaugleitung suction line
Anschaffungskosten cost, initial cost, capital investment/outlay
Anschaffungswert initial/original value
Anschauungsmuster inspection sample *(e.g. of a moulding)*
Anschlag stop
Anschlagbolzen stop pin
Anschlagleisten stop bars
Anschlagplatte stop plate
anschleifen to sand (down), to rub down
anschlußfertig ready to be connected
Anschlußflansch connecting flange
Anschlußkabel connecting cable
Anschlußleistung connected load, power rating
Anschlußnippel connecting nipple
Anschlußpaket interface package *(me)*
Anschlußprogramm interface routine *(me)*
Anschlußspannung connected voltage
Anschlußstecker connecting plug
Anschlußwert connected/installed load
Anschlußwert, elektrischer connected load
Anschlußwert, gesamtelektrischer total connected/installed load
anschmirgeln to sand (down), to rub down
anschneiden to gate *(im)*: **Es ist ratsam die Werkzeughöhlung in der Trennfläche anzuschneiden**; it is advisable to gate into the mould cavity along the parting line
Anschnitt gate *(im)*: *The translator may occasionally find that an author has used* **Anschnitt** *instead of* **Anguß***, as in this example:* **Die Zeit d5 ist so zu wählen, daß der Anschnitt des Teils am Ende von d5 sicher erstarrt ist**. *In view of the reference to the moulded part, and to the fact that the so-called* **Anschnitt** *solidifies, it is obvious that the author should have said* **Anguß***, and the word must therefore be translated as* sprue.
Anschnitt, seitlicher side gate, edge gate *(im)*
Anschnittbereich gate area *(im)* **im Anschnittbereich** near the gate
Anschnittbohrung gate *(im)*
Anschnittbuchse sprue bush *(im)*
Anschnittebene gating surface *(im)*
Anschnittferne, in 1. away from the gate. 2. away from the sprue *(see explanatory note under* **angußfern***)*
Anschnittfestlegung deciding the number of gates and their positions *(im)*
Anschnittgestaltung gate design *(im)*
Anschnittgröße gate size *(im)*
Anschnittkanal 1. runner *(if referring to a single-cavity mould)*. 2. main runner *(if referring to a multi-cavity mould)*. 3. feed channel *(if in a general context)*
Anschnittkanal, punktförmiger pin gate *(im)*
Anschnittkegel sprue *(im)*
Anschnittlage positon of the gate *(im)*
Anschnittmarkierung gate mark *(im)*
anschnittnah near the gate *(im)*
Anschnittnähe, in 1. near the gate. 2. near the sprue *(see explanatory note under* **angußnah***)*
Anschnittöffnung gate *(im)*
Anschnittpartie gate area, gate region *(im)*
Anschnittpunkt 1. gate, feed point, injection point *(im) (if referring to a machine)*. 2. gate mark *(im) (if referring to a moulded part)*. For translation examples see under **Angußstelle**
Anschnittquerschnitt 1. gate. 2. gate gross-section: **ein großer Angußquerschnitt** a large gate
Anschnittstelle 1. gate, feed point, injection point *(im) (if referring to a machine)*. 2. gate mark *(im) (if referring to a moulded part)*. For translation examples see under **Angußstelle**.
Anschnittstern 1. feed system, runner system *(im)*. 2. radial system of runners *(im)*. 3. sprues and runners *(sr) (when referring to thermoplastic scrap)*
Anschnittweg runner *(im)*
Anspeisung feed
Ansprechempfindlichkeit response sensitivity
ansprechen to respond
Ansprechgeschwindigkeit speed of response
Ansprechschwelle response/reaction threshold
Ansprechzeit response time
Anspringtemperatur kick-off temperature *(of a catalyst)*
anspritzen to gate (into) *(im)*
Anspritzen seitliches 1. side feed. 2. edge gating *(im)*
Anspritzkanal runner *(im)*
Anspritzkegel sprue *(im)*
Anspritzling sprue *(im)*
Anspritzmöglichkeit possibility of injecting/gating (into something): **...Freizügigkeit in der Formgestaltung durch wahlweiss Anspritzmöglichkeit in die Formtrennebene oder zentral** ...design latitude since the material can either be injected into the mould parting surface or centrally
Anspritzpunkt 1. gate, feed point, injection point *(im) (if referring to a machine)*. 2. gate mark *(im) (if referring to a moulded part)*. For translation examples see under **Angußstelle**
Anspritzstelle 1. gate, feed point, injection point *(im) (if referring to a machine)*. 2. gate mark *(im) (if referring to a moulded part)*. For translation examples see under **Angußstelle**

Anspritztiefe feed channel, feed channel length *(im)*. **große Anspritztiefe** long feed channel; **bei einer Anspritztiefe von 150mm** if the feed channel is 150 mm long
Anspritzung gating *(im)*
Anspritzung, axiale central gating, centre feed *(im)*
Anspritzung, direkte direct feed *(im)*
anspruchsvoll exacting, demanding
anstellbar adjustable
anstellen to adjust, to align *(e.g. calendar rolls)*
Anstellgeschwindigkeit adjustment speed, speed of adjustment
Ansteuerung actuation
Anstieg 1. slope *(of curve)* 2. rise, increase
Anstrich coat *(of paint)*, finish
Anstrichaufbau paint film structure/composition
Anstrichfarbe paint
Anstrichfarbenzusatz paint additive
Anstrichfilm paint film
Anstrichformulierung 1. paint. 2. paint formulation
Anstrichlack paint
Anstrichmittel paint
Anstrichmittelsystem paint system
Anstrichschäden paint film defects
Anstrichsektor paint sector/industry, coatings industry
Anstrichstoff paint
Anstrichsystem paint, surface coating compound
anstrichtechnisch *relating to paint films:* **die anstrichtechnischen Eigenschaften hängen weitgehend von der...** the properties of the paint film largely depend on...
Anstrichträger substrate *(already painted or to be painted)*, painted surface
Anströmkanal 1. runner *(if referring to a single cavity mould)*. 2. main runner *(if referring to a multi-cavity mould)*. 3. feed channel *(if in a general context)*
Anströmleitung feed line
Anteil 1. share, quota. 2. content
anteilige Mitverwendung incorporation of small amounts
anteilmäßig quantitative, quantity-wise
Antenne antenna, aerial
Anthracensulfonsäure anthracene sulphonic acid
anthrazitfarbig anthracite-coloured
Antiabsetzeigenschaften anti-settling properties
Antiabsetzmittel anti-settling agent
Antiabsetzwirkung anti-settling effect
antiadhäsiv non-stick
Antiadhäsivmittel (mould) release agent
Antiausschwimmittel anti-flotation agent
Antibeschlageigenschaften anti-fogging properties
Antibeschlagtest anti-fogging test
Antibewuchsfarbe anti-fouling paint/composition
Antiblockeffekt anti-blocking effect
Antiblockeigenschaften anti-blocking properties
Antiblockmittel anti-blocking agent
Antiblocksystem anti-blocking agent
Antiblockwirkung anti-blocking effect
Antidegradans anti-degradant/-oxidant
Antidriftregelung anti-drift control mechanism
Antidröhnbeschichtung anti-resonance coating, resonance-deadening coating
Antidröhnmasse resonance-deadening compound
antielektrostatisch antistatic
Antiflammittel flame retardant
Antifoulingagens anti-fouling agent
Antifoulinganstrich anti-fouling coating/finish
Antifoulinganstrichstoff anti-fouling paint/composition
Antifoulingausrüstung anti-fouling finish
Antifoulingfarbe anti-fouling paint
Antifoulingmittel anti-fouling agent
Antifoulingsystem anti-fouling paint/composition
Antifoulingwirkung anti-fouling effect
Antigleitbeschichtung anti-slip/non-skid coating
Antigraffitizusatz anti-graffiti additive
Antihaftbelag non-stick coating
Antihaftbeschichtung non-stick coating
Antihaftbeschichtungsmaterial non-stick coating compound
Antihafteffekt non-stick effect
Antihafteigenschaften non-stick properties
Antihaftfähigkeit non-stick properties
Antihaftmittel non-stick agent
Antihaftschicht non-stick coating
Antihaftüberzug non-stick coating
Antihautmittel anti-skinning agent
Antikicker inhibitor
Antiklebeffekt non-stick effect
Antiklebewirkung non-stick effect
Antikorrosionspigment anti-corrosive pigment
Antimonpentoxid antimony pentoxide
Antimontrioxid antimony trioxide
Antimonverbindung antimony compound
Antioxidans antioxidant
Antioxidationsmittel anti-oxidant
Antioxidationssystem antioxidant blend
antioxidativ anti-oxidative
Antiozonans anti-ozonant
Antirutschbeschichtung non-slip/non-skid coating
Antirutschlack non-slip/non-skid paint
Antischaummittel antifoam (agent)
Antischimmelmittel fungicide
Antislipmittel non-slip agent
Antistatiksystem antistatic agent
Antistatikum antistatic (agent)
Antistatikumkonzentrat antistatic concentrate
antistatisch antistatic **antistatisch ausgerüstet** antistatic
Antrieb, motorischer motor drive

Antriebsauslegung drive unit design
Antriebsdrehmoment drive torque
Antriebseinheit drive unit
Antriebshydraulik hydraulic drive (unit)
Antriebsleistung drive power, drive rating *(of motor)*, drive load
Antriebsmotor drive motor
Antriebsrad driving wheel
Antriebsregler drive control unit
Antriebsschaltkreis drive switch circuit
Antriebstrommel powered drum
Antriebswelle drive shaft
antrocknen to become touch dry
Antrocknung surface drying, becoming dry to the touch
Anvulkanisationsbeständigkeit scorch resistance
Anvulkanisationstendenz scorch tendency, tendency to scroch
Anvulkanisationsverhalten scorch behaviour
Anvulkanisationszeit scorch value/time
Anvulkanisieren scorching, premature vulcanisation
anwählbar selectable
anwählen to select
Anwärmbedingungen preheating/heating-up conditions
anwärmen to preheat, to heat up
Anwärmphase preheating/heating-up phase
Anwärmprozeß preheating/heating-up operation
Anwärmzeit preheating/heating-up time
anwenderfreundlich user-friendly *(me)*
Anwenderindustrie user industry
Anwenderprogramm application/user program *(me)*
Anwendersoftware application software *(me)*
Anwendungen applications
Anwendungsbandbreite range of applications
Anwendungsbedingungen service conditions
Anwendungsbeispiele application examples, typical applications
Anwendungsbereich range of applications
anwendungsbezogen application-related
Anwendungsbreite range of applications
anwendungsfertig ready-to-use
Anwendungsgebiet application area, field of application
anwendungsgerecht application-oriented
Anwendungsgrenztemperatur maximum service/operating temperature
Anwendungsmöglichkeiten application possibilities, possible applications
anwendungsorientiert application-oriented
Anwendungsprogramm application/user program *(me)*
Anwendungsrichtlinien application guidelines
Anwendungsschwerpunkte main/principal applications, most important applications
Anwendungsspektrum range of applications
Anwendungstechnik 1. application research 2. application research department 3. application technology
anwendungstechnisch applicational, application-oriented **anwendungstechnische Abteilung** Application Development/Research Department. *The word is sometimes used as padding, when it should be ignored in translation, e.g.* **anwendungstechnische Lackeigenschaften** paint properties
Anwendungstemperatur service/working/operating temperature
Anwendungstemperaturgrenze service/working/operating temperature limit **obere/untere Anwendungstemperaturgrenze** maximum/minimum operating temperature
anzeigbar displayable *(me)*
Anzeige display *(me)*
Anzeige, digitale 1. digital display. 2. digital display unit *(me)*
Anzeigebaugruppe display module *(me)*
Anzeigebildschirm screen, video screen *(me)*
Anzeigefeld display console *(me)*
Anzeigegerät indicator
Anzeigeinstrument indicator
Anzeigelampe indicator/warning light
Anzeigenaktion advertising campaign
Anzeigenleuchte indicator/warning light
Anzeigetafel display console *(me)*
Anzeigezeile display line *(me)*
Anziehungskräfte forces of attraction
Anzugsdrehmoment tightening torque
AO-Anteil active oxygen content
Apfelsinenschaleneffekt orange peel effect
Apfelsinenschalenstruktur orange peel effect
apolar non-polar
Apparatebau plant construction
apparativ *relating to equipment*: **apparativer Aufwand** amount of equipment needed/required
Apparatur apparatus, equipment
Applikationsmethode method of application
Applikationstechnik method of application
applizierbar, leicht easy to apply
Applizierbarkeit ease of application
Appretur *(textile)* finish
Appreturmittel *(textile)* finishing agent
Approximationsgleichung approximation equation
äquimolar equimolecular
Äquipotentialbedingungen equipotential conditions
äquivalent equivalent
Äquivalentgewicht equivalent weight
Äquivalentmasse equivalent weight
Aramidfaser aramid fibre
Aramidfaserlaminat aramid fibre laminate
aramidfaserverstärkt aramid fibre reinforced
Aräometer hydrometer
arbeitend, periodisch operating in cycles
Arbeiter manual/blue-collar worker
Arbeiterbelegschaft manual/blue-collar workers
Arbeitsablauf 1. process. 2. process sequence. 3. (operational) procedure

Arbeitsablaufsteuerung 1. process control. 2. process control unit
Arbeitsaufnahme energy input, absorption of energy
Arbeitsaufnahmevermögen energy-absorbing capacity, capacity/ability to absorb energy
Arbeitsaufwand 1. labour costs 2. amount of work: **Reduzierung des Arbeitsaufwandes** less work
arbeitsaufwendig labour-intensive
Arbeitsbedingungen working/operating conditions
Arbeitsbreite effective width
Arbeitsbühne operating platform
Arbeitsdrehzahl 1. operating speed. 2. operating screw speed *(e)*
Arbeitsdruck working/operating pressure
Arbeitsfeld field of activity
Arbeitsfolge working/operating sequence
Arbeitsgang operation **in einem Arbeitsgang** in one operation
Arbeitsgangfolge working/operating sequence
Arbeitsgemeinschaft working party
Arbeitsgeschwindigkeit working speed, operating speed
Arbeitsgruppe working party
Arbeitshygiene workshop/factory hygiene: **Maßnahmen zur Arbeitshygiene** handling precautions
arbeitshygienisch *relating to workshop or factory hygiene*: **Die arbeitshygienischen Auswirkungen von Bleistabilisatoren sind seit vielen Jahren bekannt** The effects of lead stabilisers on workers' health have been known for many years ; **arbeitshygienische Probleme** health problems
Arbeitsinhalt energy content
arbeitsintensiv labour-intensive
Arbeitskennlinie operating curve
Arbeitskreis working party, study group
Arbeitslänge 1. effective length. 2. effective screw length *(e)*
Arbeitsleistung 1. performance, efficiency, capacity *(e.g. of a machine)* 2. output *(of power)* 3. input *(of power)* 4. rating *(electrical)* 5. energy, power
Arbeitsmodell working model
Arbeitsnehmer, gewerbliche blue-collar/manual workers
Arbeitspackung pre-weighed pack
Arbeitspferd workhorse
Arbeitsplatz 1. work place. 2. workstation *(me)*
Arbeitsplatzkonzentration, maximale maximum allowable concentration, MAC
Arbeitsprinzip operating principle, principle of operation
Arbeitsproduktivität productivity
Arbeitsprogramm program *(me)*
Arbeitsprozeß process, operation
Arbeitspunkt operating point
Arbeitsradius operating radius
Arbeitsraum workroom, workshop

Arbeitsrichtung machine direction
Arbeitsschlitten moving carriage
Arbeitsschutzbekleidung industrial protective clothing
Arbeitsschutzbrille (safety) goggles
Arbeitsschutzkleidung industrial protective clothing
Arbeitsschutzrichtlinien works safety guidelines
Arbeitssicherheit safety at work
Arbeitsspeicher working store, intermediate store, work area *(me)*
Arbeitsstation *the place where a particular operation is carried out, e.g.* moulding station, thermoforming station, *etc. depending on the context*
Arbeitsstellung operating position
Arbeitsstoffverordnung German regulations covering dangerous substances *(for translation example see under* **ArbStoffV**.*)*
Arbeitstag working day
Arbeitstakt moulding cycle
Arbeitstechnik working/production technique
Arbeitsteilung task sharing, division of labour
Arbeitstemperatur working/operating temperature
Arbeitsvorbereitung work planning
Arbeitsvorgänge operations
Arbeitsweise operation, method of operation: **adiabatische Arbeitsweise** adiabatic operation
Arbeitswerte production variables/data *(e.g. film gauge and width, frost line temperature etc.)*
Arbeitszyklus moulding cycle
ArbStoffV. *abbr. of* **Arbeitsstoffverordnung**, German regulations covering dangerous substances: **Diese Produkte sind keine gefährlichen Arbeitsstoffe im Sinne der ArbStoffV.** These products do not constitute a hazard within the meaning of the German regulations covering dangerous substances
Architektur architecture *(me)*
Archivierung archiving, filing *(me)*
arithmetischer Mittelwert arithmetic mean
Armaturen fittings
Armaturenbrett dashboard, instrument panel, fascia
Armaturentafel dashboard, instrument panel, fascia
armiert reinforced
Armierung reinforcement
Armierungsblech reinforcing metal sheet
Armierungseffekt reinforcing effect
Armierungseisen reinforcing steel (rods)
Armierungsgewebe reinforcing fabric
Armierungsmaterial reinforcing material
Armierungsstahl reinforcing steel (rods)
Armlehne armrest
aromadicht odourproof
Aromadurchlässigkeit odour permeability
Aromasperre odour barrier

Aromat 1. aromatic solvent. **2.** aromatic compound
aromatenfrei free from aromatic compounds
Aromatengehalt aromatic content
aromatisch aromatic
arretierbar lockable
arretieren 1. to lock *(a piece of equipment in position).* **2.** to stop
Arretiervorrichtung locking mechanism
Arretierzylinder locking cylinder
Art und Umfang nature and extent
artfremd dissimilar *(materials, e.g. PE and PA)*
artgleich similar *(materials, e.g. two grades of PE)*
Artikel- *see* **Formteil-**
artverwandt related *(materials, e.g. PE and PP)*
Arylalkylmercaptan aryl alkyl mercaptan
arylalkylmodifiziert aryl alkyl-modified
Arylalkylphosphat aryl alkyl phosphate
Arylalkylphthalat aryl alkyl phthalate
Arylalkylverbindung aryl alkyl compound
Arylcarbonsäure aryl carboxylic acid
Arylgrundkörper aryl group
Arylgruppe aryl group/radical
Arylphenol aryl phenol
Arylphosphit aryl phosphite
arylsubstituiert aryl-substituted
Arylverbindung aryl compound
Arzneistoff pharmaceutical preparation, medicament
A-Säule A pillar *(front pillar holding the windscreen)*
Asbest asbestos
Asbestersatz asbestos substitute
Asbestfaser asbestos fibre
asbestfrei asbestos-free
Asbestgewebe asbestos cloth
Asbestplatte asbestos sheet
Asbestzement asbestos cement
Asche ash
Aschegehalt ash content
ASM *abbr. of* **Alterungsschutzmittel,** antioxidant
Assembler assembler, assembly program *(me)*
Assemblerbefehl assembly instruction *(me)*
Assemblerprogramm assembly program *(me)*
Assemblersprache assembly language *(me)*
astatisch astatic
asymptotisch asymptotic
ataktisch atactic
Atemmaske face mask
Atemschutzgerät breathing apparatus
Atemschutzhaube smoke hood
Atemschutzmaske face mask
Äth- *see* **Eth-**
Atlasbindung satin weave *(txt)*
Atmosphärenbedingungen atmospheric conditions
Atmosphärendruck atmospheric pressure
Atmosphärendruckgießen casting at atmospheric pressure
atmosphäril atmospheric

Atmosphärilien atmospheric influences
atmosphärisch atmospheric
atmungsaktiv breathable
Atmungsaktivität breathability
Atmungsorgane respiratory organs
Atom atom
Atomabstand atomic distance
atomar atomic
Atombindung atomic bond
Atomgewicht atomic weight
Atomgruppe group of atoms
Atomgruppierung atomic grouping
ATS Airbus Test Specification
Attraktionskräfte forces of attraction
Attritor attrition mill
ATV *abbr. of* **Auftragsverwaltung,** job management *(me)*
Ätzbad pickling bath
ätzen 1. to etch. **2.** to pickle *(e.g. metal, prior to bonding or painting)*
ätzend corrosive
Ätzflüssigkeit pickling solution
Ätzkali caustic potash
Ätzlösung pickling solution
Ätznatron caustic soda
Ätzung pickling
Audiokassette audio cassette
Audioplatte, digitale compact disc, CD
auditieren to inspect, to audit
Auf- und Abbewegungen up and down movements
aufarbeiten 1. to reprocess, to reclaim. **2.** to recondition. **3.** to work up
Aufbau 1. construction. **2.** superstructure. **3.** constitution *(chemical)*
Aufbau, chemischer chemical structure/constitution
Aufbau, konstruktiver constructional features
Aufbaumerkmale structural features/characteristics
Aufbauzeit build-up period *(e.g. of pressure)*
aufbereiten 1. to compound *(e.g. PVC).* **2.** to reprocess, to reconstitute, to reclaim, to recycle *(e.g. plastics scrap);* **das aufbereitete Produkt** the reconstituted product. **3.** to make, to prepare **Zum Aufbereiten von Plastisolen werden PVC-Typen mit pastenbildenden Eigenschaften verwendet** Plastisols are made from special paste-making PVC resins
Aufbereitungsaggregat 1. compounding unit. **2.** reprocessing unit
Aufbereitungsanlage 1. compounding line. **2.** reprocessing line
Aufbereitungs-Doppelanlage 1. twin compounding unit. **2.** twin reprocessing unit
Aufbereitungsextruder compounding extruder
Aufbereitungsmaschine 1. compounding unit. **2.** reprocessing unit
Aufbereitungsstraße 1. compounding line. **2.** reprocessing line
Aufbereitungssystem 1. compounding system. **2.** reprocessing system

Aufbereitungsteil compounding section
Aufbereitungsverfahren 1. compounding (process). 2. reprocessing
aufblähbar expandable, foamable
aufblähen to expand
aufblasbar inflatable
Aufblasbedingungen blow-up conditions *(bm, bfe)*
Aufblasdorn blowing mandrel/spigot, inflating mandrel/spigot *(bm)*
Aufblasdruck inflating pressure, blow-up pressure *(bm, bfe)*
aufblasen to blow up, to inflate *(bm, bfe)*
Aufblasluft blowing air, inflation air *(bm, bfe)*
Aufblastemperatur blow-up temperature, inflation temperature *(bm, bfe)*
Aufblasverhältnis blow-up ratio *(bm, bfe)*
Aufblasvorrichtung inflation device
Aufblaszone bubble expansion zone *(bfe)*: *(see explanatory note under* **Schlauchbildungszone**)
Aufblätterung delamination
aufbrennen to suck dry *(paint on a porous surface)*
aufbringen to apply
aufbügelbare Futterstoffe iron-on interlinings
Aufenthaltsdauer residence/dwell time
Aufenthaltzeit residence time, dwell time
Auffahren opening *(of a mould)*
Auffahrversuch crash test
Auffangvorrichtung drip tray
Auffangwanne drip tray
Aufgabegut 1. feedstock. 2. compound being charged to a machine: **granulat- oder pulverförmiges Aufgabegut** granular or powdered compound
Aufgabematerial 1. feedstock. 2. compound being charged to a machine *(for translation example see under* **Aufgabegut**)
Aufgabenstellung task
Aufgabeöffnung feed throat/opening *(e)*
Aufgabeschacht feed throat/opening *(e)*
Aufgabeschnecke feed screw
Aufgabeschurre feed chute
Aufgabetrichter (feed/material) hopper *(e, im)*
aufgebläht expanded, foamed
aufgelegte Kraft applied force
aufgelegte Last aplied stress/load
aufgelistet listed
aufgelöst dissolved
aufgenommene Leistung power/energy input, power/energy used
aufgenommene Wirkleistung effective power/energy input
aufgepfropft grafted
aufgerauht roughened
aufgeschäumt expanded, foamed
aufgeschlämmt made into a slurry
aufgeschmolzen fused, melted, molten
aufgewirbelt fluidised *(e.g. powder in flame spraying)*
Aufgliederung breakdown *(of figures)*

Aufhängung suspension *(of a machine unit)*
Aufheizgeschwindigkeit heating-up rate
Aufheizmethode method of heating
Aufheizrate heating-up rate, rate of heating
Aufheizstation heating-up station
Aufheizzeit heating-up period
Aufheller brightener
Aufhellung brightening, lightening
Aufhellvermögen brightening power
aufklappbar hinged *(i.e. so that it can be raised or opened)*
Aufkleber sticker
Aufkommen generation *(from the verb to generate, e.g. of plastics waste)*
aufladen to charge *(electrically)*
Aufladung, elektrostatische electrostatic charging
Auflage 1. condition, requirement 2. support
Auflagefläche supporting/mounting/contact surface
Auflagegewicht application weight *(e.g. PVC paste on fabric)*
Auflagen, brandschutztechnische fire safety requirement/regulations
Auflagenhöhen production runs
Auflager support
Auflagerdistanz distance between supports
Auflegemethode hand lay-up technique *(grp)*
aufleuchten to light up
Auflicht incident light
Auflichtmikroskop light microscope
Auflösung resolution
Auflösungsvermögen resolving power
Aufmachung style
Aufnahme, elektronenmikroskopische electron micrograph
Aufnahme, lichtmikroskopische photomicrograph
Aufnahme, rasterelektronenmikroskopische scanning electron micrograph
Aufnahme, röntgenmikroskopische X-ray micrograph
Aufnahmefähigkeit absorptive capacity
Aufnahmevermögen absorptive capacity
aufnehmen to use, to consume *(energy). The word has many other non-technical meanings which should be looked up in a general dictionary*
Aufnehmer sensor, transducer, probe
Aufnehmerelement transducer element
Aufnehmerempfindlichkeit transducer sensitivity
Aufnehmergehäuse transducer housing
Aufnehmerkabel transducer cable
Aufnehmermeßbrücke transducer bridge
Aufnehmersignal transducer signal
aufpfropfen to graft
aufpolymerisieren to graft
Aufprallenergie impact energy
aufprallfreundlich impact resistant
Aufprallgeschwindigkeit impact speed, speed of impact

Aufpralltest impact test
Aufpreis extra/additional cost
aufrauhen to roughen
Aufreißband tear strip
Aufreißkraft 1. mould opening force *(im)*. 2. die opening force *(e)*
Aufriß elevation
Aufrollung wind-up (unit)
Aufruf retrieval *(me)*
aufrufen to retrieve *(data)(me)*
Aufsatztrichter (feed/material) hopper *(e, im)*
aufschäumbar expandable
aufschäumen to foam
Aufschäumtemperatur foaming temperature
Aufschäumung foaming, expansion
Aufschäumvorgang foaming process
Aufschlagsmittelpunkt centre of impact
Aufschlagtest impact test
Aufschlämmung slurry, suspension
aufschließbar easy to break down *(e.g. PVC particles)*
Aufschließen breaking down, opening up *(e.g. PVC particles)*
Aufschlüsselung breakdown *(e.g. of production figures)*
Aufschmelzbereich homogenising/transition section
Aufschmelzeinheit plasticating/plasticising unit
aufschmelzen to melt
Aufschmelzextruder compounding extruder
Aufschmelzleistung melting efficiency
Aufschmelzverhalten melting characteristics
Aufschmelzzone transition/homogenising section *(of screw) (e)*
Aufschwung upswing
Aufsicht 1. top view 2. control, supervision
Aufsichtsrat board of directors
Aufsichtsratmitglied member of the board of directors
Aufspannbohrungen mould attachment holes *(im)*
Aufspannelemente clamping elements
Aufspannfläche platen area *(im) (for translation example see under* **Werkzeugaufspannfläche***)*
Aufspannmaße mould fixing dimensions *(im)*
Aufspannplan mould fixing details, mould fixing diagram, standard platen details *(im)*
Aufspannplatte platen *(im)*
Aufspannplatte, auswerferseitige moving platen *(im)*
Aufspannplatte, bewegliche moving platen *(im)*
Aufspannplatte, düsenseitige fixed platen, stationary platen *(im)*
Aufspannplatte, feststehende fixed platen, stationary platen *(im)*
Aufspannplatte, schließseitige moving platen *(im)*
Aufspannplatte, spritzseitige fixed platen, stationary platen *(im)*

Aufspleißen fibrillation *(a method of making film fibres from film tape, consisting of splitting the tape lengthways)*
Aufspulgeschwindigkeit reeling speed
Aufstabilisierung boosting the stabiliser
Aufstampfen tamping
aufstapeln to stack
Aufsteckschnecke variable-design/-geometry screw *(see also exlanatory note under* **Einstückschnecke***)*
aufsteigende Feuchtigkeit rising damp
Aufstellfläche floor space *(required to install a machine)*
Aufstellung installation *(of a machine)*: **Die Aufstellung der Maschine erfolgt auf ebenen Boden ohne besondere Fundamente** the machine can be placed on any level floor, no special foundations being necessary
auftauen to thaw (out)
Auftausalz road salt
Aufteilung division
Auftrag 1. project,job *(me)*. 2. application *(of paint, adhesive etc.) (for other, non-technical uses of this word see standard dictionaries)*
auftragen 1. to apply 2. to plot *(a curve)*
Auftraggeber person/firm placing an order
Auftragnehmer person/firm accepting an order
Auftragsaggregat applicator *(e.g. of adhesive)*
Auftragsart method of application
Auftragsdatei job (data) file *(me)*
Auftragsdaten job data *(me)*
Auftragsdatenblatt job data sheet *(me)*
Auftragsdateneingabe job data input *(me)*
Auftragseingang incoming orders, orders received
Auftragsfolge job sequence *(me)*
Auftragsgerät applicator
Auftragsgewicht rate of application *(e.g. of paint or adhesive, usually expressed in g/m^2)*
Auftragsmaschine applicator *(e.g. of adhesive)*
Auftragsmenge amount applied, rate of application *(e.g. paint, adhesive, etc.)*
Auftragsmethode method of application
Auftragsnummer job number *(me)*
Auftragsprotokoll job record *(me)*
Auftragsverfahren method of application
Auftragsverwaltung job management *(me)*
Auftragswalze applicator roll
Auftragswerk applicator unit
Auftragszustand job status *(me)*
Auftreffkraft impact force
Auftreffpunkt point of impact
Auftreffstelle point of impact
Auftreibkraft 1. mould opening force *(im)*. 2. die opening force *(e)*
Auftrennung separation
Auftriebsverfahren hydrometer method *(of determining density)*
Aufwand 1. cost, expense, expenditure. 2. complexity, effort. *This is not an easy word to deal with, and translators should read through*

the examples given to get the feel of it: **Mit relativ geringem Aufwand ist es möglich...** it is relatively easy to...; **Der maschinelle Aufwand ist beträchtlich** machine costs are considerable OR: a great deal of equipment is needed; **Sie erfordern einen erheblichen finanziellen Aufwand** they require considerable financial outlay; **Entgasungssilos erfordern einen hohen Investitions und Betriebsaufwand** devolatilising silos are costly as well as being expensive to operate; **Die einzelnen Methoden unterscheiden sich dabei im Aufwand und in der Genauigkeit** the various methods differ in complexity and accuracy; **trotz des technischen Aufwandes dieser Maschine** despite the technical complexity of this machine; **mit vertretbarem Aufwand** at reasonable cost, without much difficulty; **Eine kontinuierliche Wanddickenmessung mit speziellen Geräten ist wegen des hohen Aufwands micht üblich** it is not usual to measure wall thicknesses continuously with special instruments because of the problems/difficulties/effort involved; **mit erhöhtem hydraulischen Aufwand** with more complex hydraulic equipment; **Der Nachteil bei dieser Anlage ist ein erhöhter Überwachungs- und Bedienungsaufwand** the disadvantage of this plant is that it is more difficult to monitor and operate; **Dies läßt sich ohne großen Aufwand simulieren** This can be easily simulated; **Mit kleinem Aufwand lassen sich sehr große Förderleistungen erreichen** very high throughputs are easily achieved; **...muß durch einen erheblichen Aufwand erkauft werden** ...can be achieved only at considerable cost; **ohne großen steuerungstechnischen Aufwand** without the use of complex control systems; **Der apparative Aufwand ist größer** more equipment is needed; **Der Platzbedarf und der Bedienungsaufwand sind bei diesem System etwas größer** this system requires a little more space and slightly more labour
Aufwand-Nutzen-Relation cost-benefit ratio
Aufwärtsbewegung upward movement
Aufwärtstrend upward trend
Aufweitungszone bubble expansion zone *(bfe) (see explanatory note under* **Schlauchbildungszone)**
aufwendig 1. costly, expensive, time consuming. 2. elaborate, complex, complicated, difficult, cumbersome: **Dieses Verfahren ist aber technisch und preislich aufwendig** this process is, however, technically complicated and expensive; **Die Messung des Druckgefälles ist zu aufwendig** it is too difficult to measure the pressure gradient; **konstruktiv aufwendig** complex in design
Aufwendungen expenditure, amount of money spent
∗**Aufwickelanlage** wind-up, wind-up unit
Aufwickeleinrichtung wind up, wind-up unit
Aufwickelmaschinenbaureihe range of (film) winders, range of wind-up equipment
Aufwickelspannung reeling tension
Aufwickelstelle wind-up station
Aufwickelvorrichtung wind-up (unit)
Aufwicklung wind-up, wind-up unit
aufzeichnen to record
Aufzeichnungsgerät recording instrument/device
Auf-Zu-Bewegung open-close movement
Augenfacharzt eye specialist
ausbalanciert balanced
Ausbau 1. removal 2. expansion
ausbauen 1. to remove, to dismantle. 2. to extend, to enlarge, to expand
ausbaufähig expandable
Ausbaufähigkeit expandability, possibility of expansion
Ausbeulung buckling
Ausbeute yield
Ausbleichen fading
ausblühfrei free from efflorescence (*cf.* **Salzausblühungen**)
Ausblühung 1. bloom. 2. efflorescence
Ausbluten bleeding *(of pigments or dyes)*
Ausbringungszone metering section/zone *(e)*
Ausdampfen evaporation
Ausdehnung 1. expansion 2. size. *Although the second interpretation is most unusual, it has been encountered in this context, e.g.* **Formteilausdehnung**, *q.v.*
Ausdehnungsgefäß expansion tank
Ausdehnungskoeffizient expansion coefficient, coefficient of expansion
Ausdehnungskoeffizient, kubischer coefficient of cubical/volume expansion
Ausdehnungskoeffizient, linearer coefficient of linear expansion
Ausdehnungskoeffizient, thermischer coefficient of thermal expansion
Ausdehnungsverhalten expansion behaviour
ausdiffundieren to diffuse out *(of something)*
Ausdrehmechanik unscrewing mechanism
Ausdruck print-out *(me)*
Ausdrück- *see* **Auswerfer-**
ausdrucken to print out *(me)*
Ausdünnung thinning-out
auseinanderlaufend diverging
auseinandernehmen to take apart
ausfahren to move out
Ausfall failure, breakdown
ausfällen to precipitate
Ausfallmuster reference sample

∗ *Words starting with* **Aufwickel-**, *not found here, may be found under* **Wickel-**

Ausfallöffnung space for part ejection *(im)*
Ausfallproduktion production of rejects/reject mouldings
Ausfallrutsche (delivery) chute
Ausfallschacht (delivery) chute
Ausfallsicherung part delivery safety mechanism *(im)*
Ausfallüberwachung ejection control
Ausfallwaage check weigher
Ausfallzeit downtime
Ausformdüse (extrusion) die
Ausformextruder extruder, normal extruder *(to denote an extruder which shapes profiles (***ausformen***). The word* **Ausform-** *has been used to distinguish the machine from an* **Aufbereitungs-** *or compounding extruder)*
Ausformschräge draft, draw *(im)*
Ausformwerkzeug die *(e)*
ausfugen to grout
Ausfuhr export
Ausführung 1. construction, design. 2. model **Der ABC Extruder ist die kleinste Ausführung aus der DEF Reihe** The ABC extruder is the smallest model in the DEF range
Ausführungsmöglichkeiten types of construction/design: **Bild 2 zeigt drei Ausführungsmöglichkeiten von Schnecken** Fig. 2. shows three different types of screw
Ausgabe output *(me)*
Ausgabebaugruppe output module *(me)*
Ausgabedaten output data *(me)*
Ausgabeeinheit output unit/device *(me)*
Ausgabegerät output unit/device *(me)*
Ausgabeliste output record *(me)*
Ausgabemöglichkeit output facility *(me)*
Ausgabeseite output side *(me)*
Ausgabestein output module *(me)*
Ausgang output *(me)*
Ausgangsacrylnitrilgehalt original acrylonitrile content
Ausgangsbauart original design
Ausgangsbedingungen starting conditions
Ausgangsbefehl output instruction *(me)*
Ausgangsdaten output data *(me)*
Ausgangsfestigkeit original/initial strength
Ausgangsfeuchte original/initial moisture content
Ausgangsgeometrie original shape
Ausgangsgewicht original/initial weight
Ausgangsglanz initial gloss
Ausgangsglanzwert initial gloss
Ausgangsgröße output variable *(me)*
Ausgangshärte original hardness
Ausgangsimpedanz output impedance
Ausgangskanal output channel *(me)*
Ausgangskarte output card *(me)*
Ausgangskomponenten original components
Ausgangskontrolle end control
Ausgangskonzentration original/initial concentration
Ausgangslänge original/initial length

Ausgangsmaße original dimensions
Ausgangsmeßlänge original/initial (measured) length
Ausgangsmodell master model
Ausgangsmodul output module *(me)*
Ausgangsmonomer starting monomer
Ausgangsmonomergehalt original/initial monomer content
Ausgangsposition original position
Ausgangsprobe original sample
Ausgangsprodukt starting product
Ausgangspunkt starting point
Ausgangsschnecke basic screw
Ausgangssignal output signal
Ausgangsspannung output voltage
Ausgangsspannungsbereich output voltage range
Ausgangsstellung original position
Ausgangsstoff starting material
Ausgangsstrom output current
Ausgangsstyrolgehalt original/initial styrene content
Ausgangstemperatur 1. outlet temperature. 2. original/starting temperature
Ausgangsviskosität original/initial viscosity
Ausgangswert original figure
Ausgangszustand original state/condition
Ausgasung gas release/emission
Ausgasungsverhalten gas release behaviour
ausgebaut dismantled *(e.g. screw)*
ausgedruckt printed-out *(me)*
ausgeglichen balanced
ausgehärtet cured
ausgekleidet lined
ausgelegt, falsch wrongly designed, badly designed
ausgeliert 1. gelled *(UP, EP resins)*. 2. fused *(PVC paste)*
Ausgelierung 1. gelation *(of UP, EP resins)*. 2. fusion *(of PVC paste)*
ausgereift perfected, technically developed
ausgerichtet 1. aligned, oriented *(machine parts, fibres etc.)*
ausgerüstet, antistatisch antistatic
ausgerüstet, flammhemmend flame retardant
ausgerüstet, selbstverlöschend self-extinguishing
ausgeschäumt foam-filled
ausgeschwenkt swung out/open *(e.g. machine unit on a hinged flange)*
ausgewogen 1. weighed out. 2. balanced *(e.g. properties)*
ausgewuchtet balanced *(e.g. machine part)*
Ausgiebigkeit yield
ausgießen to fill *(with casting resin)*
Ausgleichbehälter expansion tank
ausgleichen 1. to relieve *(stresses)*. 2. to equalise/compensate/adjust/balance
Ausgleichputz surfacing plaster
Ausgleichsgerade error-adjusted straight line
Ausgleichsmasse stopper, knifing filler, surfacer

Ausgleichszone homogenising zone *(e)*
Aushängen drawdown, sag *(e, bm)*: ...**ein Aushängen des Schmelzeschlauches** parison drawdown, parison sag
aushärtbar, peroxidisch peroxide-crosslinking/-curing
aushärtbar, rasch fast curing
Aushärtdauer cure time
aushärten 1. to cure 2. to set *(e.g. cement)*
Aushärteofen curing oven
Aushärteverhalten curing behaviour/characteristics
Aushärtezyklus curing cycle
Aushärtung curing, crosslinkage, setting
Aushärtungsbedingungen curing conditions
Aushärtungsgeschwindigkeit curing rate
Aushärtungsgrad degree of cure
Aushärtungsreaktion curing reaction
Aushärtungstemperatur curing temperature
Aushärtungsverzögerungsstoff cure retarder
Aushärtungszeit cure time
Aushärtungszyklus curing cycle
Aushöhlung hollow, recess
Auskleidung lining
Auskleidungsfolie liner
Auskolken erosion
Auskratzer scraper
Auskreidung chalking
auskristallisieren to crystallise out
ausladend projecting
ausländisch foreign
Auslandsbeteiligungen foreign holdings
Auslandsinvestitionen foreign investments
Auslandsumsatz foreign turnover
Auslängen drawdown, sag
Auslastung utilisation (of available capacity): **Die Maschinen arbeiten im Dreischichtbetrieb bei einer Auslastung von 80%** the machines operate on a three-shift basis, being run to 80% of their capacity
Auslaufbecher flow cup
Auslaufgehäuse discharge section *(e)*
Auslaufkanal 1. outlet 2. die opening/orifice *(e)*
auslaufseitig at the delivery/discharge end
Auslauftemperatur outlet temperature
Auslaufwalzen discharge rolls, delivery rolls
Auslaufzeit flow time *(in viscosity determination)*
Auslaugebeständigkeit extraction resistance
auslaugen to leach out
Auslegefehler design error/fault
Auslegekriterien design criteria/principles
Auslegung lay-out, design
Auslegung, konstruktive design, design features
Auslegung, verfahrenstechnische technical layout, technical design
Auslegungsdaten design data
Auslegungshilfen design aids
Auslegungskriterien design criteria/principles
Auslegungsphilosophie design philosophy
Auslegungsrechnung design calculation
Auslegungsrichtlinien design guidelines

Auslegungsvorschlag suggested layout/design
Auslenkung deflection
Auslenkungssignal deflection signal
auslesen to read-out *(me)*
auslösen to initiate *(e.g. a reaction, process etc.)*
Auslösung initiation *(e.g. of a reaction, process etc.)*
Ausnutzungsgrad productivity
Auspreßrichtung extrusion direction
Auspreßschnecke discharge screw, delivery screw
Auspuffkrümmer exhaust manifold
ausregeln to eliminate, *(a fault by operating controls)*
Ausrichteffekt orientation effect
Ausrichteverfahren alignment
Ausrichtung alignment, orientation
ausrüsten 1. to equip *(a machine)*. 2. to apply *(e.g. a size, finish etc.)*. 3. to incorporate *(e.g. a plasticiser, antistatic agent etc.)* (see explanatory note under **Ausrüstung**): This verb can sometimes be omitted, e.g. **antistatisch ausgerüstet** antistatic
Ausrüstung 1. equipment. 2. application *(e.g. of a size or finish)*. 3. incorporation *(e.g. of a plasticiser, antistatic agent etc.)*: Versions 2 and 3 are intended to convey the meaning of the word rather than indicate how it should be translated. Translation should generally be rather free, as in this example: ...**zur antistatischen Ausrüstung von HDPE** to make HDPE antistatic OR: to impart antistatic properties to HDPE; Sometimes, the word should be omitted, e.g. **PP Elastomerblends können auch in antistatischer Ausrüstung geliefert werden** antistatic PP elastomer blends are also available
Ausrüstung, maschinelle equipment
Ausrüstung, serienmäßige standard equipment
Ausrüstungsumfang range of equipment
Aussage information
aussagefähig informative
Aussagefähigkeit informative value
Aussagekraft informative value
aussagekräftig meaningful, reliable
Aussagewert informative value
ausschalten 1. to switch off. 2. to exclude, to eliminate
Ausschaltung switching-off, turning off
Ausscheidung separation
Ausschraubkern unscrewing core *(im)*
Ausschraubvorrichtung unscrewing mechanism/device
Ausschreibung invitation to tender
Ausschuß 1. waste, scrap, rejects, reject mouldings. 2. committee
ausschußfrei scrap-free
Ausschußminimierung reducing waste/scrap to a minimum
Ausschußprofile profile scrap

Ausschußquote number of rejects, scrap rate:
  geringere Ausschußquoten fewer rejects, reduced scrap rate
Ausschußrate scrap rate, number of rejects
Ausschußrohre pipe scrap
Ausschußstücke reject mouldings
Ausschußteile reject mouldings
Ausschußware reject articles/goods/mouldings
Ausschußzahlen number of rejects, scrap rate *(for translation example see under* **Ausschußquote***)*
Ausschüttung distribution *(of dividends)*
ausschwenkbar swivel-out
ausschwenkbar, seitlich can be swivelled/swung out sideways
Ausschwimmen floating
Ausschwimmen, horizontales floating
Ausschwimmen, vertikales flooding
Ausschwimmverhütungsmittel anti-floating agent
Ausschwitzen exudation *(of plasticiser)*
Aussehen appearance
Außenabmessungen external dimensions
Außenanstrich exterior finish
Außenanwendung exterior application
außenbeheizt externally heated
Außenbeleuchtung exterior lighting
Außenbeschichtung 1. exterior paint 2. exterior finish
Außenbeständigkeit outdoor weathering resistance
Außenbewitterungsbeständigkeit outdoor weathering resistance
Außenbewitterungsversuch outdoor weathering test
Außendurchmesser outside diameter *(of pipes)*
Außeneinsatz exterior/outdoor use
Außenfarbe exterior paint
Außenfaser outer fibre
Außenflügel exterior casement *(of window)*
außengekühlt cooled from the outside
Außengewinde external thread
Außengleitwirkung external lubricating effect
Außenhandel foreign trade
Außenhandelsbilanz foreign trade balance
Außenhaut outer skin/layer
Außenkalibrieren external calibration/sizing *(e)*
Außenkammer outer chamber/compartment
Außenkontur outer contours
Außenkühlluftstrom 1. external cooling air stream. 2. amount of external cooling air. 3. external cooling air
Außenlack exterior paint
Außenlagerung outdoor weathering/exposure
Außenluft outside air, incoming air
Außenluftkühlsystem external air cooling system *(bfe)*
Außenluftkühlung 1. external air cooling. 2. external air cooling system
Außenmantel outer sheath/covering
Außenmaße external/outside dimensions
Außenoberfläche outer/exterior surface

Außenradius outside radius
Außenschale outer shell
Außentemperatur outside temperature
Außenverwendung exterior/outside use
äußere Gleitwirkung external lubrication
äußere Weichmachung external plasticisation
äußeres Gleitmittel external lubricant
außereuropäisch overseas
äußerlich external
außermittig off-centre
Aussparung recess
Ausspritzvorrichtung purging mechanism *(im)*
ausspülen to purge *(see explanatory note under* **spülen***)*
Ausstoß 1. output *(general term)*. 2. output rate *(continuous processes such as extrusion, film blowing, calendering, etc.)*
Ausstoßdruck 1. extrudate delivery pressure *(e)*. 2. ejector force *(im)*
Ausstoßen ejection *(im)*
Ausstoßer ejector *(im)*
Ausstoßer- *see* **Auswerfer-**
Ausstoßextruder melt extruder *(see explanatory note under* **Austragsextruder***)*
Ausstoßgeschwindigkit output rate
Ausstoßgleichmäßigkeit uniform delivery *(of extrudate)*
Ausstoßkennlinie output rate curve *(e)*
Ausstoßleistung output, output/delivery rate
Ausstoßleistungsminderung reduction in output: **mit spürbaren Ausstoßleistungsminderungen** with markedly reduced outputs
Ausstoßmenge output, output/delivery rate
Ausstoßschnecke discharge screw
Ausstoßschwankungen output fluctuations
Ausstoßstift ejector pin, knock-out pin *(im)*
Ausstoßteil metering section/zone *(e)*
Ausstoßtemperatur temperature at the delivery end, output temperature, discharge temperature *(e.g. extrudate temperature at the die)*
Ausstoßverringerung reduced output(s)
Ausstoßvolumen output volume
Ausstoßweg ejector stroke *(im)*
Ausstoßwerte outputs
Ausstoßzone metering section/zone *(e)*
Ausstoßzylinder ejector cylinder *(im)*
Austauchzeit speed of withdrawal *(from dip coating tank)*
austauschbar exchangeable, interchangeable
Austauschbarkeit interchangeability
Austauschstoff substitute, alternative material
austenitisch austenitic
Austragsbereich metering section/zone *(e)*
Austragsextruder melt extruder *(this is usually a single-screw extruder fed with polymer melt straight from the reactor. The word derives from* **austragen***, i.e. removal (of material from the reaction vessel). Also called* **Schmelzeextruder***)*
Austragsleistung output rate

Austragspumpe extraction pump
Austragsschnecke discharge screw
Austragsteil 1. discharge section. 2. metering section/zone *(e)*. 3. die *(e)*
Austragszone metering section/zone *(e)*
Austrieb flash
Austriebsspindel ejector spindle
Austrittsbohrung 1. outlet 2. die gap/orifice
Austrittsdüse die *(e)*
Austrittsdüsenspalte die gap-orifice
Austrittsgeschwindigkeit delivery rate *(e.g. of an extrudate or melt from a die)*
Austrittslippen die lips *(e)*
Austrittsöffnung 1. outlet. 2. die gap/orifice *(e)*. 3. gate *(im)*
Austrittsspalt die gap/orifice *(e)*
Austrittsspaltweite die gap width *(e)*
Austrittstemperatur outlet temperature
Ausvulkanisieren complete vulcanisation
Auswahlkriterium selection criterion *(plural: criteria)*
Auswanderung migration
auswechselbar interchangeable, exchangeable
Auswerfen ejection *(im)*
Auswerfer ejector *(im)*
Auswerferbewegung ejector movement *(im)*
Auswerferbohrung ejector bore *(im)*
Auswerferbolzen ejector bolt *(im)*
Auswerferdämpfung ejector damping mechanism *(im)*
Auswerfereinrichtung ejector mechanism *(im)*
Auswerferendstellung ejector end position *(im)*
Auswerferfreistellung 1. ejector release mechanism *(im)* 2. ejector release
Auswerfergeschwindigkeit ejector speed *(im)*
Auswerfergrundplatte ejector retaining plate *(im)*
Auswerferhalteplatte ejector plate *(im)*
Auswerferhilfe ejection aid *(im)*: **Ein verstellbarer Druckluftzylinder an der beweglichen Aufspannplatte dient als Auswerferhilfe** An adjustable compressed air cylinder on the moving platen helps to eject the moulding
Auswerferhub ejector stroke *(im)*
Auswerferhubbegrenzung 1. ejector stroke limitation. 2. ejector stroke limiting device *(im)*
Auswerferhülse ejector bush *(im)*
Auswerferkolben ejector rod *(im)*
Auswerferkraft ejector force *(im)*
Auswerfermarkierung ejector mark *(im)*
Auswerferpaket ejector system *(im)*
Auswerferplatte ejector plate, ejector plate assembly *(im)*
Auswerferplattensicherung ejector plate safety mechanism *(im)*
Auswerferring ejector ring *(im)*
Auswerferrücklauf ejector return movement
Auswerferrücklaufgeschwindigkeit ejector return speed
Auswerferrückzugkraft ejector retraction force *(im)*

Auswerferseite ejector half, moving mould half *(im)*
auswerferseitig on the moving mould half, on the ejector side of the mould *(im)*
auswerferseitige Werkzeughälfte ejector mould half, moving mould half *(im)*
Auswerferstange ejector rod *(im)*
Auswerferstempel ejector pin, knock-out pin *(im)*
Auswerfersteuerung ejector control mechanism *(im)*
Auswerferstift ejector pin, knock-out pin *(im)*
Auswerferstößel ejector rod *(im)*
Auswerfersystem ejector mechanism *(im)*
Auswerferteller ejector plate, ejector plate assembly *(im)*
Auswerferüberwachungssteuerung ejector monitoring device *(im)*
Auswerferventil ejector valve *(im)*
Auswerfervorlauf ejector forward movement
Auswerfervorlaufgeschwindigkeit ejector forward speed
Auswerferweg ejector stroke
Auswertung 1. evaluation, assessment interpretation *(e.g. of test results)*. 2. utilisation, exploitation *(e.g. of an invention)*
Auswurf ejection *(im)*
Ausziehen sheeting-out *(e.g. PVC, rubber etc. on a mill)*
Ausziehgrad depth of draw *(t)*
Auszug, wäßriger aqueous extract
autark independent, self-sufficient
Autodachhimmel (car) roof liner
Autodecklack automotive finish/paint
Autoelektrik automotive electrical system
Autohimmel (car) roof liner
Autoindustrie car/motor/automotive industry
Autokarosse car body
autokatalytisch autocatalytic
Autoklav autoclave
Autolack automotive paint
Automat automatic machine/equipment
Automatikbetrieb automatic operation
Automatikwerkzeugschutz automatic mould safety device
Automation automation
Automatisationsgrad degree of automation
automatisiert automated, automatic
Automatisierung automation
Automatisierungsaufgabe automation task
Automatisierungsaufwand degree of automation: **Ein höherer Automatisierungsaufwand kann vorgesehen werden** a greater degree of automation can be provided
automatisierungsfreundlich easily automated, easy to automate
Automatisierungsgrad degree of automation
Automatisierungstechnik automation technology/engineering
Automobilanwendungen automotive applications

Automobilbau car construction
Automobilbereich, im in the automotive field
Automobilhersteller car manufacturer
Automobilindustrie car/motor/automotive industry
Automobilkarosserie car body
Automobillack automotive paint/enamel
Automobillenkrad steering wheel
Automobilsektor motor industry
Automobiltechnik automotive engineering
Automobilzulieferbetrieb supplier to the motor industry
Automotor car engine
autonom autonomous
Autooxidans autooxidant
Autoradiographie autoradiography
Autoreparaturlack automotive refinishing paint
Autoserienlack standard automotive paint
Autounterbodenschutzmasse underbody sealant
Autoxidation autoxidation
autoxidierbar autoxidisable
autoxidierend autoxidising
AWETA *abbr. of* **anwendungstechnische Abteilung**, Application Research Department
Axialkolbenmotor axial piston motor
Axialkolbenpumpe axial piston pump
Axialkräfte axial loads
Axiallagergruppe thrust bearing unit
Axiallagerung thrust bearing (unit)
Axialnuten longitudinal grooves
Axialspiel axial clearance
Axialstrom axial melt stream
Axialströmung axial flow
Axialvermischung axial mixing, longitudinal mixing
Azealinsäureester azelate
Azelainsäure azelaic acid
Azelat azelate
azeotrop azeotropic
azimutal azimuthal
Azodicarbonamid azodicarbonamide
Azogruppe azo group
Azoinitiator azo initiator
Azoisobutyronitril azoisobutyronitrile
Azomonomer azo monomer
Azopigment azo pigment
Azoverbindung azo compound
azovernetzt azo-crosslinked/cured

B

Babysauger baby's teat
Backenabstand distance between the jaws
Backenwerkzeug split mould *(im)*
Backform baking tin
Backwaren bakery goods
Badesalz bath salts
Badevorhang shower curtain
Badewanne bath, bath tub
badnitriert saltbath nitrided
Bagatellschäden minor damage
Bahn web, length *(of sheeting)*: **Eine Verbindung der Einzelbahnen ist nicht erforderlich** there is no need to join the individual lengths of sheeting
bahnenförmig in web form: **bahnenförmige Materialien** continuous webs
Bahnführung web guide
Bahngeschwindigkeit web speed
Bahnkantensteuerung 1. film edge control. 2. film edge control device
Bahnkesselwagen rail tanker
Bahnmaterialien continuous webs
Bahnsilo rail tanker
Bahnspannung 1. web tension. 2. web tensioning device/mechanism
Bahnspannungskontrolle 1. web tension control. 2. web tension control unit/mechanism
Bahnspannungsregelung 1. web tension control. 2. web tension control mechanism/unit
Bahnsteuereinrichtung web guide
Bahnwächter web monitor, web monitoring device
Bahnzugkraft web tension
Bahnzugkraftmeßstation web tension measuring unit
Bahnzugkraftmeßvorrichtung web tension measuring device
Bajonettverschluß bayonet coupling/joint
Bakterienbefall bacterial attack
Bakterizid bactericide
bakterizid bactericidal
Balg bellows
Balkenanzeige bar display *(me)*
Balkencode bar code *(me)*
Balkendiagramm bar graph
Balkengraphik bar graph
Balkenverteiler manifold block *(e, im)*
Balkon balcony
Balkonbrüstung balcony parapet
Ball ball
Ballen 1. roll face *(c)* 2. bale
Ballenbreite roll face width *(c)*
Ballenlänge roll face width *(c)*
Ballenrand roll periphery *(c)*
Balligkeit convex shape
Ballungsgebiet catchment area
Ballungsraum catchment area
Band tape, web
Bandablegeeinheit web depositing unit

Bandablegung 1. web deposition. 2. web depositing unit
Bandabschlagsystem strip pelletising system *(sr)*
bandangeschnitten film gated *(im)*
Bandanguß film gate *(im)*
Bandanschnitt film gate *(im)*
Bandanschnitt, mittiger central film gate *(im)*
Bandanschnitt, ringförmiger ring gate *(im)*
Bandanschnitt, seitlicher lateral film gate *(im)*
Bandbeschichtungsverfahren coil coating (process)
Bändchen film tape, film yarn
Bändchengewebe woven film tape
Bändchenreckanlage film tape stretching plant/unit
Banderole sleeve
Banderoliermaschine sleeve wrapping machine
Bandführung web guide
Bandgeschwindigkeit web speed
Bandgranulator dicer, strip pelletiser *(sr)*
Bandgranulator mit Stufenschnitt stair step dicer *(sr)*
Bandgranulierstraße strip pelletising line *(sr)*
Bandkassettengerät cassette recorder *(me)*
Bandlackierverfahren coil coating
Bandmischer ribbon mixer
Bandsäge band saw
Bandschleifmaschine abrasive belt grinder
Bandschleifpoliermaschine abrasive belt polisher
Bandschneider strip pelletiser, dicer *(sr)*
Bandstahl strip steel
Bandstahlschnittwerkzeug strip steel cutting tool
Bandzug web tension
Bandzugmeßstation web tension measuring unit
Bankkredit bank loan
Barcode bar code *(me)*
Barcodeleser bar code reader *(me)*
Bareinlage cash subscription
Barium-Cadmium-Stabilisator barium-cadmium stabiliser
Bariummetaborat barium metaborate
Bariumoctoat barium octoate
Bariumseife barium soap
Bariumstearat barium stearate
Bariumsulfat barium sulphate
Barometerdruck barometric pressure
Barriereeigenschaften barrier properties
Barrierefolie barrier film
Barrierekunststoff barrier plastic
Barrierepigment barrier pigment
Barriereschicht barrier coating
Barrierewerte barrier properties
Barrierewirkung barrier effect
Baryt barytes, barium sulphate
Base base, alkali
Basisausführung basic design/model
Basisausrüstung basic equipment
basisch basic

Basiseinheit basic unit
Basisharz base resin
Basismaschine basic machine
Basismaterial base material
Basispolymer base polymer
Basisprogramm basic program *(me)*
Basisprüfung basic test
Basisrezeptur basic formulation
Basiswerkstoff base material
Basizität basicity
Batchanlage batch-type plant, discontinuous plant
Batchverarbeitung discontinuous/batch processing
Batteriegehäuse battery box
batteriegepuffert battery-buffered *(me)*
Batteriekasten battery box
Batterieladegerät battery charger
Batterieseparator battery separator
Batteriestromversorgung, mit battery powered
Bauart 1. method/type of construction: **Schlauchköpfe mit Pinole und Spaltverstellung stellen die einfachste Schlauchkopfbauart dar** side-fed parison dies with a die gap adjusting mechanism are the simplest type of construction; **Einschneckenextruder konventioneller Bauart** conventional single-screw extruder(s). 2. made by *(referring to a particular machine)* **Bauart Reifenhäuser** made by Reifenhäuser
Bauart, einfache simply constructed
Bauart, geschlossene totally enclosed *(machine)*
Bauart, offene consisting of separate units: **Extruder offener Bauart, d.h. mit separatem Getriebe und neben dem Extruder stehenden Antriebsmotor** extruder with a separate drive and adjacently placed drive motor
bauaufsichtliche Bestimmungen building regulations
Bauchemie *A buzz word intended to convey the idea of synthetic, i.e. chemical products and their relation to the building industry. Since it is untranslatable, it must be treated as synonymous with* **Bauindustrie**. *The resultant translation will be perfectly acceptable, as the following examples show.* **Diese Produkte werden in zahlreichen Bereichen, wie der Klebstoff-und Lackindustrie, der Bauchemie usw. eingesetzt.** These products are used in many fields, e.g. the adhesives and coatings industry, the building sector etc. **Diese Kunstharze sind die Basis für die Herstellung hochwertiger Produkte für die Bauchemie.** These synthetic resins form the basis of high quality products for the building industry.
bauchemisch *Like the noun from which it is derived,* (**Bauchemie**, *q.v.*) *this word cannot be translated and must therefore be circumscribed, e.g.* **Fast alle modernen**

bauchemischen Produkte müssen bestimmten Normen entsprechen. Practically all modern synthetic products for the building industry must meet certain standards.
Baueinheit unit, module
Bauelement component
Bauemulsion emulsion for the building sector, building-grade emulsion
Bauen, textiles production of air-supported structures
Baufehler construction fault
Baufolie sheeting/film for the building sector/industry
baugleich having the same construction
Baugröße size *(of a machine)*
Baugruppe 1. module *(me)*. 2. structural unit *(of a machine)*
Baugruppen-Heißkanalsystem modular hot runner system *(im)*
Baugruppenträger rack *(me)*
Bauindustrie building industry/sector, construction industry
Bauingenieur structural engineer
baukastenartig modular
Baukastenbauweise unit construction system, modular construction system
Baukastenelement module, unit *(of a machine)*
Baukastenmaschine machine built up from modules/units, modular machine
baukastenmäßig modular
Baukastennormalien standard mould units
Baukastenprinzip modular principle/system
Baukastenprogramm modular range
Baukastenreihe modular range *(of machines)*
Baukastenschnecke modular screw, screw built up from modules/units
Baukastenschneckensatz modular screw assembly
Baukastenspritzgießmaschine modular injection moulding machine
Baukastensystem modular system/principle: **Ein Universal-Baukastensystem für rheologische Untersuchungen** a universal, modular instrument for rheological tests; **Die Meßinstrumente sind nach dem Baukastensystem aufgebaut** the measuring instruments have been designed on the modular principle
Baukleber structural adhesive
Baukonjunktur economic situation in the building sector
Baukörper component, part
Baulänge, geringe short
Baulänge, kurze short
Baumalerlack decorator's paint
Bäumanlage beaming unit *(txt)*
Baumaße dimensions
Bäumchenbildung water treeing *(an electrical phenomenon)*
Baumwollfeingewebe fine cotton fabric
Baumwollfeinstgewebe extremely fine cotton fabric
Baumwollflocken cotton flock/linters
Baumwollgewebe cotton fabric
Baumwollgrobgewebe coarse cotton fabric
Baumwollinters cotton linters
Baumwollsamenölfettsäure cottonseed oil fatty acid
Baumwollstoff cotton fabric
bauphysikalisch physical *(in a building context)* **die bauphysikalischen Werte des Steines** the physical properties of stone. *See also appendix, example 43*
Baureihe range, series **eine neue Extruder-Baureihe** a new range of extruders
Bausatz unit, module
Bausatzschnecke modular screw, screw built up from modules/units
Bauschäden structural damage
Bausektor building sector/industry
Bauserie range, series
Baustahl structural steel
Baustein building block, module, unit *(e.g. of a machine)*
bausteinartig modular
Bausteinelektronik modular electronic system
Bausteingetriebe modular drive
Bausteinsystem modular system/principle
Baustelle building site
Baustellen, auf on site
Baustoff building material
Bautechnik constructional engineering
Bauteil component part
Bauteile, elektronische electronic components/devices
Bauteile, normalisierte standard mould units
Bauteilgestaltung component design
Bauteilversagen component failure
Bautendichtungsmasse structural sealant
Bautenfarbe exterior/masonry paint
Bauteninnenlack interior paint
Bautenlack exterior/masonry paint
Bautenschutz building conservation, conservation of buildings
Bautenschutzmittel masonry water repellent
Bauweise method of construction: **Die Maschine zeichnet sich durch eine sehr einfache Bauweise aus** the machine is very simply constructed
Bauweise, flache horizontal *(in construction) (see translation examples under* **Bauweise, horizontale***)*
Bauweise, geschlossene totally enclosed *(machine)*
Bauweise, hohe upright *(in construction)*
Bauweise, horizontale horizontal *(in construction)*: **Schneckenspritzgießmaschine in horizontaler Bauweise** horizontal screw injection moulding machine

Bauweise, offene consisting of separate units *(machine) (see translation example under* **Bauart, offene***)*
Bauwerksfuge expansion joint
Bauwesen building/construction industry
Bauwirtschaft building industry/sector, construction industry
BBP *abbr. of* **Butylbenzylphthalat**, butyl benzyl phthalate
BD butadiene
BDE *abbr. of* **Betriebsdatenerfassung** production data collection/acquisition *(me)*
BDE-Anlage production data collecting system *(me)*
BDE-Daten production data *(me)*
BDES *abbr. of* **Betriebsdatenerfassungssystem** production data collecting system *(me)*
BDE-System production data collecting system *(me)*
BDE-Terminal production data terminal *(me)*
BDE-Zentrale central data processor *(me)*
BE *abbr. of* backscattered electrons
beansprucht stressed
Beanspruchung stress, load(ing), wear (and tear)
Beanspruchung, chemische chemical attack
Beanspruchung, elektrische electrical loading
Beanspruchung, mechanische mechanical loading
Beanspruchung, ruhende static stress
Beanspruchung, schlagartige 1. impact stress. 2. sudden stress
Beanspruchung, stoßartige impact stress
Beanspruchung, thermische thermal loading
Beanspruchung, wechselnde cyclic stress/loading
Beanspruchung, zügige dynamic stress
Beanspruchung, zulässige safe working stress *(of pipes)*
Beanspruchungsart type of stress
Beanspruchungsausschlag load amplitude
Beanspruchungsbedingungen conditions of use, service conditions
Beanspruchungsdauer stress duration, loading time
Beanspruchungsgeschwindigkeit stressing rate
Beanspruchungshäufigkeit stress/loading frequency
Beanspruchungshöhe stress level, amount of stress applied
Beanspruchungsparameter stress/loading parameter
Beanspruchungsrichtung stress direction
Beanspruchungsverhältnisse conditions of use, service conditions
Beanspruchungszeit stress duration, loading time
bearbeitbar machinable
Bearbeitbarkeit machinability, machining properties
bearbeiten to machine
Bearbeitung fabrication *(mainly of plastics sheet, pipes etc.)* machining, shaping
Bearbeitung, mechanische machining
Bearbeitung, spanabhebende machining
Bearbeitung, spanende machining
Bearbeitung, spangebende machining
Bearbeitung, spanlose thermoforming, shaping, forming
Bearbeitungstoleranzen machining tolerances
Becher tub, cup
Becher, offener open cup *(in flash point determination)*
Becherglas beaker
Bedampfen 1. metallisation. 2. steam treatment
Bedarfsanstieg rise in demand
Bedarfsartikel consumer goods
Bedarfsgegenstände consumer goods
Bedarfsgegenständegesetz consumer goods regulations
Bedarfsgüter consumer goods
Bedienbarkeit, leichte easy to operate: **Diese Maschine zeichnet sich durch leichte Bedienbarkeit aus** this machine is easy to operate
Bedienbefehl operating instruction *(me)*
Bediener operator
Bedienerführung operator prompt *(me)*
bedienergeführt operator prompted *(me)*
Bedienfunktion operating function
Bedienkomfort ease of operation, operator comfort *(for translation example see under* **Bedienungskomfort***)*
Bedienkonsole 1. control desk/panel/console *(general term).* 2. operator panel *(me)*
Bedienperson operator
Bedientableau 1. control panel *(general term).* 2. operator panel *(me)*
Bedientafel 1. control panel *(general term).* 2. operator panel *(me)*
Bedientaste control key
Bedienterminal operator terminal *(me)*
Bedienungsablauf operating procedure
Bedienungsanleitung operating instructions
Bedienungsaufwand *effort required to operate a machine:* **geringer Bedienungsaufwand** easy to operate
Bedienungsautomatik automatic operating mechanism
Bedienungselemente controls, control elements
Bedienungsfehler error in operation: **Selbst bei Bedienungsfehlern werden Maschinenbeschädigungen vermieden** Even if a mistake is made in operating the machine, it will not be damaged. OR: even if the machine is operated incorrectly, it will not suffer damage
Bedienungsfeld 1. control panel *(general term).* 2. operator panel *(me)*
bedienungsfreundlich 1. easy to operate/use. 2. easily accessible; **Die Steuerungen sollten bedienungsfreundlich untergebracht**

werden the controls should be easily accessible. **Genauso bedienungsfreundlich ist der Ausbau des Werkzeugs.** Dismantling the mould is just as easy OR equally simple. 3. user-friendly *(me)*
Bedienungsfreundlichkeit ease of operation, operator-friendliness *(see translation example under* **Bedienbarkeit, leichte***)*
Bedienungsfront 1. control panel *(general term)*. 2. operator panel *(me)*
Bedienungsgegenseite the side opposite the operator's side
Bedienungsgeräte controls
Bedienungskomfort ease of operation, operator comfort: **gesteigerter Bedienungskomfort** easier to operate OR: greater operator comfort
Bedienungsmann (machine) operator
Bedienungsmannschaft machine operators, operating personnel
Bedienungsorgane controls
Bedienungsperson operator
Bedienungspersonal machine operators, operating personnel
Bedienungspult 1. control desk/console/panel *(general term)*. 2. operator panel *(me)*
Bedienungsschalter operating switch
Bedienungsschutztür operator's guard door
Bedienungsseite operator's side *(of machine)*
Bedienungstableau control/instrument panel
Bedienungstastatur control keyboard
bedingt beständig limited resistance *(material in the presence of chemicals)*
bedingt durch caused by, due to
Bedingungen, verfahrenstechnische processing conditions
bedruckbar printable: **gut bedruckbar** easily printed
Bedruckbarkeit printability
Bedrucken printing
bedruckt printed
bedürfnisorientiert demand-oriented
beeinflußbar *that which can be influenced*: **Die Zersetzungstemperatur ist durch Kickerzusatz beeinflußbar** the decomposition temperature can be influenced by adding kickers
beeinträchtigen to impair
Beeinträchtigung impairment
befahrbar resistant to wheeled traffic
Befehl command, instruction *(me)*
Befehlsbus instruction bus *(me)*
Befehlsdecoder instruction decoder *(me)*
Befehlseingabe instruction input *(me)*
Befehlsregister instruction register *(me)*
Befehlszähler instruction counter *(me)*
Befestigungsbohrung fixing hole
Befestigungsflansch mounting flange
Befestigungsklemme fixing clamp
Befestigungslaschen fastening lugs
Befestigungspunkt fastening/fixing point
Befestigungsschraube fixing/fastening screw
Befeuchtigungseinrichtung humidifier

beflammen to apply a flame
Beflammen flame treatment *(eg. of a polyethylene surface to make it suitable for printing)*
Beflammstation flame treatment station, flame treating unit
Beflammungszeit flame impingement time
Beflocken flocking, flock spraying
beflockt flocked
Beflockungskleber flock spraying adhesive, flocking adhesive
Beförderungsvorschriften transport regulations
befriedigend satisfactory
Begasen 1. aeration. 2. mechanical blowing *(method of making foam)*. 3. introduction of gas
Begasungsanlage 1. aerator. 2. mechanical blowing unit *(which passes gas into PVC paste to produce foam)*
Begasungsschaum mechanically blown foam *(using gas or air as opposed to blowing agent)*
Begasungsverfahren mechanical blowing (process) *(method of making foam using gas or air)*
begehbar resistant to foot traffic
Begleiterscheinung side effect
Begleitstoff additive
begrenzt limited
begrenzt löslich sparingly soluble
begrenzt verträglich (with) limited compatibility
Begrenzung 1. limit switch. 2. limiting device
Begrenzungsbacken distance pieces, check plates *(c)*
begründet, theoretisch theoretically established
Behälter container
Behälterauskleidung tank lining
Behältergreifstation bottle gripping unit *(bm)*
Behälterinhalt tank contents
Behälterverkleidung tank lining
Behältnis container
Behandlung treatment
Behandlungsgut material/substance/product being treated
beheizbar heatable
Behörden authorities
beidseitig on both sides, two-sided
Beigabe addition, incorporation
Beilagsscheibe washer
beimengen to add, to incorporate
beimischen to add/mix in/incorporate
Beimischung addition, incorporation
Beipaß by-pass
Beispritzextruder ancillary/subsidiary extruder
Beistellextruder ancillary/subsidiary extruder
Beistellmaschine 1. ancillary extruder. 2. machine-side granulator, press-side granulator *(sr)*
Beistelltrockner machine-side drier
Beiwert coefficient, factor
Beizbad etching/pickling bath
Beizbehälter pickling tank

beizen to pickle *(e.g. metal in acid prior to bonding or painting)*
Beizmittel pickling solution
Beizsäure pickling acid
bekannt familiar
Bekleidungssektor clothing industry
Beladestation loading station
Belag 1. deposit. 2. coating, covering. 3. (floor) screed
Belagbildung plate-out
belastbar *that which can be subjected to stress:* **hoch belastbare Werkstoffe** materials which can be subjected to high stresses/loads; **thermisch hoch belastbare Werkstoffe** materials which can be subjected to high temperatures OR: which will withstand high temperatures
Belastbarkeit load-bearing/-carrying capacity
Belastbarkeit, thermische thermal stability, heat resistance
belastet stressed
Belastung stress, load(ing), application of stress, applied load *The word must often be translated very freely, as in the following examples:* **Lichtschutzmittel für normale Belastung** light stabiliser for normal conditions; **kurzzeitige Belastungen bei 300°C** short-term exposure to 300°C; **Verfärbung bei thermischer Belastung** discolouration when exposed to heat ; **...und die thermisch Belastung der Schmelze verringert wurde** ...and the melt was not exposed/subjected to such high temperatures; **Belastung der Umwelt** pollution of the environment, environmental pollution
Belastung, ruhende static stress
Belastung, schwellende fatigue stress
Belastung, wechselnde cyclic stress/loading
Belastung, zügige dynamic stress
Belastungsamplitude stress amplitude
Belastungsart type of stress/loading
belastungsbezogen stress-related
Belastungsdauer stress duration, loading time
Belastungseinrichtung stress/load application device
Belastungsfrequenz stress frequency
belastungsgerecht capable of withstanding applied loads, *e.g.* **belastungsgerecht Bauteile** load-bearing components. **Es ist möglich, Fasern im Bauteil belastungsgerecht auszurichten.** It is possible to align the fibres in the component in such a way that it can support applied loads.
Belastungsgeschwindigkeit stressing rate
Belastungsgrenze loading-bearing/-carrying capacity
Belastungsgröße stress level, amount of stress
Belastungshöhe stress level
Belastungskollektiv (applied) load
Belastungskurve stress application curve
belastungsorientiert capable of withstanding applied loads *(for translation examples see under* **belastungsgerecht***)*
Belastungsprüfung loading test
Belastungsrichtung direction of loading
Belastungsschritte loading stages
Belastungsspitze stress/load peak
Belastungsversuch loading test
Belastungszeit stress duration, loading time
Belastungszeit, thermische time of exposure to heat: **...zur Verringerung der thermischen Belastungszeit der Schmelze** to cut down the time during which the melt is subjected to high temperatures
Belastungszustand state of stress
Belastungszyklus stress cycle
Belebtschlammanlage activated sludge purification plant
Belegemischung coating compound
Belegmuster reference sample
Belegschaft workforce, employees
Beleimungswalze glue applicator roll
Belichtung light ageing, exposure to light
Belichtungsprüfung light exposure test
Belüftungsloch ventilation hole
Belüftungsöffnung ventilation hole
Belüftungsventil vent(ing) valve
Bemessen dimensioning
bemessen dimensioned
bemessene Leistung rating *(of an engine)*
benachbart adjacent
Benard'sche Zellen Benard's cells
Benetzbarkeit wettability
Benetzen wetting
Benetzungseigenschaften wetting properties/characteristics
Benetzungsfähigkeit wetting properties/characteristics
Benetzungsfehler wetting defect/fault
Benetzungsgeschwindigkeit wetting rate
Benetzungsmittel wetting agent
Benetzungsschwierigkeiten wetting problems
Benetzungsstörung wetting defect/fault
Benetzungsverhalten wetting properties/power
Benetzungsvermögen wetting characteristics/power
Benetzungswinkel contact angle
Bentonit bentonite
benutzerfreundlich 1. user-friendly *(me)*. 2. easy to use/operate
Benutzerfreundlichkeit user-friendliness
Benutzerführung user control
Benutzeroberfläche user interface *(me)*
Benutzerprogramm application/user program *(me)*
Benutzersoftware application/user software *(me)*
Benzin 1. benzine. 2. petrol, gasoline
Benzinbeständigkeit fuel resistance
Benzine petroleum fractions
benzinfest fuel resistant
Benzinkohlenwasserstoff aliphatic hydrocarbon
Benzinpumpe fuel pump

Benzintank petrol/fuel tank
Benzinverbrauch petrol/fuel consumption
Benzochinon benzoquinone
Benzochlorphenol benzochlorophenol
Benzoesäure benzoic acid
Benzoguanamin benzoguanamine
Benzoguanaminharz benzoguanamine resin
Benzol benzene
Benzolextrakt benzene extract
Benzolkern benzene ring
Benzolkohlenwasserstoff aromatic hydrocarbon
Benzolring benzene ring
Benzolsulfonsäure benzene sulphonic acid
Benzophenon benzophenone
Benzothiazol benzothiazole
Benzothiazolabkömmling benzothiazole derivative
Benzothiazolsulfenamid benzothiazole sulphenamide
Benzothiazylsulfenamid benzothiazyl sulphenamide
Benzoylgruppe benzoyl group
Benzoylperoxid benzoyl peroxide
Benzoylperoxidpaste benzoyl peroxide paste
Benztriazol benztriazole
Benzylalkohol benzyl alcohol
Benzylbutylphthalat benzyl butyl phthalate, BBP
Benzylcellulose benzyl cellulose
Benzyldimethylamin benzyl dimethylamine
Benzyloctyladipat benzyl octyl adipate, BOA
Beobachtungsdauer observation/test period
Berechnen computation
berechnet calculated
Berechnungen, statische stress analysis
Berechnungsansatz mathematical equation
Berechnungsaufwand number of calculations needed/necessary
Berechnungsformel mathematical formula
Berechnungsgleichung mathematical equation
Berechnungsmodell mathematical model
Berechnungsprogramm computer program *(me)*
Beregnung water spray
Beregnungsvorrichtung water spraying device
Bereich zone, region, area
Bereich, im thermoelastischen when soft and flexible: **Die Rohrstücke werden bis in den thermoelastischen Bereich erwärmt** the parisons are heated until they have become soft and flexible
Bereichsterminal area terminal *(me)*
Bereichswähler range selector
Bereinigung streamlining, simplification
Bergbau mining
Bergbaubetrieb mine
Bergspitzen peaks *(of an irregular surface)*
bernsteinfarben amber (-coloured)
Bernsteinsäure succinic acid
Bernsteinsäureanhydrid succinic anhydride

Bernsteinsäuredimethylester dimethyl succinate
Bernsteinsäureester succinate
Berstdruck bursting pressure
Berstdruckfestigkeit bursting strength
Berstdruckversuch bursting pressure test
bersten to burst
Berstfestigkeit bursting strength
Berstscheibe bursting disc
Berstspannung bursting stress
Berstversuch bursting test
Berufsgenossenschaft professional association
Beruhigungszone 1. die land *(e)*. 2. relaxation zone *(part of screw)*
Berührungsdruck contact pressure
Berührungsflächen contact surfaces
berührungslos 1. solid-state, electronic. 2. non-contact, contact-less
Berührungsstelle point of contact, contact point
Berührungsthermometer contact thermometer
Berührungswinkel angle of contact, contact angle
Berührungszone contact zone
Berylliumkupfer beryllium-copper alloy
beschädigt damaged
Beschädigung damage
Beschädigungsgefahr risk of damage
Beschäftigtenzahl number of employees
Beschäftigungslage employment situation
Beschichten coating, spread-coating
Beschichtung coating
Beschichtungsanlage coating line/plant
Beschichtungsaufbau paint film composition
Beschichtungsbad electroplating bath
Beschichtungsdicke coating thickness
Beschichtungseinheit coating unit
Beschichtungseinrichtung coating machine/unit
Beschichtungsextruder coating extruder
Beschichtungsfehler coating defect
Beschichtungsfilm film, coating
Beschichtungsgewicht coating weight
Beschichtungsindustrie paint/coatings industry
Beschichtungsmaschine coating machine
Beschichtungsmasse coating compound
Beschichtungsmittel coating compound
Beschichtungsoberfläche film/coating surface
Beschichtungsqualität 1. coating quality. 2. paint film quality
Beschichtungsstoff (surface) coating compound, paint
Beschichtungsstraße coating line
Beschichtungssystem (surface) coating system
Beschichtungstechnologie surface coating technology, paint technology
Beschichtungsuntergrund substrate, surface *(already painted or to be painted)*
Beschichtungsverfahren coating process
Beschichtungsversuche 1. coating trials. 2. painting trials
Beschickeinrichtung feed equipment/unit

beschicken to feed *(a hopper, calendar or compression mould)*
Beschickseite feed side
Beschickung 1. feeding. 2. feed unit/equipment
Beschickungsaggregat feed unit
Beschickungsanlage feed unit/equipment
Beschickungsautomat automatic feeder, automatic feed unit
Beschickungseinrichtung feed equipment/unit
Beschickungsöffnung feed throat/opening *(e)*
Beschickungsschnecke feed screw
Beschickungssystem feed system
Beschleuniger accelerator
Beschleunigermenge amount of accelerator
beschleunigt accelerated
Beschleunigung acceleration
Beschleunigungsaufnehmer acceleration transducer
Beschleunigungsmoment moment of acceleration, acceleration torque
Beschleunigungssystem accelerator system
beschränkt löslich sparingly soluble
Besohlungsmaterial soling compound
Besonderheiten special/distinguishing features, salient points
Besputtern sputtering
beständig unaffected, resistant *(material in the presence of chemicals)*
Beständigkeit stability, resistance, durability
Beständigkeit, chemische chemical resistance
Beständigkeit, thermische thermal stability, heat resistance
Beständigkeitseigenschaften durability, stability (characteristics), resistance: **chemische Beständigkeitseigenschaften** chemical resistance
Beständigkeitstabelle chemical resistance table
Bestandteil constituent
Bestandteile, flüchtige volatile matter/content
Bestimmung, naßchemische wet analysis
Bestimmungen, bauaufsichtliche building regulations
Bestimmungen, lebensmittelrechtliche food regulations
Bestimmungsmethode method of determination
bestrahlen to irradiate
Bestrahlgut material/substance to be OR being irradiated
bestrahlt irradiated
Bestrahlung irradiation
Bestrahlungsdosis radiation dose
Bestrahlungsempfindlichkeit radiation sensitivity
Bestrahlungsintensität radiation intensity
Bestrahlungsvernetzung radiation cure/crosslinkage
Bestrahlungsversuch irradiation test
Betastrahlen beta rays
betätigt, zwangsläufig powered
Betätigungsfelder fields of activity

Betätigungshebel operating lever, actuating lever
Betätigungsnocken actuating cam
Beteiligung 1. holding, interest *(in a company)*. 2. participation
Beteiligungsgesellschaft associated company
Beton concrete
Betonanstrich concrete finish
Betonersatzmörtel concrete patching/repair mortar
Betonfundament concrete foundations
Betongehwegplatte concrete paving slab
Betonimprägnierung 1. concrete impregnation. 2. concrete impregnating agent
Betonklebstoff concrete adhesive
Betonoberfläche concrete surface
Betonpfahl concrete pier
Betonsanierung concrete repair/restoration
Betonsanierungsfarbe concrete repair paint
Betonschäden concrete cancer
Betonschutz concrete protection, protection of concrete
Betonschutzanstrich (protective) concrete finish
Betonschutzfarbe (protective) concrete paint
Betonversiegelung concrete sealant (composition)
Betrieb, intermittierender intermittent operation, discontinuous operation
Betrieb, permanenter continuous operation
Betriebsanlage production plant/facility
Betriebsanleitung operating instructions
Betriebsanweisung operating instructions
Betriebsart 1. type of operation *(e.g. manual or automatic)*. 2. operating mode *(me)*
Betriebsartenwahlschalter function selector switch
Betriebsaufwand operating costs
Betriebsausfall production breakdown
Betriebsbedingungen operating/service conditions
betriebsbereit ready for operation/use
Betriebsdaten production data
Betriebsdatenauswertung production data evaluation
Betriebsdatenerfassung production data collection/acquisistion *(me)*
Betriebsdatenerfassungsstation production data terminal *(me)*
Betriebsdatenprotokoll production data record
Betriebsdauer time of operation, operating period: **nach einer 2000-stündigen Betriebsdauer** after 2000 hours' operation
Betriebsdrehmoment operating torque
Betriebsdrehzahl 1. operating speed. 2. operating screw speed *(e)*
Betriebsdrehzahlbereich operating speed range
Betriebsdruck operating/working pressure
betriebseigen in-plant, in-house
Betriebsenergie *energy used or needed to operate something:* **Die Heizenergie,**

einschließlich der Betriebsenergie der Kühlanlage... the heating energy, including the energy used to operate the cooling plant...
Betriebserfahrungen practical experience
betriebsfähig in working order
Betriebsfertigungsmittel tooling aid
Betriebsgrößen production data
Betriebshygiene factory/works hygiene
betriebsintern in-house, in-plant
Betriebskalender production calender
Betriebskontrolle works control
Betriebskosten running costs, operating costs
Betriebslabor works laboratory
Betriebslebensdauer working/service life
Betriebsmann (machine) operator
Betriebsmaßstab industrial scale
Betriebsmittel equipment
Betriebsmoment operating torque
Betriebsparameter operating parameter: *The word can often be omitted, since its inclusion would produce an awkward sounding sentence, as in this example:* **Es bleibt als wesentliches Ziel die Verbesserung der Konstanz der Betriebsparameter Einspritzgeschwindigkeit und Nachdruck** the essential aim is to achieve more constant injection speed and holding pressure
betriebssicher dependable, reliable (in operation)
Betriebssicherheit dependability, reliability (in operation)
Betriebssicherung guard, safety device
Betriebssoftware operational software *(me)*
Betriebsspannung operating voltage
Betriebsspiel play, clearance
Betriebsstellung operating position
Betriebsstoff fuel, petrol
Betriebsstörung malfunction, machine malfunction
Betriebsstunden hours of operation: **Sie ist ausgelegt für 60.000 Betriebsstunden** it is designed for 60,000 hours of operation
Betriebsstundenzähler hours counter, machine hours counter
Betriebstemperatur operating/working temperature
Betriebstemperaturbereich operating temperature range
betriebsunfähig out of order
Betriebsverfassungsgesetz Law on the Constitution of Businesses
Betriebsverhalten performance *(e.g. of a machine)*
Betriebsweise method of operation: **geräuscharme Betriebsweise** quiet operation
Betriebswerte operating variables
Betriebszeit time of operation, operating period *(see translation example under Betriebsdauer)*
Betriebszustände operating conditions
Beuldruck buckling pressure

Beulfestigkeit buckling resistance
Beullast buckling stress
Beulstabilität buckling resistance
Beulsteifigkeit buckling resistance
Beurteilung assessment
Beurteilung, visuelle visual inspection/ examination
Beutelautomat automatic bag-making machine
Beutelherstellungsautomat automatic bag-making machine
Beutelmaschine bag-making machine
Beutelschweißautomat automatic bag-welding machine
Bevorratung stockpiling
bevorzugt preferred
Bewässerungsrohr irrigation pipe
beweglich mobile: **bewegliche Teile** moving parts
bewegliche Werkzeugaufspannplatte moving platen *(im)*
bewegliche Werkzeughälfte moving mould half *(im)*
bewegliche Werkzeugträgerseite moving platen *(im)*
Beweglichkeit mobility
Bewegung, Brownsche Brownian movement
Bewegung, makro-Brownsche macro-Brownian movement
Bewegung, mikro-Brownsche micro-Brownian movement
Bewegungsabläufe machine movements
Bewegungsfuge expansion joint
Bewegungsgeschwindigkeit, hohe fast cycling *(injection moulding or blow moulding machine)*
Bewegungsreibung sliding friction
Bewegungsrichtung direction of movement
Bewehrung reinforcement
Bewehrungsstahl reinforcing steel rod(s)
Bewertung assessment, evaluation
Bewertungskriterien rating/assessment criteria
Bewetterungs- *see* **Bewitterungs-**
bewittert weathered
Bewitterung weathering
Bewitterungsbeanspruchung weathering: **je nach Bewitterungsbeanspruchung** depending on weathering conditions
Bewitterungsbedingungen weathering conditions
Bewitterungsbeständigkeit weathering resistance
Bewitterungsdauer weathering period: **nach 12-jähriger Bewitterungsdauer** after 12 years' weathering
Bewitterungeigenschaften weathering characteristics
Bewitterungseinflüsse weathering influences
Bewitterungsergebnisse weathering (test) results
Bewitterungsgerät weathering instrument
Bewitterungsprüfung weathering test
Bewitterungsresistenz weathering resistance
bewitterungsstabil weather resistant

Bewitterungsstabilität weathering resistance
Bewitterungsstand weathering station
Bewitterungsstation weathering station
Bewitterungsverhalten weathering properties
Bewitterungsversuch weathering test/trial
Bewitterungszeit weathering period *(for translation example see under* **Bewitterungsdauer***)*
Bewitterungszeitraum weathering period
bewußt intentional(ly)
Beziehung relation, relationship
Bezugsdruck reference pressure
Bezugsgröße reference quantity
Bezugskurve reference curve
Bezugslehre reference jig
Bezugsmaterial reference material
Bezugsmessung reference determination
Bezugsmodell master model
Bezugspunkt point of reference
Bezugsquelle source of supply
Bezugsstoff 1. reference substance 2. upholstery material
Bezugstemperatur reference temperature
Bezugszeitraum reference period
BGA *abbr. of* **Bundesgesundheitsamt**, Federal Health Authority
Biaxial-Reckanlage biaxial stretching unit/ equipment
Bicarbonsäure dicarboxylic acid
bidestilliert double distilled
bidirektional bidirectional
biegebeansprucht under bending/flexural stress
Biegebeanspruchung flexural/bending stress
Biegebelastbarkeit flexural load-carrying capacity
biegebelastet under bending/flexural stress
Biegebelastung flexural/bending stress
Biegedehnung flexural/bending strain
Biegeelastizitätsmodul flexural modulus of elasticity
Biege-E-Modul flexural modulus of elasticity
Biegefestigkeit flexural strength, cross-breaking strength
Biegefließspannung flexural yield stress
Biegekraft bending force
Biege-Kriechmodul flexural creep modulus
Biegekriechversuch flexural creep test
Biegelehre bending fixture
Biegemodul flexural modulus
Biegemoment bending moment
Biegen bending
Biegeprobe flexural test piece, flexural specimen
Biegereißbeständigkeit flex cracking resistance
Biegerißbeständigkeit flex crack(ing) resistance
Biegerißbildung flex cracking
Biegeschablone 1. curved jig, arc-shaped jig *(stress cracking test, DIN 53449).* 2. bending jig
Biegeschälfestigkeit flexural peel strength
biegeschlaff flexible, pliable

Biegeschwellfestigkeit flexural fatigue strength
Biegeschwellversuch flexural fatigue test
Biegeschwingfestigkeit flexural fatigue strength
Biegeschwingung free vibration
Biegeschwingversuch torsion pendulum test
Biegeschwingvorrichtung torsion pendulum apparatus
Biegespannung bending/flexural stress
Biegespannungsverteilung flexural stress distribution
Biegestab flexural specimen, flexural test piece
biegesteif rigid
Biegesteifigkeit 1. flexural strength. 2. rigidity, stiffness
Biegestreifenverfahren bent strip test *(for determining stress cracking resistance)*
Biegeträger bumper bracket
Biegeumformen bending
Biegeverformbarkeit resistance to deformation through bending
Biegeverformung deformation due to bending
Biegeverhalten flexural behaviour
Biegeversuch flexural/bending test
Biegevorspannung hydraulic pull-back system *(c)*
Biegevorspannung, hydraulische hydraulic pull-back system *(c)*
Biegewechselbeanspruchung flexural fatigue stress
Biegewechselbelastung flexural fatigue stress
Biegewechselfestigkeit flexural fatigue strength
Biegewechselversuch flexural fatigue test
biegeweich flexible, pliable
Biegewinkel bending angle
Biegezug flexural tension
Biegezugfestigkeit tensile strength in bending
biegsam flexible, supple
Biegsamkeit flexibility, suppleness
biegsamkeiterhöhend flexibility-enhancing
biegsamkeiterhöhende Weichmacher plasticisers to improve/enhance flexibility
Biegung 1. bending, flexure 2. deflection
Biegung beim Bruch deflection at break
Biegungs-Verformungsverhalten flexural deformation characteristics
Bierflasche beer bottle
Bierkasten beer crate
Bifunktionalität bifunctionality
bifunktionell bifunctional
Bilanz balance sheet
Bilanzsumme balance sheet total
Bildplatte video disc
Bildpunkte dots *(me)*
Bildschirm VDU screen, screen, display *(me)*
Bildschirmanzeige visual display *(me)*
Bildschirmarbeitsplatz - VDU workstation *(me)*
Bildschirmcomputer computer with integral screen *(me)*
Bildschirmeingabemaske screen input/entry mask *(me)*
Bildschirmgerät monitor *(me)*

Bildschirmgrafik graphic display *(me)*
Bildschirminhalt display contents *(me)*
Bildschirmmarke cursor *(me)*
Bildschirmmaske screen mask *(me)*
Bildschirmmenü screen menu *(me)*
Bildschirmprogrammiergerät VDU screen programmer *(me)*
Bildschirmseite display page *(me)*
Bildschirmterminal display terminal *(me)*
Bildungsgeschwindigkeit rate of formation
Bimetallthermostat bimetallic thermostat
Bimetallzylinder 1. bimetallic cylinder *(im)*. 2. bimetallic barrel *(e)*
BImSchG abbr. of **Bundesimmissionsschutzgesetz** Federal Anti-Pollution Act
binär binary *(me)*
binär kodiert binary coded *(me)*
Binärcode binary code *(me)*
Binärdaten binary data *(me)*
Binärzahl binary number *(me)*
Bindefestigkeit bond strength
Bindegarn binder twine
Bindekraft 1. adhesive force. 2. binding power
Bindemitel, hydraulisches hydraulic adhesive/cement
Bindemittel binder
bindemittelarm low-solids
Bindemittelgehalt binder content/concentration
Bindemittelkombination binder combination
Bindemittellösung 1. (paint) vehicle/medium 2. binder solution
Bindemittelmolekül binder molecule
bindemittelreich high-solids *(paint)*
Bindenaht weld line, flow line *(mark on moulded article caused by the meeting of two flow fronts during moulding. Not to be confused with **Teilungslinie** (q.v.))*
Bindenahtfestigkeit weld line strength
Bindenahtplazierung position of the weld line *(im)*
Bindenahtzähigkeit weld strength
binderarm low-solids
Bindung 1. (type of) weave. 2. linkage, bond
Bindungsenergie bond energy
Bindungskraft bond force
Binnenmarkt home market
Binnenmarkt Europa single European market
Binnenmarkt, europäischer single European market
Binnenmarkt, gemeinsamer single European market
bioabbaubar biodegradable
Biokatalysator biocatalyst
biokompatibel biocompatible
Biokunststoff biopolymer
biologisch abbaubar biodegradable
biologisches Recycling biological recycling
biomedizinisch biomedical
Biopolymer biopolymer
bioreaktiv bioreactive
Biosphäre biosphere

Biotop biotope
Biozid biocide
Bipolarität bipolarity
Bis-Acyllactam bis-acyl lactam
Bismaleinimid bis-maleinimide
Bismaleinimidharz bismaleinimide resin
Bismaleinimid-Triazinharz bismaleinimide triazine resin
Bisphenol bisphenol
Bisphenol A-Epoxidharz bisphenol A epoxy resin
Bisphenol-A-Diglycidylether bisphenol-A-diglycidyl ether
Bisphenolharz bis-phenol resin
Bit bit *(me)*
Bitprozessor bit processor *(me)*
Bitumen bitumen
bitumenbeständig bitumen resistant
Bitumenbeständigkeit bitumen resistance
Bitumenemulsion bitumen emulsion
bitumenhaltig bituminous
Bitumenmasse bitumen compound
Bitumenpappe roofing felt
bituminös bituminous
Biuret biuret
Biuretbindung biuret linkage
Biuretbindungsanteil biuret linkage content
Biuretpolyisocyanat biuret polyisocyanate
Biuretstruktur biuret structure
Blähdruck foaming pressure
blähen to expand
Blähmittel blowing agent
Blanc fixe blanc fixe, precipitated barium sulphate
blankgeglüht bright-annealed
Blas- und Füllmaschine blow moulding and filling machine
Blasaggregat 1. blow moulding unit. 2. film blowing unit
Blasanlage 1. blow moulding line *(bm)*. 2. blown film (extrusion) line *(bfe)*
Blasautomat automatic blow moulding machine
blasbar blow mouldable
Blasbutzen flash, parison waste *(bm)*
Blascoextrusionsfolie coextruded blown film
Blascoextrusionsverfahren blown film coextrusion
Blasdorn blowing mandrel/spigot, inflating mandrel/spigot *(bm)*
Blasdornträger blowing mandrel support *(bm)*
Blasdruck blowing pressure *(bm)*
Blasdüse 2. blown film die, film blowing die *(bfe)* 2. parison die *(bm)*
Blase 1. bubble, blister. 2. film bubble *(bfe)*. 3. bladder *(e.g. of a bladder accumulator)*
Blasen 1. blow moulding. 2. blown film extrusion, film blowing
blasen, ins Freie free forming *(t)*
Blasenbildung blistering, formation of bubbles
blasenfrei 1. bubble-free, free from bubbles. 2. void free
Blasenhalslänge frost line height *(bfe)*

Blaseninnenkühlsystem internal bubble cooling system, *(bfe)*
Blaseninnenkühlung internal bubble cooling system *(bfe)*
Blaseninnenluft air inside the (film) bubble *(bfe)*
Blasenkontur film bubble contours *(bfe)*
Blasenspeicher bladder accumulator
Blasenstabilität (film) bubble stability *(bfe)*
Blasextrusion 1. extrusion blow moulding. 2. blown film extrusion, film blowing
Blasextrusionsanlage extrusion blow moulding line/plant
Blasflasche blow moulded bottle
Blasfolie blown film, tubular film *(bfe)*
Blasfolien- see **Schlauchfolien-**
Blasfolienschlauchkopf blown film die, film blowing die
Blasform blow(ing) mould *(bm)*
Blasform-, Füll- und Verschließanlage blow-fill-seal packaging line
Blasform- und Füllmaschine blow moulding and filling machine
Blasformanlage blow moulding line/plant
Blasformautomat automatic blow moulding machine
Blasformautomaten-Baureihe range of automatic blow moulding machines/equipment
Blasformen blow moulding
Blasformextruder extruder: *Although this will generally be adequate, it may sometimes be necessary to qualify the word, e.g.* **Der Vergleich von Blasformextrudern mit anderen Extrusionsmaschinen...** comparing extruders for use with blow moulding machines with other extruders...
Blasformmaschine blow moulding machine, blow moulder *(the latter expression is used where brevity is important, e.g. in headlines, captions, advertising slogans etc.)*
Blasformmasse blow moulding compound
Blasformstation blow moulding station
Blasformsystem blow moulding system
Blasformtechnik blow moulding (technology)
Blasformteil 1. blow moulding, blow moulded part. 2. blown container. 3. bottle
Blasformung blow moulding
Blasformverfahren blow moulding (process)
Blasformwerkzeug blow(ing) mould *(bm)*
blasgeformt blow moulded, blown
Blasgesenk blow mould cavity
Blasigkeit porosity
Blaskopf 1. blown film die, film blowing die *(bfe)*. 2. parison die *(bm)*. **Blaskopf** *is sometimes still used in a blow moulding context, but has now been largely replaced by* **Schlauchkopf** *and* **Schlauchwerkzeug**
Blaskopf, zentral angespritzter centre-fed blown film die
Blaskopfdüse 1. blown film die, film blowing die *(bfe)* 2. parison die *(bm)*
Blaskopfkonzeption blown film die design
Blaskörper 1. blow moulding, blow moulded part. 2. blown container. 3. bottle
Blaskörperentnahme 1. removal of blow mouldings. 2. blown part removal device/unit
Blaslinie blow moulding line
Blaslippen die lips: *Since the term is invariably used in a bfe context, there is no need to translate the first part of the word*
Blasluft blowing air, inflation air *(bm, bfe)*
Blasluftdruck blowing air pressure *(bm, bfe)*
Blasluftzufuhr blowing air supply *(bm, bfe)*
Blasmarke blow moulding grade *(of moulding compound)*
Blasmaschine blow moulding machine, blow moulder *(see explanatory note under* **Blasformmmaschine***)*
Blasmasse blow moulding compound
Blasmedium blowing medium *(bm)*
Blasnadel blowing pin, inflation needle *(bm)*
Blaspinole blowing mandrel/spigot, inflating mandrel/spigot
Blasposition blowing position
Blasprozeß 1. blow moulding (process). 2. blown film extrusion (process)
Blasschlauch 1. parison *(bm)*. 2. film bubble *(bfe)*
blaßgelb pale yellow
Blasspritzkopf blown film die, film blowing die
Blasstation blowing station *(bm, bfe)*
Blasstift blowing pin, inflation needle *(bm)*
Blasteil 1. blow moulding, blow moulded part. 2. blown container. 3. bottle
Blasteilbutzen flash, parison waste *(bm)*
Blasverarbeitung 1. blow moulding. 2. blown film extrusion, film blowing
Blasverfahren 1. blow moulding. 2. blown film extrusion, film blowing
Blasvorgang 1. blow moulding. 2. blown film extrusion, film blowing
Blasvorrichtung blowing device
Blaswerkzeug 1. blow(ing) mould *(bm)*. 2. blown film die, film blowing die *(bfe)*
Blaszeit duration of blowing, time required for blowing
blättchenförmig platelet-like
Blättchenstruktur platelet structure
Blattfeder leaf spring
blättrig flaky, platelet-like
Blattschreiber page printer *(me)*
Blauer Engel This is Germany's universally recognised symbol indicating that a product is environment-friendly. There is, as yet, no equivalent term in English, so that translators should write the German expression in inverted commas, followed by something along these lines: literally "blue angel", a label issued by the Federal Office for the Environment to denote environmentally safe products. **"Blauer Engel" Farben** environment-friendly paints.
Blaustich blue tinge
Blechbandlack coil coating paint/enamel/lacquer

Bleiabgabe lead emission
bleibend permanent *(eg. dimensional change)*
bleibende Bruchdehnung residual/irreversible/ permanent elongation at break
bleibende Dehnung residual/irreversible/ permanent elongation
bleibende Verformung residual/irreversible/ permanent deformation
Bleicarbonat lead carbonate
Bleichlorid lead chloride
Bleichromatpigment lead chromate pigment
bleifrei lead-free
Bleigehalt lead content
bleihaltig lead-containing, containing lead
Bleimaleat lead maleate
Bleinaphthenat lead naphthenate
Bleioktoat lead octoate
bleiorganisch organo-lead
Bleiphosphit lead phosphite
Bleiphthalat lead phthalate
Bleipigment lead pigment
Bleisalicylat lead salicylate
Bleischrot lead shot
Bleiseife lead soap
Bleistabilisator lead stabiliser
bleistabilisiert lead stabilised
Bleistearat lead stearate
Bleisulfat lead sulphate
Bleiverbindung lead compound
Bleiweiß white lead
Blende diaphragm
Blendenkalibrator draw plate calibrator *(e)*
Blendenpaket sizing plate assembly, draw plate assembly *(e)*
blendfrei non-reflecting
Blendpartner biend component
Blendrahmen case, casing, (outer) frame
Blendwerkstoff blend
Blindlösung reference solution
Blindprobe 1. blank test 2. reference sample
Blindversuch blank test
Blindwert blank value
blinken to flash
Blinksignal flashing light signal
Blink-Störungslampe flashing light warning signal
Blisterformmaschine blister packaging machine
Blisterpackautomat automatic blister packaging machine
Blisterpackmaschine blister packaging machine
Blisterpackung blister pack
Blisterverpackung blister pack
Blockaufbau modular construction
Blockbauweise, in modular
Blockcopolymer block copolymer
Block-Copolymerisation block copolymerisation
Blockdarstellung 1. block diagram. 2. bar graph
Blockdiagramm 1. block diagram. 2. bar graph
Blocken blocking

blockfest non-blocking
Blockfestigkeit blocking resistance
blockfrei non-blocking
blockiert blocked
Blockkraft blocking force
Blockneigung tendency to block, blocking tendency
Blockpolymerisat 1. bulk polymer. 2. block copolymer
Blockpolymerisation bulk polymerisation
Blockpunkt blocking temperature
Blockschaltbild circuit block diagram
Blockschaumstoff slabstock foam
Blockschäumung slabstock foaming
Blockschäumverfahren slabstock foaming
Blockströmung solid flow
Blocktendenz blocking tendency
Blockverhalten blocking behaviour/properties
Blockware slabstock foam
Blockweichschaum flexible slabstock foam
Blumenkübel plant tub
Blutverträglichkeit blood compatibility
BMC *abbr. of* bulk moulding compound
BMC-Formmasse bulk moulding compound, BMC
BMFT *abbr. of* **Bundesministerium für Forschung und Technologie**, Federal Ministry for Reserach and Technology
BMI *abbr. of* **Bismaleinimid**, bis-maleinimide
BMI-Harz bismaleinimide resin
BOA *abbr. of* **Benzyloctyladipat**, benzyl octyl adipate, BOA
Bobine 1. bobbin, reel *(txt)*. 2. winding mandrel
Bodenabfall base flash, tail flash *(bm)*
Bodenausgleichmasse flooring screed, jointless flooring compound
Bodenbelag 1. floorcovering. 2. (floor) screed
Bodenbelagsfirma flooring contractor
Bodenbutzen base/tail flash *(bm)*
bodeneben at floor level
Bodenentgrateinrichtung base deflashing device *(bm)*
Bodenerosion soil erosion
Bodenfläche 1. floor area. 2. floor space
Bodennahtbeutel bottom-weld bag
Bodenplatte floor slab
Bodenquetschnaht bottom weld *(bm)*
Bodensatz sediment
Bodensatzbildung sedimentation, settling out
Bodenschweißnaht bottom weld *(bm)*
Bodenspachtelmasse flooring screed, jointless flooring compound
Bodenverunreinigung soil contamination
Bogenstück elbow
Bogenwiderstand arc resistance
Bohrbild mould fixing diagram, mould fixing details, standard platen details *(see explanatory note under* **Lochbild***)*
Bohren drilling
Bohrkern core sample
Bohrlehre drill jig
Bohrmaschine power drill

Bohröl cutting oil
Bohrung bore, hole: **Seitlich am Zylinder angebrachte Bohrungen dienen zur Aufnahme von Thermofühlern** holes drilled into the side of the barrel serve to accomodate thermocouples
Bohrung, nitriergehärtete nitrided liner *(of an extruder barrel)*
Bohrungsdurchmesser bore *(of an extruder barrel)*
Bohrungskern core pin *(im)*
Bohrungsstift core pin *(im)*
Bohrungswalze drilled roll
Bolzen bolt
Bolzendornhalterung bolt-type mandrel support *(e)*
Bolzenfallversuch falling weight test
Bombage convex grinding *(c)*
bombiert convex ground *(c)*
Bootsbau boatbuilding
Borax borax
Borcarbid boron carbide
bördeln to flange
Bördelwerkzeug flangeing fixture
Borfaser boron fibre
Borsäure boric acid
Bortrifluorid boron trifluoride
Borverbindung boron compound
Böschung embankment
Böschungswinkel angle of inclination
BR *abbr. of* butadiene rubber
Branche industry
branchenspezifisch industry-related
Brandausbreitung flame spread
Brandbekämpfung fire fighting
Brandentdeckung fire detection
Brandfall, im in case of fire
Brandfortpflanzung flame spread
Brandgefahr fire risk
brandgeschützt flameproofed, flame resistant
Brandgeschwindigkeit burning rate
Brandgroßversuch large scale burning test
Brandklasse flammability rating/classification/group *(for translation example see under* **Brandschutzklasse***)*
Brandklasseneinteilung flammability rating
Brandklassifizierung flammability rating/classification/group
Brandmeldeanlage fire alarm system
Brandnebenerscheinungen secondary effects of fire
Brandneigung flammability
Brandquelle source of fire
Brandrisiko fire risk
Brandschachttest chimney test *(to determine flammability)*
Brandschachtverfahren chimney test *(to determine flammability)*
Brandschutz fire protection
Brandschutzadditiv flame retardant
brandschutzausgerüstet flameproofed, flame resistant
Brandschutzausrüstung 1. making flame retardant/resistant. 2. flame retardant (additive): **GF-UP-Platten mit Brandschutzausrüstung** flame-proofed GRP sheets; **Polystyrol mit Brandschutzausrüstung** polystyrene containing a flame retardant OR: flame retardant polystyrene; **Produkttypen mit Brandschutzausrüstung** flame retardant grades; **UP-Harze lassen sich auch mit Brandschutzausrüstung herstellen** UP resins can also be made flame resistant
Brandschutzbeschichtung flameproof coating
Brandschutzbeschichtung, dämmschichtbildende flameproof coating
Brandschutzgesetze fire regulations
Brandschutzklasse flammability rating/classification/group: **...und sind in die Brandschutzklasse B1 einzuordnen** ...and have a flammability rating of B1
Brandschutzmarke flame retardant grade
Brandschutzmaßnahmen fire precautions
Brandschutzmittel flame retardant
Brandschutznorm fire safety standard
brandschutztechnisch *relating to fire protection*: **brandschutztechnische Anforderungen** fire regulations ; **brandschutztechnische Prüfung** fire safety test; **brandschutztechnische Auflagen** fire safety regulations/requirements; **brandschutztechnische Eigenschaften** flame retardant properties
Brandschutztür fire door
Brandschutzverhalten fire retardancy
brandsicher flame resistant
Brandsicherheit fire resistance
Brandsicherheitsvorschriften fire regulations
Brandtest flammability test
Brandverhalten fire behaviour
Brandverhaltensklasse fire behaviour rating/classification
brandverhindernd fire preventing ...**daß PVC von seiner chemischen Struktur her brandverhindernd wirkt...** ...that PVC, because of its chemical structure, helps to prevent fire...
Brandversuch flammability test
Brandvorbeugung fire prevention
Brandvorschriften fire regulations
Brandweiterleitung flame spread
Brandwidrigkeit fire/flame resistance
Brauchwasser water for industrial use
braune Ware brown goods *(consumer electronics such as video recorders, camcorders, fax machines, satellite dishes etc.)*
Braunfärbung brown discolouration
Brecherplatte breaker plate *(e)*
Brechungsindex refractive index
Brechungszahl refractive index
Brechwert refractive index
Brechzahl refractive index
breiartig paste-like

Breite, flachgelegte layflat width *(bfe)*
Breiteneinsprung transverse shrinkage
Breitenkonstanz constant width: **Enge Breitenkonstanz der Folie** very constant film width
Breitenregelung 1. width control. 2. width control mechanism
Breitenregulierung 1. width control. 2. width control mechanism
Breitenschwankungen width variations
Breitensteuerung 1. width control. 2. width control mechanism
breitgefächert wide-ranging, widespread
Breithalter spreader roll *(e, bfe,c) (used to keep film uniformly flat across its width)*
Breithaltevorrichtung spreader roll unit
Breitreckmaschine transverse stretching machine
Breitschlitzdüse 1. slit die, slot die. 2. flat film (extrusion) die *(for gauges below 0.25mm)*. 3. sheet (extrusion) die *(for thicker gauges)*
Breitschlitzdüsenplatte extruded sheet
Breitschlitzdüsenverfahren slit die extrusion (process)
Breitschlitzextrusion slit die extrusion
Breitschlitzextrusionsanlage slit die extrusion line
Breitschlitzflachfolienanlage flat film extrusion line
Breitschlitzfolie extruded film/sheeting:
Breitschlitzfolie aus schlagfestem Polystyrol extruded, high-impact polystyrene sheeting
Breitschlitzfolienanlage flat film extrusion line, slit die film extrusion line
Breitschlitzfolienextrusion flat film extrusion, slit die film extrusion
Breitschlitzfolienverfahren flat film extrusion, slit die film extrusion
Breitschlitzplatte extruded sheet
Breitschlitzplattenextrusion sheet extrusion
Breitschlitzverbundfolie co-extruded flat film
Breitschlitzverfahren slit die extrusion
Breitschlitzwerkzeug 1. slit die, slot die. 2. flat film (extrusion) die *(for gauges below 0.25mm)*. 3. sheet (extrusion) die *(for thicker gauges)*
Breitstreckmaschine transverse stretching machine
Breitstreckwalze 1. transverse stretching roll *(e) (used to stretch film to orient it)*. 2. spreader roll *(see explanatory note under* **Breithalter***)*
Breitstreckwerk 1. spreader roll unit *(e,bfe,c) (see explanatory note under* **Breithalter***)*. 2. transverse stretching unit *(see explanatory note under* **Breitstreckwalze***)*
Bremsaggregat brake system
Bremsbacke brake shoe
Bremsband brake band
Bremsbelag brake lining
Bremsbelagharz brake lining resin
Bremse brake

Bremsflüssigkeit brake fluid
Bremssystem braking system/mechanism
Bremsventil braking valve
Bremsvorrichtung braking device
Bremsweg braking distance
brennbar 1. flammable. 2. combustible
Brennbarkeit 1. flammability. 2. combustibility
Brennbarkeitsklasse flammability classification/group/rating
Brennbarkeitsprüfverfahren flammability test
Brennbarkeitsstufe flammability classification/rating/group
Brennbarkeitstest flammability test
Brenndauer burning time
Brenner burner
Brenngeschwindigkeit burning rate
Brenngeschwindigkeit burning rate
Brennstelle burn mark
Brennstoff fuel
Brennstoff, fossiler fossil fuel
Brennstofftank fuel tank
Brennstrecke burning distance
Brennverhalten fire behaviour
brillant sparkling
Brillanz sparkle
Brillengestell spectacle frame
Brinellhärte Brinell hardness
bromfrei bromine-free
bromhaltig bromine-containing, brominated
bromiert brominated
Bromkresolgrün bromocresol green
Bromverbindung bromine compound
Bronze bronze
Brookfield-Viskosimeter Brookfield viscometer
Broschüre booklet, brochure
Brownsche Bewegung Brownian movement
Bruch fracture, failure
Bruch, duktiler ductile fracture/failure
Bruch, Durchbiegung beim deflection at break
Bruch, katastrophaler catastrophic failure
bruchanfällig fragile, liable to break/crack
Bruchanfälligkeit fragility, tendency to crack/break
Brucharbeit fracture energy
Bruchausbreitung crack propagation
bruchauslösend fracture/initiating
Bruchbiegespannung bending/flexural stress at break
Bruchbild fracture photomicrograph
Bruchdehnung elongation at break, ultimate elongation
Bruchdehnung, bleibende residual/irreversible/permanent elongation at break
Bruchdehnungsänderung change in elongation at break
Brucheinleitung crack initiation
bruchempfindlich fragile, liable to break/crack
Bruchempfindlichkeit fragility, liability to break
Bruchenergie fracture energy
Bruchfestigkeit 1. breaking strength. 2. bursting strength *(e.g. of a blow moulded container)*. 3. fracture resistance

Bruchfläche fracture surface
Bruchfortschritt crack propagation
Bruchgrenze breaking limit
brüchig brittle
Bruchkraft breaking stress
Bruchlast breaking stress
Bruchlastspielzahl number of breaking stress cycles
Bruchmechanik fracture mechanics
bruchmechanisch fracture-mechanical
Bruchmechanismus failure mechanism
Bruchmorphologie fracture morphology
Bruch-Schwingspielzahl number of vibrations to failure
bruchsicher break resistant
Bruchsicherheit break resistance
Bruchspannung 1. breaking stress, tensile stress at break, ultimate tensile stress. 2. rupture voltage *(in electrophoretic paint deposition)*
Bruchstauchung compression at break
Bruchstück fragment
Bruchverhalten fracture behaviour
Bruchvorgang fracture process
Bruchzähigkeit fracture toughness
Brückenanschluß bridge abutment expansion joint
Brückenlager bridge bearing pad
Brückenpfeiler bridge pier
Brückenplatte bridge deck
Brückenübergänge nosings and cover plates for bridge expansion joints
Brückenzufahrt bridge approach
Brunnentopf wellpoint
Brüstungsplatte parapet cladding panel
Brüstungsverkleidung parapet cladding
Bruttogewicht gross weight
Bruttosozialprodukt gross national product
Bruttozusammensetzung gross/overall composition
B-Säule B pillar *(pillar to which the seatbelt is fixed)*
BT-Harz bis-maleinimide triazine resin
Bubblepackung bubble pack
Buchdruck letterpress printing
Buchse bush, bushing, socket
Bügeleisen electric/domestic iron
Bügellänge die land *(e)*
Bügelstrecke die land *(e)*
Bügelzone die land *(e)*
Bugspoiler front spoiler
Bühne (machine) platform
Bündelkabel bunched cables
Bundesgesundheitsamt Federal Health Authority
Bundesimmissionsschutzgesetz Federal Anti-Pollution Act, Federal Pollution Control Act
Bundesimmissionsschutzverordnung Federal Anti-Pollution Act, Federal Pollution Control Act
Bundesländer, alte the former W. Germany
Bundesländer, neue the former E. Germany

bündig flush
Bunsenbrenner Bunsen burner
Bunsenflamme Bunsen flame
bunt brightly coloured, multi-coloured
Buntmetall non-ferrous metal
Buntmetall-Legierung non-ferrous metal alloy
Buntpigment 1. pigment. 2. coloured pigment *(where a distinction is made between* **Buntpigmente** *and* **Weißpigmente**, *q.v)*
buntpigmentiert containing coloured pigments *(see explanatory notes under* **Buntpigment***)*
Bürette burette
Bürogeräte office equipment
Büromaschinen office/business machines
Büromöbel office furniture
Büroraum office
Bürostuhl office chair
Bürste brush
Bürsteffekt brushed surface finish
Bürstenapplikation application by brush, brush application
Bus-Leitung bus, highway *(me)*
Busschnittstelle bus interface *(me)*
Butadien butadiene
Butadienacrylnitril butadiene acrylonitrile
Butadien-Acrylnitrilcopolymer butadiene-acrylonitrile copolymer/rubber, nitrile rubber
Butadien-Acrylnitrilkautschuk butadiene-acrylonitrile rubber, nitrile rubber
Butadienkautschuk butadiene rubber
Butadien-Nitrilkautschuk butadiene-acrylonitrile rubber, nitrile rubber
Butadien-Styrolkautschuk butadiene-styrene rubber
Butan butane
1,4-Butandiol 1,4-butane diol
Butandiol butane diol
Butandioladipat butane diol adipate
Butandioldimethacrylat butane diol dimethacrylate
Butanol butanol
Buttersäure butyric acid
Buttersäureanhydrid butyric anhydride
Butylacetat butyl acetate
Butylacrylat butyl acrylate
Butylalkohol butyl alcohol
Butylbenzoesäure butyl benzoic acid
Butylbenzol butyl benzene
Butylbenzyladipat butyl benzyl adipate
Butylbenzylphthalat butyl benzyl phthalate, BBP
Butylbrenzkatechin butyl pyrocatechol
Butyldiglykolacetat butyl diglycol acetate
Butylether butyl ether
Butylglycidylether butylglycidyl ether
Butylglykol butyl glycol
Butylgruppe butyl group
Butylhydroperoxid butyl hydroperoxide
butyliert butylated
Butylkautschuk butyl rubber
Butylmethacrylat butyl methacrylate
Butylmethylphenylsiliconharz butyl methyl

phenyl silicone resin
Butylmethylsiliconharz butyl methyl silicone resin
Butylperbenzoat butyl perbenzoate
Butylperoctoat butyl peroctoate
Butylperoxid butyl peroxide
Butylphenylglycidylether butylphenylglycidyl ether
Butylstearat butyl stearate
Butyltitanat butyl titanate
Butyltrichlorsilan butyl trichlorosilan
Butylzinncarboxylat butyl-tin carboxylate
Butylzinnmercaptid butyl-tin mercaptide
Butylzinnstabilisator butyl-tin stabiliser
Butylzinnverbindung butyl-tin compound
Butyraldehyd butyraldehyde
Butzen flash, parison waste *(bm)*
Butzenabfall flash, parison waste *(bm)*
Butzenabschlageinrichtung deflashing device *(bm)*
Butzenabtrennung 1. flash trimming. 2. flash trimming mechanism *(bm)*
Butzenabtrennvorrichtung flash trimmer *(bm)*
Butzenbeseitigung flash removal *(bm)*
Butzenkammer flash chamber *(bm)*
Butzenmaterial flash, parison waste *(bm)*
Butzenminimierung minimising the amount of flash produced *(bm)*
Butzentrenner flash trimmer *(bm)*
BV-Klasse *abbr. of* **Brandverhaltensklasse**, fire behaviour rating/ classification
Bypasshahn by-pass valve
Bypassventil by-pass valve
B-Zustand B-stage

C

CAD 1. *abbr. of* computer aided design 2. *abbr. of* computer aided drafting
CAD-Arbeitsplatz CAD workstation *(me)*
CAD/CAM-Technik computer aided design and manufacturing, CAD/CAM
CAD-Konstruktion computer aided design, CAD *(me)*
CAD-Konstruktionsprogramm CAD program *(me)*
Cadmium cadmium
Cadmiumabgabe cadmium emission
cadmiumfrei cadmium-free
Cadmiumgelb cadmium yellow
Cadmiumlaurat cadmium laurate
Cadmiumorange cadmium orange
Cadmiumpigment cadmium pigment
cadmiumpigmentiert cadmium pigmented
Cadmiumrot cadmium red
CAD-Paket CAD package *(me)*

CAD-Rechentechnik computer aided design, CAD *(me)*
CAD-Softwarepacket CAD software package *(me)*
CAE *abbr. of* computer aided engineering *(me)*
CAED *abbr. of* computer aided engineering and design *(me)*
CAE-Paket CAE package *(me)*
CAE-Programm CAE program, computer-aided engineering program *(me)*
CAE-Technik computer aided engineering, CAE
Calcit calcite
Calciumborat calcium borate
Calciumcarbonat calcium carbonate
Calciummontanat calcium montanate
Calciumoktoat calcium octoate
Calciumoxid calcium oxide
Calciumseife calcium soap
Calciumsilikat calcium silicate
Calciumstearat calcium stearate
Calcium/Zink-Stabilisierung calcium-zinc stabiliser
calziniert calcined
Calzinierung calcination
CAM *abbr. of* computer aided manufacturing *(me)*
CAM Rechentechnik computer aided manufacturing, CAM *(me)*
Campher camphor
Camphersäure camphoric acid
cancerogen carcinogenic
CAP *abbr. of* computer aided planning *(me)*
Caprolactam caprolactam
Caprolactamring caprolactam ring
Caprolactamschmelze caprolactam melt
CAQ *abbr. of* computer aided quality control/ assurance *(me)*
CAQ-Anlage computer aided quality control/ assurance system
Carbamat carbamate
Carbamidharz urea resin
Carbamidsäure carbamic acid
Carbamidsäurechlorid carbamic chloride
Carbaminsäure carbamic acid
Carbamylsulfenamid carbamyl sulphenamide
Carbonamidgruppe carbonamide group
Carbonat carbonate
Carbonatisierung carbonation
Carbonatisierungsbremse carbonation inhibitor
Carbonfaser carbon fibre
Carbonsäure carboxylic acid
Carbonsäureamid carboxylamide
Carbonsäureanhydrid carboxylic acid anhydride
Carbonsäureester carboxylate, carboxylic acid ester
Carbonylgruppe carbonyl group
Carbonylverbindung carbonyl compound
Carboxylat carboxylate
Carboxylendgruppe terminal carboxyl group
Carboxylgruppe carboxyl group

carboxylgruppenaufweisend containing carboxyl groups
carboxylgruppenhaltig containing carboxyl groups
carboxyliert carboxylated
Carboxylierungsgrad carboxyl (group) content
Carboxylsäure carboxylic acid
Carboxymethylcellulose carboxymethylcellulose
carboxymethyliert carboxymethylated
Carnaubawachs carnauba wax
Casein casein
CASS *abbr. of* computer assisted solvent selection *(me)*
ca.-Wert approximate figure
Ca/Zn stearat calcium-zinc stearate
C-C-Bindung C-C/carbon-carbon bond
CC-Ruß conductive channel black
Cd-frei cadmium-free
CD-Platte compact disc, CD
CD-Plattenspieler compact disc player, CD player
Celluloid celluloid
Cellulose cellulose
Celluloseacetat cellulose acetate
Celluloseacetobutyrat cellulose acetate butyrate
Celluloseacetopropionat cellulose acetate propionate
Cellulosederivat cellulose derivative
Celluloseester cellulose ester
Cellulosefaser cellulose fibre
Cellulosehydrat cellulose hydrate
Cellulosenitrat cellulose nitrate
Cellulosepropionat cellulose propionate
Cellulosetriacetat cellulose triacetate
Cernaphthenat cerium naphthenate
C-Faser carbon fibre
CFK *abbr. of* **kohlenstoffaserverstärkte Kunststoffe**, carbon fibre reinforced plastics, CRP
CFK-Prepregmaterial carbon fibre prepreg
CF-Ruß conductive furnace black
Charakterisierung characterisation
charakterisieren to characterise, to describe
Charge batch
Chargengröße batch size
Chargenmenge batch size
Chargennummer batch number
Chargenprozeß batch/discontinuous process
Chargenprüfung batch control
Chargenqualität batch quality
Chargenschwankungen batch variations
Chargenunterschiede batch variations/differences
chargenweise batchwise, discontinuous(ly)
Charpygerät Charpy impact tester, Charpy apparatus
Charpy-Methode Charpy test, Charpy (flexural) impact test
Checkliste check list
Chelatbildner chelator

Chelator chelator
chelierend chelating
Chemieanlage chemical plant
Chemieanlagenbau chemical plant construction
Chemieapparatebau chemical equipment construction
Chemiefaser synthetic/man-made fibre
chemiefaserarmiert synthetic fibre reinforced
Chemiefasergewebe synthetic/man-made fabric
Chemiewerkstoff polymer material, synthetic resin, plastic
Chemiewerkstoffe, technische engineering plastics/polymers
Chemikalienangriff chemical attack
chemikalienbeansprucht exposed/subjected to chemical attack
Chemikalienbeanspruchung exposure to chemical attack
Chemikalienbelastung exposure to chemicals
chemikalienbeständig chemical resistant
Chemikalienbeständigkeit chemical resistance
Chemikaliendämpfe chemical fumes/vapours
Chemikalieneinfluß effect of chemicals
Chemikalieneinwirkung chemical attack: **Widerstandsfähigkeit gegen Chemikalieneinwirkung** resistance to chemicals, resistance to chemical attack
chemikalienfest chemical resistant
Chemikalienlagerung immersion in chemicals
chemikalienresistent chemical resistant
Chemikalienresistenz chemical resistance
Chemikalientauglichkeit chemical resistance
Chemikalienversprödung chemical embrittlement
Chemikalienzeitstandverhalten long-term chemical resistance
Chemiker chemist
Chemiluminiszenz chemiluminescence
chemisch chemical
chemisch getrieben chemically blown
chemisch neutral chemically inert
chemische Beanspruchung chemical attack
chemische Beständigkeit chemical resistance
chemische Umsetzung chemical reaction
chemische Verfahrenstechnik chemical process engineering
chemische Verwertung chemical recycling
chemischer Aufbau chemical structure
chemischer Schaum chemically blown foam
chemischer Umsatz chemical reaction
chemisches Recycling chemical recycling
Chemischreinigung dry cleaning
Chemismus chemical action
Chemisorption chemisorption
Chillrollabzug chill roll take-off unit
Chillrollanlage cast film extrusion line, chill roll casting/extrusion line
Chillrollfilmgießanlage cast film extrusion line, chill roll casting/extrusion line
Chillroll-Flachfolienextrusionsanlage cast film extrusion line, chill roll casting/extrusion line
Chillrollfolie cast film

Chillrollfolienanlage cast film extrusion line, chill roll casting/extrusion line
Chillrollverfahren cast film extrusion, chill roll casting/extrusion
Chillrollwalze chill roll
Chillrollwalzengruppe chill roll unit
Chinolin quinoline
Chinon quinone
chinonvulkanisiert quinone-vulcanised
chirurgisch surgical
Chlor chlorine
Chloracetamid chloroacetamide
Chloralkansäure chloroalkane acid
Chloratom chlorine atom
Chlorbenzol chlorobenzene
Chlorbutylkautschuk chlorinated butyl rubber
Chlorderivat chlorine derivative
Chlorethan chloroethane
Chlorgehalt chlorine content
chlorhaltig chlorine-containing, chlorinated
chloriert chlorinated
Chlorierung chlorination
Chlorierungsgrad degree of chlorination, chlorine content
Chlorit chlorite
Chlorkautschuk chlorinated rubber
Chlorkautschukanstrich chlorinated rubber coating/finish
Chlorkautschukfarbe chlorinated rubber paint
Chlorkautschuklack chlorinated rubber paint
Chlorkohlenwasserstoff chlorinated hydrocarbon
Chlormethylstyrol chloromethyl styrene
Chloroprenkautschuk chloroprene rubber
Chloroprenpolymerisat chloroprene polymer
Chlorparaffin chlorinated paraffin
Chlorphenol chlorophenol
Chlorpolyethylen chlorinated polyethylene
Chlorpolyolefin chlorinated polyolefin
Chlorpropylsilan chloropropyl silane
Chlorsilan chlorosilane
chlorsulfoniert chlorosulphonated
Chlorverbindung chlorine compound
Chlorwasserstoff hydrogen chloride
Chlorwasserstoffgas hydrogen chloride
Chlorwasserstoffsäure hydrochloric acid
cholesterisch cholesteric
Chromatogramm chromatogram
Chromatographie chromatography
chromatographisch chromatographic
Chromatpigment chrome pigment
Chromgelb chrome yellow
Chromophor chromophore
chromophor chromophoric
Chromoxidgrün chrome green
Chromsäure chromic acid
Chromschwefelsäurelösung chromic acid solution
CIM *abbr. of* computer integrated manufacturing *(me)*
CIM-Baustein CIM module *(me)*
CIM-Fertigungsbetrieb CIM plant *(me)*

CIM-Konzept CIM system *(me)*
Cirka-Gewicht approximate weight
cis-trans-Isomerie cis-trans isomerism
CIWH *abbr. of* computer integrated warehousing *(me)*
CKW *abbr. of* **Chlorkohlenwasserstoff**, chlorinated hydrocarbon
CM *symbol for* chlorinated polyethylene
CMC carboxymethyl cellulose
CMHR-Schaum combustion modified high resilience foam
CNC-geregelt CNC-controlled, computer numerically controlled
CNC-gesteuert CNC-controlled, computer numerically controlled *(me)*
CNC-Maschine CNC machine, computer numerically controlled machine *(me)*
CNC-System CNC system, computer numerical control system
CN-Gruppe cyanide group
CO 1. *symbol for* epichlorohydrin rubber 2. *chemical symbol for* carbon monoxide
Coaleszenz coalescent
coaleszierend coalescent
Coaten *German version of the word* "coating"
Cobalt- *see* **Kobalt-**
Co-Beschleuniger cobalt accelerator
Co-beschleunigt cobalt accelerated
Cocosölfettsäure coconut oil fatty acid
Codiereinrichtung coding equipment *(me)*
codiert coded *(me)*
Codierung coding *(me)*
Codimer co-dimer
coextrudierbar coextrudable
coextrudierende Flachfolienanlage flat film coextrusion line
coextrudierende Schlauchfolienanlage blown film coextrusion line
coextrudiert coextruded
Coextrusionsanlage coextrusion line
Coextrusionsbeschichtung coextrusion coating
Coextrusionsbeschichtungsanlage coextrusion coating line
Coextrusionsbeschichtungsverfahren coextrusion coating (process)
Coextrusionsblasanlage 1. coextrusion blow moulding plant *(bm)*. 2. blown film coextrusion line *(bfe)*
Coextrusionsblasfolienanlage blown film coextrusion line
Coextrusionsblasformanlage coextrusion blow moulding plant
Coextrusionsblasformmaschine coextrusion blow moulding machine
Coextrusionsblasformverfahren coextrusion blow moulding (process)
Coextrusionsbreitschlitzfolienanlage flat film coextrusion line
Coextrusionsschlauchkopf parison coextrusion die *(bm)*
Coextrusionsverfahren coextrusion (process)

Coflockulation coflocculation
coflockulierend cofloccultating
Coil-Coating-Decklack coil coating paint/enamel
Coil-Coating-Grundierung coil coating primer
Coil-Coating-Industrielack industrial coil coating paint/enamel/lacquer
Coil-Coating-Lack coil coating paint/enamel/lacquer
Cokondensat co-condensate
cokondensiert co-condensed
Collodiumwolle nitrocellulose, collodion cotton
coloristisch coloristic
Comonomer co-monomer
Compiler compiler *(me)*
Compilersprache compiler language *(me)*
Compoundieranlage compounding plant/line
Compoundieren compounding
Compoundierextruder compounding extruder
Compoundiermaschine compounder, compounding unit
Compoundierspielraum scope for compounding
Compoundierstraße compounding line
Compoundierung compounding
Computerausdruck computer print-out *(me)*
Computerband computer tape
Computereinsatz the use of computers
Computergehäuse computer housing
computergeregelt computer controlled
computergesteuert computer controlled *(me)*
computergestützt computer aided
computergestütztes Fertigen computer aided manufacturing, CAM
computergestütztes Konstruieren computer aided design, CAD
computerintegrierte Fertigung computer integrated manufacturing, CIM
Computerisierung computerisation
Computerprogramm computer program
Computerschnittstelle computer interface
Computersimulation computer simulation
Computertechnik computer technology/engineering
Computertechnologie computer technology/engineering
computerüberwacht computer controlled/supervised
computerunterstützt computer aided
computerunterstützte Qualitätssicherung computer aided quality assurance, CAQ
Co-Octoat cobalt octoate
COOH-Gehalt carboxyl (group) content
Copolyamid copolyamide
Copolyester copolyester
Copolymer copolymer
copolymer copolymeric
Copolymerfolie copolymer film
Copolymerharz copolymer resin
Copolymerisat copolymer
Copolymerisatharz copolymer resin
Copolymerisation copolymerisation

copolymerisierbar copolymerisable
copolymerisieren to copolymerise
Copräzipitat co-precipitate
Copräzipitieren co-precipitation
Coronabehandlung corona treatment
Coronabeständigkeit corona resistance
Coronaentladung corona discharge
Coronagenerator corona generator
Coronavorbehandlung corona pretreatment
Coronavorbehandlungsanlage corona pretreating unit
Co-Sikkativ cobalt drier
Costabilisator co-stabiliser
Covernetzer co-curing/-vulcanising agent
Covulkanisat co-vulcanisate
Covulkanisation co-vulcanisation
CP *abbr. of* cross-polarisation
CPU *abbr. of* central processing unit *(me)*
CPVC-Anstrich chlorinated PVC finish
CR 1. *abbr. of* chloroprene rubber 2. *abbr. of* controlled rheology
Craze craze
crazeauslösend craze-initiating
Crazebildung craze formation, crazing
Crazefeld craze zone
Crazeinitiierung craze initiation
Crazewachstum craze propagation
Crazezone craze zone
Crotonaldehyd crotonaldehyde
Cryogenkühlrohr cryogenic cooling pipe
C-Säule C pillar *(rear window pillar in a car)*
CSI *abbr. of* closed systems interconnection *(me)*
CSM *symbol for* chlorosulphonated polyethylene
CTAB cetyl trimethylammonium bromide
CuBe copper-beryllium
Cumaron coumarone
Cumaron-Indenharz coumarone-indene resin
Cumolhydroperoxid cumol hydroperoxide
Cursor cursor *(me)*
CV-Anlage continuous vulcanisation plant
CV-Belag cushioned vinyl floorcovering
c_w-Wert drag coefficient/factor
Cyanacrylatklebstoff cyanoacrylate adhesive
Cyanacrylsäure cyanacrylic acid
Cyanacrylsäureester cyanoacrylate, cyanoacrylic acid ester
Cyanatgruppe cyanate group
Cyansäurechlorid cyanuric chloride
Cyanursäure cyanuric acid
Cyanwasserstoff hydrogen cyanide
Cyclisierung cyclisation
Cycloakylgruppe cycloalkyl group
cycloaliphatisch cycloaliphatic
Cycloalkylrest cycloaklyl radical
cycloaromatisch cycloaromatic
Cyclohexan cyclohexane
Cyclohexanol cyclohexanol
Cyclohexanon cyclohexanone
Cyclohexanonharz cyclohexanone resin
Cyclohexanonperoxid cyclohexanone peroxide
Cyclohexen cyclohexene

Cyclohexylvinylether cyclohexyl vinyl ether
Cyclokautschuk cyclo-rubber, cyclised rubber
Cycloolefin cycloolefin
cyklisch cyclic

D

D/A Wandler digital-analog converter *(me)*
Dach mit Flachneigung low-pitched roof
Dachbahn roofing sheet/membrane
Dachbeschichtungsmasse roofing compound
Dachdämmung roof insulation
Dachdichtungsbahn roofing sheet/membrane
Dachfenster skylight, dormer window
Dachfolie roofing sheet/membrane
Dachhaut roofing sheet/membrane
Dachhautentlüfter roof ventilator
Dachhimmel (car) roof liner, headlining
Dachisolierung roof insulation
Dachrinne roof gutter
Dachspoiler roof spoiler
DAM *abbr. of* Diaminophenylmethan, diaminophenyl methane
Dämmaterial insulating material
Dämmdicke insulation thickness
Dämmeffekt insulating effect
Dämmeigenschaften insulating properties
Dämmplatte insulating sheet/panel
Dämmschicht insulating layer
dämmschichtbildende Brandschutzbeschichtung flameproof coating
Dämmschichtbildner substance which forms an insulating layer
Dämmschichtdicke insulation thickness
Dämmstoff insulating material
Dämmstoffmaterial insulating material
Dämmung insulation
Dämmvermögen insulating properties
Dämmwirkung insulating effect
Dammzone restricted flow zone
Dampf 1. vapour *(solvents or water)*. 2. steam *(only water)*
Dampfabschaltautomatik steam-actuated switch-off mechanism
Dampfalterung steam ageing
Dampfautoklav autoclave
Dampfbehandlung steam treatment
dampfbeheizt steam heated
dampfbremsende Schicht vapour barrier
dampfdicht vapour proof/impermeable
Dampfdichtheit vapour impermeability
Dampfdiffusionswiderstand vapour diffusion resistance
Dampfdruck 1. vapour pressure. 2. steam pressure

Dampfdurchlässigkeit vapour permeability
dämpfend damping
Dampferzeuger steam generating unit
Dampferzeugung 1. generation of steam. 2. steam generating unit
dampfförmig in vapour form: **dampfförmige Phase** vapour phase
dampfgeheizt steam heated
Dampfkammer steam chamber
Dampfkasten steam chest
Dampfkessel steam boiler
Dampfphase vapour phase
Dampfphasenlöten vapour phase soldering
Dampfsperrbahn vapour barrier (sheeting)
Dampfsperre vapour barrier
Dampfteildruck partial vapour pressure
Dampfundurchlässigkeit vapour impermeability
Dämpfung damping
Dämpfungseigenschaften damping characteristics
Dämpfungseinrichtung damping mechanism
Dämpfungselement damping element
Dämpfungsflüssigkeit damping medium/fluid
Dämpfungsgehäuse silencer housing
Dämpfungsmaximum damping maximum
Dämpfungsmedium damping medium/fluid
Dämpfungsmodul damping modulus
Dämpfungsverhalten damping behaviour
Dämpfungsvermögen damping characteristics
Dampfvulkanisation steam vulcanisation
Dampfvulkanisieranlage steam vulcanising plant
Darstellung, in schematischer schematically
Darstellung, schematische schematic diagram
Datei data file *(me)*
Dateistruktur data file structure *(me)*
Daten data *(Although data is plural, it is treated as a singular noun in computer terminology)*
Daten, toxikologische toxicological data/information
Datenabfrage data retrieval *(me)*
Datenablage data filing system, data store *(me)*
Datenabnahme data collection/acquisition/gathering *(me)*
Datenabspeicherung data storage *(me)*
Datenanalysator data analyser *(me)*
Datenanzeige 1. data display. 2. data display unit *(me)*
Datenarchivierung data filing *(me)*
Datenaufbereitung data preparation *(me)*
Datenaufschreibung data recording *(me)*
Datenausgabe data output *(me)*
Datenaustausch data exchange *(me)*
Datenauswertung data evaluation *(me)*
Datenbank data bank, database *(me)*
Datenbankdatei data file *(me)*
Datenbankmanagementsystem databank management system *(me)*
Datenbaseninhalt database contents *(me)*
Datenbasis database *(me)*
Datenbestände databases *(me)*
Datenbreite word length *(me)*

Datenbus data bus *(me)*
Datendarstellung data presentation *(me)*
Datendatei data file *(me)*
Datendokumentation data documentation *(me)*
Datendurchsatz data throughput *(me)*
Dateneingabe data input *(me)*
Datenerfassung data gathering/acquisition/ capture *(me)*
Datenerfassungsanlage data acquisition unit *(me)*
Datenerfassungsgerät data gathering device/ unit/equipment, data acquisition device/unit/ equipment *(me)*
Datenerfassungsstation data acquisition unit *(me)*
Datenerhebung data collection **Die Datenerhebung erfolgte durch persönliche Interviews** Data was collected during personal interviews
Datenfeld data field *(me)*
Datenfernübertragung remote data transfer/ transmission *(me)*
Datenfile data file *(me)*
Datenfluß data flow/stream *(me)*
Datenflut data deluge, flood of data *(me)*
Datengeber data generator *(me)*
Datenhaltung data retention *(me)*
Dateninhalt data content *(me)*
datenkompatibel data-compatible *(me)*
Datenkompression data compression *(me)*
Datenkorrektur data correction *(me)*
Datenleitung data line *(me)*
Datenmenge amount of data
Datenorganisation data organisation *(me)*
Datenreduktion data reduction *(me)*
Datenregister data register *(me)*
Datensammelsystem data collecting/acquisition system *(me)*
Datensammlung data collection/acquisition/ gathering *(me)*
Datenschnittstelle data interface *(me)*
Datenschutz data protection *(me)*
Datensicherheit data reliability *(me)*
Datensicherung data protection *(me)*
Datensichtgerät data display unit, visual display unit *(me)*
Datenspeicher data bank/store *(me)*
Datenspeichermöglichkeit data storage facility *(me)*
Datenspeicherplatte data storage disk *(me)*
Datenspeicherplatte, optische optic(al) disk *(me)*
Datenspeicherung 1. data storage. 2. data bank/store *(me)*
Datenstation data station *(me)*
Datenstrom data flow/stream *(me)*
Datentechnik data processing
Datenträger data carrier *(me)*
Datentransfer data transfer/transmission *(me)*
Datentransport data transfer/transmission *(me)*
Datenübertragung data transfer/transmission *(me)*

Datenverarbeitung data processing *(me)*
Datenverarbeitung, elektronische electronic data processing, EDP *(me)*
Datenverarbeitungsanlage data processing unit *(me)*
Datenverarbeitungseinheit data processing unit *(me)*
Datenverarbeitungsmöglichkeiten data processing facilities *(me)*
Datenverbund data link *(me)*
Datenverdichtung data compaction *(me)*
Datenverlust data loss *(me)*
Datenverwaltung data management *(me)*
Datenvorrat data store *(me)*
Datenwortbreite word length *(me)*
Datumsanzeige data display
Datumsidentifikation date identification *(me)*
dauerantistatisch permanently antistatic
Dauerbeanspruchbarkeit resistance to long-term stress, resistance to sustained loads
Dauerbeanspruchung long-term stress
Dauerbelastung long-term stress
Dauerbetrieb, im in continuous use/operation
Dauerbetriebsgrenztemperatur maximum continuous operating temperature
Dauerbetriebstemperatur continuous service temperature
Dauerbiegefestigkeit long-term flexural strength
Dauerbiegewechselfestigkeit flexural fatigue strength
Dauerbruch fatigue failure
Dauerdruckbelastung continuous compressive stress
Dauerdurchschlagfestigkeit long-term dielectric strength
Dauereinsatz continuous operation/use
dauerelastisch permanently flexible
Dauerelastizität long-term, permanent flexibility
Dauerfestigkeit 1. fatigue strength. 2. durability
Dauerfestigkeitsversuch fatigue test
dauerflexibel permanently flexible
Dauerflexibilität long-term flexibility
Dauergebrauch continuous use
Dauergebrauchseigenschaften long term performance
Dauergebrauchstemperatur continuous working/operating temperature, long-term service temperature
Dauergebrauchstüchtigkeit long-term serviceability
dauerhaft permanent, durable, lasting, long-lasting
Dauerhaftigkeit permanence, durability
dauerklebrig permanently tacky
Dauer-Knickversuch folding endurance test
Dauerkontakt permanent/continuous contact
Dauerkühlung continuous cooling
Dauerlagerung 1. long-term ageing. 2. prolonged immersion
Dauerlast continuous/permanent/sustained load
Dauerleistungsbedarf continuous power consumption

dauerplastisch permanently flexible
Dauerproduktion continuous production
Dauerprüfspannung long-term test voltage
Dauerschlagzähigkeit long-term impact strength
Dauerschwellbelastung fatigue stress
Dauerschwellfestigkeit fatigue/endurance limit
Dauerschwingfestigkeit fatigue strength
Dauerschwingfestigkeit fatigue strength
Dauerschwingverhalten fatigue behaviour
Dauerschwingversuch fatigue test
Dauerstandfestigkeit 1. creep strength. 2. creep rupture strength *(of pipes)*
Dauerstandversuch creep test
Dauerstrom constant current
Dauertemperatur sustained temperature
Dauertemperaturbelastung long-term exposure to high temperatures
Dauertemperaturbelastungsbereich 1. long-term heat resistance. 2. long-term thermal stabilty
Dauertemperaturbeständigkeit long-term heat resistance, long-term thermal stability
Dauertemperaturgrenze long-term temperature limit
Dauertest 1. creep test 2. long-term test
Dauertorsionsbiegebeanspruchung long-term torsional bending stress
Dauerverformung permanent deformation
Dauerverhalten long-term performance/behaviour
Dauerversuch long-term test
Dauerwalztest long-term milling test
Dauerwärmebelastbarkeit long-term heat resistance
dauerwärmebeständig *having long-term heat resistance:* **Der Kautschuk ist dauerwärmebeständig bei 250°C und darüber** the rubber will withstand temperatures of 250°C and over under conditions of continuous use
Dauerwärmebeständigkeit long-term heat resistance, long-term thermal stability
Dauerwärmeformbeständigkeit long-term heat resistance, long-term thermal stability: **Die Grenztemperaturen als Maß der Dauerwärmeformbeständigkeit betragen bei allen Produkten mindestens 150°C** all products will withstand temperatures of up to 150°C in continuous use
Dauerwärmelagerung long-term heat ageing
Dauerwärmestabilität long-term heat resistance, long-term thermal stability
dauerwärmestandfest *having long-term heat resistance. For translation example see under* **dauerwärmebeständig**
Dauerwechselfestigkeit resistance to long-term alternating stress
Dauerwitterungsstabilität long-term weathering resistance
DB *abbr. of* **Datenbank**, data bank, database
DBE *abbr. of* dibasic ester

DC *abbr. of* **Dünnschicht-Chromatographie**, thin-layer chromatography
DC-Analyse thin-bed chromatography
DCUP dicumyl peroxide
DDC-Regelung direct digital control *(me)*
DDC-Temperaturregelkreis direct digital temperature control circuit
DDC-Temperaturregelung direct digital temperature control
DDK *abbr. of* **dynamische Differenzkalorimetrie**, dynamic differential calorimetry
DD-Lack polyurethane paint
DE *abbr. of* **Defo-Elastizität**, Defo elasticity
Decarboxylierung decarboxylation
Deckanstrich top/finishing coat
Deckelfaß lidded drum
deckend opaque
Deckenleuchte ceiling light
Deckfähigkeit hiding power, opacity *(of a pigment or paint)*
Deckkraft hiding power *(of a pigment)*
Decklack paint
Decklackierung top coat
Decklage overlay *(grp)*
Deckschicht 1. gel coat *(grp)*. 2. top coat *(e.g. of a PVC paste)*. 3. outer layer *(eg. of a sandwich structure)*
Deckschichtharz gel coat resin *(grp)*
Deckschichtoberfläche gel coat surface *(grp)*
Deckstrich top coat
Deckstrichpaste top coating paste
Deckung opacity *(of a pigment)*
Deckungsgrad hiding power, opacity *(of a pigment or paint)*
Deckvermögen hiding power, opacity *(of a pigment or paint)*
Defekt defect
defekt defective, faulty
definiert specified, definite, clearly defined, definite
definitionsgemäß by definition
Deflokulation deflocculation
deflokuliert deflocculated
Defo-Elastizität Defo elasticity
Defo-Gerät Defo plastometer
Defo-Härte Defo hardness
Deformation deformation
Deformationsenergie deformation energy
Deformationsfestigkeit deformation resistance
Deformationsgeschichte deformation history
Deformationsgeschwindigkeit 1. rate of deformation. 2. shear rate
Deformationsgleichgewicht deformation equilibrium
Deformationsgrenze deformation limit
Deformationsmechanismus deformation mechanism
Deformationsreversibilität elastic recovery
Deformationsverhalten deformation behaviour
Deformationswiderstand deformation resistance

deformierbar deformable
Deformierbarkeit deformability
deformieren to deform
Degradation degradation
DEHA *abbr. of* **Diethylhexyladipat**, diethylhexyl/dioctyl adipate, DOA
dehnbar extensible
dehnbares Polystyrol expandable polystyrene
Dehnbarkeit 1. extensibility, elasticity. 2. elongation at break
Dehnbeanspruchung 1. strain 2. elongation, extension
Dehnbolzen thermal expansion piece
dehnfähig elastic, extensible
Dehnfähigkeit 1. extensibility, elasticity. 2. elongation at break
Dehnfolie stretch wrapping film
Dehnfuge expansion joint
Dehngeschwindigkeit straining rate, rate of elongation
Dehngrenzlinie creep curve
Dehnmeßstreifen strain gauge
Dehnspannung offset yield stress
Dehnströmung stretching flow
Dehnung 1. strain. 2. elongation, extension
Dehnung bei Streckgrenze elongation at yield
Dehnung bei Streckspannung elongation at yield stress, yield strain, tensile strain at yield
Dehnung beim Bruch elongation at break
Dehnung, bleibende residual/irreversible elongation
Dehnungausschlag strain amplitude
Dehnungsaufnehmer strain transducer/sensor
Dehnungsgeschwindigkeit rate of elongation/straining
Dehnungsgrad amount of strain/elongation
Dehnungsgrenze elastic limit
dehnungsinduziert strain-induced
Dehnungsinkrement strain increment
Dehnungsmeßstreifen strain gauge
Dehnungsspannung offset yield stress
Dehnungsverformung strain deformation
Dehnungsverhalten extensibility
Dehnungsvorgeschichte previous strain history
Dehnviskosität stretching viscosity
DEHP *abbr. of* **Diethylhexylphthalat**, diethylhexyl/dioctyl phthalate, DOP
Dehydratation dehydration
dehydratisierend dehydrating
Dehydratisierung dehydration
dehydrierend dehydrogenating
Dehydrierung dehydrogenation
Dehydrochlorierung dehydrochlorination
deionisiert de-ionised
Dekade decade
Dekadenschalter decade switch
Dekadenzähler decade counter
Dekahydronaphthalin decahydronaphthalene
Dekompressionsschnecke vented screw *(e)*
Dekompressionsventil relief valve
Dekompressionszone decompression section, devolatilising section, vent zone *(e)*

Dekontamination decontamination
dekontaminierbar decontaminable
dekorativ decorative
Dekorativlack decorative paint
Dekorfilm decorative film
Dekorfolie decorative film/sheeting
dekorgebend decorative
Dekorlaminat decorative laminate
Dekorpapier decorative paper
Dekorschichtstoff decorative laminate
Dekrement decrement
Delaminierung delamination
Delta-Anguß film gate *(im)*
Demontage dismantling
demontagefreundlich easy to dismantle
demontierbar removable
demontiert dismantled
Dendrit dendrite
dendritisch dendritic
Dentalabdruckmasse dental impression paste
Depolymerisation depolymerisation
Deponie waste/rubbish tip, waste/rubbish dump, landfill site/tip
Deponie, geordnete controlled tipping *(of waste)*
Deponiekapazität landfill site capacity
Deponiekosten dumping costs
Deponierung dumping, tipping, landfilling
Deponiesickerwasser leachate
Depotverfahren sandwich construction
Derivat derivative
Dermatosegefahr risk of dermatitis
Desagglomeration deagglomeration
desaktivieren to deactivate
Desaktivierung deactivation
Desinfektionsmittel disinfectant
desorbiert desorbed
Desorientierung disorientation
Desorption desorption
Desorptionsenergie desorption energy
Desorptionskurve desorption curve
destabilisiert destabilised
Destillat distillate
Destillation distillation
Destillationsgerät distillation apparatus
Destillationskolonne distillation column
destilliert distilled
detailgetreu faithful (to the original) in every detail
Detailinformationen detailed information
DETDA *abbr. of* **Diethyltoluylendiamin**, diethyl toluylene diamine
Detergentienfestigkeit detergent resistance
Detergentienlösung detergent solution
deuteriert deuterated
deutlich distinct(ly)
Devisenbedarf need for foreign exchange
Dextrin dextrin
dezentral decentralised
Dezentralisierung decentralisation
Dezimalbruch decimal fraction
DH *abbr. of* **Defo-Härte**, Defo hardness

Dia *abbr. of* **Diapositiv**, slide
Diacetonalkohol diacetone alcohol
Diacetylgruppe diacetyl group
Diacrylphthalat diacryl phthalate
Diacylperoxid diacyl peroxide
Diagnosefunktion diagnostic function *(me)*
Diagnoseprogramm diagnostics program *(me)*
Diagnosesystem diagnostics system *(me)*
Diagnosetext diagnostic text *(me)*
Diagrammpapier graph paper
Diagrammschreiber 1. chart recorder. 2. (graph) plotter *(me)*
Dialkyldichlorsilan dialkyl dichlorosilane
Dialkyldithiophosphat dialkyl dithiophosphate
Dialkyldithiophosphorsäuresalz dialkyl dithiophosphate
Dialkylester dialkyl ester
Dialkylperoxid dialkyl peroxide
Dialkylperoxidgruppe dialkyl peroxide group
Dialkylsulfid dialkyl sulphide
Dialkylverbindung dialkyl compound
Dialkylzinnbismerkaptid dialkyltin-bis-mercaptide
Dialkylzinndichlorid dialkyltin dichloride
Dialkylzinndimerkaptid dialkyltin dimercaptide
Dialkylzinnlaurat dialkyltin laurate
Dialkylzinnmaleat dialkyltin maleate
Dialkylzinnmerkaptid dialkyltin mercaptide
Dialkylzinnmerkaptochlorid dialkyltin mercaptochloride
Dialkylzinnstabilisator dialkytin stabiliser
Dialkylzinnthioglykolsäureester dialkyltin thioglycolate
Dialkylzinnverbindung dialkyltin compound
Diallylphthalat diallyl phthalate
Diallylphthalatpreßmasse diallyl phthalate moulding compound
Dialog, im in the interactive/conversational mode *(me)*
Dialogbetrieb interactive/conversational mode *(me)*
Dialogführung interactive/conversational mode *(me)*
dialoggeführt interactive, conversational *(me)*
Dialogverkehr interactive/conversational mode *(me)*
Diamin diamine
Diaminodiphenylmethan diaminodiphenyl methane
Diaminodiphenylsulfon diaminodiphenyl sulphone
Diaminododecan diaminododecane
Diaminohexan diaminohexane
Diaminomethylacrylat diaminomethyl acrylate
Diaminophenylmethan diaminophenyl methane
Diaminothiodiazol diamino-thiodiazole
Dianhydrid dianhydride
Diaprojektor slide projector
Diarylperoxid diaryl peroxide
Diatomenerde diatomaceous earth, kieselguhr
Diazodicarbonamid diazodicarbonamide
Diazoniumverbindung diazonium compound
Diazotierung diazotisation
Diazoverbindung diazo compound
dibasisch dibasic
Dibenzodioxin dibenzodioxin
Dibenzofuran dibenzofurane
Dibenzyladipat dibenzyl adipate
Dibenzylamin dibenzylamine
Dibenzyltoluol dibenzyl toluene
Diblockcopolymer diblock copolymer
Dibutyladipat dibutyl adipate
Dibutylmaleat dibutyl maleate
Dibutylphthalat dibutyl phthalate, DBP
Dibutylsebacat dibutyl sebacate, DBS
Dibutylzinndilaurat dibutyltin dilaurate
Dibutylzinnmaleinat dibutyltin maleinate
Dibutylzinnmerkaptid dibutyltin mercaptide
Dibutylzinnmerkaptoverbindung dibutyltin mercapto compound
Dibutylzinnstabilisator dibutyltin stabiliser
Dibutylzinnthioacetat dibutyltin thioacetate
Dibutylzinnthioglykolat dibutyltin thioglycolate
Dicarbonsäure dicarboxylic acid
Dicarbonsäureanhydrid dicarboxylic acid anhydride
Dicarbonsäureester dicarboxylate, dicarboxylic acid ester
Dicetylperoxydicarbonat dicetylperoxy dicarbonate
Dichlorbenzol dichlorobenzene
Dichlordiphenylsulfon dichlorodiphenyl sulphone
Dichlormethan dichloromethane
Dichlorpolyether dichloropolyether
dicht 1. dense *(texture)*. 2. tight, watertight, leakproof
Dichtbacken distance pieces, check plates *(c)*
Dichte density
Dichte, relative relative density, specific gravity
Dichtegradient density gradient
Dichtfläche sealing face
dichtgepackt tightly packed
Dichtigkeit tightness, imperviousness
Dichtigkeitsprüfsystem tightness checking system
dichtkämmend closely intermeshing *(screws) (e)*
Dichtkanten sealing edges
Dichtmittel sealant
Dichtprofil self-sealing profile, self-wiping profile *(e)*
Dichtprofilschnecken intermeshing screws with a self-sealing/-wiping profile *(e)*
Dichtring gasket
Dichtstoff sealant
Dichtung seal, gasket
Dichtungsbahn waterproof(ing)) membrane/sheet
Dichtungsband sealing tape
Dichtungsmasse sealant
Dichtungsmittel sealant
Dichtungsring gasket
Dichtungssatz seal assembly

Dickbettfliesenkleber thick-bed tile adhesive
Dickbettkleber thick-bed adhesive *(for fixing tiles)*
Dickbettverfahren thick-bed method
Dicke thickness
Dicke, verpreßte pressed thickness *(e.g. of a multi-layer laminate)*
Dickenabweichungen 1. thickness variations. 2. gauge variations *(of film)*
Dickenkalibrierung 1. thickness calibrations. 2. thickness calibration unit
Dickenmeßeinrichtung thickness gauge
Dickenmeßgerät thickness gauge
Dickenschwankungen 1. thickness variations. 2. gauge variations *(of film)*
Dickentoleranz 1. thickness tolerance. 2. gauge tolerance *(of film)*
Dickenverteilung thickness distribution
dickflüssig high-viscosity, viscous
Dickflüssigkeit high viscosity
Dickschichtanstrich high-build finish
Dickschichtfarbe high-build paint
Dickschichtlasur high-build glaze
Dickschichtsystem high-build system
dickwandig thick-walled
Dicumylperoxid dicumyl peroxide
Dicyandiamid dicyandiamide
Dicyclohexyl-Peroxydicarbonat dicyclohexylperoxy dicarbonate
Dicyclohexylphthalat dicyclohexyl phthalate
Dicyclopentadien dicyclopentadiene
Dicyclopentylacrylat dicyclopentyl acrylate
Dielektrikum dielectric
dielektrisch dielectric
dielektrische Verlustzahl loss index
dielektrischer Verlustfaktor dissipation factor
Dielektrizitätskonstante dielectric constant, relative permittivity
Dielektrizitätskonstante, relative relative permittivity, dielectric constant
Dielektrizitätszahl dielectric constant, relative permittivity
Dien diene
Dienkautschuk diene rubber
Dienolfettsäure dienol fatty acid
Dienstleistung service
Dienstprogramm service program/routine *(me)*
Diensynthesekautschuk diene rubber
Dieselkraftstoff diesel fuel
Dieselmotor diesel engine
Diester diester
Diethylanilin diethyl aniline
Diethylenglykol diethylene glycol
Diethylenglykoladipat dieythylene glycol adipate
Diethylenglykoldimethacrylat diethylene glycol dimethacrylate
Diethylenglykolmonomethacrylat diethylene glycol methacrylate
Diethylentriamin diethylene triamine
Diethylethanolamin diethyl ethanolamine
Diethylether (diethyl) ether

Di-2-ethylhexyladipat dioctyl adipate, di-2-ethylhexyl adipate, DOA
Di-2-ethylhexylazelat dioctyl azelate, di-2-ethylhexyl azelate, DOZ
Di-2-ethylhexylfumarat dioctyl fumarate, di-2-ethylhexyl fumarate
Di-2-ethylhexylmaleinat dioctyl maleinate, di-2-ethylhexyl maleinate
Di-2-ethylhexylphthalat dioctyl phthalate, di-2-ethylhexyl phthalate, DOP
Di-2-ethylhexylsebacat dioctyl sebacate, di-2-ethylhexyl sebacate, DOS
Di-2-ethylhexylterephthalat dioctyl terephthalate, di-2-ethylhexyl terephthalate, DOTP
Diethylphthalat diethyl phthalate, DEP
Diethylsebacat diethyl sebacate, DES
Diethyltoluylendiamin diethyl toluylenediamine
Differentialdichtung differential gear seal
Differentialgleichung differential equation
Differentialkalorimetrie differential calorimetry
Differentialthermoanalyse differential thermoanalysis, DTA
Differentialthermoanalysengerät differential thermal analyser
differentialthermoanalytisch differential-thermoanalytical
Differentialthermometrie differential thermometry
Differentialwaage differential scales
Differenzdruck pressure difference, differential pressure
Differenzdruckumformer differential pressure transducer
Differenzierung differentiation
Differenzkalorimetrie differential calorimetry
Differenzleistungskalorimetrie differential scanning calorimetry, DSC
Differenzspektrometrie differential spectrometry
Differenztemperatur temperature differential
Differenzwärmestromkalorimetrie differential heat flow calorimetry
diffundieren to diffuse
Diffusion diffusion
Diffusionsbarriere diffusion barrier
diffusionsdicht diffusion resistant
Diffusionsfähigkeit permeability
diffusionsfest diffusion resistant
Diffusionsfestigkeit diffusion resistance
Diffusionsgesamtwiderstand total diffusion resistance
Diffusionsgeschwindigkeit diffusion rate, rate of diffusion
Diffusionsgesetz diffusion law
diffusionshemmend diffusion-inhibiting
Diffusionskleben diffusion bonding
Diffusionskoeffizient diffusion coefficient, diffusivity
Diffusionskonstante diffusion constant
Diffusionspumpe diffusion pump
Diffusionsrate diffusion rate

diffusionsregulierend diffusion-controlling
Diffusionstiefe diffusion depth
Diffusionsverhalten diffusion behaviour
Diffusionsvorgang diffusion process
Diffusionsweg diffusion distance
Diffusionswiderstand diffusion resistance
Diffusionswiderstandszahl diffusion resistance coefficient
Diffusionszeit diffusion time
Difluoracrylat difluoroacrylate
difunktionell bifunctional
digital digital
Digital-Analog-Wandler digital-analog converter *(me)*
Digitalanzeige digital display *(me)*
Digitalausgang digital output *(me)*
Digitaldrucker digital printer *(me)*
digitale Audioplatte compact disc, CD
Digitaleingang digital input *(me)*
Digitaleinstellung digital setting
Digitalhydraulik digital hydraulics, digital hydraulic system
Digitalisierer digitiser *(me)*
Digitalisiertablett digitising tablet *(me)*
Digitalisierung digitisation *(me)*
Digitalisierungsgerät digitising unit, digitiser *(me)*
Digitalmengenblock digital volume control unit
Digitalrechner digital computer *(me)*
Digitalschallplatte compact disc, CD
Digitalschalter digital switch *(me)*
Digitalschalterreihen rows of digital switches *(me)*
Digitalsignal digital signal *(me)*
Digitalsteuerung 1. digital control. 2. digital control unit
Digitalventil digital valve
Digitalwaage 1. digital scales 2. digital balance
Digitalzeichenmaschine digital drawing instrument
Digitalzeitmeßgerät digital timer
Digitalzeituhr digital timer
Diglycidylether diglycidyl ether
Diglycidylverbindung diglycidyl compound
Dihexyladipat dihexyl adipate
Dihydrochinolin dihydroquinoline
2,6-Dihydroxynaphthalin 2,6-dihydroxynaphthalene
Dihydroxypolydimethylsiloxan dihydroxypolydimethyl siloxane
Dihydroxyverbindung dihydroxy compound
Diisocyanat di-isocyanate
Diisodecyladipat di-isodecyl adipate, DIDA
Diisodecylphthalat di-isodecyl phthalate, DIDP
Diisononyladipat di-isononyl adipate, DINA
Diisononylphthalat di-isononyl phthalate, DINP
Diisooctylphthalat di-isooctyl phthalate, DIOP
Diisopropylether di-isopropyl ether
Diisotridecylphthalat di-isotridecyl phthalate
Dikresylphenylphosphat dicresylphenyl phosphate
dilatant dilatant

Dilatanz dilatancy
Dilatation expansion, dilation
Dilatometer dilatometer
dilatometrisch dilatometric
Dilauroylperoxid dilauroyl peroxide
Dilaurylthiodipropionat dilauryl thiodipropionate
dimensioniert, großzügig generously dimensioned
Dimensionierungsprogramm dimensioning program *(me)*
Dimensionsänderung change in dimensions
dimensionsbeständig dimensionally stable
Dimensionsbeständigkeit dimensional stability
Dimensionsgenauigkeit dimensional accuracy
dimensionslos dimensionless
dimensionsstabil dimensionally stable
Dimensionsstabilität dimensional stability
Dimer dimer
dimer dimeric
Dimerisat dimer
Dimerisation dimerisation
Dimerisationsprodukt dimer
dimerisiert dimerised
Dimerisierung dimerisation
Dimerisierungsvorgang dimerisation reaction
Dimersäure dimeric acid
Dimethoxyethylphthalat dimethoxyethyl phthalate
Dimethylanilin dimethyl aniline
Dimethylbenzylamin dimethyl benzylamine
Dimethylcyclohexylphthalat dimethylcylclohexyl phthalate
Dimethylenetherbrücke dimethylene ether bridge
Dimethylformamid dimethyl formamide
Dimethylglkyolphthalat dimethyl glycol phthalate
Dimethylglyoxim dimethyl glyoxime
Dimethylolharnstoff dimethylol urea
Dimethylolphenolharz dimethylolphenol resin
Dimethylphenol dimethyl phenol
Dimethylphthalat dimethyl phthalate, DMP
Dimethylpolysiloxan dimethyl polysiloxane
Dimethylsulfat dimethyl sulphate
Dimethylsulfoxid dimethyl sulphoxide
Dimethylterephthalat dimethyl terephthalate
Dimethylvinylsilanol dimethylvinyl silanol
Dinatriumsalz disodium salt
Di-n-butylzinnmaleinat di-n-butyltin maleinate, dibutyltin maleinate
Di-n-butylzinnthioglykolat di-n-butyltin thioglycolate, dibutyltin thioglycolate
DIN-Klima DIN standard conditioning atmosphere *(see explanatory note under* **Normklima**)
Dinonyladipat dinonyl adipate, DNA
Dioctyladipat dioctyl adipate, DOA
Dioctylazelat dioctyl azelate, DOZ
Dioctylfumarat dioctyl fumarate
Dioctylmaleinat dioctyl maleinate
Dioctylphthalat dioctyl phthalate, DOP

Dioctylsebacat dioctyl sebacate, DOS
Dioctylzinnmerkaptid dioctyltin mercaptide
Dioctylzinnthioglykolat dioctyltin thioglycolate
Diode diode
Diol diol
Diolacrylat diol acrylate
Diolefinkautschuk diolefin rubber
Diolefinpolymerisat diolefin polymer
Diolmethacrylat diol methacrylate
Diorganosilan diorganosilane
Diorganozinnverbindung di-organotin compound
Dioxan dioxan
Dioxin dioxin
Dioxinausstoß dioxin emission
Diphenylalkylmercaptan diphenylalkyl mercaptan
Diphenylamin diphenylamine
Diphenylester diphenyl ester
Diphenylether diphenyl ether
Diphenyl-(2-ethylhexyl)-phosphat diphenyloctyl phosphate, diphenyl 2-ethylhexyl phosphate
Diphenylguanidin diphenylguanidine
Diphenylharnstoff diphenyl urea
Diphenylkresylphosphat diphenylcresyl phosphate
Diphenylmethandiisocyanat diphenylmethane di-isocyanate, MDI
Diphenyloctylphosphat diphenyloctyl phosphate
2,2'-Diphenylolpropan 2,2'-diphenylolpropane
Diphenylphthalat diphenyl phthalate
Diphenylsulfongruppe diphenylsulphone group
Diphenylthioharnstoff diphenyl thiourea
diphylbeheizt Dowtherm-heated, using Dowtherm as a heat transfer fluid: *(Dowtherm is the trade name of a eutectic mixture of diphenyl and its oxide)*
Dipol dipole
Dipolbindung dipole linkage
Dipolcharakter dipole character
Dipol-Dipol-Anziehung dipole-dipole attraction
Dipolkräfte dipole forces
Dipolmolekül dipole molecule
Dipolmoment dipole moment
Dipropylenglykol dipropylene glycol
Dipropylenglykolmethylether dipropylene glycol methyl ether
Direktabschlagsystem die face pelletising system, hot cut pelletising system
Direktanspritzung direct gating/feed *(im)*
Direktanspritzung, angußlose direct gating/feed *(im)*
direktbegast mechanically blown *(expanded by introducing gas as opposed to chemical blowing using a blowing agent)*
Direktbegasungsanlage mechanical blowing unit *(see explanatory note under* **direktbegast***)*
Direktbegasungsverfahren mechanical blowing (process) *(method of making foam)*

Direktdosierung direct addition
Direkthinterschäumung direct foam backing
Direkttitration direct titration
Direktwickler centre-drive winder
Direktzugriff direct access *(me)*
Disäure di-acid
Disilyltetrasulfid disilyl tetrasulphide
Diskette diskette *(me)*
Diskettenlaufwerk diskette drive *(me)*
Diskettenmassenspeicher diskette mass/bulk storage unit *(me)*
Diskettenverwaltung diskette management *(me)*
diskontinuierlich discontinuous(ly)
Diskontinuität discontinuity
diskret separate, discrete
Dispergieradditiv dispersing agent
Dispergieraggregat stirrer, mixer, dispersing unit
Dispergieraufwand *This indicates the problems associated with dispersion.* **Feinteilige Ruße erfordern einen höheren Dispergieraufwand als grobteilige.** Finely divided carbon blacks are more difficult to disperse than coarse ones.
dispergierbar dispersible
Dispergierbarkeit dispersibility
dispergieren to disperse
Dispergierenergie dispersion energy
Dispergierfähigkeit dispersibility
Dispergiergrad degree of dispersion
dispergierhart difficult to disperse
Dispergierhilfe dispersion aid
Dispergierhilfsmittel dispersing agent
Dispergierleistung dispersing efficiency
Dispergiermedium dispersion medium
Dispergiermittel dispersing agent
Dispergierprobleme dispersing problems
Dispergierschwierigkeiten dispersion problems
dispergierte Phase disperse phase
Dispergierung dispersion
disperse Phase disperse phase
Dispersierwirkung dispersing effect
Dispersion dispersion
Dispersionklebstoff dispersion/emulsion/latex adhesive, water-based adhesive
Dispersionsanstrichmittel emulsion paint
Dispersionsanstrichstoff emulsion paint
Dispersionsbindemittel dispersion binder
Dispersionsfarbe emulsion paint
Dispersionsflügel rotor blade
Dispersionsharz redispersible resin
Dispersionshilfsstoff dispersing agent
Dispersionskleber dispersion/water-based/emulsion/latex adhesive
Dispersionslack emulsion paint
Dispersionslackfarbe emulsion paint
Dispersionsmittel dispersing agent
Dispersionspulver redispersible powder
Dispersionsstabilisator dispersion stabiliser
dispersionsstabilisierendes Additiv dispersing agent

Dispersionsträger dispersion medium
Displayanzeige display *(me)*
Disproportionierung disproportionation
Disproportionierungsreaktion disproportioning reaction
Dissipationswärme heat of dissipation
dissipiert dissipated
Dissolver high speed stirrer/mixer, agitator. Since a **Dissolver** *is mainly used to mix and disperse, rather than to dissolve, it is best to use the terms given here, rather than* dissolver
Dissoziation dissociation
Dissoziationsenergie dissociation energy
Dissoziationskonstante dissociation constant
dissoziieren to dissociate
Distanzblock distance block/piece
Distanzbolzen threaded spacer, distance piece
Distanzhalter spacer, distance piece
Distanzplatten spacer plates/platens
Distanzring spacer ring
Distanzscheibe spacer disc
Distanzstück distance piece
Distributionskette distribution chain
disubstituiert di-substituted
Disulfid disulphide
disulfidisch disulphide
Di-tert.-butylperoxid di-tertiary butyl peroxide, DTBB
Dithiocarbamat dithiocarbamate
Dithiocarbaminsäure dithiocarbamic acid
Dithiophosphat dithiophosphate
Diurethandimethacrylat diurethane dimethacrylate
divergent divergent, diverging
divergierend diverging
Diversifikation diversification
Diversifizierung diversification
DI-Wasser deionised water
DKP-Schnecke *abbr. of* Dekompressionsschnecke, vented screw
DMA 1. *abbr. of* dynamic-mechanical analysis 2. *abbr. of* dynamic-mechanial analyser
DMC *abbr. of* dough moulding compound
DMS *abbr. of* **Dehnungsmeßstreifen**, strain gauge
DMTA 1. *abbr. of* dynamic-mechanical-thermal analysis 2. *abbr. of* dynamic- mechanical-thermal analyser
DOA *abbr. of* **Dioctyladipat**, dioctyl adipate
Dodecalactam dodecalactam
Dodecansäure dodecanoic/lauric acid
Dodecansäureanhydrid dodecanoic/lauric anhydride
Dodecylbenzol dodecyl benzene
Dodecylbernsteinsäureanhydrid dodecyl succinic anhydride
Dodecylmercaptan dodecyl mercaptan
Dodecylmonoethanolamin dodecyl monoethanolamine
Dokumentation documentation *(me)*
Dokumentieren documentation *(me)*
Dolomit dolomite

Domäne domain
Donator donor
Doppelaufwickler twin winder, twin wind-up (unit)
Doppelbandabzug twin-belt take-off (unit/system)
Doppelbindung double bond
Doppelbindungsanteil double bond content
doppelbindungsfrei free from double bonds
Doppelbindungsgehalt double bond content
Doppelblaskopf twin-die film blowing head *(bfe)*
Doppelbrechung double refraction, birefringence
Doppeldiskettenlaufwerk twin-diskette drive *(me)*
Doppeldosierschnecke twin feed screw
Doppeleinlaufschnecke twin feed screw
Doppel-Floppy-Disk-Laufwerk twin floppy disk drive *(me)*
doppelgängig two-start, double flighted *(screw)*
Doppelkehlnaht double internal angle weld, double inside corner weld
Doppelkniehebel double toggle
Doppelkniehebelmaschine double toggle machine
Doppelkniehebelschließeinheit double toggle clamp unit *(im)*
Doppelkniehebelschließsystem double toggle clamping system/mechanism *(im)*
Doppelkniehebelsystem double toggle system/mechanism
doppelkonisch double-conical *(screw)*
Doppelkonusschnecke double-conical screw
Doppelkopf twin die (extruder) head
Doppelkopfblasfolienanlage twin-die film blowing line
doppellagig two-layer
Doppellaufwerk twin drive *(me)*
Doppelmantel double-walled jacket
Doppelmischkopf twin mixing head
Doppelplattenwerkzeug two-plate mould, two-part mould, single-daylight mould
Doppelprägewerk twin embossing unit
Doppelrücken secondary backing web *(of carpet)*
Doppelschieberwerkzeug two-part sliding split mould *(im)*
Doppelschlauchkopf twin parison die *(bm)*
Doppelschnecke twin screw
Doppelschneckenausführung twin screw design
Doppelschneckencompounder twin screw compounder
Doppelschneckenentgasungsextruder vented twin screw extruder
Doppelschneckenextruder twin screw extruder
Doppelschneckenextrusion twin screw extrusion
Doppelschneckengranulieranlage twin screw pelletising line *(sr)*
Doppelschneckenkneter twin-screw compounder/plasticator

Doppelschneckenlaborextruder laboratory twin screw extruder
Doppelschneckenmaschine 1. twin screw extruder. 2. twin screw compounder
Doppelschneckenpresse twin-screw extruder
Doppelschneckenprinzip twin screw principle
Doppelschneckenseitenextruder auxiliary twin screw extruder
Doppelschneckenzylinder twin screw barrel
Doppelschreiber twin-channel chart recorder
doppelseitig on both sides, double sided
Doppelsiebkopf twin screen pack *(e)*
Doppelspritzkopf twin die (extruder) head *(e)*
Doppelstegdornhalter(ung) twin spider-type mandrel support *(bm)*
Doppelstegplatte twin-wall sheet
Doppelstrangwerkzeug twin-orifice die *(e)*
doppeltlogarithmisch log-log
Doppelvakuumtrichter twin vacuum hopper
Doppelverglasung double glazing
Doppelverteilerkanal 1. double runner *(im)*. 2. twin manifold *(e)*
Doppel-V-Kerbe double-V notch
Doppelwalze two-roll mill
Doppelwalzenextruder roller die extruder *(an extruder whose die is formed of two rolls)*
doppelwandig double-walled
Doppelwerkzeug 1. twin die (extruder) head *(e)*. 2. two-cavity mould *(im)*
Doppelwickler twin winder/wind-up (unit)
Doppelzylinder twin barrel *(e)*
Dorn mandrel *(e,bm)*
Dornbiegefestigkeit mandrel flexibility
Dornbiegeversuch mandrel flex test
Dornhalteplatte mandrel holding/locating plate *(e)*
Dornhalter mandrel support *(e,bm)*
Dornhalter mit versetzten Stegen spider with staggered legs *(e,bm)*
Dornhalterblaskopf centre-fed blown film die *(bfe)*
Dornhalterkonstruktion mandrel support design/construction
Dornhalterkopf centre-fed die *(e,bm) (in contrast to the **Pinolenkopf** (q.v.), this design incorporates a spider or a breaker plate to support the mandrel)*
Dornhalterkopf, zentralgespeister centre-fed die *(e,bm) (see explanatory note under **Dornhalterkopf**)*
Dornhaltermarkierungen spider lines *(e,bm)*
Dornhalterschlauchkopf centre-fed parison die *(bm)*
Dornhalterspritzkopf centre-fed die *(e,bm) (see explanatory note under **Dornhalterkopf**)*
Dornhaltersteg spider leg *(e, bm)*
Dornhalterung mandrel support (system) *(e, bm)*
Dornhalterwerkzeug centre-fed die *(e,bm) (see explanatory note under **Dornhalterkopf**)*
Dornsteghalter spider-type mandrel support, spider *(e,bm)*

Dornsteghalterung spider-type mandrel support, spider *(e,bm)*
Dornstegmarkierungen spider lines *(e,bm)*
Dornträger mandrel support *(e,bm)*
Dornträgersteg spider leg *(e,bm)*
Dose can
Dosier- und Mischmaschine metering and mixing unit
Dosieraggregat 1. metering unit. 2. feed/dispensing unit
Dosieranlage 1. metering equipment. 2. feed/dispensing equipment
Dosierautomat 1. automatic metering unit. 2. automatic feeder/dispenser
Dosierbehälter feed tank
Dosiereinheit 1. metering unit. 2. feed/dispensing unit
Dosiereinrichtung 1. metering equipment. 2. feed/dispensing equipment
Dosierelement 1. metering unit. 2. feed/dispensing unit
dosieren 1. to meter. 2. to feed, to dispense, to add, to incorporate. *This word should never be translated as* to dose, *which is used only in a medical context. Unfortunately, the word* dosing *is often encountered in translations of German texts but to use it in connection with machines is bad English.*
Dosiergenauigkeit metering accuracy: **Dosiergenauigkeiten zwischen 0.5 und 2% sind zu erreichen** it is possible to achieve metering with an accuracy of 0.5 - 2%
Dosiergerät 1. metering unit. 2. feed/dispensing unit
Dosierhub metering stroke
Dosierkammer feed compartment
Dosierkapazität metering capacity
Dosierkolben feed ram
Dosierleistung metering performance
Dosiermenge metered amount, required amount: **Alle Dosiermengen können getrennt für beide Komponenten eingestellt werden** all the required amounts can be set separately for each component
Dosiermischmaschine metering-mixing unit
Dosiermodus metering mode
Dosierphase plasticising time *(im)*
Dosierpumpe 1. metering pump. 2. feed pump
Dosierschieber metering valve
Dosierschnecke feed screw *(e)*
Dosierstation 1. metering station. 2. feed station
Dosiersystem 1. metering system. 2. feed system
Dosiertakt feed(ing) cycle
Dosiertrichter (feed/material) hopper *(e, im)*
Dosierung 1. metering: **Die Dosierung der Schmelze erfolgt volumetrisch** the melt is metered volumetrically. 2. feeding, adding: **Die direkte Dosierung von Füllstoffen zu Polyolefinen...** direct addition of fillers to polyolefins.... 3. amount (added): **Von großer Bedeutung ist die Dosierung des**

Dosierungenauigkeit   Drehmomentschlüssel

Weichmachers the amount of plasticiser used is very important; **Die Verträglichkeit der Weichmacher mit PVC wird durch eine maximale Dosierung begrenzt** the compatibility of plasticisers with PVC is limited by the maximum amount that can be added; **Bei Dosierungen oberhalb der Verträglichkeitsgrenze neigt der Weichmacher zum Ausschwitzen** if the amount of plasticiser added exceeds the compatibility limit, it will tend to exude; **...während die Härte mit steigenden Dosierungen von Füllstoff zunimmt...** whilst the hardness increases with increasing amounts of filler. 4. metering unit. 5. feed unit. 6. dosage
Dosierungenauigkeit inaccurate metering
Dosierverzögerung delayed feed
Dosiervolumen shot volume *(im)*
Dosiervorrichtung 1. metering unit 2. feed unit
Dosierwaage weigh feeder
Dosierwalze feed roll
Dosierweg metering stroke *(im)*
Dosierwerk 1. metering unit. 2. feed unit
Dosierzeit plasticising time *(im)*
Dosierzylinder 1. metering cylinder. 2. feed cylinder
Dosis dose, dosage
Dotieren doping
dotiert doped
Doubliereinrichtung contact laminator/laminating unit
doublieren to laminate
Doublierkalander contact laminating calender
DPCF *abbr. of* **Diphenylkresylphosphat**, diphenyl cresyl phosphate, DPCP
DPOF *abbr. of* **Diphenyloctylphosphat**, diphenyl octyl phosphate, DPOP
Drahtbürste wire brush
Drahtführungsspitze torpedo tip: *This is the removable tip of a* **Drahtummantelungspinole** *(q.v.) which ensures that the wire being covered is fully centred. Since the word is used only in descriptions of wire-covering crossheads, its function will be obvious, so that the English version given is adequate.*
Drahtgewebe wire gauze, wire cloth *(mf)*
Drahtgewebefilter wire gauze screen/filter *(mf)*
Drahtgitter wire netting
Drahtisolierlinie wire covering line *(e)*
Drahtisolierung wire insulation
Drahtlack wire enamel
Drahtnetz wire mesh/gauze
Drahtrichtevorrichtung wire aligning device *(e)*
Drahtschweißen hot air/gas welding: **Draht** here refers to **Schweißdraht** *(q.v.) and can be ignored in translation, since the use of welding rod automatically implies hot air/gas welding. Translators should beware of a literal translation, which would be meaningless as well as misleading.*

Drahtsieb wire gauze/mesh
Drahtsiebboden wire mesh screen *(mf)*
Drahtummantelung 1. wire covering *(when referring to the operation of covering wire)* 2. wire insulation *(when referring to the end product)*
Drahtummantelungsanlage wire covering line
Drahtummantelungsdüsenkopf wire covering die *(e)*
Drahtummantelungskopf wire covering die *(e)*
Drahtummantelungspinole wire covering torpedo *(e)*
Draht-und Kabelummantelungen cable sheathing and wire insulation
Drainagerohr drainpipe
Dralldorn rotating mandrel
Drallwinkel helix angle
Dränrohr drainpipe
Dränung drainage
drapierbar, leicht easily draped
drapieren to drape
Drapierfähigkeit draping qualities
Draufsicht view/seen from above
DRAW *abbr. of* direct read after write *(me)*
Drechslerbank lathe
Drehachse axis of rotation
Drehbank lathe
drehbar rotatable
Drehbarkeit *denotes that a machine or part thereof can be rotated or turned round*: **Die Dreh- und Schwenkbarkeit der Schließeinheit ermöglicht es...** since the clamping unit can be rotated and tilted, it is possible to...
Drehbewegung rotary movement
Dreheinschlag twist wrap
Drehen turning
drehend rotating
Dreherbindung true Leno weave *(txt)*
Drehextruder rotary extruder
Drehfenster side-hung window
Drehflügel turning casement
Dreh-Kippfenster tilt-and-turn window
Dreh-Kipptür tilt-and-turn door
Drehknopf knob
Drehkolbenpumpe rotary pump
Drehmasse torsional mass
Drehmoment torque
Drehmomentabfall torque decrease
Drehmomentanstieg torque increase
Drehmomentaufnehmer torque sensor/transducer
Drehmomentdefizit insufficient torque
Drehmomenteinstellung 1. torque adjustment. 2. torque adjusting mechanism
Drehmomenterhöhung torque increase
Drehmomentgrenze maximum torque
drehmomentkonstant with constant torque
Drehmomentkorrektur torque correction
Drehmomentmaximum maximum torque
Drehmomentrheometer torque rheometer
Drehmomentschlüssel torque wrench

Drehmomentsensor torque sensor/transducer
Drehmomentveränderung change in torque
Drehmomentverstärker torque amplifier
Drehmomentverteilergetriebe torque dividing drive
Drehmomentwaage torque meter
Drehrichtung direction of rotation
Drehschwingung torsional vibration
Drehsinn direction of rotation
Drehspulrotor moving coil rotor
Drehstab torsion bar
Drehstahl turning tool
Drehstrom three-phase current
Drehstromgetriebemotor three-phase geared motor
Drehstrommotor three-phase motor
Drehteller rotary table/platform
Drehtisch rotary table/platform
Drehtischbauweise rotary table design, carousel-type design
Drehtischmaschine rotary table machine, carousel-type machine
Drehtisch-Spritzgußmaschine rotary table injection moulding machine, carousel-type injection moulding machine
Drehtischsystem rotary table system/arrangement
Drehverschluß screw cap
Drehwiderstand rheostat
Drehwinkel angle of rotation
***Drehzahl** 1. number of revolutions. 2. speed *(of a rotating element)*. 3. screw speed
Drehzahl, regelbare variable speed
Drehzahländerung 1. change in speed. 2. change in screw speed
Drehzahlanzeige speed indicator
Drehzahlbereich 1. speed range. 2. screw speed range
Drehzahleinstellung 1. speed setting. 2. speed setting mechanism
Drehzahlerhöhung speed increase
Drehzahlerniedrigung speed reduction/decrease
Drehzahlgeber speed transducer/sensor
Drehzahlgenauigkeit accurate speed
Drehzahlgrenze 1. maximum speed. 2. maximum screw speed
Drehzahlistwert actual speed
Drehzahlkonstanz constant speed
Drehzahlmesser speed counter
Drehzahlprogramm speed program
Drehzahlprogrammablauf speed profile
Drehzahlregelbereich 1. speed range. 2. screw speed range
Drehzahlregelung 1. speed control. 2. speed control system/mechanism/device *(for translation example see **Drehzahlsteuerung**)*
Drehzahlregler speed regulator, speed control device
Drehzahlregulierung 1. speed control. 2. speed control system/mechanism/device *(for translation example see **Drehzahlsteuerung**)*
Drehzahlreserve speed reserve
Drehzahlschwankungen speed variations
Drehzahlsollwert required speed
Drehzahlsteigerung 1. speed increase. 2. screw speed increase: **Bild 3 zeigt die lineare Zunahme des Ausstoßes bei Drehzahlsteigerung** fig. 3. shows the linear increase in output with increasing screw speed
Drehzahlsteigerungsrate rate of speed increase
Drehzahlsteuerautomatik automatic speed control system/mechanism/device
Drehzahlsteuerung 1. speed control. 2. speed control system/mechanism/device: **Die Drehzahlsteuerung erfolgt vielfach mit Hilfe des Antriebsmotors** the screw speed is often controlled by means of the drive motor
Drehzahlstufen speed stages
drehzahlunabhängig independent of speed
drehzahlvariabel (of) variable speed
drehzahlveränderlich (of) variable speed
dreiachsig triaxial
dreiadrig triple-core *(cable)*
dreibasisch tribasic
dreibindig trivalent
Dreiblockcopolymer triblock copolymer
dreidimensional three-dimensional
dreieckig triangular
dreifach 1. triple, treble, threefold 2. three-ply *(e.g. paper sacks)*
Dreifachbindung triple bond
Dreifachform three-cavity/-impression mould
Dreifachkopf triple-die (extruder) head
Dreifachschnecke triple screw
Dreifachverglasung triple glazing
Dreifachverteilerkanal triple runner *(im)*
Dreifachwerkzeug three-cavity/-impression mould
dreigängig three-start, triple-flighted *(screw)*
Dreigangschaltgetriebe three-speed control mechanism
Dreikanaldüse three-channel nozzle *(im)*
Dreikanalwerkzeug triple-manifold die *(e)*
Dreiphasensystem three-phase system
dreiphasig three-phase
Dreiplattenabreißwerkzeug three-plate mould *(im)*
Dreiplattenmehrfachwerkzeug three-plate multi-cavity/-impression mould *(im)*
Dreiplattenpreßwerkzeug three-plate compression mould
Dreiplattenschließeinheit three-plate clamp(ing) unit *(im)*
Dreiplattenschließsystem three-plate clamping mechanism *(im)*
Dreiplattenwerkzeug three-plate mould, three-part mould, double-daylight mould *(im); see explanatory note under **Etage**)*
Dreiplatzanordnung three-station design

* Words starting with **Drehzahl**-, not found here, may be found under **Geschwindigkeits**-

Dreipunktauflage three-point support *(eg. of a test specimen)*
Dreipunktbelastung three-point loading
Dreipunktbiegeversuch three-point bending test
Dreipunktbiegung three-point bending
Dreipunktregler three-point controller
Dreischichtbetrieb three-shift operation
Dreischichtcoextrusion three-layer coextrusion
Dreischichtdüse three-layer (coextrusion) die *(e)*
Dreischichtextrusionskopf three-layer extruder/die head
Dreischichtfolienblaskopf three-layer blown film die *(bfe)*
Dreischichthohlkörper three-layer blow moulding
Dreischichtquerspritzkopf three-layer crosshead (die)
Dreischneckenextruder triple screw extruder
Dreistufenprozeß three-stage process
dreistufig three-stage
dreiviertelfett 75% oil length
Dreiwalze triple roller, triple roll mill
Dreiwalzenglättkalander three-roll polishing stack
Dreiwalzenglättwerk three-roll polishing stack
Dreiwalzen-I-Kalander three-roll vertical/superimposed calender
Dreiwalzenkalander three-roll calender
Dreiwalzenkalander, Schrägform three-roll offset calender
Dreiwalzenmaschine triple roller, triple roll mill
Dreiwalzennmühle triple roll mill, triple roller
Dreiwalzenstuhl triple roller, triple roll mill
Dreiwegehahn three-way tap/stopcock
Dreiwegemengenregelventil three-way flow control valve
Dreiwegestromregelung 1. three-way flow control. 2. three-way flow control unit
Dreiwegeventil three-way valve
dreiwertig 1. trivalent. 2. trihydric *(if an alcohol)*
Dreizonenplastifiziereinheit three-section plasticising unit
Dreizonenschnecke three-section screw *(e)*
Drift drift
driftfrei drift-free
Drossel flow restrictor, flow restriction device
Drosselfeld restricted flow zone *(e)*
Drosselgitter flow restrictor grid
Drosselkennzahl pressure flow-drag flow ratio *(e)*
Drosselklappe throttle valve
Drosselkörper flow restrictor, flow restriction device
Drosselorgan flow restrictor, flow restriction device
Drosselquotient pressure flow-drag flow ratio *(e)*
Drosselrückschlagventil throttle-check valve
Drosselschieber throttle valve
Drosselspalt flow restrictor gap *(e)*

Drosselstelle restricted flow zone
Drosselsteuerung throttle control
Drosselung flow restrictor
Drosselventil throttle valve
Drosselvorrichtung flow restrictor, flow restriction device
Drosselwirkung flow restriction effect
Druck 1. pressure, compression 2. printing
Druckabbau pressure decrease/reduction
Druckabfall pressure decrease/drop, decrease in pressure, loss of pressure, reduction in pressure
Druckabfallphase pressure decreasing phase: **während der Druckabfallphase** whilst the pressure is decreasing
druckabhängig pressure-dependent, depending on the pressure
Druckabschneidung pressure cut-out
Druckanstieg pressure increase, increase in pressure
Druckanzeige pressure gauge
Druckanzeigegerät pressure gauge
Druckaufbau pressure build-up
Druckaufbauphase pressure build-up phase: **während der Druckaufbauphase** whilst the pressure is building up
Druckaufbauvermögen capacity/ability to build up pressure
Druckaufbauzeit pressure build-up period
Druckaufgabe application of pressure
Druckaufnehmer pressure transducer/sensor
Druckaufnehmerabdruck mark left by the pressure transducer
Druckaufzeichnung recording of pressure
Druckausgleichskammer surge chamber
Druckbeanspruchbarkeit compression resistance
druckbeansprucht under compressive stress, under pressure
Druckbeanspruchung compressive stress, pressure: **...bei Raumtemperatur unter Druckbeanspruchung** ...at room temperature under pressure
Druckbedarf pressure requirement(s)
Druckbedarfsberechnung pressure requirement calculation
Druckbegrenzung pressure limitation
Druckbegrenzungsventil pressure release/relief valve
Druckbehälter pressurised vessel/tank
druckbelastbar resistant to compression, resistant to compressive stress
Druckbelastbarkeit compressive strength
druckbelastet under compressive stress, under pressure
Druckbelastung compressive stress, pressure
Druckbild printed image
Druckdifferenz pressure difference, difference in pressure
Druckdose pressure transducer
Druckeigenspannungen internal compressive stresses

Druckeinbruch drop in pressure
Druckeinstellorgan pressure adjusting device/mechanism
Druckeinstellventil pressure adjusting valve
Druckeinwirkung action/effect/application of pressure
Druck-Elastizitätsmodul modulus of elasticity in compression
Druck-E-Modul modulus of elasticity in compression
druckempfindlich pressure sensitive
Druckentlastung pressure release/relief
Druckentlastungsventil pressure release/relief valve
Druckentspannung pressure relief/release
Druckentspannungssystem pressure release/relief system
Drucker printer *(me)*
Druckerhöhungspumpe booster pump
Druckerschnittstelle printer interface *(me)*
Druckfarbe printing ink
Druckfarbenbindemittel printing ink binder
Druckfarbensektor printing ink industry
Druckfeder compression spring
druckfest 1. resistant to pressure. 2. resistant to compression
Druckfestigkeit compressive/crushing strength
Druckfestigkeitsentwicklung development of compressive strength
Druckfestigkeitsprüfmaschine compressive/crushing strength test machine
Druckfilm printed image
Druckfilter pressure filter
Druckfluß pressure flow
Druckflüssigkeit hydraulic fluid
Druckfortpflanzung pressure propagation
Druckfühler pressure transducer/sensor
Druckführung 1. pressure profile. 2. pressure control *(for details of how* profile *and* control *are used, see the translation examples under* **Temperaturführung***)*
Druckgasbehälter compressed gas cylinder
Druckgeber pressure transducer
Druckgefälle pressure gradient
Druckgefäß pressurised vessel/tank
Druckgelierverfahren pressure gelation (process)
Druckgleichgewicht equilibrium pressure
Druckgradient pressure gradient
Druckgrenze pressure limit, maximum pressure
Druckguß (pressure) diecasting
Druckgußform (pressure) diecasting mould
Druckgußwerkzeug (pressure) diecasting mould
Druckhalten maintaining the pressure
Druckhaltezeit holding pressure time *(im)*
Druckindustrie printing industry
druckinduziert pressure-induced
Druckistwert actual pressure
Druckkalibrierung 1. air pressure calibration/sizing. 2. air pressure calibrator/sizing unit *(e)*
Druckkennlinie pressure curve

Druckkessel pressure vessel
Druckkissen pressure pad
Druckknopf push button
Druckknopfschalter push-button switch
Druckknopfstation push-button console
Druckknopfsteuerung 1. push-button control. 2. push-button control unit
Druckkolben pressure ram
Druckkorrektur pressure correction
Druckkraft compressive force
Drucklager thrust bearing (unit)
Drucklagerung thrust bearing (unit)
Drucklast compressive stress, pressure
Druckleitung 1. pressure line. 2. delivery line *(of hydraulic system)*
Drucklimit pressure limit
drucklos pressure-less, at normal/atmospheric pressure
Drucklossintern free sintering *(ptfe)*
Druckluft compressed air
Druckluftauswerfer pneumatic ejector *(im)*
druckluftbetrieben pneumatically operated
Druckluftfördergerät pneumatic conveyor
Druckluftformmaschine compressed air forming machine *(t)*
Druckluftformung compressed air forming *(t)*
Druckluftkalibrierhülse compressed air calibrating/sizing sleeve *(e)*
Druckluftkalibrierung 1. air pressure calibration/sizing. 2. air pressure calibrating/sizing unit *(e)*
Druckluftspritzpistole compressed air spraygun
Druckluftverformen compressed air forming *(t)*
Druckluftzylinder compressed air cylinder
Druckmanometer manometer
Druckmaximum peak pressure, maximum pressure
Druckmedium hydraulic fluid
Druckmeßarmatur pressure transducer/sensor
Druckmeßaufnehmer pressure transducer/sensor
Druckmeßdose pressure transducer
Druckmeßeinrichtung pressure gauge
Druckmeßgeber pressure gauge
Druckmeßgerät pressure gauge
Druckmeßstelle pressure measuring point
Druckmeßsystem pressure measuring system
Druckmeßumformer pressure transducer
Druckmessung 1. pressure gauge. 2. measurement of pressure
Druckminderer pressure release/relief valve
Druckminderung reduction in pressure
Druckminderventil pressure release/relief valve
Druckmittelakkumulator hydraulic accumulator
Druckmittelspeicher hydraulic accumulator
Druckmodul compressive modulus
Druckobergrenze maximum pressure
Drucköl hydraulic oil
Druckölspeicher hydraulic accumulator
Druckölung 1. pressurised oil lubrication. 2. pressurised oil lubricating system

Druckpegel pressure (level)
Druckplastifizierung pressure plasticisation/plastication
Druckplatte pressure pad
druckpolymerisiert high pressure polymerised
Druckprobe compression test piece
Druckprofil pressure profile
Druckprogramm pressure program
Druckprüfung compression test
Druckpulsationen pressure fluctuations
Druckpumpe pump *(since all pumps generate pressure the first part of the word need not be translated)*
Druckreaktor pressure reactor
Druckreduzierung pressure reduction
Druckreduzierventil pressure release/relief valve
Druckregelgerät pressure controller/regulator, pressure control device
Druckregelkreis 1. pressure control circuit *(general term)*. 2. closed-loop pressure control circuit *(if the text differentiates between* **steuern** *and* **regeln***)*
Druckregelung 1. pressure control. 2. closed-loop pressure control. 3. pressure controller. 4. closed-loop pressure control system *(use 2. and 4. if the text differentiates between* **steuern** *and* **regeln***)*
Druckregelventil pressure control valve
Druckregulierventil pressure control valve
Druckreserve pressure reserve
Druckring thrust ring
Druckrohr pressure pipe
Druckrohrleitung pressure pipeline
Druckschalter push-button switch
Druckscherversuch compressive shear test
Druckschlauch pressure hose/tubing
Druckschraube thrust screw
Druckschreiber pressure recorder, pressure recording device
Druckschrift leaflet, publication
Druckschwankungen pressure variations/fluctuations
Druckschwellbereich region of repeated compressive stress
Druckschwellwert threshold pressure
Druckschwingungen pressure variations/fluctuations
Drucksensor pressure transducer/sensor
Druckservoventil pressure servo-valve
Drucksollwert required pressure, set pressure
Drucksonde pressure transducer/sensor
Druckspannung compressive stress
Druckspeicher pressure accumulator
Druckspeicherbetrieb operation with a pressure accumulator: **Der Druckspeicherbetrieb bietet dann wirtschaftliche Vorteile...** the use of a pressure accumulator offers economic advantages if...
Druckspitze 1. peak pressure, maximum pressure. 2. pressure peak

druckspitzenlos without pressure peaks
Druckstandfestigkeit compression set
Drucksteifigkeit compressive strength
Druckstempel piston, ram, plunger
Drucksteuerkreis 1. pressure control circuit *(general term)*. 2. open-loop pressure control circuit *(if the text differentiates between* **steuern** *and* **regeln***)*
Drucksteuerung 1. pressure control. 2. open-loop pressure control. 3. pressure controller. 4. open-loop pressure control system *(use 2. and 4. if the text differentiates between* **steuern** *and* **regeln***)*
Druckstoß 1. sudden pressure increase, pressure surge. 2. water hammer
Druckströmung pressure flow
Druckstufe I injection pressure *(im)*
Druckstufe II holding pressure *(im)*
Drucktaste push button
drucktastengesteuert push-button controlled
Drucktastenschalter push-button switch
Drucktastensteuerung 1. push-button control. 2. push-button control unit
Druckübertragung pressure transfer
Drucküberwachung 1. monitoring pressure 2. pressure monitor
Druckumformen pressure forming
Druckumformer pressure transducer
Druckumschaltung pressure change-over, change-over from injection to holding pressure *(im)*
Druckumschaltzeit pressure change-over time *(im)*
druckunabhängig independent of pressure
Druckunabhängigkeit independence of pressure
Druckunterschied pressure difference, difference in pressure
Druckuntersuchung compressive test
Druckventil pressure control valve
Druckverformung compressive deformation
Druckverformungsrest compression set
Druckverformungsverhalten compressive deformation (behaviour)
Druckverhalten compressive behaviour
Druckverlauf 1. pressure profile. 2. changes in pressure
Druckverlaufkurve pressure curve
Druckverlust pressure loss/drop, loss of pressure
Druckversuch compression test
Druckverteilung pressure distribution
Druckwaage pressure regulator
Druckwalze 1. printing roller/cylinder 2. pressure roller
Druckwasser pressurised water
Druckwassergerät pressurised water unit
Druckzone compression zone
Druckzylinder hydraulic cylinder
Dry-Blendmasse dry blend
Dry-Blendmischung dry blend

DSB *abbr. of* **Dämmschichtbildner**, substance which forms an insulating layer
DSB-Material substance which forms an insulating layer
DSC-Methode diffential scanning calorimetry
DSD *abbr. of* **Duales System Deutschland** Dual System
Duales System Dual System *(This forms part of the German packaging directive which came into force in January 1993 and obliges packaging producers and distributors to set up their own system for collecting and recycling used transit and post-consumer packaging materials. The full German name is* **Duales System Deutschland***)*
DTA *abbr. of* **Differential-Thermoanalyse**, differential thermal analysis
duktil ductile
duktiler Bruch ductile fracture/failure
duktiles Versagen ductile fracture/failure
Duktilität ductility
dunkelfarbig dark coloured
dunkelpigmentiert dark coloured
Dunkelverfärbung darkening
Dunkelwerden darkening
Dünnbettfliesenkleber thin-bed tile adhesive
Dünnbettkleber thin-bed adhesive *(for fixing tiles)*
Dünnbettverfahren thin-bed method
dünnflüssig low-viscosity, runny
Dünnflüssigkeit low viscosity
Dünnfolie thin film
Dünnschichtchromatographie thin-layer chromatography
dünnschichtchromatographisch thin-layer chromatographic
Dünnschichtentgasungsmischer thin-film degassing mixer
dünnschichtig thin: **dünnschichtiger Überzug** thin coating/film
Dünnschichtlasur thin glaze
Dünnschichtverdampfer thin-film vaporiser
Dünnschliff microsection
Dünnschnitt microtome section
dünnwandig thin-wall
Duplikatmodell duplicate model
Durchbiegung deflection
Durchbiegung beim Bruch deflection at break
Durchbruch 1. breakthrough. 2. opening, aperture
Durchbruchfeldstärke disruptive strength
Durchbruchfestigkeit rupture resistance
durchdringendes Netzwerk interpenetrating network
Durchdringung penetration
Durchdrückpackung push-/press-through pack
Durchdrückverpackung push-/press-through pack
Durchdrückzähigkeit puncture resistance
Durchfeuchtung damp penetration
Durchfluß 1. flow rate. 2. throughput
Durchflußfühler flow sensor, flow sensing element
Durchflußgeschwindigkeit flow rate
Durchflußkühlung 1. continuous flow cooling. 2. continuous flow cooling system
Durchflußleistung 1. flow rate. 2. throughput
Durchflußmenge 1. flow rate. 2. throughput
Durchflußmengenmesser flowmeter
Durchflußmeßgerät flowmeter
Durchflußmessung measurement of flow rate
Durchflußschwankungen flow/throughput variations
Durchflußsteuerventil flow control valve
Durchflußstrom 1. flow. 2. flow rate
Durchflußwiderstand flow resistance
Durchflußzeit flow time *(in viscosity determinations)*
Durchführbarkeit practicability, feasibility
Durchführung 1. bushing, duct 2. execution *(of an operation, plan etc.)*
Durchgangsfläche, freie effective filter area *(mf)*
Durchgangswiderstand volume resistance: *The word is sometimes used instead of* **spezifischer Durchgangswiderstand,** *in which case it should be translated accordingly.* **Durchgangswiderstand** *is expressed in Ohms,* **spezifischer Durchgangswiderstand** *in Ohms/cm.*
Durchgangswiderstand, spezifischer volume resistivity
durchgefärbt self-coloured, with moulded-in colour
durchgeschnitten fully flighted *(screw)*
durchgetrocknet completely dry
durchgezogen continuous *(curve)*
Durchhaltevermögen endurance
Durchhang drawdown, sag *(for translation example see under* **Aushängen***)*
Durchhängen drawdown, sag *(for translation example see under* **Aushängen***)*
Durchhärtung (full) cure
durchlässig permeable
Durchlässigkeit permeability
Durchlässigkeitswert permeability coefficient
Durchlaufanlage continuous plant, continuously operating plant
Durchlaufanzeige flow indicator
Durchlaufgeschwindigkeit throughput rate
Durchlauflackieranlage continuous painting line
Durchlaufmenge 1. flow rate. 2. throughput
Durchlaufmischer continuous mixer
Durchlaufofen tunnel oven
Durchlicht transmitted light
Durchlichtelektronenmikroskop transmission electron microscope, TEM
Durchlichtmikroskop optical microscope
Durchlüftung ventilation
Durchmesser diameter
Durchmesserverteilung particle size distribution
Durchsatz 1. throughput *(mostly of liquids through a pipe for example)*. 2. output, output

Durchsatz, volumetrischer / Düsenaustritt

rate. *Version 2. is preferable in connection with polymer processing*
Durchsatz, volumetrischer volume throughput, volumetric flow rate
Durchsatzeinbuße reductions in output
Durchsatzleistung output, output rate
Durchsatzmaximierung output maximisation: **zum Zweck der Durchsatzmaximierung** to achieve maximum outputs
Durchsatzmenge output, output ratae
Durchsatzrate output rate
Durchsatzreduktion reduction in output: **Aus diesem Grund ist eine Durchsatzreduktion zu erwarten** for this reason, outputs are likely to be lower
Durchsatzregulierung 1. flow control. 2. flow control device
Durchsatzsteigerung increase in output: **Durch diese Maßnahmen wurden die in Tabelle 3 aufgeführten Durchsatzsteigerungen erreicht** these measures resulted in the increased outputs shown in table 3
durchsatzunabhängig independent of throughput
durchscheinend translucent
Durchschlag 1. strike-through, fabric penetration *(penetration of substrate, e.g. fabric, by PVC paste)*. 2. breakdown *(electrical)*
Durchschlagfeldstärke breakdown field strength
Durchschlagfestigkeit 1. dielectric strength. 2. penetration resistance *(e.g. of safety glass)*
Durchschlagspannung breakdown voltage
Durchschlagversuch breakdown test
Durchschnittskosten average costs
Durchschnittsleistung average output
Durchschnittsprobe representative sample
Durchsenkung deflection
durchsetzen, sich to gain acceptance
durchsichtig transparent
Durchsichtigkeit transparency
Durchspritzverfahren sprueless injection moulding, hot-runner injection moulding
Durchstichfestigkeit puncture resistance
Durchstoßapparatur penetration test instrument/apparatus
Durchstoßarbeit penetration energy
Durchstoßfestigkeit puncture resistance
Durchstoßofen tunnel oven
Durchstoßprüfung penetration test
Durchstoßversuch penetration test
durchtränkt impregnated
Durchtränkung impregnation
durchtrocknen to hard-dry, to become completely dry

durchvernetzt fully crosslinked
Durchvulkanisation complete vulcanisation
durchvulkanisiert fully vulcanised/cured
Duromer thermoset material
duromer thermoset
Duroplast thermoset (material): *The word duroplastic, often encountered in translations, is not English and should never be used*
Duroplastformmasse thermoset moulding compound
Duroplastformmassen-Sortiment range of thermoset moulding compounds
Duroplastformteil thermoset moulding
duroplastisch thermoset
Duroplastmasse thermoset moulding compound
Duroplastschnecke thermoset screw
Duroplastspritzgießmaschine thermoset injection moulding machine
Duroplastspritzgießverfahren thermoset injection moulding (process)
Duroplastverarbeitung processing of thermosets
Duschwanne shower tray
Düse 1. die *(e)*. 2. nozzle *(im)*
Düse, offene free-flow nozzle *(im)*
Düse, verlängerte extended nozzle, long-reach nozzle *(im)*
∗**Düsenabhebegeschwindigkeit** nozzle retraction speed *(im)*
Düsenabhebeweg nozzle retraction stroke *(im)*
Düsenabhebung nozzle retraction *(im)*
Düsenabhub nozzle retraction *(im)*
Düsenablagerungen deposits formed in the die *(e)*
Düsenabmessungen 1. die dimensions *(e)*. 2. nozzle dimensions *(im)*
Düsenanfahrgeschwindigkeit nozzle approach speed *(im)*
Düsenanlage nozzle contact *(im)*
Düsenanlagedruck nozzle contact pressure *(im)*
Düsenanlagekraft nozzle contact pressure *(im)*
Düsenanlegebewegung nozzle forward movement *(im)*
Düsenanpreßdruck nozzle contact pressure *(im)*
Düsenanpreßkraft nozzle contact pressure *(im)*
Düsenanpressung 1. nozzle contact. 2. nozzle contact pressure *(im)*
Düsenanpreßzylinder nozzle advance cylinder *(im)*
Düsenarten 1. types of nozzle *(im)*. 2. types of die *(e)*
Düsenausgang die opening/orifice *(e).*
Düsenauslegung die design *(e)*
Düsenaustritt 1. die opening/orifice *(e)*. 2. nozzle aperture *(im)*. 3. *the word is also used*

∗ *Words starting with* **Düsen-***, not found here, may be found under* **Werkzeug-***. Certain words starting with* **Düsen-***, for which only an injection moulding version is given, may occur in an extrusion context and vice versa. In such cases, "nozzle" should be replaced by "die", and vice versa.*

to denote melt coming out (**austreten**) of the die: **Es ist zu beachten, daß der Vorformling nach dem Düsenaustritt bis zum zehnfachen seines Volumens aufschäumt** it should be noted that the parison expands to ten times its volume as it leaves the die
Düsenaustrittsöffnung die gap/orifice *(e)*
Düsenaustrittsspalt die gap/orifice *(e)*
Düsenaustrittstemperatur extrudate delivery temperature
Düsenbauart 1. nozzle design *(im)*. 2. die design *(e)*. 3. type of nozzle *(im)*. 4. type of die *(e)*
Düsenbaustoff 1. nozzle material *(im)*. 2. die material *(e)*
Düsenblock nozzle block *(im)*
Düsenbohrung 1. nozzle aperture *(im)*. 2. die gap/orifice *(e)*
Düsenbreite die width *(e)*
Düsendeformation nozzle deformation *(im)*
Düsendruck die pressure *(e)*
Düsendurchmesser die diameter *(e)*
Düseneinsatz die insert
Düseneinstellung 1. die adjustment. 2. die adjusting mechanism *(e)*
Düseneintrittdruck die head pressure *(e)* *(melt pressure as material enters the die)*
Düsenfahrgeschwindigkeit nozzle speed *(im)*
Düsenfluß amount of melt passing through the die *(e)*
Düsenform 1. type of die *(e)*. 2. type of nozzle *(im)*
Düsenhalteplatte nozzle carriage *(im)*
Düsenhalterung die mount *(e)*
Düsenheizband nozzle heater band *(im)*
Düsenheizkörper nozzle heater *(im)*
Düsenheizung nozzle heater *(im)*
Düsenhub nozzle stroke *(im)*
Düsenjustlereinrichtung die adjusting mechanism *(e)*
Düsenkalotte nozzle seating *(im)*
Düsenkanal manifold *(e)*
Düsenkennlinie die characteristic (curve) *(e)*
Düsenkennwert die constant
Düsenkennzahl die constant *(e)*
Düsenkonstruktion 1. nozzle design *(im)*. 2. die design *(e)*
Düsenkopf 1. extruder head, die head. 2. die *(see explanatory note under* **Kopf***)*
Düsenkörper die body *(e)*
Düsenkraftstoff jet fuel
Düsenlippe, einstellbare adjustable/flexible lip *(e)*
Düsenlippe, feste fixed lip *(e)*
Düsenlippen die lips *(e)*
Düsenlippenspalt die gap/orifice *(e)*
Düsenlippenspaltweite die gap width *(e)*
Düsenlippenverstellung 1. die lip adjustment *(e)* 2. die lip adjusting mechanism *(e)*
Düsenmund 1. die gap/orifice *(e)*. 2. nozzle orifice *(im)*

Düsenmundstück (outer) die ring *(e)*
Düsennadel valve pin *(im)*
Düsennähe, in 1. near the nozzle *(im)*. 2. near the die *(e)*
Düsenöffnung 1. nozzle aperture/orifice *(im)*. 2. die gap/orifice *(e)*
Düsenöffnungsdruck die opening force *(e)*
Düsenparallelführung die land *(e)*
Düsenplatte 1. die plate *(e)*. 2. fixed platen, stationary platen *(im)*
Düsenquellung die swell *(e)*
Düsenring die ring *(e)*
Düsensatz die assembly *(e)*
Düsenschnellabhebung high speed nozzle retraction mechanism *(im)*
Düsenschwellung die swell *(e)*
Düsenseite fixed mould half, stationary mould half *(im)*
düsenseitig on the fixed mould half, on the stationary mould half *(im)*
düsenseitige Werkzeughälfte fixed/stationary mould half *(im)*
Düsensitz nozzle seating *(im)*
Düsenspalt 1. die gap/orifice *(e)*. 2. nozzel clearance *(im)*
Düsenspaltbreite die gap width *(e)*
Düsenspaltverstellung 1. die gap adjustment. 2. die gap adjusting device *(e)*
Düsenspaltweite die gap width *(e)*
Düsenspitze nozzle point *(im)*
Düsenstandzeit (working) life of the die: **es wird eine größere Düsenstandzeit erreicht** the die will last longer
Düsenteil die section *(e)*
Düsenteller die plate *(e)*
Düsentreibstoff jet fuel
Düsentrockner spray dryer
Düsenumfang die circumference *(e)*
Düsenversatz nozzle displacement/misalignment *(im)*: **...um einen Düsenversatz zu verhindern** ...to prevent the nozzle becoming misaligned
Düsenverschluß shut-off nozzle *(im)*
Düsenvorlaufgeschwindigkeit nozzle advance speed *(im)*
Düsenwechsel replacing the die, changing the die: **bei einem erforderlichen Düsenwechsel** if the die needs changing
Düsenweg nozzle stroke *(im)*
Düsenwerkstoff 1. nozzle material *(im)*. 2. die material *(e)*
Düsenwerkzeug (extrusion) die
Düsenwiderstand die resistance *(e)*
Düsenzapfen sprue *(im)*
Düsenzentrierung die centring device *(e)*
Düsenziehverfahren pultrusion *(grp)*
DV-Aufgabe data processing task *(me)*
DVR *abbr. of* **Druckverformungsrest**, compression set
dynamisch dynamic
dynamisch-mechanische Analyse dynamic-mechanical analysis

Dynstatgerät Dynstat impact tester, Dynstat apparatus

E

E/A *abbr. of* **Eingang/Ausgang**, input-out-put *(me)*
E/A-Karte input-output card *(me)*
EBC *abbr. of* electron beam curing
eben plane
Ebene plane
EB-Verfahren electron beam curing
Echtheit fastness *(e.g. of pigment)*
Echtzeit real time *(me)*
Echtzeiteigenschaften real time properties
Echtzeitmeßmethode real time method of determination
Echtzeitmessung real time measurement
Echtzeitsteuerung real time control *(me)*
Echtzeitverarbeitung real time processing *(me)*
Ecke corner
Ecken, tote dead spots
Eckfestigkeit corner strength *(of a PVC window frame)*
Eckverbindung corner joint
ECO *symbol for* epichlorohydrin rubber
Edelgas inert gas
Edelmetallkatalysator precious metal catalyst
Edelputz facing plaster
Edelstahl stainless steel
Edelstahlmaschengewebe 1. stainless steel wire mesh. 2. stainless steel filter screen *(mf)*
Editiermöglichkeiten editing options *(me)*
Editierung editing *(me)*
EDM *abbr. of* electric discharge machining
EDRAW *abbr. of* erasable direct read after write *(me)*
EDS-Analyse electron dispersive spectroscopy
EDV *abbr. of* **elektronische Datenverarbeitung** electronic data processing, EDP *(me)*
EDV-Anlage computer system *(me)*
EDV-gestützt computer aided
EDV-kontrolliert computer controlled
EDV-Programm computer program *(me)*
EDV-Rechenprogramm computer program *(me)*
EDV-System electronic data processing system *(me)*
EDV-unterstützt computer aided
EDV-Unterstützung computer aid/support/assistance
EDX *abbr. of* **energiedispersive Röntgenanalyse** energy-dispersive X-ray analysis
EEPROM *abbr. of* electrically erasable programmable read only memory *(me)*
effizient efficient

Effizienzsteigerung increased efficiency
EFTA-Staaten EFTA countries, countries of the European Free Trade Association
Egalisierungsmasse levelling compound
EG-Binnenmarkt single European market
E-Glas low-alkali glass
EG-Mitgliedsländer EEC member countries
EG-Schnecke *abbr. of* **Entgasungsschnecke**, vented screw
Eichfaktor calibration factor
Eichkurve calibration curve
Eichmarke (calibration) mark
Eichprogramm standardising/calibrating program
Eichung calibration, standardisation
Eichungsfaktor calibrating factor
Eichvorrichtung calibrating device/unit
Eierschachtel egg box
Eigenfarbe inherent colour
Eigenfestigkeit inherent strength
Eigenfrequenz natural frequency
Eigengeruch characteristic smell
Eigenkapital own capital/resources *(of a company)*
Eigenklebrigkeit inherent tack
Eigenkontrolle in-house control
Eigenschaften, filmmechanische mechanical film properties
Eigenschaften, kennzeichnende characteristic features/ properties
Eigenschaften, schaumzerstörende antifoam properties
Eigenschaften, thermische thermal properties
Eigenschaften, toxikologische toxicological properties
Eigenschaftsänderung change in properties
Eigenschaftsanforderungen property requirements
Eigenschaftsanisotropie anisotropic properties
eigenschaftsbezogen property-related
Eigenschaftsbild (general) properties: **elektrisches Eigenschaftsbild** electrical properties
Eigenschaftseinbuße deterioration of properties
Eigenschaftskennwerte properties
Eigenschaftskombination combination of properties
Eigenschaftsmerkmale properties, characteristics
Eigenschaftsniveau properties: **hohes Eigenschaftsniveau** excellent properties
Eigenschaftsprofil (range of) properties
Eigenschaftsrichtwerte typical properties
Eigenschaftsschwankungen property variations
Eigenschaftsspektrum range of properties
Eigenschaftstabelle table of properties
eigenschaftsverbessernd property-enhancing
Eigenschaftsvergleich comparision of properties
Eigenschaftswerte properties
eigensicher intrinsically safe

Eigenspannung   **Einfachkopf**

Eigenspannung internal stress
Eigenspannungsfeld internal stresses
eigenspannungsfrei free from internal stresses
Eigenspannungszustand state of internal stress
Eigenstabilität inherent stability
eigenständig independent
Eigenverstärkung self-reinforcement *(e.g. by causing morphological changes in the polymer through compressing or stretching the melt)*
Eignungsprüfung suitability test
Eilgang, im fast
Eilgangzylinder high speed cylinder *(im)*
einachsig uniaxial
einarbeiten to incorporate, to mix in
Einarbeitung 1. mould cavity *(im)*. 2. incorporation
Einatmen inhalation
Ein-Ausgabebaugruppe input-output module *(me)*
Ein-Ausgabeeinheit input-output unit *(me)*
Ein-Ausgabeeinrichtung input-output device *(me)*
Ein-Ausgabekanal input-output channel *(me)*
Ein-Ausgabestein input-output module *(me)*
Ein-Aus-Schalter on-off switch
einaxial uniaxial
Einbau fitting, installation
einbauen to fit, to install
einbaufertig ready for installation
Einbauhöhe mould height/space *(im)*
Einbauküche fitted kitchen
Einbaukühlmöbel fitted refrigerators
Einbaumaße 1. mould mounting dimensions, platen dimensions *(im)*. 2. installtion dimensions *(of a machine)*
Einbaumethode method of installation
Einbaurichtlinien fitting instructions
Einbetteil insert
Einbettmasse embedding/potting compound
Einbettmaterial embedding/potting compound
Einbettung embedding/potting
Einbettungsmittel potting/embedding compound
Einbeulen denting, indentation
einbindig monovalent
Ein-Bit-Prozessor one-bit processor *(me)*
Einblattfeder single-leaf spring
Einblickfenster sight glass, window
Einbrennalkyd stoving alkyd
Einbrennalkydharz stoving alkyd resin
einbrennbar stovable
Einbrennbedingungen stoving/baking conditions
Einbrennbereich stoving/baking temperature range
Einbrenndauer stoving/baking time
Einbrenndecklack stoving/baking paint
Einbrennen stoving, baking
einbrennfest resistant to stoving (temperatures)
Einbrennfüller stoving/baking filler
Einbrenngeschwindigkeit stoving rate

Einbrenngrundierung stoving/baking primer
Einbrenngrundlack stoving/baking primer
Einbrennlack stoving enamel/paint/lacquer
einbrennlackiert stove enamelled
Einbrennlackierung stoved/baked finish
Einbrennofen stoving oven
Einbrennrückstand stoving residue
Einbrennstrukturlack textured stoving paint
Einbrennsystem stoving enamel/paint/lacquer
Einbrenntemperatur stoving/baking temperature
Einbrenntemperaturbereich stoving/baking temperature range
Einbrennversuch stoving test/trial
Einbrennzeit stoving/baking time
Einbruch fall, drop, decline, reduction, slump *(in prices, demand, sales etc.)*
einbügelbar iron-on *(interlinings)*
Einbügelstoff iron-on interlining
eindampfed to evaporate
Eindampfung evaporation
eindicken to thicken
Eindickmittel thickener, thickening agent
Eindickungsaktivität thickening effect
Eindickungsmittel thickener, thickening agent
eindiffundieren to diffuse into *(a material)*
Eindiffusion diffusion into *(a material)*
eindimensional one-dimensional
eindispergieren to mix in
eindosieren to feed, to add, to charge *(material to an extruder, injection moulding machine etc.)*
Eindringamplitude penetration amplitude
Eindringen penetration
eindringen to penetrate
Eindringgeschwindigkeit speed of penetration, penetration rate
Eindringgrad degree of penetration
Eindringkoeffizient penetration coefficient
Eindringtiefe 1. depth of penetration. 2. depth of indentation *(when measuring ball identation hardness)*
Eindringvermögen penetrating power
Eindruckfestigkeit indentation resistance
Eindruckhärte indentation resistance
Eindruckkörper indenter
Eindrucktiefe depth of indentation
Eindruckversuch identation test
Eindruckweg depth of indentation
Eindruckwiderstand indentation resistance
Einetagenwerkzeug two-plate mould, two-part mould, single-daylight mould *(im) (see explanatory note under* **Etage***)*
einfach, verfahrenstechnisch technically simple
Einfachbindung single bond
Einfachextrusionskopf single-die extruder head
Einfachform single-cavity/-impression mould
Einfachkniehebel toggle lever
einfachkonisch conventional conical *(screw) (e)*
Einfachkopf single-die extruder head

Einfachschlauchkopf single-parison die *(bm)*
Einfachschnecke single screw *(as opposed to* **Doppelschnecke** *(q.v.))*
Einfachschubschnecke single reciprocating screw
Einfachspritzgießwerkzeug single-cavity/-impression injection mould
Einfachwerkzeug 1. single-cavity/-impression mould. 2. single-die extruder head *(e)*
einfahren to move in, to move into position
Einfahrgeschwindigkeit forward speed *(im)*
Einfallstelle sink mark *(im)*
Einfärbbarkeit pigmentability **universelle Einfärbbarkeit** all-round pigment tolerance
Einfärbegerät pigment mixing unit
einfärben to colour, to pigment
Einfärbung 1. colouring, pigmentation. 2. colour: *Für transparente Einfärbungen sind lösliche Farbstoffe üblich transparent colours are obtained by using soluble dyes*
Einflußfaktor influencing factor
Einflußgröße 1. influencing factor. 2. independent variable
Einflußhöhe degree of influence
Einflußparameter influencing factor
Einfrierbereich 1. frost line *(bm)*. 2. glass transition range
einfrieren 1. to freeze *(melt in a mould) (im)*. 2. to solidify, to set
Einfriergrenze frost line *(bfe)*
Einfriergrenzenabstand frost line height *(bfe)*
Einfriertemperatur glass transition temperature
Einfriertemperaturbereich glass transition range
Einfrierungstemperatur glass transition temperature
Einfuhr import
Einführungsphase introductory phase/period
Einführungssortiment introductory range
Einfuhrzoll import duty
Einfüllabschnitt feed section/zone *(e)*
Einfüllbereich feed section/zone *(e)*
Einfüllgehäuse (feed/material) hopper *(e,im)*
Einfüllöffnung feed opening/port/throat *(e)*
Einfüllschacht feed opening/port/throat *(e)*
Einfüllteil feed sction/zone *(e)*
Einfülltrichter (feed/material) hopper *(e,im)*
Einfüllzeit injection time, mould filling time *(im)*
Einfüllzeitende end of mould filling time *(im)*
Eingabe input *(me)*
Eingabebaugruppe input module *(me)*
Eingabebaustein input module *(me)*
Eingabedaten input data *(me)*
Eingabeeinheit input unit *(me)*
Eingabefehler input error(s) *(me)*
Eingabegerät input device *(me)*
Eingabemaske input mask *(me)*
Eingabemenü input menu *(me)*
Eingabemodus input mode *(me)*
Eingabemöglichkeiten input options *(me)*
Eingabeseite input side *(me)*
Eingabestation input station *(me)*
Eingabestein input module *(me)*
Eingabetablett input tablet *(me)*
Eingabetastatur input keyboard *(me)*
Eingabewert input variable *(me)*
Eingang input *(me)*
eingängig single-start/-flighted *(screw)*
Eingangsbefehl input instruction *(me)*
Eingangsdaten input data *(me)*
Eingangsdrehzahl 1. initial speed. 2. initial screw speed *(e)*
Eingangsempfindlichkeit input sensitivity
Eingangsgröße input variable *(me)*
Eingangsimpedanz input impedance
Eingangskabel input cable
Eingangskanal input channel *(me)*
Eingangskarte input card *(me)*
Eingangskontrolle 1. incoming goods control. 2. incoming goods control section
Eingangskreis input circuit
Eingangsladung input charge
Eingangsmassetemperatur melt temperature at the feed end *(e.g. of an extruder)*
Eingangsmodul input module *(me)*
Eingangsseite input side *(me)*
Eingangssignal input signal *(me)*
Eingangsspannung input voltage
Eingangsspannungsbereich input voltage range
Eingangsteil feed section/zone *(e)*
Eingangstemperatur inlet temperature
Eingangstransistor input transistor
Eingangswiderstand input resistance
Eingangszone feed section/zone *(e)*
eingebbar input: **eingebbarer Wert** input value
eingeben to input *(me)*: *Although this sounds odd to the ear unaccustomed to computer jargon, it is perfectly correct usage*
eingebördelt flanged
eingebrannt stoved, baked
eingefallene Stellen sink marks *(im)*
eingefärbt coloured, pigmented
eingefärbt, gedeckt pigmented *(the use of pigments automatically causes opacity so that the word* **gedeckt** *can be ignored in translation)*
eingefärbt, in der Masse self-coloured
eingefettet greased
eingeformt moulded-in
eingefroren frozen-in *(stresses or orientation)*
eingekapselt encapsulated, embedded
eingekerbt notched
eingelagert embedded
eingerastet engaged
eingeschwenkt in position *(part of a machine)*
Eingeschwindigkeitsschließeinheit single-speed clamp(ing) unit *(im)*
eingespannt clamped
eingespeist, seitlich side-fed *(e,bm,bfe)*
eingespeist, zentral centre-fed *(e,bm,bfe)*
eingespritzt moulded-in *(e.g. inserts)*
eingestellt formulated: *Although this represents the meaning of the word, it is often omitted in*

translation, e.g.: **Das Produkt ist wärmestabil eingestellt** the product is heat stabilised; **antistatisch eingestellt** antistatic
eingestellt, hart with a low plasticiser content *(PVC paste or compound)*
eingestellt, weich with a high plasticiser content *(PVC paste or compound)*
eingreifend intermeshing *(screws) (e)*
eingreifend, voll closely intermeshing *(screws) (e)*
Eingriffsbereich intermeshing zone *(between twin screws) (e)*
Eingußloch pouring hole
Einheit unit
Einheit, verfahrenstechnische processing unit
Einkammertrichter single-compartment hopper
Einkanaldüse single-manifold die *(e)*
einkapseln to encapsulate, to pot, to embed
Einkaufspreis purchase price
Einkaufszentrum shopping centre
Ein-Kavitätenwerkzeug single-cavity/-impression mould
Einkomponentenkleber one-pack adhesive
Einkomponentenklebstoff one-pack adhesive
Einkomponentenmasse one-component/-pack compound
Einkomponenten-Schaumspritzgießverfahren one-component structural foam moulding (process)
Einkomponentensilikonkautschuk one-pack silicone rubber
Einkomponenten-Spritzgießverfahren conventional injection moulding. *It is not necessary here to translate the* **Einkomponenten** *part because conventional injection moulding is never anything else.*
Einkomponentensystem 1. one-component system. 2. one-pack system
einkomponentig one-component/one-pack
Einkopfanlage single-die extrusion line
Einkopfbedienung one-man operation
Einkreiskühlsystem single-circuit cooling system
Einkreissystem single-circuit system
Einkristall single crystal
Einkristallplättchen single-crystal platelet
Einlage insert
Einlagerung immersion
Einlagerungsversuch immersion test
Einlagerungszeit time of immersion *(of test specimen in a medium)*, immersion period
Einlagevlies interlining (material)
einlagig single-layer
Einlaßgrund(ierung) low viscosity primer *(especially for very absorbent surfaces like concrete, etc.)*
Einlaßventil inlet valve
Einlauf 1. inlet. 2. feed throat/opening/port *(e)*
Einlaufgehäuse (feed/material) hopper *(e,im)*
Einlaufhülse feed bush
Einlaufmassetemperatur melt temperature at the feed end *(e.g. of an extruder)*

Einlauföffnung feed opening/port/throat *(e)*
Einlaufrohr feed pipe
einlaufseitig at the feed end
Einlaufstück feed section/zone *(e)*
Einlaufstutzen inlet port *(e.g. for cooling water)*
Einlaufteil feed section/zone *(e)*
Einlauftemperatur 1. inlet temperature *(e.g. of water or oil)*. 2. feed temperature *(e.g. of melt)*
Einlauftrichter (feed/material) hopper *(e,im)*
Einlaufwalze feed roll
Einlegeautomat insert-placing robot
Einlegeroboter insert-placing robot
Einlegeteil insert
Einleitung 1. introduction. 2. initiation
einlesen to key in *(me)*
Einmalgebrauchserzeugnisse disposables
einmalig 1. non-recurring. 2. unique
Einmannbedienung one-man operation
Einmannbetrieb one-man business
einmischen to mix in, to incorporate
Einphasenstruktur single-phase structure
Einphasensystem single-phase system
einphasig single-phase
einpolymerisiert incorporated by polymerisation
Einpreßschnecke stuffing screw
Einpreßteil insert
Einpunktmessung single-point determination
einregeln to adjust *(by operating controls)*
Einreißfestigkeit tear resistance
Einreißversuch tear resistance test
Einreißwiderstand tear resistance
Einrichtblätter setting-up instructions
Einrichten setting *(of a machine)*
Einrichter machine setter
Einrichtezeit setting time *(for a machine)*
Einrichtung arrangement, device, apparatus, equipment
Einriß crack, tear
einrühren to stir in
Einsatz 1. use, employment *(of machines, labour etc.)*. 2. insert
Einsatzbedingungen conditions of use
einsatzbereit ready for use
Einsatzbreite range of uses/applications
Einsatzdeterminanten decisive factors for using...
einsatzfähig serviceable, fit for use, in working order
Einsatzgebiet field/area of use, application
einsatzgehärtet case hardened
Einsatzgrenztemperatur maximum working/operating temperature
Einsatzkonzentration concentration, amount used/added *(e.g. catalyst, filler etc.)*
Einsatzmenge amount used, dosage
Einsatzmöglichkeiten possible uses/applications, application possibilities
Einsatzmonomer monomer (used): **Die Polymerisationswärme wird durch Kühlung des Lösemittels und der Einsatzmonomeren abgeführt** The heat of

polymerisation is removed by cooling the solvent and the monomers
Einsatzplatte insert plate *(im)*
Einsatzring insert ring *(im)*
Einsatzschwerpunkt principal application/use, most important application
Einsatzspektrum range of uses/applications
einsatzspezifisch application-oriented
Einsatzstahl case-hardened steel
Einsatzstück insert
Einsatztemperatur working/operating temperature range
Einsatztemperaturbereich working/operating temperature range
Einsatzzweck application, end use
einschalig single-shell
Einschaltautomatik automatic switch-on mechanism
einschalten to switch/turn on
Einschaltreihenfolge switch-on sequence
Einscheibensicherheitsglas single-layer safety glass
Einschichtdüse single-layer die *(e)*
Einschichtenbetrieb 1. one-shift operation. 2. one-shift plant
Einschichtextrusion single-layer extrusion
Einschichtfolie single-layer film
einschichtig single-layer
Einschichtlack one-coat paint
Einschicht-PS-Schaumhohlkörper single-layer expanded polystyrene container
Einschichtsystem one-coat system
Einschichtwerkzeug single-layer die *(e)*
einschiebbar slide-in
Einschiebehorde slide-in tray
Einschlagfolie wrapping film
einschließlich, bis up to and including
Einschlüsse occlusions
Einschnecke single screw *(as opposed to* **Doppelschnecke** *(q.v.))*
Einschneckenaggregat single-screw extruder
Einschneckenanlage 1. single-screw extrusion line. 2. single-screw extruder
Einschneckenanordnung single-screw arrangement *(e)*
Einschneckenentgasungsextruder vented single-screw extruder
Einschneckenextruder single-screw extruder
Einschneckenextrusion single-screw extrusion
Einschneckenextrusionsanlage 1. single-screw extrusion line. 2. single-screw extruder
Einschneckenhochleistungsextrusionsanlage high speed single-screw extrusion line
Einschneckenhochleistungsmaschine high speed single-screw extruder
Einschneckenkneter single-screw compounder/plasticator
Einschneckenkurzextruder short single-screw extruder
Einschneckenmaschine 1. single-screw machine *(general term)*. 2. single-screw extruder. 3. single-screw compounder. 4. single-screw devolatiliser. 5. single-screw injection moulding machine
Einschneckenplastifizieraggregat single-screw compounder/plasticator
Einschneckenplastifizierextruder single-screw compounding/plasticating extruder
Einschneckenpresse single-screw extruder
Einschneckenprinzip single-screw principle
Einschneckenspritzgießmaschine single-screw injection moulding machine
Einschneckenstrangpresse single-screw extruder
Einschneckensystem single-screw system
Einschneckenzylinder single-screw barrel *(e)*
Einschnitt incision
Einschnürung constriction
Einschraubdüse screw-in nozzle
Einschraubwiderstandsthermometer screw-in resistance thermometer
Einschreiblaser(strahl) input recording laser
Einschrumpfen shrink wrapping
Einschub slide-in module *(me)*
Einseitenbodenklebstoff flooring adhesive for one-sided application
Einseitenkleber adhesive for one-sided application
Einseitenklebstoff adhesive for one-sided application
einseitig on one side, one-sided
einsetzbar suitable
einsetzbar, universell universal, general purpose *(machine)*
Einspannbacke clamping jaw
Einspanndruck clamping pressure
Einspannklemme clamp
Einspannlänge clamping distance *(of test specimen)*
Einspannvorrichtung clamping device *(for test specimen)*
Einsparung saving *(e.g.* **Energieeinsparung, Stabilisatoreinsparung** *etc.)*
Einspeisebohrung feed opening/port/throat
Einspeisekanal feed channel
einspeisen to feed, to add, to charge *(material to an extruder, injection moulding machine, etc.)*
Einspeisepunkt feed point
Einspeiseschnecke feed screw
Einspeisestelle feed point
Einspeisevorrichtung feed system/mechanism
Einspeisung feed, feed system
Einspeisungsschwierigkeiten feed problems
Einspindelschnecke single screw
✱**Einspritzaggregat** injection unit *(im)*
Einspritzarbeit injection force
Einspritzbaustein injection module *(im)*
Einspritzbedingungen injection conditions *(im)*
Einspritzbewegung injection stroke *(im)*

✱ *Words starting with* **Einspritz-***, not found here, may be found under* **Spritz-**

Einspritzdruck injection pressure *(im)*
Einspritzdüse injection nozzle *(im)*
Einspritzeinheit injection unit *(im)*
einspritzen to inject
Einspritzende end of injection: **unmittelbar for dem Einspritzende** immediately before injection has been completed
einspritzfertig ready for injection
Einspritzgeschwindigkeit injection speed/rate, rate of injection
Einspritzgeschwindigkeitsverlauf injection speed profile
Einspritzgewicht shot weight *(im)*
Einspritzhub injection stroke *(im)*
Einspritzhydraulik hydraulic injection unit *(im)*
Einspritzkolben 1. injection plunger *(im)*. 2. transfer plunger *(tm)*
Einspritzkraft injection force *(im)*
Einspritzleistung injection capacity *(im)*
Einspritzleistung, installierte installed injection capacity *(im)*
Einspritzmenge 1. injection rate. 2. amount of material injected
Einspritzmöglichkeiten injection options
Einspritzöffnung gate *(im)*
Einspritzphase injection phase *(im)*
Einspritzprogramm injection programme *(im)*
Einspritzpunkt gate, feed point, injection point *(im)*
Einspritzrate injection speed/rate, rate of injection *(im)*
Einspritzseite 1. fixed mould half, stationary mould half *(im)*. 2. injection unit *(im)*
Einspritzsteuereinheit injection control unit
Einspritzstrom injection speed/rate, rate of injection
Einspritztemperatur injection temperature
Einspritzung 1. injection. 2. gating: **Einspritzung in die Trennebene** gating at the mould parting line
Einspritzvolumen injection volume, shot volume *(im)*
Einspritzvolumenstrom injection speed/rate, rate of injection *(im)*
Einspritzvorgang injection (process/operation)
Einspritzvorrichtung injection device
Einspritzweg injection stroke *(im)*
Einspritzzeit injection time, mould filling time *(im)*
Einspritzzylinder injection cylinder
Einsprung necking *(of film)*
Einstampfen tamping
Einstationenanlage single-station machine
Einstationenblasformautomat single-station automatic blow moulding machine
Einstationenblasformmaschine single-station blow moulding machine
Einstationenblasmaschine single-station blow moulding machine
Einstationenextrusionsblasformanlage single-station extrusion blow moulding machine
Einstationenmaschine single-station machine

einsteckbar plug-in
einstellbar variable, adjustable
einstellbar, digital digitally adjustable
einstellbar, stufenlos steplessly variable, infinitely variable
Einstellbarkeit adjustability
Einstelldaten (machine) settings
Einstelldatenabspeicherung 1. the storage of machine settings. 2. machine settings store *(me)*
Einstelleinrichtung adjusting/setting mechanism
Einstellelemente controls
Einstellenaufwicklung single-station winder/wind-up (unit)
Einstellfehler setting error
Einstellfeinheit setting accuracy
Einstellgrößen (machine) settings
Einstellmöglichkeiten (machine) setting options
Einstellparameter (machine) settings
Einstellprotokoll (machine) setting record
Einstellschraube adjusting screw
Einstellung 1. setting, adjustment *(of a machine or instrument)*. 2. discontinuance *(of a process or operation)*. 3. formulation: **Phosphatweichmacher in schwerentflammbaren Einstellungen...** phosphate plasticisers in flame retardant formulations
Einstellung, harte low-plasticiser formulation *(PVC paste or compound)*
Einstellung, weiche high-plasticiser formulation *(PVC paste or compound)*
Einstellvorrichtung adjusting/setting mechanism
Einstellwerte (machine) settings
Einstichpyrometer needle pyrometer
Einströmdruck (mould) filling pressure
Einströmgeschwindigkeit (mould) filling speed
Einstückschnecke one-piece screw *(as opposed to a **Schaftschnecke** (q.v.) which is a screw assembled from different sections pushed on to a shaft)*
Einstufenblasverfahren single-stage blow moulding process
Einstufenextrusionsstreckblasen single-stage extrusion stretch blow moulding
Einstufenprozeß single-stage process
Einstufenverfahren single-stage process
einstufig single-stage
Eintauchversuch immersion test
Eintauchzeit immersion time/period
einteilig one-part
Eintopfsystem one-pack/one-component system
eintrennen to treat/spray with (mould) release agent: **Die Werkzeugoberfläche muß deshalb mit einem Trennmittel eingetrennt werden** The cavity surface must therefore be treated with a release agent
Eintrennung treatment with (mould) release agent: **Die Eintrennung der Form mittels**

Trennmittel kann drastisch reduziert werden the amount of release agent used can be drastically reduced
Eintrennvorgang application of (mould) release agent
Eintrittstemperatur inlet temperature
Einwaage amount weighed out *(material or sample)*, amount of substance
einwandfrei, physiologisch non-toxic
einwandfrei, physiologisch nicht toxic
einwandfrei, toxikologisch non-toxic
einwandig single-shell
Einwegartikel disposable article *(plural: disposables)*
Einwegbehälter returnable/disposable/one-trip container
Einwegfaß returnable/disposable/one-trip drum
Einwegflasche returnable/disposable/one-trip bottle
Einwegspritze disposable (hypodermic) syringe
Einwegtasse disposable cup
Einwegverpackung returnable/disposable/one-trip pack(aging)
Einwellenextruder single-screw extruder
Einwellenmaschine 1. single-screw machine *(general term)*. 2. single-screw extruder. 3. single-screw compounder. 4. single-screw devolatiliser. 5. single-screw injection moulding machine
Einwellenschnecke single screw
Einwellensystem single-screw system
einwellig single-screw: **einwellige Austragsschnecke** single-screw delivery unit
einwertig 1. monovalent, univalent. 2. monohydric *(if an alcohol)*
einwiegen to weigh out
Einwirkungszeit exposure time/period *(e.g. to chemicals)*
Einzeladditive separate additives *(as opposed to additive blends)*
Einzelaggregat single unit: **Im folgenden wird kurz auf die Einzelaggregate einer automatischen Rohrfertigungsstraße eingegangen** we shall now briefly describe the various units which make up an automatic pipe production line
Einzelanfertigung one-off production, production of single/ individual units
Einzelantrieb separate/independent drive: **Moderne Kalander haben für jede Walze Einzelantrieb** in modern calenders, each roll is driven separately
Einzelbaustein separate module
Einzel-E-Moduln separate elastic moduli
Einzelfertigung one-off production, production of single/individual units
Einzelform separate/individual mould
Einzelformnestzentrierung individual cavity centring device
Einzelfunktionen separate functions
Einzelgeräte individual/single/serarate units
Einzelkerne separate cores *(im)*

Einzelkosten itemised cost
Einzelmessungen separate determinations
Einzelparameterregelkreis single-parameter control circuit
Einzelpartikel individual particle
Einzelpunktanschnitt single pin gate *(im)*
Einzelregelkreis separate control circuit
Einzelregler individual/separate controller, individual/separate control unit
Einzelschnecke single screw *(as opposed to* **Doppelschnecke**, *q.v.)*
Einzelstückfertigung one-off production, production of single/individual units
Einziehen feeding
Einziehverhalten feed performance/characteristics
einzigartig unique
Einzonenschnecke one-section screw
Einzug 1. feed section/zone *(e)*. 2. feed end *(e)*. 3. feed throat/opening *(e)*. 4. feeding *(of material to a machine)*
Einzugsbereich 1. feed section/zone *(e)*. 2. catchment area
Einzugsbuchse feed bush
Einzugsgangtiefe feed section flight depth *(e)*
Einzugsgeschwindigkeit feed rate
Einzugsnutbuchse grooved feed bush
Einzugsöffnung feed opening/port/throat *(e)*
Einzugsschnecke feed screw
Einzugsschwierigkeiten feed problems
Einzugsshilfe feeding aid
Einzugstasche feed pocket
Einzugsteil feed section/zone *(e)*
Einzugsverhalten feed performance/characteristics
Einzugsverhältnisse feed conditions
Einzugsvermögen feed capacity
Einzugsvorrichtung feed device
Einzugswalze feed roll
Einzugswerk feed mechanism
Einzugszone feed section/zone *(e)*
Einzugszonenabschnitt feed section/zone *(e)*
Einzugszonenbereich feed section/zone *(e)*
Einzugszonengangtiefe flight depth in the feed zone/section *(e)*
Einzugszonenteil feed section/zone *(e)*
Einzweckausführung 1. special purpose design. 2. special purpose machine
Einzweckextruder special purpose extruder
Eisen iron
Eisenbahnkesselwagen rail tanker
Eisenbahntransport rail transport
Eisenblech iron/steel sheet
Eisenglimmer micaceous iron ore
eisenhaltig containing iron
Eisen-Konstantanthermoelement iron-constantan thermocouple
Eisenmetall ferrous metal
Eisenoxidbraun brown iron oxide
Eisenoxidgelb yellow iron oxide
Eisenoxidpigment iron oxide pigment
Eisenoxidrot red iron oxide

Eisenoxidschwarz black iron oxide
Eisenspuren traces of iron
Eisessig glacial acetic acid
Elastifikator impact modifier, toughening agent
elastifizieren to elasticise, to make elastic, to plasticise
elastisch 1. flexible, elastic, resilient
elastische Rückdeformation elastic recovery
elastische Rückfederung elastic recovery
Elastizität flexibility, elasticity
Elastizitätsgrenze elastic limit
Elastizitätsmodul modulus of elasticity, elastic modulus, Young's modulus
Elastizitätsverhalten resilience
Elastizitätsvermögen 1. flexibility 2. elasticity *(see explanatory note under **elastisch**)*
Elastizitätszahl elasticity number
Elastomer elastomer
elastomer elastomeric
Elastomerblend elastomer blend
Elastomerlegierung elastomer blend/alloy
elastomermodifiziert elastomer-modified
Elastomerphase elastomer phase
elektrisch electric(al)
elektrisch abgesichert electrically interlocked *(e.g. safety guard)*
elektrische Beanspruchung electrical loading
elektrische Entladung electric discharge
elektrische Feldstärke electric field strength
elektrische Leistung power
elektrische Werte electrical properties
elektrischer Widerstand electrical resistance
elektrisches Treeing water treeing *(an electrical phenomenon)*
Elektroantriebsmotor electric drive motor
Elektroanwendungen electrical applications
Elektroausrüstung electrical equipment
Elektrobauteil electrical component
Elektrobeheizung electric heating
elektrobraun clcctrical-grade brown
elektrochemisch electrochemical
elektrochemische Korrosion electrochemical corrosion
Elektrodenanordnung electrode arrangement
elektroerosiv hergestellt spark eroded
Elektrogerät electrical appliance
Elektrogeräteindustrie electrical appliance industry
elektrohydraulisch electrohydraulic
Elektroindustrie electrical industry
Elektroinstallationskanal (cable) conduit
Elektroinstallationsrohr electric wiring conduit
Elektroisolierband insulating tape
Elektroisolierfolie electrical insulating film
Elektroisolierharz electrical insulating resin
Elektroisolierlack electrical insulating varnish
Elektroisoliermasse electrical insulating compound
Elektroisolierrohr electrical conduit
Elektroisolierstoff electrical insulating material
Elektroisolierteil electrical insulating component
Elektroisolierverhalten electrical insulating properties
Elektrokabel electric cable
elektrokinetisch electrokinetic
Elektrokontrollschrank control cabinet
Elektrokorrosion electrolytic corrosion
Elektrolaminat printed circuit board material
Elektrolumineszens electroluminescence, electrofluorescence
Elektrolyt electrolyte
elektrolythaltig containing electrolyte
elektrolytisch electrolytic
elektrolytische Korrosion electrolytic corrosion
elektrolytische Korrosionswirkung electrolytic corrosion
elektromagnetisch electromagnetic
elektromagnetische Abschirmung electromagnetic screening/shielding
elektromagnetische Interferenz electromagnetic interference, EMI
Elektromotor electric motor
elektromotorisch electric, electrical
Elektron electron
elektronegativ electronegative
Elektronegativität electronegativity
Elektronenbestrahlung electron irradiation
Elektronendonator electron donor
Elektronenmikroskop electron microscope
Elektronenmikroskopie electron microscopy
elektronenmikroskopisch electron-microscopic
elektronenmikroskopische Abbildung electron micrograph
elektronenmikroskopische Aufnahme electron micrograph
Elektronenpaar electron pair
Elektronenrückstreuung electron back scattering
Elektronenschale electron shell
Elektronenstoß electron collision
Elektronenstrahl electron beam
elektronenstrahlgehärtet electron beam cured
elektronenstrahlhärtbar electron beam curable
Elektronenstrahlhärtung electron beam curing
Elektronenstrahloszillograph electron beam oscillograph
Elektronenstrahlresonanzspektroskopie electron beam resonance spectroscopy
elektronenstreuend electron-scattering
Elektronenstreuung electron scattering
Elektronentransfer electron transfer
Elektronenwolke electron cloud
Elektronik electronics, electronic system
Elektronikbaustein electronic module
Elektronikindustrie electronics industry
Elektronikplatine printed circuit board
Elektroniksteuerung 1. electronic control. 2. electronic control system
elektronisch electronic
elektronische Bauteile electronic components/devices
elektronische Datenverarbeitung electronic data processing
elektronische Spielgeräte electronic games

Elektropapier mica paper
Elektrophoreselack electrophoretic/ electrodeposition paint
Elektrophoresis electrophoresis
elektrophoretisch electrophoretic
Elektropreßspan electrical grade presspahn
Elektroschaltschrank control cabinet
Elektroschichtpreßstoff printed circuit board material
Elektroschlackenraffinierung electrolytic refining
Elektroschutzrohr cable conduit
Elektrostatikspritzen electrostatic spraying
elektrostatisch electrostatic
elektrostatische Aufladung electrostatic charging
Elektrosteuerung electric controls, electric control system
Elektrotauchbad electrodeposition/ electrophoretic bath
Elektrotauchlack electrophoretic/ electrodeposition paint
Elektrotauchlackierung electrodeposition/ electrophoretic painting
Elektrotechnik electrical engineering
elektrotechnisch electrical: **elektrotechnischer Isolierstoff** electrical insulating material
Elektroteile electrical parts/ components/fittings
Elektrowerkzeug power tool
Elementaranalyse elementary analysis
Elementarfaden monofilament
Elementarfaser fibril, elementary fibre
Elementarteilchen elementary particle
Elementarteilchenstrahlung elementary particle radiation
Elementbauweise modular construction, unit construction
Elementbauweise, offene consisting of separate units *(see translation example under* **Bauart, offene***)*
Elevationswinkel angle of elevation
Eliminierung elimination
Ellipse ellipse
elliptisch elliptic(al)
eloxiert anodised
Eluat extract
Elution elution
Emaillackfarbe enamel
Emaille enamel
Emballage package, packaging container
Emballageindustrie packaging industry
EMI *abbr. of* **elektromagnetische Interferenz**, electromagnetic interference
EMI-Abschirmung electromagnetic/EMI shielding/screening
Emission emission
emissionsarm low-emission
emissionsfrei emission-free
Emissionsgrenze emission limit
Emissionsgrenzwert emission limit
Emissionskoeffizient emission coefficient

Emissionsminderungsmaßnahmen measures to reduce emissions
Emissionsrate emission rate
E-Modul elastic modulus, modulus of elasticity
E-Motor electric motor
Empfehlung recommendation
empfindlich delicate, sensitive
Empfindlichkeit sensitivity
Empfindlichkeitsänderung sensitivity shift
Empfindlichkeitsschwelle sensitivity threshold
empfohlen suggested, recommended
empirisch empirical
Emulgator emulsifier, emulsifying agent
emulgatorarm low-emulsifier, with a low emulsifier content
Emulgatorart type of emulsifier
Emulgatorgehalt emulsifier content
Emulgatorgemisch emulsifier blend
emulgatorhaltig containing emulsifier
Emulgatormenge amount of emulsifier
emulgatorreich high-emulsifier, with a high emulsifier content
Emulgatorreste emulsifier residues
Emulgatorsystem emulsifier system
emulgierbar emulsifiable
emulgieren to emulsify
Emulgiersystem emulsifier system
Emulgierwirkung emulsifying effect
Emulsion emulsion
Emulsionshilfsmittel emulsifier, emulsifying agent
Emulsionshomopolymerisat emulsion homopolymer, emulsion polymer
Emulsionskleber dispersion/emulsion/latex adhesive, water-based adhesive
Emulsionspolymerisat emulsion polymer
Emulsionspolymerisation emulsion polymerisation
Emulsionspolymerisationsverfahren emulsion polymerisation (process)
emulsionspolymerisiert emulsion polymerised
Emulsions-PVC emulsion PVC
Emulsionstyp emulsion PVC, emulsion polymer
Emulsionsverfahren emulsion polymerisation
Endabnehmer end user
Endanschläge stops
Endanwendung end use, application
Endartikel end product
Endauswahl final choice
Endeigenschaften end properties
Enderzeugnis end product
Endfestigkeit final strength
Endfeuchte final moisture content
Endgruppe terminal/end group
Endkonsistenz final consistency
Endkontrolle final control/check
Endkonzentration final concentration
Endkunde end user
Endlagendämpfung end-of-travel damping mechanism
endlich finite
Endlosfaden continuous filament

Endlosfaser continuous filament/strand
Endlosfasermatte continuous strand mat
Endlosfaserverbund continuous strand composite
Endlosglasfasern continuous (glass) strands
Endlosmatte continuous strand mat
Endoskopie endoscopy
endotherm endothermic
Endothermie endothermic character
Endposition final/end position
Endprodukt end product
Endprodukteigenschaften end product properties
Endschalter limit switch
endständig terminal
Endtemperatur final temperature
Endtermin final target date
Endverbraucher end user
energetisch *relating to energy*; **energetische Wiederverwertung, energtisches Recycling** energy recovery/recycling; **vom energetischem Standpunkt...** from the energy point of view
energetische Wiederverwertung energy recovery/recycling
energetisches Recycling energy recovery/recycling
Energie energy, power
Energie, aufgenommene energy input, energy used
Energie, kinetische kinetic energy
Energieabrechnung energy calculation
Energieabsorber energy absorber
energieabsorbierend energy absorbent
Energieabsorption energy absorption
Energieanlieferungsrate energy input rate
energiearm low-energy
Energieaufnahme energy input/absorption, energy used
energieaufnehmend energy-absorbing
Energieaufwand amount of energy required
energieaufwendig energy-intensive, requiring a lot of energy
Energiebedarf energy requirements
Energiebilanz energy balance
Energiedichte energy density
energiedispersiv energy-dispersive
Energiedissipation energy dissipation
energiedissipierend energy-dissipating
Energiedosis energy dose
Energieeinleitung energy input
Energieeinsparung energy saving: **Dies entspricht einer Energieeinsparung von etwa 45%** this is equivalent to a saving in energy of about 45%
energieelastisch energy-elastic
Energieelastizität energy elasticity
Energieemission energy release/emission
Energieersparnis energy saving
Energieerzeugung production of energy
Energiefreisetzung energy release
Energiefreisetzungsrate energy release rate

Energiegehalt energy content
Energiegewinnung production of energy
Energiegleichung energy equation
energiegünstig using less energy: **Das Zweikreis-System ist wesentlich energiegünstiger** the two-circuit system uses far less energy
Energiehaushalt energy balance
Energieinhalt energy content
energieintensiv energy-intensive
Energieknappheit energy shortage
Energiekosten energy costs
Energiemehraufwand increased energy requirements: **Der Energiemehraufwand beträgt etwa 15%** about 15% more energy is required
Energienutzung energy recovery
Energiequelle energy source, source of energy
Energieraufnahmevermögen energy-absorbing capacity, capacity/ability to absorb energy
energiereich energy-rich
Energierrecycling energy recovery/recycling
Energierückführung energy recovery/recycling
Energierückgewinnung energy recovery
energiesparend energy saving
Energiesparung energy saving
Energiespeicherung energy storage, storage of energy
Energiespender energy source, source of energy
Energiestrom amount of energy
Energieträger, fossiler fossil fuel
Energieübertragung energy transfer
Energieumsatz energy conversion
Energieumsetzung energy conversion
Energieumwandlungsprozeß energy conversion process
Energieverbrauch energy consumption
Energieverbraucher energy consumer
Energieverlust energy loss
Energieversorgung 1. energy supply. 2. energy supply system
Energieversorgungsleitung power supply line
energieverzehrend energy-consuming
Energievorräte energy reserves
Energiezufuhr energy/power supply: *This word must sometimes be translated very freely - indeed ignored altogether - as in this example*: **Infolge zu starker Energiezufuhr wird die Schmelze stark überhitzt, thermisch geschädigt und abgebaut** If the melt is overheated, it will be charred and degraded. *Although this departs considerably from the original, it conveys its meaning exactly*
engkämmend closely intermeshing (twin screws)
engmaschig close-mesh
engmaschig vernetzt closely cross-linked
Engpaß bottleneck
engtoleriert close-tolerance *(e.g. mouldings)*
engvernetzt closely crosslinked
Enolether enol ether

Enstaubungssystem dust removal system
Entbutzeinrichtung deflashing device *(bm)*
Entbutzen deflashing, flash removal
Entbutzstation deflashing station *(bm)*
Entbutzvorrichtung deflashing device *(bm)*
entdröhnend sound insulating
entfernen to remove
entfetten to degrease
Entfettungsanlage degreasing plant
Entfettungsmittel degreasing agent
Entfeuchtung dehumidification
entflammbar flammable
Entflammbarkeit flammability
Entflammbarkeitsklasse flammability rating
Entformbarkeit demouldability, release properties, ease of demoulding: **Die Entformbarkeit der Formteile ist ausgezeichnet** the parts are extremely easy to demould
Entformbarkeitszeit vulcanising/cure time *(e.g. for silicone rubber moulds)*
entformen to demould
Entformen, spritzteilschonendes demoulding without damaging the moulded part
Entformungsdruck demoulding pressure
Entformungshilfe demoulding aid, (mould) release agent
Entformungshilfsmittel (mould) release agent
Entformungskonizität draft, draw *(im)*
Entformungskraft demoulding force: **Dies gilt besonders für Spritzlinge mit Hinterschneidungen, die höhere Entformungskräfte erfordern** this particularly applies to injection mouldings with undercuts, which require more force to demould them
Entformungsmittel (mould) release agent
Entformungsprinzip method of demoulding
Entformungsrichtung direction of demoulding
Entformungsschräge draft, draw *(im)*
Entformungsschwierigkeiten demoulding problems
Entformungssteifigkeit stiffness on demoulding
Entformungssystem demoulding system/mechanism
entformungstechnisch *relating to demoulding*: **entformungstechnisch kompliziert** difficult to demould
Entformungstemperatur demoulding temperature
Entformungsweg demoulding stroke
Entformungszeitpunkt moment of demoulding
Entformzeit demoulding time
Entgasen 1. degassing, devolatilisation *(e.g. of moulding powders)*. 2. venting *(e.g. of extruder or mould)*. 3. removal *(e.g. of gases from polymer)*: **Entgasen von Ethylen aus Hochdruck-Polyethylen** removal of ethylene from low density polyethylene**entgast** 1. degassed, deaerated *(compound or melt from which volatiles such as moisture, solvent or unreacted monomer have been removed)*. 2. vented *(extruder or mould)*
entgast 1. degassed, deaerated *(compound or melt from which volatiles such as moisture, solvent or unreacted monomer have been removed)*. 2. vented *(extruder or mould)*
∗**Entgasungsaggregat** 1. vented unit. 2. devolatilising unit, devolatiliser
Entgasungsanlage degassing/devolatilising unit
Entgasungsausrüstung venting system: *This word may require some care in translation, as the following example shows*: **Entgasungsausrüstung mit L/D Verhältnissen von etwa 20**. *Since the L/D ratio invariably refers to a screw, the word must obviously be translated as* vented screw
Entgasungsbedingungen degassing/devolatilisation conditions
Entgasungsbereich vent zone, devolatilising section *(e)*
Entgasungsbohrung vent
Entgasungsdom vent
Entgasungseffekt degassing/devolatilising effect
Entgasungseinheit 1. vented unit. 2. devolatilising unit, devolatiliser
Entgasungseinrichtung venting/degassing/devolatilising system: **Doppelschneckenextruder mit Entgasungseinrichtung** vented twin screw extruder
Entgasungsergebnis degassing/devolatilising efficiency
Entgasungsextruder vented/devolatilising extruder
Entgasungsintensität degassing/venting intensity
Entgasungskamin vent
Entgasungskammer degassing/devolatilising chamber
Entgasungsleistung degassing/venting efficiency
Entgasungsmaschine 1. vented extruder. 2. vented injection moulding machine. 3. devolatiliser. 4. devolatilising compounder
Entgasungsmöglichkeiten venting facilities: **Erforderlich für optimale Extrusionsergebnisse sind ausreichende Entgasungsmöglichkeiten und die richtige Temperierung** the essential conditions for the best extrusion results are adequate venting facilities (OR: an adequate venting system) and correct temperature control
Entgasungsoberfläche exposed melt surface *(i.e. during degassing)*
Entgasungsöffnung vent
Entgasungsplastifiziereinheit vented plasticising unit *(im)*
Entgasungsplastifizierung 1. vented plasticisation. 2. vented plasticising unit *(im)*

∗ *Words starting with* **Entgasungs-**, *not found here, may be found under* **Entlüftungs-**

Entgasungsqualität venting efficiency
Entgasungsschacht vent
Entgasungsschnecke vented screw
Entgasungsschneckenpaar vented twin screws *(e)*
Entgasungsschuß vent section/zone, devolatilising section *(e)*
Entgasungssilo degassing/devolatilising silo
Entgasungsstufen degassing/devolatilising stages
Entgasungsstutzen vent
Entgasungssystem venting/degassing/ devolatilising system
Entgasungsteil vent section
Entgasungstrichter vented hopper, vacuum hopper
Entgasungsvorrichtung venting/degassing/ devolatilising device/system
Entgasungswirkung venting/degassing effect
Entgasungszapfen vent
Entgasungszone vent zone, devolatilising section *(e)*
Entgasungszylinder 1. vented barrel *(e)*. 2. vented cylinder *(im)*
entgegengesetzt rotierend counter-rotating
entgiften to decontaminate
Entgratautomat automatic deflashing unit
Entgraten deflashing
Entgratprozeß deflashing (operation)
Entgratroboter deflashing robot
Entgratungsabfälle flash
Entgratungsstation deflashing unit
Entgratungsvorrichtung deflashing device
Entgratvorgang deflashing (operation)
Entgratwerkzeug deflashing tool
Enthalpie enthalpy
entionisiert deionised
entknäuelt disentangled
Entknäuelung disentanglement
Entkopplung decoupling
Entladeenergie discharge energy
Entladeschieber discharge valve
Entladestation unloading station
Entladung discharge
Entladung, elektrische electric discharge
Entladungsgeschwindigkeit discharge rate
Entladungslampe discharge lamp
entlasten to relieve, to release
Entlastung stress removal
Entlastungskurve stress removal curve
Entlastungsvorrichtung load/stress removal device/mechanism
Entlastungszeit stress removal time
entleeren to empty, to drain
Entleerüberwachung discharge monitor
Entleerung emptying
Entleerungsöffnung outlet, discharge opening
Entleerungsstutzen outlet, discharge opening
Entleerungsvorgang emptying/discharging operation

entlüften 1. to evacuate. 2. to vent, to deaerate. 3. to allow (solvent) to evaporate
Entlüftungsbaustein vent unit
***Entlüftungsbohrung** vent
Entlüftungskanal venting channel
Entlüftungslamelle venting lamella
Entlüftungsnute vent groove
Entlüftungsöffnung vent
Entlüftungsschlitz venting slit, vent groove
Entlüftungsspalt venting slit, vent groove
Entlüftungsstift vent(ing) pin
Entlüftungssystem venting/degassing system
Entlüftungsvorrichtung venting/degassing device
entmineralisiert demineralised
Entmischung separation *(e.g. of PVC and plasticiser)*
Entmischungserscheinungen (signs of) separation
Entnahmeautomat parts-removal robot, demoulding robot
Entnahmedrehtisch rotary take-off unit
Entnahmegerät parts-removal robot, demoulding robot
Entnahmemaske gripper
Entnahmeroboter parts-removal robot, demoulding robot
Entnahmestation parts-removal station *(im)*
Entnahmevorrichtung parts-removal robot, demoulding robot
Entnetzungsvorgang dewetting (process)
Entpalettisierungssystem pallet emptying system/unit
entriegelt unlocked
Entropie entropy
Entropieabnahme entropy decrease
entropieelastisch entropy-elastic, rubbery, rubbery-elastic
Entropieelastizität entropy-elasticity, rubber-like elasticity
Entropiezunahme entropy increase
Entrostung removal of rust
entsalzt desalinated
Entschalungsmittel mould oil
Entschäumer defoaming agent, defoamer, antifoam (agent)
Entschäumertropfen defoamer droplet
Entschäumerwirkung defoaming effect
entschäumungsaktiv antifoam
Entschäumungsmittel defoaming agent, defoamer, antifoam (agent)
Entscheidungsgremium decision-making body
Entscheidungshilfe decision aid *(me)*
Entscheidungskriterien decision-making criteria
Entscheidungsmerkmale decision-making features
Entscheidungsprozeß decision-making process
Entscheidungstabelle decision table *(me)*
Entschlaufung disentanglement

---

* *Words starting with **Entlüftungs-**, not found here, may be found under **Entgasungs-***

entschlichtet de-sized
Entschwefelungsanlage desulphurisation plant
Entsorgung 1. discharge. 2. disposal *(e.g. of effluent, waste, etc.)*
Entsorgungsanlage waste tip/dump, waste disposal site, landfill site
entsorgungsfreundlich easily/readily disposed of
Entsorgungskonzept disposal concept
Entsorgungsleitung discharge line
Entsorgungslinie waste disposal chain
Entsorgungsunternehmen waste disposal contractor
Entspannung 1. relief, relaxation *(of stresses or pressures)*. 2. decompression, devolatilisation *(of melts)*
Entspannungsverdampfung flash vaporisation
Entspannungsversuch stress relaxation test
Entspannungsvorgang 1. relaxation (process) *(of stresses)*. 2. decompression, devolatilisation
Entspannungszone vent zone, devolatilising section *(e)*
Entstatisierungseinrichtung destatisising unit
Entstauben dust removal
Entwässerung 1. drainage. 2. removal of water
entwerfen to design
Entwicklung, verfahrenstechnische technical development(s)
Entwicklungarbeit(en) development work
Entwicklungsaktivitäten development activities
Entwicklungsaufwand development costs
Entwicklungsländer developing/ underdeveloped countries
Entwicklungsmöglichkeiten development potential/possibilities
Entwicklungsphase development phase
Entwicklungsprodukt development product
Entwicklungstendenzen development trends
Entwicklungszeit development period
Entwurfsdaten design data
Entwurfsphase design stage
Entwurfsstadium design stage
Entzunderung descaling
Entzündlichkeit flammability
Entzündungspunkt ignition point
Entzündungsquelle source of ignition
Entzündungstemperatur ignition temperature
Entzündungswiderstandsindex ignition resistance index
entzündungswidrig flame resistant
Entzündungswidrigkeit flame resistance
Enzym enzyme
enzymatisch enzymatic
EO *abbr. of* **Ethylenoxid**, ethylene oxide
EP-Anstrichmittel epoxy paint
EP-Beschichtung 1. epoxy coating 2. epoxy paint
EPC-Ruß easy procesing channel black
EPDM *symbol for* ethylene propylene-diene rubber
EP-Harz epoxy resin

EPIC-Harz epoxy-isocyanate resin (**EPIC** *is a combination of* **EP**oxy *and* **IsoC**yanat*)*
Epichlorhydrin epichlorohydrin
Epichlorhydrinkautschuk epichlorohydrin rubber
Epichlorkautschuk epichlorohydrin rubber
epidemiologisch epidemiological
Epihalogenhydrin epihalogen hydrin
epitaktisch epitaxial
EP-Laminat epoxy laminate
EPM *symbol for* ethylene-propylene ruber
EP-Mörtel epoxy-based mortar
E-Polymerisationsverfahren emulsion polymerisation (process)
Epoxi- *see* **Epoxy-** *or* **Epoxid-**
Epoxid epoxide, epoxy
Epoxidacrylat epoxy acrylate
Epoxidäquivalent epoxy equivalent
Epoxidäquivalentgewicht epoxy equivalent
Epoxidation epoxidation
Epoxidbeschichtung 1. epoxy coating 2. epoxy paint
Epoxidformmasse epoxy moulding compound
epoxidfunktionell epoxy-functional
Epoxidgehalt epoxy/epoxide value
Epoxidgießharz epoxy casting resin
Epoxidgießharzisolator cast epoxy insulator
Epoxidgruppe epoxy/epoxide group
Epoxidharz epoxy/epoxide resin
Epoxidharzanstrichsystem epoxy paint, epoxy coating system
Epoxidharzbeton epoxy concrete
Epoxidharzester epoxy ester
Epoxidharzestrich epoxy screed
Epoxidharzform epoxy mould
Epoxidharzformmasse epoxy moulding compound
Epoxidharzformstoff cured epoxy resin
Epoxidharzglashartgewebe glass-epoxy laminate
Epoxidharzglashartgewebe epoxy glass cloth laminate
Epoxidharzgruppe epoxy/epoxide group
Epoxidharzhärter epoxy hardener
Epoxidharzhartpapier epoxy paper laminate
epoxidharzimprägniert epoxy (resin) impregnated
Epoxidharzklebstoff epoxy adhesive
Epoxidharzlack epoxy paint
Epoxidharzlösung epoxy resin solution
epoxidharzmodifiziert epoxy modified
Epoxidharzpreßmasse epoxy moulding compound
Epoxidharzschaum epoxy foam
Epoxidharzsystem epoxy system
epoxidharzvergütet epoxy resin modified
Epoxidharzversiegelung epoxy sealant
epoxidiert epoxidised
Epoxidlaminat epoxy laminate
Epoxidpreßmasse epoxy moulding compound
epoxidtypisch typical of epoxy resins
Epoxidumhüllung epoxy encapsulation

Epoxidverbindung epoxy/epoxide compound
Epoxidweichmacher epoxy plasticiser
Epoxidwert epoxy/epoxide value
Epoxidzweikomponentenklebstoff two-pack epoxy adhesive
Epoxyacrylat epoxy acrylate
Epoxyalkoxysilan epoxy alkoxysilane
Epoxydierung epoxidation
Epoxygießharz epoxy casting resin
Epoxygruppe epoxy/epoxide group
Epoxyharzschichtstoff epoxy laminate
Epoxymethacrylat epoxy methacrylate
Epoxysilan epoxysilane
Epoxyverschnitt epoxy blend
EP-Reaktionsharz epoxy/epoxide resin
EP-Reaktionsharzmasse catalysed epoxy resin *(see explanatory note under* **Reaktionsharz***)*
Eprom-Speicher EPROM *(me) (abbr. of erasable and programmable read-only memory)*
EP-Spachtel epoxy-based knifing filler, epoxy-based stopper
E-PVC *abbr. of* **Emulsions-PVC,** emulsion PVC
EP-Werkstoff epoxy material
Erdalkalicarbonat alkaline earth carbonate
Erdalkalimetall alkaline earth metal
Erdalkalioxid alkaline earth oxide
Erdalkalisulfat alkaline earth sulphate
Erdatmosphäre earth's atmosphere
Erdbau earthwork
Erdbeschleunigung acceleration due to gravity
erden to earth
Erdfarbe earth pigment
Erdgas natural gas
Erdkabel underground cable
Erdölfördertechnik petroleum extraction technology/engineering
Erdölförderung petroleum extraction
Erdölfraktion petroleum fraction
Erdölharz petroleum resin
Erdölindustrie petroleum industry
Erdölreserven oil reserves
Erdölressourcen petroleum resources
Erdpotential earth potential
erdverlegt underground, buried *(e.g. pipes)*
Erdverlegung, für die for burying underground, for underground installation
Erfahrungsaustausch exchange of ideas
erfahrungsgemäß from previous experience, in our experience, we have found that...
Erfahrungswert known value
erfaßbar determinable
erfassen to register, to determine, to measure, to pick up, to include, to cover
Erfassung 1. determination, measurement. 2. acquisition, gathering *(of data) (me)*
Erfassungssystem measuring system
erfreulich encouraging *(results, developments etc.)*
Ergänzungsaggregat supplementary unit
Ergänzungseinheit supplementary unit
Ergebnis 1. result. 2. profit, earnings

Ergebnisunterschiede differences in the results obtained
Ergiebigkeit yield
Ergonomie ergonomics
ergonomisch ergonomic
Erhebungsformular questionnaire
erhöht increased
erhöht schlagzäh with enhanced impact resistance/strength, high-impact
erhöhte Temperatur elevated temperature
Erholungskurve recovery curve
Erholungsprozeß recovery process
Erholungsverhalten recovery behaviour
Erholungszeit recovery period
Erholungszeitspanne recovery period
Erichsendehnbarkeit Erichsen distensibility
Erichsentiefung Erichsen indentation
Erinnerungseffekt memory effect
erkalten to cool down
Erkenntnisse findings
Erlenmeyerkolben Erlenmeyer flask
Ermüdung fatigue
Ermüdungsbeanspruchung fatigue stress
Ermüdungsbelastung fatigue stress
Ermüdungsbeständigkeit fatigue resistance
Ermüdungsbruch fatigue failure/fracture
Ermüdungseigenschaften fatigue properties
ermüdungsfest fatigue resistant
Ermüdungsfestigkeit fatigue strength
Ermüdungsprüfung fatigue test
Ermüdungsriß fatigue crack
Ermüdungsrißausbreitung fatigue crack propagation
Ermüdungsrißbildung fatigue cracking
Ermüdungsrißwachstum fatigue crack growth
Ermüdungsschutzmittel fatigue inhibitor, anti-fatigue agent
Ermüdungsverhalten fatigue behaviour
Ermüdungsversagung fatigue failure
Ermüdungsversuch fatigue test
Ermüdungswiderstand fatigue resistance
Ermüdungswiderstandsfähigkeit fatigue resistance
Erniedrigung reduction
Erodieren spark erosion, electric discharge machining, EDM
Erosion 1. erosion. 2. spark erosion, electric discharge machining, EDM
Erosionsbeständigkeit erosion resistance
Erosionsfestigkeit erosion resistance
Erosionsgrad amount of erosion
Erosionsoberfläche spark eroded surface
Erosionsverschleiß erosive wear
erosiv erosive
erprobt proven
Erprobung trial
Erprobungsergebnis test result
Erprobungsphase trial period
Ersatzprodukt alternative product, substitute
Ersatzstoff alternative material, substitute
Ersatzteile spares, spare parts

Ersatzteilkosten cost of spare parts **geringe Ersatzteilkosten** spare parts are cheap
Ersatzteilliste list of spare parts
Ersatztreibmittel alternative blowing agent
Ersatzzylinder replacement barrel *(e)*/cylinder *(im)*
Erscheinungsbild appearance
Erschütterung shock, vibration
erschütterungsempfindlich shock-sensitive
erschütterungsfest shock/vibration resistant
Erschütterungsfestigkeit shock/vibration resistance
erschütterungsfrei vibration-free
ersetzen to replace
Erstanwender first-time user
erstarren to freeze *(melt in a mould) (im)*, to solidify, to set
Erstarrungsdauer 1. cooling time. 2. setting time *(im)*
Erstarrungsgeschwindigkeit setting speed
Erstarrungslinie frost line *(bfe)*
Erstarrungslinienhöhe frost line height *(bfe)*
Erstarrungsphase setting phase *(im)*
Erstarrungspunkt freezing/crystallising/ soldifying point
Erstarrungsschrumpf shrinkage on solidification
Erstarrungszeit cooling time, setting time *(im)*
Erstausrüstung original equipment
Erstrezept starting formulation
Erstwerkzeugkosten initial tooling costs
Ertrag profits
Ertragskraft earning power
Erwärmung heating-up
Erwärmungsgeschwindigkeit heating-up rate
Erwärmungsphase heating-up phase
Erwärmungszeit heating-up period
erwartungsgemäß as expected
Erweichung softening
Erweichungsbereich softening range
Erweichungsgebiet softening range
Erweichungspunkt softening point
Erweichungstemperatur softening point
erweiterungsfähig expandable
Erw.Pkt *abbr. of* **Erweichungspunkt**, softening point
Erzeugnisgruppe product group, group of products
ESCR *abbr. of* environmental stress cracking resistance
ESCR-Daten stress crack resistance data/ information
ESG *abbr. of* **Einscheibensicherheitsglas**, single-layer safety glass
ESH *abbr. of* **Elektronenstrahlhärtung**, electron beam curing
ESH-Technologie electron beam curing *(technology)*
ESR-Spektroskopie electron beam resonance spectroscopy
Essigester ethyl acetate
Essigestertest ethyl acetate test

Essigsäure acetic acid
Essigsäureanhydrid acetic anhydride
Essigsäurebutylester butyl acetate
Essigsäureethylester ethyl acetate
Essigsäurevinylester vinyl acetate
Esteralkoholrest ester alcohol radical
Estercarbonylgruppe ester-carbonyl group
Estergehalt ester content
Esterglied ester segment
Estergruppe ester group
Estergruppierung ester group
esterlöslich ester-soluble
Esterweichmacher ester plasticiser
Estrich (floor) screed
Estrich, schwimmender floating screed/floor
Estrichmörtel screed mortar
Etage daylight *(im)*: *It should be remembered that* **Etage** *is the space between two mould parts, which is why, for example, a* **Zweietagenwerkzeug** *is equivalent to a* **Dreiplattenwerkzeug**
Etagenbauweise multi-daylight design: **Spritzgießwerkzeug in Etagenbauweise** multi-daylight injection mould
Etagenpresse multi-daylight press
Etagenspritzen 1. multi-daylight injection moulding *(im)*. 2. multi-daylight transfer moulding *(tm)*
Etagenspritzgießwerkzeug multi-daylight injection mould
Etagenwerkzeug multi-daylight mould *(im,tm)*, stack mould
Etatjahr financial year
Ethan ethane
Ethanol ethanol
Ethanolamin ethanolamine
ethanolisch ethanolic
ethanolische Lösung ethanol solution
ethanollöslich ethanol-soluble
Ethanollöslichkeit solubility in ethanol
Ethanolverdünnbarkeit dilutability with ethanol
Ether ether
Etheralkohol ether alcohol
Etheramin ether amine
Etherbindung ether linkage
Etherbrücke ether bridge
Etherdiol ether diol
Ethereinheit ether group
Etherextrakt ether extract
Etherglied ether segment
Ethergruppe ether group
Etherverknüpfung ether linkage
Ethoxygruppe ethoxy group
ethoxyliert ethoxylated
Ethoxypropylacetat ethoxypropyl acetate
Ethylacetat ethyl acetate
Ethylacrylat ethyl acrylate
Ethylalkohol ethyl alcohol
Ethylcellulose ethyl cellulose
Ethylchlorid ethyl chloride
Ethylcyanacrylat ethyl cyanoacrylate
Ethylen ethylene

Ethylen-Acrylatcopolymer ethylene-acrylate copolymer
Ethylen-Acrylatkautschuk ethylene-acrylate rubber
Ethylen-Acrylsäurecopolymer ethylene-acrylic acid copolymer
Ethylenanteil ethylene content
Ethylen-Butylacrylatcopolymerisat ethylene butyl acrylate copolymer
Ethylenchlorid ethylene chloride
Ethylenchlortrifluorethylen ethylene chlorotrifluoroethylene
Ethylendicarbonsäure ethylene dicarboxylic acid
Ethylenediamin ethylene diamine
Ethylengas ethylene
Ethylengehalt ethylene content
Ethylenglykol ethylene glycol
Ethylenglykoldimethacrylat ethylene glycol dimethacrylate
Ethylenglykolether ethylene glycol ether
Ethylenglykolmonoalkylether ethylene glycol alkyl ether
Ethylenglykolmonobutylether ethylene glycol butyl ether
Ethylenglykolmonoethylether ethylene glycol ethyl ether
Ethylenglykolpolyadipat ethylene glycol polyadipate
ethylenhaltig containing ethylene
ethylenisch ethylenic
Ethylenkette ethylene chain
Ethylenoxid ethylene oxide
Ethylenoxidgruppe ethylene oxide group
Ethylen-Propylen-Dienkautschuk ethylene-propylene-diene rubber, EPDM
Ethylen-Propylenelastomer ethylene-propylene rubber, EPDM
Ethylen-Propylenkautschuk ethylene-propylene rubber, EPDM
Ethylen-Tetrafluorethylencopolymer ethylene-tetrafluoroethylene copolymer
Ethylenverteilung ethylene distribution
Ethylen-Vinylacetatcopolymerisat ethylene-vinyl acetate copolymer, EVA
Ethylen-Vinylacetatkautschuk ethylene-vinyl acetate rubber
Ethylen-Vinylalkoholcopolymerisat ethylene-vinyl alcohol copolymer
Ethylglykol ethyl glycol
Ethylglykolacetat ethyl glycol acetate
Ethylgruppe ethyl group
Ethylhexanol ethyl hexanol
Ethylhexylacrylat ethylhexyl acrylate
Ethylhexylglycidylether ethylhexyl glycidyl ether
Ethylsilikat ethyl silicate
Etikett label
Etikettieren labelling
europäischer Binnenmarkt single European market
eutektisch eutectic

EVAC ethylene-vinyl acetate, EVA
evakuieren to evacuate
eventuell possibly, perhaps *The word is sometimes used as padding, when it should be omitted, e.g.* **Nach einer sich eventuell anschließenden Kühlzeit kann das Spritzgießwerkzeug geöffnet werden.** *The mould is ALWAYS allowed to cool down before opening - there is no "eventuell" about it. The translation should therefore be:* The mould is then allowed to cool, and opened
EWG *abbr. of* **Europäische Wirtschaftsgemeinschaft**, European Economic Community, EEC
EWG-Länder EEC countries, common market countries
Exaktheit precision, accuracy, exactness
ex-geschützt explosion-proof
exotherm exothermic
Exothermie exothermic character
expandierbar expandable
expandiert expanded, foamed
Expansionsdruck expansion pressure
Expansionsphase expansion phase
Expansionsvermögen expansion capacity
Experimentierbedingungen experimental conditions
explosibel explosive
explosionsartig explosive
Explosionsdarstellung exploded-view diagram
explosionsfähig explosive
explosionsgefährlich potentially explosive, representing an explosive hazard
explosionsgeschützt explosion-proof
Explosionsgrenze explosive limit
Exponat exhibit
exponentiell exponential
exponiert exposed
Exportauflagen export requirements
Exportaussichten export prospects
Exportgeschäft export business
Exportquote export share
Exposition exposure
Expositionsbedingungen conditions of exposure, exposure conditions
Expositionsdauer time/period of exposure *(e.g. of a test specimen)*
Expositionszeit time/period of exposure *(.eg. of a test specimen)*
Ex-Schutzbereich explosion-proof area
Exsiccator, Exsikkator desiccator
Extender extender
Extenderfüllstoff filler
Extender-PVC extender PVC
Extenderweichmacher (plasticiser) extender
Externspeicher external store *(me)*
Extinktion extinction
Extinktionskoeffizient extinction coefficient
extrahierbar extractable
Extrahierbarkeit extractability
Extrahieren extraction
extrahiert extracted

Extraktion extraction
extraktionsbeständig extraction resistant
Extraktionsbeständigkeit extraction resistance
Extraktionsflüssigkeit extractant
Extraktionsgeschwindigkeit extraction rate
Extraktionsmittel extractant
Extraktionsrückstand extraction residue
extraktionsstabil extraction resistant
Extraktionsverhalten extraction behaviour
Extrapolation extrapolation
extrapoliert extrapolated
Extremforderungen extreme demands
Extremtemperaturen temperature extremes
Extrudat extrudate
Extruder extruder
Extruder, reversierender reciprocating extruder
Extruderabschnitte parts of the extruder: **die einzelnen Extruderabschnitte** the different parts of the extruder
Extruderanfahrladen extruder start-up waste
Extruderanlage extrusion line
Extruderanschluß extruder connection
Extruderantrieb extruder drive
Extruderaufhängung extruder suspension (unit)
Extruderausstoß extruder output (rate)
Extruderbauart(en) type(s) of extruder
Extruderbaufirma extruder manufacturer
Extruderbaulänge extruder length
Extruderbaureihe extruder range, range of extruders
Extruderdiagramm extruder diagram *(output vs. pressure curve)*
Extruderdrehzahl screw speed *(e)*
Extruderdrehzahlsollwert required screw speed
Extruderdüseneinheit (extrusion) die assembly
Extruderdüsenkopf extruder head, die head
Extrudereinfüllbereich feed section/zone *(e)*
Extrudereinfüllöffnung feed opening/port/throat *(e)*
Extrudereinfülltrichter (feed/material) hopper *(e,im)*
Extrudereinheit extruder
Extrudereinstellwerte extruder settings
Extrudereinzugsbereich feed section/zone *(e)*
Extruderfabrikat make of extruder
Extruderfahrwagen extruder carriage
Extruderfertigung extrusion
Extruderfolie extruded film
Extrudergehäuse (extruder) barrel
Extrudergetriebe extruder drive
Extruderhalle extrusion shop
Extruderhöhe extrusion height, screw centre height *(see explanatory note under* **Extrudierhöhe***)*
Extruderkaskade cascade extruder
Extruderkennlinie extruder characteristic (curve)
Extruderkennliniengleichung extruder output equation
Extruderkonstruktion extruder design

Extruderkonzept 1. extruder design. 2. extruder *(see explanatory note under* **Konzept***)*
Extruderkonzeption extruder design
Extruderkopf extruder head, die head
Extruderkühlung 1. extruder cooling: **Neuerdings wird zur Extruderkühlung nicht nur Luft, sondern zusätzlich auch noch Wasser verwendet** a recent development is cooling the extruder with water in addition to using air cooling. 2. extruder cooling system
Extruderlänge 1. extruder length. 2. screw length: **Bei einem Verarbeiten von Elastomeren reicht eine Extruderlänge von 16D aus** an extruder with a 16D screw is adequate for extruding elastomers
Extruderleistung 1. extruder output. 2. extruder efficiency
Extruderlochplatte breaker plate *(e)*
Extrudermundstück (outer) die ring *(e)*
Extrudernachfolge extruder downstream equipment, post-extrusion equipment
Extrudernachfolgemaschine extruder downstream equipment, post-extrusion equipment
Extruderprozeß extrusion (process)
Extruderschnecke (extruder) screw
Extruderspeisung extruder feed: **Da die Extruderspeisung über einen Trichter erfolgt...** since the extruder is fed from a hopper...
Extruderstraße extrusion line
Extrudersystem extruder system
Extrudertrichter (feed/material) hopper *(e)*
Extrudertyp type of extruder
Extruderverarbeitung extrusion
Extruderverhalten extruder performance
Extruderwerkzeug (extrusion) die
Extruderzubehör extruder accessories
Extruderzylinder (extruder) barrel
extrudierbar extrudable
Extrudierbarkeit extrudability
Extrudierbedingungen extrusion conditions
extrudieren to extrude
Extrudieren extrusion
Extrudiergeschwindigkeit extrusion speed/rate
Extrudierhöhe extrusion height, screw centre height *(height from floor level to centre line of extruder barrel)*
Extrudierwerkzeug (extrusion) die
Extrusion extrusion
Extrusion, stetige continuous extrusion
Extrusion, taktweise intermittent extrusion
Extrusionblasverfahren extrusion blow moulding (process)
Extrusionsabfälle 1. extrusion scrap. 2. extruder start-up waste
Extrusionsanlage extrusion line/plant
Extrusionsartikel extruded products, extrusions
Extrusionsaufgaben extrusion tasks
extrusionsbedingt caused by extrusion *(e.g. stresses)*
Extrusionsbedingungen extrusion conditions

Extrusionsbeschichten extrusion coating
Extrusionsbeschichtungsanlage extrusion coating line/plant
Extrusionsbeschichtungslinie extrusion coating line
Extrusionsblasanlage extrusion blow moulding plant/line
Extrusionsblasen extrusion blow moulding
Extrusionsblasformanlage extrusion blow moulding plant/line
Extrusionsblasformen extrusion blow moulding
Extrusionsblasformmaschine extrusion blow moulding machine, extrusion blow moulder
Extrusionsblasformprozeß extrusion blow moulding (process)
Extrusionsblasformverfahren extrusion blow moulding (process)
extrusionsblasgeschäumt extrusion blow foam moulded: *Although this is acceptable, the translator would be better advised to base his translation on the interpretation given under* **TSB** *(q.v.) e.g.* **extrusionsblasgeschäumte PS Becher** cups made by extrusion blow moulding expandable polystyrene
Extrusionsblaskopf parison die: *The first part of the word need not be translated since* **-kopf** *automatically implies extrusion* blow moulding expandable polystyrene
Extrusionsblasmaschine extrusion blow moulding machine, extrusion blow moulder
Extrusionsblastechnik extrusion blow moulding (technology/process): **Aus den geschilderten Merkmalen der Extrusionsblastechnik...** from the above described features of the extrusion blow moulding process...; **Die in letzter Zeit erzielten Fortschritte in der Extrusionsblastechnik...** the progress made in recent years in extrusion blow moulding (technology)...
Extrusionsblasverfahren extrusion blow moulding (process)
Extrusionsblaswerkzeug parison die *(bm)*
Extrusionsbreitschlitzdüse 1. slit die, slot die 2. flat film (extrusion) die *(for gauges below 0.25mm)* 3. sheet (extrusion) die *(for thicker gauges)*
Extrusionsdruck extrusion pressure
Extrusionsdüse (extrusion) die
Extrusionseinheit extruder
extrusionsfähig extrudable
Extrusionsfolie extruded film/sheeting
Extrusionsformanlage combined extrusion-thermoforming line
extrusionsgeblasen extrusion blow moulded
Extrusionsgeschwindigkeit extrusion speed/rate, rate of extrusion
Extrusionsgießverfahren cast film extrusion, chill roll casting/extrusion (process)
Extrusionshalle extrusion shop
Extrusionshöhe extrusion height, screw centre height *(see explanatory note under* **Extrudierhöhe***)*

Extrusionskapazität extrusion capacity
Extrusionskopf extruder head, die head
Extrusionsleistung extruder output/capacity
Extrusionslinie extrusion line
Extrusionsmarke extrusion grade *(of moulding compound)*
Extrusionsmaschine extruder
Extrusionsmasse extrusion compound
Extrusionsmischung extrusion compound
Extrusionsmundstück (outer) die ring *(e)*
Extrusionsqualität 1. extrusion quality. 2. extrusion grade *(of compound)*
Extrusionsquellung die swell
Extrusionsrichtung extrusion direction
Extrusionsrunddüse round section die
Extrusionsschnecke (extruder) screw
Extrusionsschweißen extrusion welding
Extrusionsschweißgerät extrusion welding unit
Extrusionsspinnanlage spinneret extrusion line
Extrusionsspinneinheit spinneret extrusion unit
Extrusionsspinnprozeß spinneret extrusion (process)
Extrusionsstaudruck back pressure *(e)*
Extrusionsstraße extrusion line
Extrusionsstreckblasen extrusion stretch blow moulding
Extrusionsstreckblasformen extrusion stretch blow moulding
Extrusionssystem extrusion system
Extrusionstechnik extrusion technology, extrusion: **Die Herstellung von Schlauchfolien ist eines der bedeutendsten Gebiete der Extrusionstechnik** the manufacture of tubular film is one of the most important branches of extrusion technology; **Die Extrusionstechnik dient zur Fertigung von Profilen, Rohren usw.** Profiles, pipes etc. are made by extrusion
Extrusionstrichter (feed/material) hopper *(e,im)*
Extrusionstyp extrusion grade *(of compound)*
Extrusionsverarbeitung extrusion
Extrusionsverfahren extrusion (process)
Extrusionsverhalten extrusion performance
Extrusionsverlauf extrusion process/operation: **Der Extrusionsverlauf gliedert sich in zwei Abschnitte** the extrusion process is divided into two stages
Extrusionsversuch extrusion trial
Extrusionsviskosimeter extrusion rheometer/plastimeter
Extrusionsvorgang extrusion (process)
Extrusionswerkzeug (extrusion) die
Extrusionszylinder (extruder) barrel
Exzenterhub eccentric stroke/movement
E-Zugsmodul tensile modulus of elasticity

F

F + E *abbr. of* **Forschung und Entwicklung**, research and development
Fabrikationshalle 1. production shed. 2. moulding shop, extrusion shop
Fabrikationskontrolle production control
Fabrikationsprogramm 1. production schedule/plan/programme. 2. product range, range of products
Fabrikationsprüfung production control
Fabrikationsstätte factory, plant, production facility
Fabrikationsüberwachung production control
Fabrikhalle factory shed
Fach (mould) cavity, impression *(im) (see explanatory note under* **Formhohlraum***);*
  -**fach Heißkanalanguß** (*e.g.* **4-, 6-, 8-,** *etc.* **fach**) - point hot runner feed system (*e.g.* 4-, 6-, 8- point hot runner feed system)
Fachausschuß technical committee
Fachberater technical adviser
Fächelscheißen hot air/gas welding **Fächeln** *refers to the stroking or fanning action of the welding torch, i.e. the movement of the nozzle across the polymer surface to prevent charring. There is no need to translate the word, since* **fächeln** *invariably forms part of hot gas welding.*
Fachhochschule technical university
Fachleute experts
Fachliteratur technical literature
Fachmann expert
Fachnormenausschuß technical standards committee
fadenartig thread-like
Fadenlegetechnik filament winding
Fadenmolekül thread-like molecule
Fadenzeit fibre time *(PU foaming)*
Fadenziehen 1. stringing *(im)*. 2. cobwebbing *(lacquers)*
FAF *abbr. of* **flexibel automatisierte Fertigungsanlage**, flexibly automated production line
Fahrbahn roadway
Fahrbahndecke road surface
Fahrbedingungen 1. processing conditions. 2. driving conditions
Fahrbelag surfacing *(of roads, bridges etc.)*
Fahrbewegung movement *(of a machine unit)*
Fahrerhaus driver's cab
Fahrgastraum passenger compartment
Fahrgastzelle passenger compartment
Fahrgeschwindigkeit speed, machine speed:
  Fahrgeschwindigkeiten beim Schließen und Öffnen mould opening and closing speeds
Fahrgeschwindigkeitsüberwachung (machine) speed control system
Fahrgestell chassis
Fahrkomfort driving comfort
Fahrrad bicycle
Fahrradfelge bicycle wheel rim
Fahrradrahmen bicycle frame
Fahrweise method of operation
Fahrwerk chassis
Fahrzeug vehicle
Fahrzeugbau vehicle construction
Fahrzeugelektrik car's electrical system
Fahrzeuginnenausstattung vehicle interior trim
Fahrzeuginnenraum vehicle interior
Fahrzeugkarosserie vehicle body
Fahrzeugkonstrukteur car designer
Fahrzeuglack automotive paint
Fahrzeugrahmen chassis
Fahrzeugvorbau front end
Fallbolzen falling weight
Fallbolzenprüfgerät falling weight tester
Fallbolzenversuch falling weight test
Fallbruchfestigkeit drop impact strength
Falldorntest falling dart test
Fallfilmverdampfer falling film vaporiser
Fallgewicht falling weight
Fallgewichtsprüfung falling weight test
Fallhammer falling weight
Fällhilfsmittel precipitating agent
Fallhöhe drop height
Fällmittel precipitating agent
Fallrohr downpipe
Fällung precipitation
Fällungsmittel precipitating agent
Fällungsparameter precipitation conditions
Fällungspolymerisation precipitation polymerisation
Fallversuch drop impact test
Faltbiegeversuch folding endurance test
Faltenbalg bellows
Faltenfilter fluted filter
faltenfrei crease-free, wrinkle-free, free from creases
faltenlos crease-free, wrinkle-free, free from creases
Faltkern collapsible core
falzen to bend, to fold, to flex
FAR *abbr. of* Federal Aviation Regulations
Faraday'scher Käfig Faraday cage
Farbabweichungen colour deviations
Farbänderung change in colour
Farbanstrich paint film, coating
Farbanteil pigment content
Farbbatch pigment paste/concentrate
farbbeständig colourfast
Farbbeständigkeit colour stability, colourfastness
Farbbildschirm colour screen *(me)*
Farbbrillanz brilliant/sparkling colour
Farbdia (colour) transparency
Farbdichte colour intensity
Farbdruckvorrichtung colour printing unit
Farbe 1. colour. 2. paint
Farbechtheit colour fastness
Farbenhersteller paint manufacturer
Farbenindustrie coatings/paint industry

Farbgebung colouring, imparting colour
Farbgranulat masterbatch
Farbgraphikbildschirm colour graphic display screen *(me)*
Farbgraphiksystem colour grphics system *(me)*
Farbhomogenität uniform colour
Farbintensität colour intensity
farbintensiv intensely coloured
Farbkennzeichnungsmaschine colour coding machine
Farbkodierung colour coding
Farbkonstanz colour uniformity, uniform/consistent colour
Farbkonzentrat pigment concentrate
Farbkraft 1. pigment strength. 2. colour intensity
farblos colourless
Farblosigkeit colourlessness, absence of colour
Farbmetrik colorimetry
farbmetrisch colorimetric
Farbmittel colouring agent, colourant *(which can be a pigment or a dye)*
Farbnachstellung colour matching
Farbnester pigment agglomerates
Farbpalette colour range, range of colours
Farbpaste pigment paste
Farbpastenharz pigment paste resin, resin suitable for pigment pastes
Farbpigment 1. pigment. 2. coloured pigment *(when a distinction is made between* **Farbpigmente** *and* **Weißpigmente,** *q.v.)*
Farbpigmentpaste pigment paste
Farbpulver powdered pigment
Farbrasterbildschirm colour raster display (screen) *(me)*
Farbrasterterminal colour raster terminal *(me)*
Farbreste paint residues
Farbruß carbon black
Farbschlieren coloured streaks
Farbschwankungen colour variations, variations in colour
Farbsichtgerät colour visual display unit, colour VDU *(me)*
Farbskala colour scale
Farbsortiment colour range
Farbspritzanlage paint spraying unit
farbstabil colourfast
Farbstabilität colourfastness
farbstark intensely coloured
Farbstärke colour intensity
Farbstoff dye, dyestuff: *Occasionally, the word is carelessly and wrongly used in place of* **Pigment.** *In an article on PVC pastes, for example, the phrase* **Füll- und Farbstoffe** *must obviously be rendered as* fillers and pigments
Farbstoffnester local pigment concentrations
Farbstoffpaste pigment paste
Farbtafel colour chart
Farbteig pigment paste
farbtief deep in colour
Farbtiefe depth of colour
Farbton shade, hue
Farbtonabweichung colour variation
Farbtonänderung colour change, change in colour
farbtonbeständig colourfast
Farbtonbeständigkeit colour fastness/stability
Farbtonkarte colour chart
Farbtonkonstanz colour uniformity, uniform/consistent colour
Farbtonnuancierung tinting
Farbtonspektrum colour range, range of colours
Farbtontiefe colour intensity/saturation
Farbtonverschiebung colour shift
Farbübergang change in colour, colour change
Farbumschlag colour change, change in colour
Farbunterschied difference in colour
Farbveränderung colour change, change in colour
Farbverteilung pigment dispersion
Farbvertiefung deepening in colour
Farbwiedergabe colour rendition
Farbzahl, Gardner Gardner colour number/value
Fase chamfer
fasen to chamfer
Faser fibre, filament
Faser, polymeroptische optic(al) fibre
Faserabschnitte fibre offcuts
Faseranlage fibre extrusion line
Faseranordnung fibre orientation
Faseranteil fibre content
Faserausrichtung fibre orientation
Faserbündel fibre strand
Faserextrusionsanlage fibre extrusion line
faserförmig fibrous
Faserfüllstoff fibrous filler
Fasergewebe woven fabric
Faserharzspritzanlage spray lay-up plant/equipment *(grp)*
Faserharzspritzen spray lay-up *(grp)*
faserig fibrous
Faser-Kunststoffverbund fibre-plastic composite
Faserlänge fibre length
Faserleder - reconstituted leather
Faser-Matrixgrenzschicht fibre-matrix boundary/interface
Faser-Matrixhaftung fibre-matrix adhesion
Faseroptik fibre optics
Faseroptiksensor optic(al) fibre probe
Faserorientierung fibre orientation
Faserrichtung fibre orientation
Faserschnittmatte chopped strand mat *(grp)*
Faserschutt fibre sweepings *(i.e. waste fibres less than 0.15 mm long)*
Faserspritzanlage spray lay-up plant/equipment *(grp)*
Faserspritzen spray lay-up *(grp)*
Faserspritzlaminat spray-up laminate *(grp)*
Faserstrang fibre strand
Fasersuspension fibre suspension
Faserverbund fibre composite

Faserverbundbauteil fibre composite part/component
Faserverbundwerkstoff fibre composite
Faserverlauf fibre orientation
faserverstärkt fibre reinforced
Faservlies bonded/non-woven fabric
Faservolumenanteil fibre volume content
Faserwellen local fibre accumulations
Faserwickelverfahren filament winding
fasrig fibrous
Faß drum
Fassade facade, frontage
Fassadenanstrich exterior finish
Fassadenbau curtain wall construction
Fassadenfarbe exterior paint
Fassadenplatte facade panel
Fassadenverkleidung facade cladding, exterior wall cladding
Faßblasmaschine machine for blow moulding drums
fasserrelevant *relating to fibres*: **faserrelevante Kenndaten** fibre properties/characteristics
fassoniert shaped, contoured
Fassungsvermögen capacity
Faßwickler drum winder
fäulnisfest rot-proof
Faustformel rule of thumb
Faustregel rule of thumb
FCKW *abbr. of* **Fluorchlorkohlenwasserstoff**, chlorofluorocarbon, CFC
FCKW-frei CFC-free
FEA *abbr. of* **Finite-Element-Analyse**, finite element analysis
FE-Analyse finite element analysis
FE-Berechnung finite element calculation
Feder spring
federbelastet spring loaded
federbetätigt spring actuated
Federeigenschaften resilience
Federgehäuse spring housing
Federhaus spring housing
Federkennlinie hysteresis curve
Federpaket spring assembly
Federstahl spring steel
Federstahlmesser spring steel blade
Federsteifigkeitsrate spring rate
Federungseigenschaften resilience
Federweg spring deflection
Federwert spring rate/constant
Federwirkung resilience
FEF-Ruß fast extrusion furnace black
Fehlbedienung faulty operation
Fehleinstellungen wrong settings
Fehler 1. fault, error. 2. flaw, defect *(in a material)*
Fehler, optischer surface defect
Fehleranalyse fault analysis
Fehleranzeige malfunction indicator, alarm signal
Fehlerdiagnose error diagnostics *(me)*
Fehlerdiagnoseprogramm error diagnostics program/routine *(me)*

Fehlerdiagnosis error diagnostics *(me)*
Fehlererkennbarkeit defect recognisability
Fehlererkennbarkeitsgrenze defect recognisability limit
fehlerfrei 1. error-free. 2. free from defects, faultless, perfect
Fehlergrenze limit/margin of error
fehlerhaft faulty, incorrect, wrong, defective
Fehlerkatalog list of defects
Fehlerlokalisierung fault localisation *(me)*
Fehlermeldung error message
Fehlerort fault location
Fehlerortsignalisierung fault location indicator
Fehlerquadrate, Summe der kleinsten sum of least squares
Fehlerquelle source of error
Fehlersuchliste trouble-shooting chart
Fehlersuchprogramm fault location program *(me)*
Fehlfunktion malfunction
Fehlinformation wrong information
Fehlinterpretation misinterpretation, wrong interpretation
Fehlinterpretation misinterpretation
Fehlkonstruktion faulty design
Fehlleistung malfunction
Fehlmanipulierung mishandling
Fehlmessung measurement error
Fehloperation malfunction
Fehlschlüsse wrong conclusions
Fehlstart false start
Fehlstelle flaw, defect
Fehlzeit downtime
Feinabgleichpotentiometer fine-tuning potentiometer
Feinbohrarbeiten precision drilling
Feindispergierung fine dispersion
feindispers fine-particle
feindispers verteilt finely dispersed
Feindreharbeiten precision turning
feineinstellbar accurately adjustable
Feineinstellung fine adjustment
Feinfilter-Siebeinrichtung fine filtration unit *(mf)*
Feinfraktion fine fraction
feinfühlig sensitive *(instrument)*: **feinfühlige Steuerung** sensitive control
feingenutet finely grooved
feingeschliffen precision ground
Feingewebe fine fabric
Feinjustierung fine adjustment
feinkörnig fine-particle
Feinkörnigkeit fine-particle character
Feinkreide finely powdered chalk
feinkristallin finely crystalline
Feinmahlaggregat fine grinding machine, pulveriser, pulverising unit
Feinmahlen fine grinding, pulverisation
feinmaschig fine-mesh
Feinmechanik precision engineering
feinmechanisch precision **feinmechanische Teile** precision parts
Feinmeßmanometer precision manometer

Feinmessung precision measurement/
determination
Feinnuten fine grooves
feinporig fine-cell *(foam)*
Feinpositioniersystem precision positioning
mechanism/system
feinpulverig finely powdered
feinpulverisiert finely powdered
Feinregulierung fine adjustment
Feinschicht gel coat *(grp)*
Feinschleifarbeiten precision grinding
Feinschliff microtome section
Feinsieb fine-mesh screen *(mf)*
Feinstaub fine dust
feinstgemahlen very finely ground
Feinstgewebe extremely fine fabric
Feinstölfilterung extremely fine-mesh oil filter
Feinstpartikel extremely fine particle
feinstpoliert highly polished
Feinstregulierung extremely fine adjustment
Feinstruktur microstructure
feinstverteilt extremely finely divided
feinteilig fine-particle
Feinteiligkeit fine-particle character
Feinvakuum medium vacuum
Feinvermahlen fine grinding, pulverisation
Feinverstellung fine adjustment
feinverteilt 1. finely divided. 2. finely dispersed
Feinwaage precision balance
Feinwerktechnik precision engineering
feinwerktechnisch precision
  feinwerktechnische Zahnräder precision
  gears
feinzellig fine-cell *(foam)*
Feinzermahlung fine grinding
feinzerteilt finely divided
Fe-Ko Thermoelement iron-constantan
thermocouple
Feld, magnetisches magnetic field
Feldspat fel(d)spar
Feldstärke field strength
Feldstärke, elektrische electric field strength
Feldversuch field trial
Felge (car wheel) rim
Fell sheeted-out compound, web, strips, milled
sheet *(c)*: **Das gelierte Material wird auf
einem Walzwerkzeug zum Fell ausgezogen**
the gelled compound is sheeted out on the mill
Fellwendevorrichtung *This is a device which
cuts up the sheeted-out compound into strips
and returns these to the mill. There appears to
be no succinct English equivalent*
FEM *abbr. of* finite element method
FEM-Analyse finite element analysis
FEM-Berechnung finite element calculation
FE-Metall ferrous metal
FEM-Netz finite element network
FE-Modell finite element model
FEM-Programm finite element program
FEM-Rechnung finite element calculation
FEM-Verfahren finite element method
FE-Netzwerk finite element network

Fensteranschluß jamb fixing
Fensterbank window sill
Fensterbankprofil window sill profile
Fensterdichtung window seal
Fensterflügel casement
Fensterformulierung window profile compound
Fensterglasplatte window pane
Fensterhochleistungsprofil heavy-duty window
profile
Fensterläden shutters
Fensterprofil window profile
Fensterprofilrezeptur window profile
(compound) formulation
Fensterrahmen window frame
Fensterstock window casing, outer window
frame
fernbedient remote-operated
Ferneinstellung 1. remote control adjustment/
setting. 2. remote control adjusting/setting
mechanism
ferngesteuert remote-controlled
Fernkontrolle 1. remote control. 2. remote
control unit
Fernmeldekabel telecommunications cable
Fernmeldetechnik telecommunications
(engineering)
Fernost Far East
Fernsehapparat television set
Fernsehgerät television set
Fernsprechkabel telecommunication cable
fernsteuerbar remote-controlled
Fernsteuerung 1. remote control. 2. remote
control unit
Fernwärme district heating
Fernwärmeleitung district heating line
Fernwärmerohr district heating pipe
ferroelektrisch ferroelectric
Fertigartikel end product, finished article/
product
Fertigen, computergestütztes computer aided
manufacturing, CAM
Fertigen, computerintegriertes computer
integrated manufacturing, CIM
Fertigerzeugnis end product, finished article/
product
Fertiggelieren fusion *(of PVC paste)*
Fertiggericht ready meal
Fertighimmel prefabricated roof liner
Fertigmischung ready-to-use mix
Fertigpigmentpaste ready-made pigment paste
Fertigprodukt end product, finished article/
product
Fertigprodukteigenschaften end product
properties
Fertigschäumen final foaming
Fertigteil 1. part, moulded part, moulding, end
product, finished article. 2. prefabricated
component
Fertigteil- *see* **Formteil-**
Fertigteilbau industrialised building methods
Fertigung 1. manufacture *(of finished products)*.
2. production *(of finished or semi-finished*

products) 3. moulding process: **Einfluß der Fertigung auf die Formteileigenschaften** effect of the moulding process on part properties
Fertigungsabfall production waste/scrap
Fertigungsablauf production/manufacturing process
Fertigungsanlage production line
Fertigungsautomatisierung production automation
Fertigungsbedingungen production conditions
Fertigungsbetrieb 1. factory, works. 2. moulding shop
Fertigungscharge production batch
Fertigungseinheit production unit
Fertigungseinrichtungen production equipment
Fertigungsextruder production extruder
Fertigungsfehler production fault
Fertigungsfluß production cycle
Fertigungsgemeinkosten production overheads
Fertigungsgeschwindigkeit production rate/speed
Fertigungsgüte product quality
Fertigungshalle production shed
Fertigungsinsel production unit
Fertigungskontrolle production control
Fertigungskosten production/manufacturing costs
Fertigungslinie production line
Fertigungsmittel 1. production equipment. 2. mould. 3. machine
Fertigungsnebenkosten extra/additional production costs
Fertigungspanne production breakdown
Fertigungsparameter 1. production/manufacturing conditions. 2. production variable
Fertigungsplanung 1. production planning. 2. production planning department
Fertigungsprogramm 1. range of products/machines/equipment. 2. production plan/schedule/programme
Fertigungsschrumpf moulding shrinkage *(see explanatory note under* **Verarbeitungsschwindung***)*
Fertigungsstätte production facility, factory, plant
Fertigungssteuerung 1. production control. 2. production control system. 3. production control department
Fertigungsstraße production line
Fertigungsstrecke production line
fertigungstechnisch *relating to manufacture or production. Can sometimes be translated as* technical: **Wenn der seitliche Bandanschnitt aus fertigungstechnischen Gründen nicht angewandt werden kann** if technical considerations prohibit the use of lateral film gating
Fertigungstechnologie production technology

Fertigungstoleranz 1. production tolerance. 2. moulding tolerance
Fertigungsüberwachung 1. production control. 2. production control system. 3. production control department
Fertigungsunterlagen production records
Fertigungsverfahren manufacturing/production process
Fertigungsverlauf production/manufacturing process
Fertigungsversuch production trial
Fertigungsversuchsanlage experimental production plant
Fertigungszeichnung production drawing
Fertigungszeit production time
Fertigungszelle production unit
Fertigungszyklus 1. production cycle. 2. moulding cycle
fertigvernetzt fully cured/crosslinked
Fertigwaren finished products/goods
Fertigwarenlager finished goods store
fest solid, firm
Festanschlag fixed stop
Festblende fixed diaphragm
feste Werkzeugträgerseite fixed/stationary platen *(im)*
festfressen to seize (up)
Festgehalt solids content
festgelegt specified, prescribed
Festharz solid resin
Festharzgehalt solid resin content
Festharzkombination solid resin blend
Festigkeit strength
Festigkeit, mechanische mechanical strength
Festigkeitsabfall decrease in strength
Festigkeitsabnahme decrease in strength
Festigkeitsanstieg increase in strength
Festigkeitseigenschaften strength
Festigkeitseinbuße loss of strength
Festigkeitserhöhung increase in strength
Festigkeitsminderung decrease in strength
festigkeitsoptimiert high strength
Festigkeitssteigerung increase in strength
Festigkeitsträger reinforcement, reinforcing material
Festigkeitsuntersuchung measurement/determination of strength
Festigkeitsverlust decrease in strength
Festkautschuk solid rubber
Festkörper solid
Festkörperanteil solids content
Festkörpergehalt solids content
Festkörperoberfläche solid surface
Festkörperreibung solid friction
festkörperreich high-solids
Festkörperzustand solid state
festlegen to lay down, to stipulate, to establish
Festmenge fixed amount
Festmesser bed/stationary knife *(sr)*
Festphase solid phase
Festphasenkondensation solid phase condensation

Festphasenpyrolyse solid phase pyrolysis
Festplatte hard/fixed disk *(me)*
Festplattenlaufwerk hard/fixed disk drive *(me)*
Festplattenspeicher fixed magnetic disk *(me)* *(see explanatory note under* **Magnetplattenspeicher***)*
Festpunkt fixed point
FE-Struktur finite element structure
Festsilikonkautschuk solid silicone rubber
Feststampfen tamping
feststehende Werkzeugaufspannplatte fixed/stationary platen *(im)*
feststehende Werkzeughälfte fixed/stationary mould half *(im)*
Feststellschraube locking screw
Feststoff solid, solid material/matter
Feststoffabfall solid waste
Feststoffansammlungen solids accumulations
Feststoffanteil solids content
Feststoffbereich 1. feed section/zone *(e)*. 2. solid phase
Feststoff-Förderung 1. solids conveying. 2. solids conveying system
Feststoffgehalt solids content
Feststoffkern solid core (of material)
Feststoffpartikel solid particle
Feststoffzone solid phase: *If the word is seen to contribute nothing to the general meaning, it should be omitted in translation, e.g.* **In den letzten Jahren hat sich die Problematik des Verschleißes in der Feststoffzone von Kunststoffverarbeitungsmaschinen zusehends verschärft** The problems caused by wear in plastics processing equipment have greatly increased in recent years
festverankert firmly fixed
festverdrahtet hard-wired
Festwalze fixed roll
Festwert fixed value
Festzustand solid state
fett 1. greasy 2. long-oil
Fettalkohol fatty alcohol
Fettalkoholsulfat fatty alcohol sulphate
Fettamid fatty amide
fettbeständig grease resistant
Fettbeständigkeit grease resistance
fettdicht greaseproof
fettes Alkydharz long-oil alkyd (resin)
fettfrei free from grease
fettgedruckt printed in bold type
fettig greasy
fettlösend grease-dissolving
Fettlösungsmittel grease solvent
Fettpresse grease gun
Fettreste grease residues
Fettsäure fatty acid
Fettsäureamid fatty acid amide
Fettsäurederivat fatty acid derivative
Fettsäureester fatty acid ester
fettsäuremodifiziert fatty acid modified
Fettsäuremonoester fatty acid mono-ester
Fettsäureseife fatty acid soap

Fettschmierpresse grease gun
Fettseife fatty acid soap
Fettzentralschmierung central grease lubricating system
feucht 1. damp. 2. humid
Feuchte 1. moisture *(in a substance)*. 2. damp *(in the atmosphere)*. 3. humidity *(of the air)*
Feuchteadsorption moisture adsorption
Feuchteaufnahme moisture absorption
Feuchtebeständigkeit moisture resistance
Feuchtebestandteil moisture content
Feuchtediffusion moisture diffusion
Feuchtediffusionskoeffizient moisture diffusion coefficient
Feuchteeinfluß effect of moisture
Feuchteeinwirkung effect of damp: **gegen Feuchteeinwirkung geschützt** protected against damp
Feuchteempfindlichkeit sensitivity/susceptibility to moisture
Feuchtegehalt 1. moisture content. 2. humidity *(of the air)*
Feuchtehaushalt moisture equilibrium
Feuchteklima damp/humid conditions
Feuchteschutz protection against damp
Feuchteschwankungen variations/fluctuations in moisture content
feuchteunempfindlich unaffected by moisture
Feuchtewert 1. moisture content. 2. humidity *(of the air)*
Feuchtfestigkeit moisture resistance, resistance to damp (conditions)
Feuchtformmasse 1. dough moulding compound, DMC. 2. bulk moulding compound, BMC
Feuchthaltemittel humectant
feuchtheiß hot and humid
Feuchtigkeit 1. moisture *(in a substance)*. 2. damp *(in the atmosphere)*. 3. humidity *(of the air)*
Feuchtigkeit, aufsteigende rising damp
Feuchtigkeitsaufnahme moisture absorption
Feuchtigkeitsbelastung *this indicates the presence of moisture/damp;* **bei hoher Feuchtigkeitsbelastung** under very damp conditions
feuchtigkeitsbeständig moisture resistant
feuchtigkeitsdicht moisture proof
Feuchtigkeitsdurchlässigkeit moisture permeability
feuchtigkeitsempfindlich affected by moisture
Feuchtigkeitsempfindlichkeit moisture sensitivity
Feuchtigkeitsgehalt 1. moisture content. 2. humidity *(of the air)*
feuchtigkeitshaltend moisture-retaining
feuchtigkeitshärtend moisture-curing
feuchtigkeitsleitend moisture-transmitting
Feuchtigkeitsniveau 1. moisture content. 2. humidity *(of the air)*
Feuchtigkeitsresistenz moisture resistance
Feuchtigkeitsspuren traces of moisture

feuchtigkeitsundurchlässig impermeable to moisture, moisture proof/impermeable
Feuchtigkeitsundurchlässigkeit moisture impermeability
feuchtigkeitsunempfindlich unaffected by moisture
Feuchtigkeitsverlust moisture loss
feuchtigkeitsvernetzend moisture-curing
Feuchtpolyester 1. polyester dough moulding compound, polyester DMC. 2. polyester bulk moulding compound, polyester BMC
Feuchtpolyesterformmasse 1. polyester dough moulding compound, polyester DMC. 2. polyester bulk moulding compound, polyester BMC
Feuchtpolyesterpreßmasse 1. polyester dough moulding compound, polyester DMC. 2. polyester bulk moulding compound, polyester BMC
Feuchtpreßmasse dough moulding compound, DMC, bulk moulding compound, BMC
Feuchtraumleuchte damp-proof lamp
Feuchtschranktest humidity cabinet test
Feucht-Warm-Lagerung ageing under warm, humid conditions
feuerabweisend fire/flame resistant
feuerbeständig fire/flame resistant, flameproof
feuerfest refractory
Feuerfeststoff refractory (material)
feuergefährlich flammable
feuerhemmend fire/flame retardant
Feuerrisiko fire risk
feuersicher fire/flame resistant
Feuersicherheit fire safety
feuersicherheitlich relating to fire safety:
  feuersicherheitliche Prüfung fire safety test
Feuersicherheitsempfehlungen fire safety recommendations
feuerverzinkt hot galvanised
Feuervorschriften fire regulations
Feuerweiterleitung flame spread
Feuerwiderstandsdauer fire endurance
feuerwiderstandsfähig fire/flame resistant
Feuerwiderstandsfähigkeit fire/flame resistance
Feuerwiderstandsverhalten fire resistance
FFL *abbr. of* **flexible Fertigungslinie**, flexible production line
F-Form four-roll inverted L-type calender
FF-Ruß fine furnace black
FFS *abbr. of* **flexibles Fertigungssystem**, flexible production system
FFZ 1. *abbr. of* **flexible Fertigungszelle**, flexible production unit. 2. *abbr. of* **flexibles Fertigungszentrum**, flexible production centre
Fibrille fibril
fibrillenartig fibril-like
Fibrillieren fibrillation *(a method of making film fibres from film tape, consisting essentially of splitting the tape lengthways)*
fibrogen fibrous

FID *abbr. of* **Flammenionisationsdetektor**, flame ionisation detector
Filamentgarn filament yarn
filigran filigree
Film film
Filmanguß film gate *(im)*
Filmanschnitt film gate *(im)*
Filmaussehen film appearance
Filmballon film bubble *(bfe)*
Filmbildehilfsmittel film forming aid
filmbildend film forming
Filmbildner film former
Filmbildnermaterial film forming material/substance
Filmbildung film formation
Filmbildungsbedingungen film forming conditions
Filmbildungsdauer time needed to form a film
Filmbildungshilfsmittel film forming aid
Filmbildungsmechanismus film forming mechanism
Filmbildungstemperatur film forming temperature
Filmdefekt film defect
Filmdicke film thickness
Filmeigenschaften film properties/characteristics
Filmflexibilität film flexibility
Filmhärte film hardness
filmmechanische Eigenschaften mechanical film properties
Filmoberfläche film surface
Filmoberflächeneigenschaften film surface characteristics
Filmoptik film appearance
Filmqualität film quality
Filmscharnier integral hinge
Filmstärke film thickness
Filmstörungen film defects
Filmstruktur film structure
Filmziehgerät film casting instrument
Filterapparatur filtration unit
Filterautomatik automatic filter unit *(mf)*
Filtereinsatz filter insert *(mf)*
Filterelement filter element/medium *(mf)*
Filterelementhalteplatte filter element support *(mf)*
Filterfläche screen area *(mf)*
Filtergerät filtration unit
Filtergewebe 1. filter fabric. 2. screen, filter screen *(mf)*
Filtergewebepaket screen pack *(mf)*
Filterhilfsmittel filtration aid *(mf)*
Filterkammer filtration chamber *(mf)*
Filterkarte filter card *(me)*
Filterkartusche cartridge filter *(mf)*
Filterkerze cartridge/candle filter *(mf)*
Filterkerzenpaket candle filter assembly
Filterkorb cartridge filter *(mf)*
Filterkuchen filter cake *(mf)*
Filtermaske face mask
Filtermedium filter medium/element *(mf)*

Filtermittel filter medium/element/screen *(mf)*
Filterpaket screen pack *(e)*
Filterplatte filter plate *(mf)*
Filterpresse filter press *(mf)*
Filterrückstand filtration residue
Filterscheibe filter disc
Filtersieb screen *(mf)*
Filtertemperatur filtration temperature
Filtertuch filter cloth *(mf)*
Filterung filtration, screening
Filtration filtration, screening
Filtrationsverhalten filtration characteristics
Filtrierbarkeit filtrability
Filtrieren filtration, screening
Finanzkraft financial strength
Fingerkühlung cooling by means of long, slender cooling channels *(method of cooling long core inserts in injection moulding)*
Fingernageltest fingernail test
Fingerspitzengefühl intuition, instinct, sixth sense
Finite-Elementnetz finite element network *(me)*
firmeneigen in-house, in-company:
  firmeneigene Prüfverfahren the company's own test methods
firmenintern in-house, in-company
Firmenleistung company's contribution *(e.g. to welfare, insurance etc.)*
Firmenliteratur company literature
Firmenmerkblätter manufacturers' technical literature
Firmenstrategie company policy
Firmenstruktur company structure
Firnis varnish, lacquer
Fischaugen fish eyes
Fischkasten fish box
Fischschwanzdüse fishtail die *(e)*
Fischschwanzkanal fishtail manifold *(e)*
fix fixed
Fixierstift fixing pin
Fixkosten overheads
Fixlänge standard length
Fixpunkt fixed point
FK *abbr. of* **faserverstärkter Kunststoff**, fibre reinforced plastic
F-Kalander four-roll inverted L-type calender
FKM-Mischung fluororubber mix
FKV *abbr. of* **Faser-Kunststoff-Verbundstoff**, fibre-plastic composite
FKV-Blattfeder composite leaf spring
FKW *abbr. of* **Fluorkohlenwasserstoff**, fluorinated hydrocarbon
Flachbettplotter flat-bed plotter *(me)*
Flachbodentank flat-bottom tank
Flachdach flat roof
Flachdüse flat film (extrusion) die
Fläche, projizierte projected area
flächenartig flat
Flächenbeflammung application of flame to a flat surface
flächenbündig flush with the surface
flächendeckend 1. nationwide, covering the whole country, OR the whole area. 2. total, comprehensive **Die Produktionslinie weist als Besonderheit ein flächendeckendes Kontroll- und Informationssystem auf.** The special feature of the production line is an extensive/comprehensive control and information system
Flächeneinheit unit area
Flächenfilter surface filter
flächenförmig flat, two-dimensional
Flächengebilde flat material *(e.g. paper, film, fabric etc.)*
Flächengewicht weight per unit area
flächengleich having the same surface area
Flächenpressung surface pressure
Flächenscherteil faceted smear section *(of screw)*
Flächenschwerpunkt centre of gravity of the surface
Flächenträgheitsmoment moment of inertia
flächenversetzt staggered in relation to the surface
Flachfolie flat film
Flachfolienanlage flat film extrusion line
Flachfolienanlage, coextrudierende flat film coextrusion line
Flachfoliendüse flat film (extrusion) die
Flachfolienextruder flat film extruder
Flachfolienextrusion flat film extrusion
Flachfolienextrusionseinheit flat film extruder
Flachfolienmaschine flat film extruder
Flachfolienstreckverfahren flat film orientation (process)
Flachfolienverfahren flat film extrusion (process)
Flachfolienwerkzeug flat film (extrusion) die
flachgängig shallow-flighted *(screw)(e)*
Flachgarn film tape/yarn
flachgedrückt flattened, compressed
flachgelegte Breite layflat width *(bfe)*
flachgeschnitten shallow-flighted *(screw)* *(e)*
Flachgewebe flat-woven fabric
flächig flat, two-dimensional
flächiger Abtrag surface abrasion
Flachlegebleche collapsing boards *(bfe)*
Flachlegebreite layflat width *(bfe)*
Flachlegebretter collapsing boards *(bfe)*
Flachlegeeinrichtung collapsing boards *(bfe)*
flachlegen to collapse *(film bubble) (bfe)*
Flachlegewinkel bubble collapsing angle *(bfe)*
Flachlegung 1. collapsing *(of film bubble)*. 2. collapsing boards *(bfe)*
Flachliegebreite layflat width *(bfe)*
Flachprobe flat test piece/specimen
Flachprofildüse flat-profile die *(e)*
Flachstab flat test specimen/piece
Flammbarriere fire barrier
Flammenausbreitung flame spread;
  Flammenausbreitung an der Oberfläche surface flame spread
Flammenausbreitungsgeschwindigkeit rate of flame spread

Flammenausbreitungsrichtung direction of flame spread
Flammenionisationsdetektor flame ionisation detector
flammfest flame resistant
Flammfestausrüsten flame/fire proofing
Flammfestigkeit flame/fire resistance
flammgehemmt flame retardant
flammhemmend flame retardant
flammhemmend ausgerüstet flame retardant
flammhemmende Additive flame retardants
Flammhemmer flame retardant
Flammhemmung flame retardancy: **Die Kunststoffen zur Flammhemmung zugesetzten Additive** the flame retardants added to plastics
Flammklasse flammability classification/group/rating
Flammpolieren flame polishing
Flammpoliergerät flame polisher
Flammpunkt flash point
Flammruß furnace/lamp black
Flammschutz flame/fire proofing: **Flammschutz für Kunststoffe und Chemiefasern** flame retardants for plastics and synthetic fibres
Flammschutzadditiv flame retardant
flammschutzausgerüstet flameproofed
Flammschutzausrüstung making flame retardant, imparting flame retardant properties: **Kunststoffe können mit einer permanenten Flammschutzausrüstung versehen werden** polymers can be made permanently flame retardant; **mit Flammschutzausrüstung** containing flame retardant
Flammschutzeffekt flame retardant effect
Flammschutzeigenschaften flame retardant properties
Flammschutzfarbe flame retardant paint
Flammschutzkomponente flame retardant
Flammschutzmittel flame retardant
Flammschutzmittelzusatz flame retardant
Flammschutzsystem flame retardant
Flammschutzwirkung flame retardant effect
flammsicher flameproof, flame resistant
Flammspritzen flame spraying
Flammsprühen flame spraying
Flammstrahlen flame treatment
flammstrahlvorbehandelt flame treated
Flammverzögerungseigenschaften flame retardant properties
Flammvorbehandlung flame pretreatment
flammwidrig flame resistant
Flammwidrigkeit flame resistance
Flanke flank
Flanke, aktive thrust/front face of flight, leading edge of flight *(e)*
Flanke, hintere rear face of flight, trailing edge of flight *(e)*
Flanke, treibende thrust/front face of flight,

Flanke, passive rear face of flight, trailing edge of flight *(e)*
leading edge of flight *(e)*
Flanke, vordere thrust/front face of flight, leading edge of flight *(e)*
Flankenabstand flight land clearance, inter-screw clearance *(e)*
Flankenspiel flight land clearance, inter-screw clearance *(e)*
Flankenwinkel flank angle
Flansch flange
Flanschanschluß flange coupling/connection
Flanschbefestigung flange mounting
Flanschverbinding flange joint
Flapperventil flap valve
Flaschenblasanlage bottle blowing plant
Flaschenblasautomat automatic bottle blowing machine
Flaschenblasmaschine bottle blowing machine
Flaschencompound bottle blowing compound
Flaschenkasten bottle crate
Flaschenmaschine bottle blowing machine
Flaschenproduktionslinie bottle blowing line
Flaschenprüfstand bottle testing machine
Flaschentransportkasten bottle crate
Flaschenverschluß bottle cap/closure
Flashverdampfung flash vaporisation
Fleckenunempfindlichkeit stain resistance
fleckig patchy
flexibel flexible
Flexibilisator flexibiliser *(used with epoxy resins)*
flexibilisiert flexibilised
Flexibilisierung flexibilisation
Flexibilisierungseffekt flexibilising effect
Flexibilität flexibility
flexible Formgebung design flexibility
Flexlippe adjustable lip *(e)*
Flexodruck flexographic printing
Flexodruckfarbe flexographic printing ink
Flexofarbe flexographic printing ink
Flickmörtel patching mortar
fliegender Stopfen floating plug
Fliehkraft centrifugal force
Fliese (floor/wall) tile
Fliesenkleber tile adhesive
***Fließanomalien** flow anomalies
Fließband 1. conveyor belt. 2. assembly line
Fließbandfertigung assembly line production
Fließbandprozeß assembly line production
Fließbandverfahren assembly line production
Fließbedingungen flow conditions
Fließbehinderung flow restriction
Fließbelag self-levelling screed
Fließbett fluidised bed
Fließbild flow diagram
Fließeigenschaften flow characteristics
Fließen, laminares laminar flow
Fließen, Newtonsches Newtonian flow
Fließen, viskoses viscous flow

---

* *Words starting with* **Fließ-**, *not found here, may be found under* **Strömungs-**

Fließexponent flow index
fließfähig free-flowing, pourable
Fließfähigkeit flow, ease of flow, flowability
Fließfiguren flow pattern
Fließfront melt front, flow front
Fließfrontgeschwindigkeit melt/flow front velocity
Fließfrontverlauf melt/flow front profile
fließgepreßt flow moulded
fließgerecht *designed or constructed to facilitate melt flow*: **fließgerechtes Breitschlitzwerkzeug** slit die designed to permit optimum melt flow
Fließgeschichte previous history *(ie. the thermal or mechanical history of a test piece or moulded part)*
Fließgeschwindigkeit 1. flow rate. 2. flow velocity *(liquids in pipes)*
Fließgesetz flow law
Fließgesetzexponent flow law index
Fließgießverfahren flow moulding (process)
Fließgrenze yield point
Fließguß flow moulding
Fließhilfe flow promoter
Fließhindernis flow obstruction
Fließindex melt flow index
Fließkanal 1. flow channel *(general term)*. 2. runner *(im)*. 3. barrel *(although the word has been encountered in this context, this is a very unusual interpretation)*
Fließkanalgestaltung 1. flow channel design. 2. runner design
Fließkomma floating decimal point
Fließkontrollmittel flow control agent
Fließkorrektur flow correction
Fließkurve flow curve
Fließlänge flow length/distance
Fließlinie weld line, flow line *(see explanatory note under* **Bindenaht**)
Fließmarkierungen 1. flow marks. 2. spider lines *(e,bm)*
Fließmittel eluent *(used in chromatography)*
Fließnaht weld line, flow line *(see explanatory note under* **Bindenaht***)*
Fließpfad 1. (melt) flow path, (melt) flow-way 2. runner *(im)* 3. flow length/ distance
Fließpfad (melt) flow path
Fließpressen flow moulding
Fließpreßmaterial flow moulding compound
Fließprozeß 1. continuous process. 2. flow *(of melt)*
Fließquerschnitt 1. flow channel *(general term)*. 2. runner *(im)*
Fließrichtung direction of flow, flow direction
Fließschatten spider lines *(e,bm)*
Fließschema flow diagram
Fließspan swarf
Fließspannung tensile stress at yield, yield stress
Fließstörungen flow irregularities
Fließstudie flow study
fließtechnisch 1. *relating to flow*:

fließtechnische Besonderheiten flow peculiarities; **Der relativ einfach gebaute und kostengünstige Pinolenkopf bringt auf Grund der seitlichen Einspeisung fließtechnisch ungünstige Voraussetzungen** the side-fed die, though of relatively simple design and inexpensive, does not provide ideal flow conditions. 2. rheological **fließtechnische Eigenschaften** rheological properties
fließtote Räume dead spots
Fließuntersuchung flow test/investigation
Fließverbesserer flow promoter
Fließverhalten flow characteristics/properties
Fließvermögen flowability
Fließvorgang flow process
Fließweg 1. (melt) flow path, (melt) flow-way. 2. runner *(im)*. 3. flow length/distance: **da der zurückgelegte Fließweg zunimmt...** since the flow distance increases....
Fließweglänge flow length/distance
Fließwegrichtung direction of flow, flow direction
Fließweg/Wanddickenverhältnis flow length-wall thickness ratio *(im)*
Fließwiderstand flow resistance
Fließwiderstandsänderungen flow resistance changes, changes in flow resistance
Fließzone craze zone
Fließzonenbildung crazing
Fließzonenmaterial crazed material
Fließzonenstruktur craze zone structure
flockenförmig flake-like
Flockulat flocculate
Flockulation flocculation
Flockulationsfreiheit freedom from flocculation
Flockulationsneigung tendency to flocculate
Flockulationsstabilität flocculation resistance
flockuliert flocculated
Flockung flocculation
Flockungsmittel flocculating agent
Flockungstendenz tendency to flocculate
Floppy-Disklaufwerk floppy disk drive *(me)*
Flotation flotation
Flüchte volatiles, volatile matter
Flüchtegehalt volatile content
fluchtend aligned
flüchtig volatile
flüchtige Bestandteile volatile matter/content
Flüchtigkeit volatility
Flüchtigkeitsgrad (degree of) volatility
Fluchtungsfehler misalignment
Flugbenzin aviation fuel
Flügel 1. flight, thread *(of a screw)*. 2. paddle, blade *(of a mixer)*. 3. casement *(of a window)*
Flügelfenster casement window
Flügelprofil casement profile
Flügelrahmen casement
Flügelzellenmotor vane-type motor
Flügelzellenpumpe vane-type pump
Flügelzellenregelpumpe variable-delivery vane-type pump

Flugtreibstoff aviation fuel
Flugzeugbau aircraft construction
Flugzeughalle aircraft hangar
Flugzeugindustrie aircraft/aviation industry
Flugzeugsitz aircraft seat
fluidisieren to fluidise
Fluidisierungstrockner fluidising dryer
Fluidität fluidity
Fluidmischer fluid mixer
Fluoracrylat fluoroacrylate
Fluoracrylsäure fluoroacrylic acid
Fluoralkoxyterpolymer fluoroalkoxy terpolymer
fluoralkylmodifiziert fluoroalkyl modified
Fluorcarbonkautschuk fluorocarbon rubber
Fluorchlorkohlenwasserstoff chlorofluorocarbon, CFC
Fluorelastomer fluororubber
Fluoreszenz fluorescence
Fluoreszenzfarbe fluorescent colour
Fluoreszenzpigment fluorescent pigment
fluoreszierend fluorescent
Fluorethylenpropylen fluorinated ethylene-propylene copolymer, FEP
Fluorgehalt fluorine content
fluoriert fluorinated
Fluorierung fluorination
fluorisiert fluorinated
Fluorkautschuk fluororubber
Fluorkohlenwasserstoff fluorinated hydrocarbon
Fluorkunstharz fluoroplastic, fluoropolymer, fluorocarbon plastic/polymer
Fluorkunststoff fluoroplastic, fluoropolymer, fluorocarbon plastic/polymer
Fluorolefin fluoro-olefin
Fluorpolymer fluoropolymer, fluorocarbon polymer
Fluorsiliconelastomer fluorosilicone elastomer
Fluorsiliconfett fluorosilicone grease
Fluorsilikon fluorosilicone
Fluorsilikonkautschuk fluorosilicone rubber
Fluorsilikonöl fluorosilicone fluid
Fluorsubstituent fluorosubstitutent
Fluorthermoplast fluorothermoplastic
Fluortrichlormethan fluorotrichloromethane
Fluß, kalter cold flow
Fluß, laminarer laminar flow
Flußdiagramm flow diagram
flüssig liquid, fluid
Flüssigadditiv liquid additive
flüssige Mittel liquid assets
flüssige Seele fluid centre; *for translation example see* **Seele, plastische**
Flüssigfolie liquid film
Flüssigharz liquid resin
Flüssigharzkombination liquid resin blend
flüssigisotrop liquid-isotropic
Flüssigkautschuk liquid rubber
Flüssigkeit liquid, fluid
Flüssigkeit, Newtonsche Newtonian liquid
Flüssigkeit, nicht-Newtonsche non-Newtonian liquid
Flüssigkeit, wärmeaustauschende heat exchanging fluid
Flüssigkeitsaufnahme absorption of liquid
Flüssigkeitsbad liquid (bath): **Warmlagerung im Flüssigkeitsbad** heat ageing in a liquid
Flüssigkeitschromatographie liquid chromatography
Flüssigkeitsdiffusion liquid diffusion
Flüssigkeitsfilm liquid film
Flüssigkeitsheizung 1. fluid heating. 2. fluid heating system
Flüssigkeitsphase liquid phase
Flüssigkeitsreibung liquid friction
Flüssigkeitssäule liquid column
flüssigkeitstemperiert *derived from* **Flüssigkeitstemperierung** *(q.v.)*: **Die Schnecken sind flüssigkeitstemperiert** the screw temperature is controlled by means of a heating-cooling fluid
Flüssigkeitstemperierung temperature control by means of a heating-cooling fluid: **Für thermisch empfindliche und wenig stabilisierte Kunststoffe eignet sich ein solcher 2-Stufen-Extruder mit Flüssigkeitstemperierung von Schnecke und Zylinder besonders gut** this type of two-stage extruder, whose barrel and screw temperatures are controlled by a heating-cooling fluid, is specially recommended for thermally sensitive plastics containing only small amounts of stabiliser
Flüssigkeitsverdrängung liquid displacement
Flüssigkristall liquid crystal
Flüssigkristallanzeige liquid crystal display, LCD
Flüssigkristallfolie liquid crystal film
flüssigkristallin liquid crystalline
flüssigkristallines Polymer liquid crystal polymer
Flüssigpolymer liquid polymer
Flüssigsilikonkautschuk liquid silicone rubber
Flüssigstickstoffkühlung 1. liquid nitrogen cooling. 2. liquid nitrogen cooling system
Flüssigwaschmittel liquid detergent
Flußkanal 1. flow channel (general term). 2. runner *(im)*
Flußkorrektur flow correction
Flußregulierung 1. flow control: **...welches ebenfalls zur Flußregulierung dient** which likewise helps to control flow. 2. flow control mechanism/system
Flußsäure hydrofluoric acid
Flußschema flow diagram
Flußwasserstoffsäure hydrofluoric acid
Fluten flow coating
foggingfrei non-fogging
Fokussierung focusing
Folgeaggregat downstream unit/equipment
Folgeanstriche subsequent coats
Folgeeinheit downstream unit/equipment
Folgeeinrichtung downstream unit/equipment
Folgegerät downstream unit/equipment

Folgemaschine downstream machine
Folgeprodukt secondary product
Folgereaktion secondary reaction
Folgeregelung sequence control
Folgeregler sequence controller
Folgesteuerung sequence control
Folie 1. film, sheeting *(if plastic)*. 2. foil *(if metal)*
Folien, papierähnliche paper-like film
Folien- und Bahnenware continuous film/sheeting
Folienabfall film scrap
Folienabziehwerk film take-off (unit)
Folienabzug film take-off (unit)
Folienanlage 1. flat film extrusion line, slit die film extrusion line. 2. film blowing line, blown film line. 3. cast film extrusion line, chill roll casting/extrusion line. 4. film production line *(general term)*
Folienaufbereitungsanlage film scrap reprocessing plant
Folienaufwickler film winder, film wind-up (unit)
Folienaufwicklung film winder, film wind-up (unit)
Folienbahn film web
Folienbahnabfall web scrap
Folienbahnführung film web guide
Folienbahnspannung 1. web tension 2. web tensioning device/mechanism
Folienballon film bubble *(bfe)*
Folienband film tape, film yarn
Folienbändchen film tape, film yarn
Folienbändchenanlage film tape production line
Folienbandextruder film tape extruder
Folienbandreckanlage film tape stretching unit
Folienbandstreckwerk film tape stretching unit
Folienbeobachtungsstand film inspection unit
Folienbeschaffenheit film quality/characteristics
Folienblasanlage film blowing line, blown film line
Folienblasbetrieb film blowing plant
Folienblase film bubble *(bfe)*
Folienblaseinheit film blowing machine
Folienblasen blown film extrusion, film blowing
Folienblaskopf blown film die, film blowing die
Folienblaskopf, seitlich eingespeister side-fed blown film die
Folienblaskopf, stegloser side-fed blown film die
Folienblasverfahren blown film extrusion, film blowing
Folienbreite, flachgelegte layflat width *(bfe)*
Folienbreitenregelung 1. film width control. 2. film width control mechanism
Foliendicke film gauge/thickness
Foliendickenabweichungen film gauge variations
Foliendickenmeßgerät film gauge measuring instrument
Foliendickenunterschiede film gauge variations
Foliendüse 1. flat film (extrusion) die. 2. blown film die

Folienextrusionsanlage flat film extrusion line, slit die film extrusion line
Folienfertigbreite width of film after trimming
Folienführung 1. film web guide. 2. guiding of the film bubble: **ungünstige Folienführung bei der Flachlegung** unsatisfactory guiding of the bubble when it is being collapsed
Folienführungswalzen film guide rolls *(bfe)*
Foliengießanlage 1. cast film extrusion line. 2. film casting line *(for making, say, cellulose acetate film from a solution)*
Foliengießen 1. cast film extrusion. 2. film casting *(see explanatory note under* **Foliengießanlage***)*
Folienhals bubble neck *(bfe)*
Folieninnenkühlvorrichtung internal bubble cooling system *(bfe)*
Folieninnnenkühlung internal bubble cooling system *(bfe)*
Folienkaschieranlage film laminating plant
Folienkompaktanlage compactly designed/constructed film extrusion line
Folienkopf 1. flat film (extrusion) die. 2. blown film die
Folienkühlung film bubble cooling system
Folienlaufrichtung, in in machine direction
Folienmechanik mechanical film properties
Folienmerkmale film properties
Folienproduktionsmaschine film production machine
Folienrandstreifen edge trim
Folienrandstreifenaufbereitung 1. edge trim reclaiming. 2. edge trim reclaim unit
Folienreckanlage film stretching plant
Folienreckvorrichtung film stretching device
Folienregranulieranlage film scrap reclaim plant
Folienschlauch film bubble *(bfe)*
Folienschlauchflachlegung collapsing boards *(bfc)*
Folienschlauchhalbmesser film bubble radius *(bfe)*
Folienschlupf sheet slippage *(during thermoforming)*
Folienschneidaggregat film slitter, film slitting unit *(used to slit film into film tapes)*
Folienschneidautomat automatic film slitter
Folienschnitzel cut-up film scrap
Folienschnitzelgranulieranlage film scrap reclaim plant
Folienspritzwerkzeug 1. flat film (extrusion) die. 2. blown film die
Folienstreifen film tape
Folientastatur membrane/touch-sensitive keyboard, keyboard pad *(me)*
Folientransport film transport
Folienverlegeeinheit film gauge equalising unit *(see explanatory note under* **Folienverlegegerät***)*
Folienverlegegerät film gauge equalising unit *(this equalises gauge variations across the width of the roll to avoid local diameter build-*

ups when operating with a stationary die head or winder)
Folienverlegung film gauge equalising unit *(see explanatory note under Folienverlegegerät)*
Folienvorbehandlunsgerät film pretreating instrument
Folienweiterverarbeitung 1. film conversion. 2. film conversion plant
Folienwerkzeug 1. flat film (extrusion) die. 2. blown film die
Folienwickel reel
Folienwickelmaschine film winder
Folienwickelsystem film winding system/mechanism/arrangement
Folienwickler film winder
Folienziehen calendering
Folienziehkalander sheeting calender
forciert trocknend force drying
Förderaggregat conveying unit/equipment
Förderband conveyor belt
Förderdruck conveying pressure
Fördergebläse material transport blower
Fördergurt conveyor belt
Fördergut material being transported/conveyed, material to be transported/conveyed
Förderkapazität pumping/conveying capacity *(of screw) (e,im)*
Förderlänge feed section/zone *(e)*
Förderleistung 1. conveying rate, throughput rate. 2. conveying capacity
Förderleitung feed pipe
Fördermenge (material) throughput, flow rate
Fördern transport, conveyance *(eg. of melt in an extruder barrel)*
Förderorgan transporting element/unit, conveying element/unit: **Das Einschneckenaggregat mit der Schnecke als Förder- und Plastizierorgan** the single-screw extruder with the screw acting as conveying and plasticising unit
Förderprobleme transport problems
Förderpumpe delivery pump
Förderrate throughput, output, throughput rate
Förderrichtung transport direction
Förderschnecke 1. screw *(general term)*: **Der Extruder mit der 25D Förderschnecke ist auf die speziellen Materialeigenschaften ausgelegt** the extruder with its 25D screw has been designed with the special material properties in mind. 2. transport/conveying screw *This is a constant pitch screw with constant root diameter, whose sole function it is to convey the melt along the barrel, in contrast to the* **Kernprogressivschnecke** *(q.v.) whose root diameter increases towards the tip and which compresses the melt. Version 2. should be used when the text describes and compares the finer points of different types of screw*
Förderschnecke, kompressionslose transport/conveying screw
Förderspitze flighted screw tip *(e)*
Förderstrecke 1. conveyor 2. conveying/transporting section
Förderstrom throughput
Förderung extraction *(e.g. of petroleum)*
Forderungen 1. demands. 2. debts
Förderungleichmäßigkeiten uneven transport
Förderungsgeschwindigkeit throughput rate
Fördervolumen volume throughput
Fördervorgang conveying process
förderwirksam with forced/positive conveying action: **Plastifizierextruder mit förderwirksamer Einzugszone** plasticating extruder with a forced conveying feed section
Förderwirksamkeit conveying effect
Förderwirkungsgrad conveying efficiency
Förderzeit metering time
Förderzone feed section/zone
Förderzyklus feed cycle
Förderzylinder barrel *(e)*
***Form** 1. mould. 2. shape
Formaldehyd formaldehyde
Formaldehydemission formaldehyde emission
Formaldehydemissionswert formaldehyde emission value
Formänderungen dimensional changes
Formänderungsfestigkeit dimensional stability, shape retention
Formartikel moulded article/part/product
formatgerecht of the right/correct size
formatiert formated *(me)*
Formatierung formating *(me)*
formatschneiden to cut to size
Formattoleranz dimensional tolerance
Formaufdrückzylinder mould opening cylinder
Formauffahrgeschwindigkeit mould opening speed *(im)*
Formaufspannfläche platen area *(see explanatory note and translation example under* **Werkzeugaufspannfläche**)
Formaufspannhöhe mould height/space *(im)*
Formaufspannmaße mould fixing dimensions
Formaufspannplatte platen *(im)*
Formaufspannplatte, düsenseitige fixed platen, stationary platen *(im)*
Formauftreibdruck mould opening force
Formautomat automatic thermoforming machine
formbar mouldable
Formbarkeit mouldability, moulding properties
formbeständig dimensionally stable
Formbeständigkeit dimensional stability, shape retention
Formbeständigkeit in der Wärme 1. deflection temperature, heat distortion temperature. 2. heat resistance: *Use 1. in tables and test descriptions, otherwise use 2*
Formbeständigkeit in der Wärme mit Vicatnadel Vicat softening point

* *Words starting with* **Form-** *or* **Formen-**, *not found here, may be found under* **Werkzeug-**

Formbeständigkeit in der Wärme nach Martens Martens heat distortion temperature
Formbeständigkeit in der Wärme nach Vicat Vicat softening point
Formbeständigkeit nach Martens Martens heat distortion temperature
Formbeständigkeitstemperatur deflection temperature, heat distortion temperature
Formbildungsprozeß moulding process/operation
Formbildungsvorgang moulding process/operation
Formdruck cavity pressure *(im)*
Formeinarbeitung (mould) cavity
Formeinbauhöhe mould height/space *(im)*
Formeinbauraum mould height/space *(im)*
Formeinheit mould assembly
Formeinrichtekarte mould setting record card
Formeinsatz mould insert *(im)*
Formel formula
Formenanschluß mould attachment
Formenbauer toolmaker, mould maker
Formenbauteile 1. mould components. 2. standard mould units
Formenbauzeit mould construction time: **Hierdurch wird die Formenbauzeit wesentlich verkürzt** in this way the mould can be made much more quickly
Formenfüllung mould filling operation *(for translation example see under* **Formfüllvorgang***)*
Formenkonstruktion mould design
Formenkorrekturmaß mould correction factor
Formenmaße mould dimensions
Formenmaterial mould making material
Formenöffnungs- und Schließspiel mould opening and closing cycle
Formenöffnungs- und Schließventil mould opening and closing valve
Formenschließgeschwindigkeit mould closing speed
Formentemperierung 1. mould temperature control. 2. mould temperature control system
Formentlüftungszeit mould venting time
Formenträger 1. mould carrier *(of a carousel-type foam moulding or injection moulding machine)*. 2. platen *(im)*
Formentrennmittel (mould) release agent
Formenwerkzeug mould
Formfahrgeschwindigkeit mould advance speed
Formfläche forming area *(t)*
Formfreisintern free sintering *(PTFE)*
Formfülldruck mould filling pressure
Formfüllgeschwindigkeit mould filling speed
Formfüllphase mould filling phase, injection phase *(im)*
Formfüllprozeß mould filling operation *(for translation example see under* **Formfüllvorgang***)*
Form-Füll-Siegelmaschine form-fill-seal machine

Formfüllüberwachung mould filling control system
Formfüllung mould filling: **Bei bestimmten Bedingungen hängt die Geschwindigkeit der Formfüllung vom Durchmesser des Punktanschnittes ab** under certain conditions, the rate at which the mould is filled depends upon the diameter of the pin gate. **Nach erfolgter Formfüllung** when the mould has been filled
Formfüllverhalten mould filling characteristics
Formfüllvolumen cavity volume *(im)*
Formfüllvorgang mould filling operation: **Der Formfüllvorgang muß in kürzester Zeit vor sich gehen** the mould must be filled as quickly as possible
Formfüllzeit mould filling time, injection time *(im)*
Formgebung 1. shape, shaping, design: **die spezielle Formgebung der Torpedospitze** the special shape of the torpedo tip. 2. moulding. 3. extrusion
Formgebung, flexible design flexibility
Formgebungsbedingungen moulding conditions
Formgebungsdruck holding pressure, dwell pressure *(im)*
Formgebungsphase holding pressure phase, dwell pressure phase *(im)*
Formgebungsprozeß moulding process
Formgebungsteil shaping/forming component: **Kernstück der Anlage ist das Profilwerkzeug. Es ist das Formgebungsteil, das den Schmelzestrang zum Profil formt** the essential part of the machine is the profile die which shapes the melt strand into the profile
Formgebungsverfahren moulding process
Formgebungswerkzeug 1. mould *(im)* 2. die *(e)*
Formgenauigkeit dimensional accuracy
formgepreßt compression moulded
formgerecht suitably shaped: **formgerechte Verpackung** suitably shaped pack
formgeschäumt foam moulded
Formgestalter mould designer
Formhälfte mould half
Formheizgerät mould heater
Formhöhenverstellung 1. mould height adjustment. 2. mould height adjusting mechanism *(for translation example see under* **Werkzeughöhenverstellung***)*
Formhohlraum (mould) cavity, impression: *Technically speaking these two words are not the same, since the impression (into which the polymer melt is injected) is made up of the cavity (female portion) and the core (male portion). Nevertheless the two words are treated as synonyms.*
Formhohlraumtiefe (mould) cavity depth
Formhohlraumvolumen (mould) cavity volume
Formhöhlung (mould) cavity, impression *(see explanatory note under* **Formhohlraum***)*

Forminhalt mould contents
Forminnendruck cavity pressure *(im)*
forminnendruckabhängig cavity pressure-dependent
Forminnendruckspitzenwert maximum cavity pressure *(im)*
forminnendruckunabhängig cavity pressure-independent
Forminnendruckverlust drop in cavity pressure *(im)*
Forminnenraum (mould) cavity, impression
Forminnenwand cavity wall
Formkavität (mould) cavity/impression
Formkern (mould) core *(im)*
formkompliziert of complex shape
Formkontur (mould) cavity/impression
Formkörper 1. casting. 2. moulding, part, moulded part
Formkosten tooling costs
Formling moulding, injection moulding
Formmasse moulding compound
Formmasse, pulverförmige dry blend
Formmassedruck melt pressure
Formmassehersteller moulding compound producer
Formmassetemperatur melt temperature
Formmassetemperaturverlauf melt temperature profile
Formmassetrichter (feed/material) hopper *(e,im)*
Formnest (mould) cavity, impression
Formnestabstände distances between (mould) cavities
Formnestbelag deposit on the cavity wall/surface
Formnestblock cavity plate *(im)*
Formnestdruck cavity pressure *(im)*
Formnesteinsatz mould insert
Formnestentlüftung mould cavity venting: **Möglichkeiten zur Formnestentlüftung** methods of venting the mould cavity
Formnestfüllvolumen cavity volume *(im)*
Formnesthohlraum (mould) cavity/impression
Formnestinhalt mould contents: **Ursache des Verzugs sind Eigenspannungen, die dadurch entstehen, daß der Formnestinhalt beim Abkühlen nicht gleichmäßig schwindet** Warping is caused by internal stresses due to uneven shrinkage of the moulded part whilst it is cooling down inside the mould
Formnestinnendruck cavity pressure
Formnestkontur (mould) cavity/impression *(im)*
Formnestoberfläche cavity surface
Formnestoberflächentemperatur cavity surface temperature
Formnestplatte cavity plate *(im)*
Formnestwand(ung) cavity wall
Formnestzahl number of cavities/impressions *(im)*
Formoberfläche mould surface, cavity surface

Formoberflächentemperatur cavity surface temperature
Formöffnungsbewegung mould opening movement
Formöffnungshub mould opening stroke
Formöffnungsrichtung mould opening direction
Formöffnungsweg mould opening stroke
Formplatte 1. mould plate *(see also explanatory note under* **Werkzeugplatte***).* 2. pattern plate
Formplatte, düsenseitige cavity plate *(im)*
Formplatte, schließseitige core plate *(im)*
Formpressen compression moulding
Formpreßmaterial compression moulding compound
Formschäumanlage foam moulding plant/equipment
Formschäumen foam moulding
Formschaumstoff moulded foam
Formschäumverfahren foam moulding (process)
Formschließaggregat (mould) clamping unit *(im)*
Formschließbewegung mould closing movement
Formschließeinheit (mould) clamping unit
Formschließgehäuse housing containing the mould clamping mechanism
Formschließhub (mould) clamping stroke
Formschließkraft 1. clamping force. 2. locking force *(im) (see explanatory note under* **Schließkraft***)*
Formschließregelung 1. mould clamping control. 2. mould clamping control mechanism
Formschließsicherung mould safety mechanism
Formschließsystem (mould) clamping mechanism *(im)*
Formschließverzögerung mould clamp delaying mechanism
Formschließzeit mould closing time
Formschließzylinder clamping cylinder *(im)*
Formschluß (mould) clamping mechanism *(im)*
Formschluß- *see* **Formschließ-**
formschlüssig positive *(connection or joint)*
formschön attractively designed, of pleasing appearance
Formschrumpf moulding shrinkage
Formschutz mould safety mechanism
Formschwindung moulding shrinkage
Formschwund moulding shrinkage
Formsohlen moulded soles
formstabil dimensionally stable
Formstabilität dimensional stability
Formstation 1. thermoforming station. 2. moulding station
Formsteifigkeit dimensional stability
Formstempel punch
Formstoff 1. cured resin *(if referring to thermosets, e.g. epoxies, polyesters, phenolics etc.).* 2. moulded material *(which can be either thermoset or thermoplastic; sometimes wrongly used instead of* **Formmasse***(q.v.))*

Formstück fitting
Formtechnik moulding technique
*****Formteil** 1. moulding, part, moulded part. 2. part of the mould
Formteilabweichungen 1. faulty mouldings: **werkzeugbedingte Ursachen der Formteilabweichungen** faulty moudlings due to defective moulds. 2. moulding faults: **Zu den Formteilabweichungen zählt ein vorstehender Angußabriß** another moulding fault is a torn, protruding sprue. 3. variations in part dimensions: **Die Formteilabweichung erfordert meist eine kostspielige Reparatur am Werkzeug** variations in part dimensions usually need expensive repairs to be done to the mould
Formteilanguß sprue *(see explanatory notes under* **Anguß** *and* **Anschnitt**)
Formteilanschnitt sprue *(see explanatory note under* **Anschnitt** *and* **Anguß** *This is another example of the word* **Anschnitt** *being used instead of the correct word, i.e* **Anguß**
Formteilausdehnung moulded-part size *(see explanatory note under* **Ausdehnung***)*
Formteilautomat automatic moulding machine
Formteilbildungsprozeß moulding process/ operation
Formteilbildungsvorgang moulding process/ operation
Formteilebene (mould) parting surface/line
Formteileigenschaften part properties, moulded-part properties
Formteilentnahmegerät parts-removal robot, demoulding robot
Formteilfläche, projizierte projected moulding area
Formteilgestalt part shape, moulded-part shape
Formteilgestaltung part design, moulded-part design
Formteilgewicht part weight, moulded-part weight
Formteilgröße part size, moulded-part size
Formteilhandhabung 1. parts handling. 2. parts handling device
Formteilherstellung moulding (process)
Formteilkonstrukteur part designer, moulded-part designer
Formteilkonstruktion part design, moulded-part design
Formteilmasse part weight, moulded-part weight
Formteilmaße part dimensions, moulded-part dimensions
Formteiloberfläche part surface, moulded-part surface
Formteilprojektionsfläche projected moulding area
Formteilqualität part quality, moulded-part quality
Formteilquerschnitt part cross-section, moulded-part cross-section
Formteilschäumautomat automatic foam moulding unit
Formteilschäumen foam moulding
Formteilschwindung part shrinkage, moulded-part shrinkage
Formteiltoleranzen part tolerances, moulded-part tolerances
Formteilung (mould) parting surface/line
Formteilvolumen part volume, moulded-part volume
Formteilwanddicke part wall thickness, moulded-part wall thickness
Formteilzeichnung part drawing, moulded-part drawing
Formtemperatur mould temperature
Formträgerplatte 1. mould carrier *(of a carousel-type foam moulding or injection moulding machine)*. 2. platen *(im)*
Formtrennaht parting line *(see explanatory note under* **Teilungslinie**)
Formtrennebene (mould) parting surface/line *(im)*
Formtrennfläche (mould) parting surface/line *(im)*
Formtrennlinie (mould) parting line/surface *(im)*
Formtrennmittel mould release agent
Formtrennwirkung mould release effect
Formtreue dimensional accuracy
Formulierung formulation
Formulierungsbeispiel typical formulation
Formulierungsbestandteile formulation components
Formulierungshinweise formulation guidelines
Formungshilfe thermoforming aid
Formungskraft thermoforming pressure
Formungstemperatur thermoforming temperature
Formverfahren thermoforming (process)
formverschäumt foam moulded
Formverschäumung foam moulding
Formversiegler mould sealant
Formverweilzeit mould residence time
Formvolumen cavity volume
Formwandung mould wall, cavity wall
Formwerkstoff mould material: **Als Formwerkstoffe eignen sich Zinklegierungen, Beryllium-Kupfer oder auch Aluminium** moulds may be made of zinc alloys, beryllium-copper as well as of aluminium
Formwerkzeug 1. mould. 2. thermoforming mould
Formwerkzeugnormalien standard mould units
Formzufahrgeschwindigkeit mould closing speed *(im)*
Formzufahrkraft clamping force *(im)*
Formzuhaltekraft locking force *(im)*
Formzuhaltung (mould) locking mechanism
Formzyklus moulding cycle
Forschung research

* *Words starting with* **Formteil-**, *not found here, may be found under* **Spritzteil-**, **Teil-**, *or* **Teile-**

Forschung, angewandte applied research
Forschungsarbeit(en) research work
Forschungsaufwand research costs/expenditure
Forschungseinrichtungen research facilities
Forschungsinstitut research institute
Fortentwicklung improvement *(for translation example see under* **Weiterentwicklung***)*
Fortluft waste air
fortschreitend progressive
Fortschrittsbericht progress report
fossiler Brennstoff fossil fuel
fossiler Energieträger fossil fuel
Foto-, foto- *see* **Photo-, photo-**
Foulardieren padding
Fourierzahl Fourier number
FPM *symbol for* fluororubber
Fragebogen questionnaire
fragmentiert fragmented
Fraktion fraction
fraktioniert fractionated
Fraktionierung fractionation
Fräsaggregat milling unit
Fräsen milling
Fräser milling cutter
Fräslehre router jig
Fräsrotor milling-type cutter *(sr)*
Freialterung outdoor weathering
freibewittert naturally weathered
Freibewitterung outdoor weathering
Freibewitterungsbeständigkeit outdoor weathering resistance
Freibewitterungsdauer outdoor weathering period/duration
Freibewitterungsprüfung outdoor/natural weathering test
Freibewitterungsstand outdoor weathering station
Freibewitterungsstation outdoor weathering station
Freibewitterungsverhalten outdoor weathering performance
Freibewitterungsversuch outdoor weathering test
Freie blasen, ins free forming *(t)*
freifließend free-flowing
freigesetzte Wärmemenge amount of heat released
freihängend freely suspended
Freiheitsgrad degree of freedom *(e.g. of a control system)*
Freilandbewitterung outdoor weathering
Freiluftbeständigkeit weathering resistance
Freiluftbetrieb outdoor use: **Versuche, diese Harze auch für elektrische Isolatoren im Freiluftbetrieb zu verwenden, scheiterten** attempts to use these resins for outdoor insulators failed
Freiluftbewitterung outdoor weathering
Freiluftbewitterungsanlage outdoor weathering station

Freilufteinsatz outdoor use *(for translation example see under* **Freiluftbetrieb***)*
Freiluftisolator outdoor insulator
Freiluftklima outdoor conditions
Freiluftprüffeld outdoor weathering station
Freiluftprüfstand outdoor weathering station
Freiluftprüfstation outdoor weathering station
Freilufttauglichkeit suitability for outdoor use
Freiluftversuch outdoor weathering test
freiprogrammierbar freely programmable
freischwingend freely vibrating/swinging
Freisintern free sintering *(ptfe)*
Freispiegelleitung non-pressure pipeline
freistehend 1. free-standing, separate *(e.g. control cabinet).* 2. cantilevered *(e.g. the clamping unit of an injection moulding machine, fixed at one end and free at the other)*
Freistrahl jet of material *(im):* **Durch diese Maßnahme wird vermieden, daß in den Formenhohlraum ein Freistrahl hineinspritzt** in this way one can prevent jetting. *Whenever the text speaks of a jet of material suddenly and quickly entering the mould cavity, the translator should use the word* jetting
Freistrahlbildung jetting *(im) (see also explanatory note under* **Freistrahl***)*
freiverlegt 1. laid above ground *(e.g. pipes).* 2. loose-laid *(e.g. roofing sheet)*
Freiverschäumung free foaming
Freiwinkel clearance/relief angle
Freizeitbereich leisure industry/sector
Freizeitboot leisure craft
Freizeitsektor leisure industry/sector
Fremdbeheizung separate heaters
Fremdbestandteile foreign matter/particles/substances
fremdbezogen purchased/obtained from outside
fremdbezogenes Programm ready-made program *(me)*
Fremddosierung separate feed/metering unit
fremde Software packaged software, ready-made software, software package *(me)*
Fremdenergie outside energy
Fremdentzündung flash ignition
Fremdentzündungstemperatur flash ignition temperature
fremdgesteuert separately controlled
Fremdharz extender/blending resin, extender polymer
Fremdkapital borrowed capital
Fremdkonstrukteur outside designer
Fremdkontrolle outside control
Fremdkörper foreign body
Fremdkörperabscheider contaminant separator
Fremdmaterial contaminant, impurity *(any material introduced from an outside source into a moulding compound for example),* foreign matter
Fremdmittel borrowed funds
Fremdpartikel foreign particle

Fremdspannung external stress
Fremdstoff foreign substance, contaminant, impurity
Fremdsubstanz foreign substance, contaminant, impurity
Fremdteile foreign matter/particles/substances
Fremdüberwachung outside control
fremdvernetzend externally crosslinking
Fremdverstärkung external reinforcement *(e.g. by means of glass or carbon fibres)*
Fremdzündung flash ignition
Frequenz frequecy
Frequenzbereich frequency range
Frequenzgang frequency response
fressen to seize (up), to jam
Friktion 1. friction. 2. friction ratio
Friktionierung frictioning *(c)*
friktionsarm low-friction
friktionsbedingt due to friction
Friktionsenergie frictional energy
Friktionsmischung frictioning compound *(c)*
Friktionsverhältnis friction ratio *(c)*
Friktionswärme frictional heat
Frischbeton fresh concrete
Frischhaltepackung airtight pack
Frischluftgerät breathing apparatus
Frischluftmenge amount of fresh air: **Es empfiehlt sich, pro Minute eine Frischluftmenge von 100 l durch den Ofen zu blasen** it is advisable to blow fresh air through the oven at the rate of 100 l/min
Frischluftzufuhr supply of fresh air, fresh air supply
frischpolymerisiert freshly polymerised
Frischrohstoff virgin material/compound
Frontalaufprall head-on collision
Frontalzusammenstoß head-on collision
Frontend front end
Frontgießharz facing resin
Frontguß facing, face casting
Fronthaube bonnet
Frontklappe bonnet
Frontplatte 1. control panel. 2. front panel
Frontscheibe windscreen
Frontschürze front apron
Frontspoiler front spoiler
Frontstoßfänger front bumper
Fronttafel control panel
Frontteil front end
Frontverkleidung front panel *(of a car)*
frostbeständig frost resistant
Frostbeständigkeit 1. frost resistance 2. freeze stability
Frostgrenze frost line *(bfe)*
Frostlinie frost line *(bfe)*
Frost-Taubeständigkeit freeze-thaw resistance
Frost-Tauwechselbelastung freeze-thaw cycle
Frostwechsel freeze-thaw cycle
Frostzone frost line *(bfe)*
Fruchtsaftverpackung fruit juice pack
Frühfestigkeit initial strength
Frühphase, in der in the early stages
Frühstadium, im in the early stages
FST-Wert flame-smoke-toxicity value
FTIR-Gerät FTIR spectroscope, Fourier transform infra-red spectrometer
FTIR-Spektroskopie FTIR spectroscopy, Fourier transform infra-red spectroscopy
FTK *abbr. of* **Flaschentransportkasten**, bottle crate
FT-Ruß fine thermal black
FTS *abbr. of* **fahrerloses Transportsystem**, driver-less transport system
Fuge joint
Fügedruck joining pressure *(w)*
Fügefläche adherend surface
Fugenabstände joint spacings
Fügenaht weld
Fugenbruch joint failure
Fugendichtungsmasse 1. jointing filler/compound 2. (tile) grout
fugenfüllend gap-filling
Fugenfüller 1. jointing filler/compound. 2. (tile) grout
fugenlos jointless
Fugenmasse 1. jointing filler. 2. (tile) grout
Fugenmörtel 1. jointing filler/compound. 2. (tile) grout
Fugenspachtel 1. jointing filler. 2. (tile) grout
Fugenvergußmasse joint filler
Fugenversiegelungsmasse joint sealant (compound)
Fügestelle joint
Fügeteil adherend
Fügeteiloberfläche adherend surface
Fühler sensor, transducer, probe
Fühlerspitze sensor tip
Fühlerstift sensor pin
Führungsbahn guideway
Führungsbolzen guide bolt
Führungsbuchse guide bush
Führungselement guide element
Führungsetagen upper echelons *(of a company)*
Führungsholm guide pin/pillar *(im)*
Führungskorb calibrating/sizing basket *(bfe)*
Führungskräfte, obere senior management
Führungsleisten guide bars
Führungsloch guide/locating hole
Führungsplatten guide plates
Führungsprozessor host/central/controlling processor, master computer *(me)*
Führungsring guide ring
Führungsrolle guide roll
Führungssäule guide pin/pillar *(im)*
Führungsschienen guide rails
Führungsstange guide rod
Führungsstift guide pin
Führungswalze guide roll
Füllautomat automatic filling machine
Füllbarkeit filler tolerance
Füllbild flow pattern *(of a melt in a mould)*
Füllbildkonstruktion preparation of a flow pattern

Füllbildmethode flow pattern method *(im)*. *This is a method used by part designers to determine the ideal number of gates in a mould, and their position, as well as determining the most suitable part geometry.*
Fülldichte apparent density
Fülldruck injection pressure *(im)*
Fülle 1. body *(of a paint or paint film)* 2. build *(of a paint film)*
Füller filler
Füllergemisch filler blend
Füllfaktor bulk factor
Füllgeschwindigkeit 1. filling speed. 2. mould filling speed *(im)*
Füllgrad 1. amount of material *(e.g. inside a mixer, a mould, in the screw flights, etc.)* 2. filler loading/content
Füllgut contents *(of a package, bottle etc.)*
Verpackungen für fetthaltige Füllgüter containers for fat-containing products
Füllguttrichter (feed/material) hopper *(e,im)*
Füllinhalt capacity *(of a bottle or container)*
füllkräftig high-build
Füllphase mould filling phase, injection phase *(im)*
Füllquerschnitt gate *(im)*
Füllraum transfer chamber *(tm)*
Füllraumwerkzeug positive mould *(cm)*
Füllschnecke feed screw
Füllstand level *(of solid or liquid inside a hopper, container etc.)*
Füllstandangeber 1. fuel gauge. 2. level indicator
Füllstandanzeige(r) 1. fuel gauge. 2. level indicator
Füllstandmelder 1. fuel gauge. 2. level indicator
Füllstandüberwachung level monitor, level monitoring unit
Füllstation filling station
Füllstoff filler
Füllstoffanteil filler content/loading *(see also explanatory note under **Pigmentanteil**)*
füllstoffarm with a low filler content
Füllstoffart type of filler
Füllstoffaufnahme filler tolerance
Füllstoffdosierung 1. addition of filler 2. amount of filler (added)
Füllstoffgehalt filler content/loading
füllstoffhaltig filled
Füllstoffkonzentration filler concentration/content
Füllstoffkorn filler particle
Füllstoffmenge amount of filler, filler content
Füllstoffnester local filler concentrations
Füllstofforientierung filler particle orientation
Füllstoffpulver powdered filler
füllstofffrei unfilled
füllstoffreich highly filled
Füllstoffsorte type of filler
Füllstoffteilchen filler particle
füllstoffverträglich filler-compatible, compatible with fillers

Füllstoffzugabe 1. addition of filler. 2. amount of filler (added): **Abhängigkeit der Dichte bei Variation der Füllstoffzugabe** effect of filler content on density
Füllstudie mould filling study
Füllsubstanz filler
fülltechnisch *relating to mould filling*: **der fülltechnisch kritischste Bereich...** the most critical region from the mould filling point of view...
Fülltrichter (feed/material) hopper *(e,im)*
Fülltrichterinhalt 1. hopper capacity. 2. hopper contents
Füllungsgrad filler content/loading
Füllvolumen 1. cavity volume *(im)*. 2. capacity
Füllwerkzeug positive mould *(cm)*
Füllzeit injection time, mould filling time *(im)*
Füllzone feed section/zone *(im)*
Füllzylinder feed cylinder
Fumarat fumarate
Fumarsäure fumaric acid
Fumarsäureester fumarate, fumaric acid ester
Fundament foundation
Fünffachanguß five-point gating *(im)*
Fünffachverteilerkanal five-runner arrangement *(im)*
Fünfpunktdoppelkniehebel five-point double toggle
Fünfpunktkniehebel five-point toggle
Fünfwalzenkalander five-roll calender
Fünfwalzen-L-Kalander five-roll L-type calender
Fungizid fungicide
fungizid fungicidal
funkenerodiert spark eroded, electric discharge machined
Funkenerosion spark erosion, electric discharge machining, EDM
funkenerosive Abtragung spark erosion, electric discharge machining, EDM
funkenerosives Schneiden spark erosion, electric discharge machining, EDM
Funkenüberschlag spark-over
Funktion function, performance, activity, duty, responsibility, task, job
Funktionalität functionality
funktionell functional
Funktionsablauf operating sequence
Funktionsablaufanzeige operating sequence display *(me)*
Funktionsanforderungen performance requirements
Funktionsanzeige function display
Funktionsausfall breakdown
Funktionsbeschreibung mode of operation
Funktionselemente control/operating elements
funktionsfähig operational
Funktionsfähigkeit serviceability
Funktionsgenerator function generator *(me)*
Funktionskarte function card *(me)*
Funktionslänge effective length
Funktionsplan functional diagram, logic diagram *(me)*

Funktionsprinzip operating principle, principle of operation
Funktionsprüfung performance test
funktionssicher reliable
Funktionssicherheit reliability, dependability *(in operation)*
Funktionsstörung malfunction
Funktionstaste function key *(me)*
Funktionstauglichkeit serviceability
funktionstüchtig serviceable
Funktionstüchtigkeit serviccability
Funktionsumfang range/number of functions: **Die "D" Maschinen zeichnen sich durch einen großen Funktionsumfang der Grundmaschine aus** the basic machines in the "D" range are capable of carrying out a large number of functions
Funktionsvorteile performance advantages
Funktionsweise method of operation
Furan furan
Furanharz furan resin
Furfurol furfurol
Furfurylalkohol furfuryl alcohol
Furnaceruß furnace black
Furnier veneer
Fußbefestigung foot mounting
Fußbodenausgleichmasse jointless flooring compound, floor screed
Fußbodenbelag floorcovering
Fußbodenheizung underfloor heating
Fußbodenheizungsrohr underfloor heating pipe
Fußbodenklebstoff flooring adhesive
Fußbodenleiste skirting board
Fußbodenmasse flooring compound
Fußbodenplatte floor tile
Fußleiste skirting board
Fußschalter foot operated switch
Futterstoff interlining
Futterstoffe, aufbügelbare iron-on interlinings
Fütterstreifen strips for feeding to the calender (or extruder)
FVK *abbr. of* **faserverstärkte Kunststoffe**, fibre reinforced plastics
FVW *abbr. of* **Faserverbundwerkstoff** fibre composite

G

Gabelstapler fork-lift truck
Galetten godets *(txt)*
Galettenstreckwerk godet roll stretch unit *(txt)*
Galvanikanlage electroplating plant/unit
Galvanikbad electroplating bath
galvanisch electrolytic
galvanisch hergestellt mady by electrodeposition

galvanisierbar electroplatable
Galvanisierflüssigkeit electroplating solution
galvanisiert 1. galvanised *(usually metals)*. 2. electroplated *(metals and plastics)*
Galvanisierungslösung electroplating solution
Galvanobeschichtung electroplating
Galvanoform electroformed mould
Galvanoformung electroforming
Galvanoplastik electrotype
Galvanoschale electroformed shell
Galvanotechnik electroplating
Gammabestrahlung gamma irradiation
Gammabestrahlungsanlage gamma irradiation plant
Gammastrahlen gamma rays
Gammastrahlenresistenz gamma ray resistance
Gammastrahlung gamma radiation
Gang 1. (screw) channel *(where the screw and barrel are treated as one unit) (e)*. 2. (screw) flight *(where the screw is referred to in isolation, i.e. without the barrel) (e)*; **5 Gänge vor dem Schneckenende** 5 flights away from the screw tip; **die ersten, tiefgeschnittenen Schneckengänge sind nur teilgefüllt** the first, deeply cut screw flights are only partly filled
Gangbreite (screw) channel width *(e)*
Gangdurchbrüche interrupted flights *(e)*
Ganggrund (screw) root surface *(e)*
gängig conventional, standard, commonly used
Gängigkeit number of starts *(e)*
Gangkammer chamber *(of screw) (e) (see explanatory note under* **Kammer***)*
Gangprofil (screw) channel profile *(e)*
gangprogressiv with gradually increasing flight depth *(screw) (e)*
Gangreserve power reserve *(e.g. of a timer or time switch)*
Gangschalthebel gear shift lever
Gangsteigung pitch *(e)*
Gangsteigungswinkel helix angle *(e)*
Gangtiefe flight depth *(e) (see explanatory note under* **Kanaltiefe***)*: **mit großer Gangtiefe** deep-flighted, with deep flights
Gangtiefenverhältnis flight depth ratio *(e)*
Gangvolumen (screw) channel volume *(e)*
Gangzahl number of flights *(e)*
Ganzkunststoffrahmen all-plastics frame
Ganzstahlmesser all-steel knife
Garantieausstoß guaranteed output
Gardinenbildung curtaining, sagging *(of paint film)*
Gardinengleiter curtain rail
Gardner Farbzahl Gardner colour number/value
Gardner Farbskala Gardner colour scale
Garn yarn
Gartenbau horticulture
Gartenmöbel garden furniture
Gartenmöbellack garden furniture paint
Gartenschlauch garden hose
Gartenstuhl garden chair

Gartenzwerg garden gnome
Gas gas
Gasabgabe gas release/emission
gasabspaltend gas-producing
Gasabspaltung gas evolution
Gasamtkosten total cost
Gasausbeute gas yield
gasbeheizt gas heated
gasbeladen containing gas, charged with gas
Gasbeladung 1. introduction of gas. 2. gas content
Gasbeton cellular/aerated concrete
Gasbläschen gas bubbles
Gasblase gas bubble
gasbremsend gas/vapour impermeable
Gaschromatogramm gas chromatogram
Gaschromatographie gas chromatography
gaschromatographisch gas chromatographic
gasdicht vapour proof, gas-tight
Gasdichtheit gas impermeability
Gasdichtigkeit gas impermeability
Gasdiffusion gas diffusion
Gasdruck gas pressure
Gasdurchlässigkeit gas permeability
Gasentladung gas discharge
gasförmig gaseous
Gasfreisetzung gas release/emission
Gasgemisch gas mixture
Gasinnendruck internal gas pressure
Gaskonstante gas constant
Gasleitung gas supply line
gasnitriert gas nitrided: *See also explanatory note under* **Nitrierschicht**
Gasnitrierung gas nitriding
Gaspedal accelerator (pedal)
Gasphase gaseous/vapour phase
Gasrohr gas pipe
Gasrohrleitung gas supply line
Gasruß channel black
Gassperreigenschaften gas impermeability
Gasundurchlässigkeit gas impermeability
Gasversorgung gas supply
Gatter creel *(txt)*
gaufrieren to emboss
GC *abbr. of* **Gaschromatographie**, gas chromatography
gealtert aged
Geber sensor, transducer, probe
Gebinde container
Gebläse blower
Gebläsegehäuse fan shroud
Gebläsekühlung 1. fan cooling. 2. cooling fan
geblasen blown, blow moulded
geblockt blocked
Gebrauchsartikel consumer goods
Gebrauchsdauer pot life
Gebrauchseigenschaften functional properties, performance characteristics
Gebrauchsfähigkeit serviceability
gebrauchsfertig ready-to-use
Gebrauchsfestigkeit final/maximum strength
Gebrauchsgüter consumer goods/articles

Gebrauchslebensdauer working/service life
Gebrauchsstabilität stability in use
gebrauchstauglich suitable for the job
Gebrauchstauglichkeit serviceability
Gebrauchstemperatur working/service/operating temperature
gebrauchstüchtig serviceable
Gebrauchstüchtigkeit serviceability
Gebrauchswert usefulness, serviceability
Gebrauchtfenster scrap windows
gecoatet *German version of the word* coated
Gedankenaustausch exchange of ideas
gedeckt opaque
gedeckt eingefärbt pigmented: *the use of pigments automatically causes opacity, so that the first word can be ignored in translation.*
nur gedeckt eingefärbt lieferbar only available in opaque colours
gedruckte Schaltung printed circuit
geeicht calibrated
Gefahrenklasse hazard classification
Gefahrgüter hazardous/dangerous goods
Gefahrguttransportvorschriften German regulations covering the transport of dangerous goods
Gefahrgutverordnung Eisenbahn German regulations covering the transport of dangerous goods by rail
Gefahrgutverordnung Straße German regulations covering the transport of dangerous goods by road
Gefälle gradient, slope
gefällt precipitated
gefedert spring-loaded
gefettet greased
geflusht *German version of the word* flushed
geformten Zustand, im in the moulded state
Gefrierpunkt freezing point
Gefrierschrank 1. freezer 2. refrigerator
Gefriertrocknung freeze drying
Gefriertruhe freezer, deep-freeze
Gefüge structure *(of a material, polymer etc.)*
Gefügeänderungen structural changes
Gefügeschädigung structural damage
Gefügestörung structural defect
Gefügeumwandlungen structural changes
Gegebenheiten conditions
Gegebenheiten, maschinentechnische technical conditions
Gegenbiegen roll bending *(c)*
Gegendralldoppelschnecke counter-rotating twin screw
Gegendrallkontraextruder contra-extruder
Gegendrallmaschine counter-rotating twin screw extruder
Gegendrallschnecke counter-rotating twin screw *(e)*
Gegendrallschneckenextruder counter-rotating twin screw extruder
Gegendrallsystem counter-rotating twin screw (system)

Gegendruck 1. back pressure *(e)*. 2. counter pressure *(general term)*
Gegendruckwalze backing roll
gegeneinanderlaufend counter-rotating, rotating in opposite directions *(twin screws)*
Gegenkraft opposing force, counterforce
Gegenlager support
gegenläufig counter-rotating, rotating in opposite directions *(twin screws)*
Gegenmaßnahme countermeasure
Gegenmesser bed knife, stationary knife *(sr)*
Gegenschnecke opposing screw *(e)*
Gegenseite opposite side
gegensinnig counter-rotating, rotating in opposite directions *(twin screws)*
Gegenstromölkühler counterflow oil cooling system, counterflow oil cooler
Gegenstromprinzip counterflow principle
Gegenüberstellung comparison
geglüht calcined
gegossen cast
gehackt chopped *(e.g. glass fibres)*
Gehalt 1. content. 2. salary
gehärtet 1. cured *(resin, rubber, etc.)*. 2. hardened *(steel)*
Gehäuse 1. casing, housing. 2. barrel *(e)*
Gehäuseabschnitt barrel section *(e)*
Gehäuseinnenwandung barrel liner *(e)*
Gehäuseschuß barrel section *(e)*
Gehäuseteile barrel sections
Gehäusewand barrel wall *(e)*
gehont honed
Gehrung mitring
Gehrungsfuge mitre (joint)
Gehrungssäge mitre saw
Gehrungsschnitt mitre cut
Geisterbild ghost image
Geisterschicht unmanned shift
gekapselt 1. embedded, encapsulated, potted. 2. enclosed *(part of a machine)*
gekerbt notched
gekickt containing kicker(s)
geknäuelt entangled *(molecules)*
gekrümmt curved
Gel gel
geladen charged
Geländefahrzeug cross-country vehicle
Gelanteil gel content
gelartig gel-like
Gelatine gelatine
gelatinisierend solvating *(plasticiser)*
Gelbfärbung yellowing, yellow discolouration
gelblich yellowish
Gelbstich yellow tinge
Gelbstichindex yellowness index
Gelbtönung yellowing
Gel-Chromatogramm gel chromatogram
Gel-Chromatographie gel chromatography
Gelcoatierung gel coat *(grp)*
Gelcoatschicht gel coat *(grp)*
Gele gel particles
Gelege bonded/non-woven fabric

Gelenkwelle cardan shaft
gelförmig gel-like
Gelgehalt gel content
Gelierdauer 1. gel time *(UP, EP resins)*. 2. fusion time *(PVC pastes)*
Gelieren 1. gelation *(UP, EP resins)*. 2. solvation *(PVC in plasticiser)*. 3. fusion *(PVC pastes)*
Words starting with **Gelier-** require special care when used in a PVC paste context, since they can mean two different things - solvation (i.e. gelling) or fusion. The former is the dissolving of PVC particles in plasticiser to form a gel-like substance, the latter is the conversion of this gel into a solid by heating to 150°C or more. Which of the two versions applies will only be apparent from the context
gelierend solvating *(plasticiser)*
Gelierfähigkeit solvating/dissolving power, gelling capacity *(of plasticiser for PVC)*
Gelierfreudigkeit readiness to gel
Geliergeschwindigkeit 1. gelation speed. 2. fusion rate
Geliergrad 1. degree of gelation. 2. degree of solvation *(PVC in plasticiser)*
Gelierhilfe gelling aid
Gelierkanal 1. gelling tunnel. 2. fusion tunnel *(for PVC pastes)*
Gelierkraft solvating/dissolving power *(of plasticiser for PVC)*
Geliermittel gelling agent
Gelierneigung tendency to gel
Gelierofen 1. gelling oven. 2. fusion oven *(for PVC pastes)*
Gelierpunkt 1. gel point. 2. solvating/dissolving temperature *(of PVC in plasticiser)*. 3. fusion temperature/point *(of PVC paste)*
Geliertemperatur 1. gel point *(of UP, EP resins)*. 2. solvating/dissolving temperature *(of PVC in plasticiser)*. 3. fusion point/temperature *(of PVC pastes)*
Gelierungsbeständigkeit gelation resistance
Gelierungserscheinungen signs of gelling
Gelierungsgrad 1. degree of gelation. 2. degree of fusion
Gelierverhalten 1. gelling/solvating behaviour. 2. fusion characteristics
Geliervermögen gelling/solvating/dissolving power
Gelierwirkung gelling/solvating effect
Gelierzeit 1. gel time *(of UP, EP resins)*. 2. fusion time *(of PVC pastes)*
Gelierzeitautomat automatic gel time tester
gelocht perforated
gelöst dissolved
Gelpartikel gel particle
Gelpermeationschromatographie gel permeation chromatography
gelpermeationschromatographisch gel permeation chromatographic(ally)
Gelpunkt 1. gel point 2. solvating temperature *(of PVC in plasticiser)* 3. fusion temperature *(of PVC paste)*

Gelspinnen gel spinning
Gelspinnverfahren gel spinning (process)
Gelteilchen gel particle
Gelzeitprüfgerät gel time tester
gemahlen ground
gemahlen, grob coarsely ground
gemäßigt moderate
Gemeinkosten overheads
gemeinsamer Binnenmarkt single European market
gemessen measured
gemischtdispers of mixed/non-uniform particle size
Gemischtfraktion mixed fraction
gemischtzellig mixed-cell *(foam)*
genarbt grained, textured
Genauigkeit accuracy, precision
Genauigkeitsgrad degree of accuracy
Genauigkeitswaage precision balance/scales
genehmungsbedürftig requiring official permission/approval
genehmungspflichtig requiring official permission/approval
Generierung generation *(me)*
genormt standardised
Genußmittel *There is no succinct, adequate equivalent for this word in English. The term covers products such as tea, coffee, tobacco, etc. which have no food value and are consumed for enjoyment. Many dictionaries give* semi-luxuries *as the English equivalent but this should be avoided since it is not English and comes nowhere near the true meaning of the German word which, literally, means "things for enjoyment". Where the text mentions* **Nahrungs- und Genußmittel,** *the best the translator can do is to write* foodstuffs and products such as tea, coffee and tobacco.
Genußmittelindustrie *see explanatory note under* **Genußmittel.** *Where the word occurs on its own it is best to translate it as* food industry. *Whilst this will not be 100% accurate, there is a chance that the word was used loosely in place of* **Nahrungsmittelindustrie.** *See also explanatory note under* **Genußmittel)**
genutet grooved
genutet, spiralförmig spirally grooved
Geometrie geometry, shape
Geometrievorschlag suggested geometry/shape
geordnet ordered *(molecular structure)*
geordnete Deponie controlled tipping *(of waste)*
gepanzert hardened, hard faced
gepfropft grafted
gepolstert padded
gepreßt compression moulded
geprimert primed
gepuffert buffered
Geradeausextrusionswerkzeug straight-through die, in-line die *(e)*
Geradeauskopf in-line die, straight-through die *(e)*
Geradeausspritzkopf straight-through die, in-line die *(e)*
Geradeauswerkzeug straight-through die, in-line die *(e)*
geradkettig straight-chain
geradlinig straight-line
Gerät 1. tool, implement, instrument, appliance, device, unit. 2. equipment, piece of equipment
gerätebedingt due to the instrument/machine/equipment
geräteextern outside the instrument/machine
Gerätefrontplatte instrument's/unit's front panel
geräteintern inside the instrument/machine
Geräterückseite back of the unit/instrument
Gerätetechnik instrumentation *(me)*
∗**geräuscharm** quiet in operation, quiet-running, low-noise, silent *(e.g. machine, pump etc.)*
Geräuschabsorption sound absorption
Geräuscharmut low noise level
Geräuschdämpfer silencer
Geräuschdämpfung noise reduction/sound deadening qualities
Geräuschdämpfungsmaßnahmen noise reduction measures
Geräuschkapsel engine sound shield
geräuschlos silent
Geräuschniveau noise level
Geräuschpegel noise level
Geräuschreduzierung noise reduction, sound deadening qualities
geräuschstark noisy
Geräuschverminderung noise reduction: **aus Gründen der Geräuschverminderung** to reduce noise
-gerecht *This suffix implies that something is appropriate to, or correct for the properties of a material or the characteristics of a process. Whilst the translator should try to convey the meaning of the word if possible, he may find that the result is a clumsy, verbose sentence. In that case he should substitute the word* correct. *Sometimes, indeed, the word makes no contribution to the meaning of a sentence and can be omitted. In some cases one can add the word* oriented, *e.g.* **kunststoffgerecht** plastics-oriented. *For translation examples see under* **materialgerecht, verarbeitungsgerecht, verfahrensgerecht** *and* **werkstoffgerecht**
geregelt, thermostatisch thermostatically controlled
gerichtet aligned, oriented
geriffelt grooved
Geruch odour, smell, aroma
geruchlos odourless
Geruchlosigkeit freedom from odour
geruchsarm low-odour
Geruchsarmut low-odour properties
Geruchsdichtigkeit odour impermeability

∗ *Words starting with* **Geräusch-**, *not found here, may be found under* **Lärm-** *or* **Schall-**

geruchsfrei odourless
Geruchsfreiheit freedom from odour
geruchsintensiv strong-smelling, smelly
geruchsneutral odourless
Geruchsneutralität freedom from odour
geruchsschwach low-odour
Geruchssperre odour barrier
Gerüst framework, skeleton
Gesamtabmessungen overall dimensions
Gesamtanschlußwert total connected/installed load
Gesamtantriebsleistung total drive power
Gesamtbelegschaft total number of employees
Gesamtbeschäftigtenzahl total number of employees
Gesamtbeurteilung overall assessment
Gesamtbleigehalt total lead content
Gesamtbrenndauer total burning time
Gesamtdeformation total deformation
Gesamtdehnung total elongation
Gesamtenergieaufnahme total energy input, total energy used
Gesamtfüllstoffverbrauch total filler consumption
Gesamtgasabgabe total gas release/emission
Gesamtheizleistung total heating capacity/units
Gesamtinvestitionen total investments
Gesamtkautschukverbrauch total rubber consumption
Gesamtkonzept overall design/concept
Gesamtlagenzahl total number of layeres
Gesamtlänge total/overall length
Gesamtleistung 1. total power. 2. total output. 3. overall performance
Gesamtleistung, installierte total connected/installed load
Gesamtleistung, installierte elektrische total connected/installed load
Gesamtleistungsbedarf total power consumption
Gesamtmenge total amount
Gesamtökobilanz overall ecological balance
Gesamtpigmentierung total pigment content
Gesamtproduktion total production
Gesamtproduktionskosten total production costs
Gesamtproduktionsleistung total output
Gesamtquerschnitt entire/complete cross-section
Gesamtschalldruckpegel 1. overall sound pressure level/intensity. 2. overall noise level. Version 1. should be used in scientific texts, version 2. in a general context, e.g. in connection with granulators and other noisy machines
Gesamtschallpegel overall noise level
Gesamtschaltzeit total switching time
Gesamtschichtdicke total film/coating thickness
Gesamtschrumpfung total shrinkage
Gesamtschwindung total shrinkage
Gesamtspaltlast total nip pressure *(c)*

Gesamtstromkosten total electricity costs
Gesamttemperaturniveau overall temperature
Gesamtunternehmungsführung corporate management
Gesamtverbrauch total consumption
Gesamtverformung total deformation
Gesamtverhalten general/overall performance
Gesamtwärmekosten total heating costs
Gesamtweichmachermenge total amount of plasticiser
Gesamtzahl total number
Gesamtzykluszeit total cycle time
gesättigt saturated
geschädigt, thermisch charred *(polymer, during processing)*
Geschäftsbedingungen conditions of sale
Geschäftsbereich division
Geschäftsbereichleiter divisional manager
Geschäftsbericht annual report
Geschäftsführung Board of Directors
Geschäftsjahr financial year
Geschäftspolitik company/corporate policy
Geschäftsvolumen business volume, volume of business
geschätzt 1. estimated 2. highly regarded
geschäumt foamed, expanded
geschirmt shielded, screened
Geschirrqualität tableware quality *(moulding powder)*
Geschirrspülmaschine diswasher
geschliffen ground
geschlossen 1. self-contained *(machine unit)*. 2. closed, shut
geschlossene Wartezeit closed assembly time
geschlossene Zeit closed assembly time
geschlossenporig closed-cell
geschlossenzellig closed-cell
Geschlossenzelligkeit 1. closed-cell character 2. closed-cell content
geschmacklos tasteless
geschmacksarm having little taste, with little taste, almost tasteless
Geschmacksfreiheit freedom from taste
geschmacksneutral tasteless
Geschmacksneutralität freedom from taste
geschmeidig soft, supple
Geschmeidigkeit softness, suppleness
geschmiert lubricated
geschmolzen molten *(especially metal)*, melted
geschnitten, durchgehend fully-flighted *(screw)* *(e)*
geschnittene Glasfasern chopped strands
geschnittenes Textilglas chopped strands
geschützt protected *(e.g. data in a memory)(me)*
geschützt, patentrechtlich patented
geschweißt welded
***Geschwindigkeitsaufnehmer** speed transducer
Geschwindigkeitsfernsteuerung 1. remote speed control. 2. remote speed control

* Words starting with **Geschwindigkeits-**, not found here, may be found under **Drehzahl-**

mechanism
Geschwindigkeitsfolge speed sequence
Geschwindigkeitsfühler speed transducer
Geschwindigkeitsgefälle 1. velocity gradient. 2. shear rate
geschwindigkeitsgeregelt speed-controlled
Geschwindigkeitsgradient 1. velocity gradient. 2. shear rate
Geschwindigkeitskennlinie velocity curve
Geschwindigkeitskonstante velocity constant
Geschwindigkeitskonstanz constant speed/velocity
Geschwindigkeitsprofil velocity profile
Geschwindigkeitsprogramm speed programme
Geschwindigkeitsregler speed regulator
Geschwindigkeitsumschaltpunkt speed change-over point *(im)*
Geschwindigkeitsumschaltung speed changing device
Geschwindigkeitsverteilung velocity distribution
Gesellschaftsmittel company funds/reserves
Gesenk 1. (mould) cavity. 2. cavity plate *(im)*
Gesenkeinsatz cavity insert *(im)*
Gesenkform mould
Gesenkhöhlung (mould) cavity
Gesenkplatte cavity plate *(im)*
Gesenkseite fixed/stationary mould half *(im)*
Gesetzesanforderungen legal requirements
Gesetzgeber legislator
Gesetzgebung legislation
Gesetzgeber legislator
Gesetzgebungsmaßnahmen legislative measures
gesetzliche Regelung law, regulation
geshreddert *German version of* shredded
Gesichtsmaske face mask
gespeichert stored
gespeist, seitlich side-fed *(e,bm,bfe)*
gespeist, zentral centre-fed *(e,bm,bfe)*
gespritzt 1. extruded. 2. moulded, injection moulded
gestaffelt, dreifach in three stages
Gestalt shape
gestalterisch *relating to design*: **gestalterische Freiheit** design freedom
Gestaltfestigkeit dimensional strength
Gestaltung design
Gestaltung, konstruktive design
Gestaltungsfreiheit design freedom
Gestaltungshinweise design guidelines
Gestaltungsmöglichkeiten design possibilities
Gestaltungsrichtlinien design guidelines
gestaltungstechnisch *relating to design: The second part of this word should always be omitted, as in these examples:* **gestaltungstechnische Richtlinien** design guidelines; **vom gestaltungstechnischen Gesichtspunkt** from the design point of view
Gestaltungsvorschlag suggested design
gestanzt punched

Gestein stone, stonework
Gesteinsmehl mineral powder
Gestell (machine) frame
gesteuert, numerisch digitally controlled
gestrainert strained, sieved
gestreckt, biaxial biaxially oriented
gestrichelt dashed *(curve)*
gestrichen coated, painted
gestuft multi-stage
gesundheitlich unbedenklich non-toxic
gesundheitliche Unbedenklichkeit non-toxicity
gesundheitsgefährdend toxic, hazardous to health
Gesundheitsgefährdung health hazard/risk, danger to health
Gesundheitsrisiko health risk
gesundheitsschädlich toxic, hazardous to health
getempert annealed, conditioned
Getränk drink, beverage
Getränkedoseninnenlack can coating lacquer
Getränkeflasche drinks pack
Getränkeverpackung drinks pack
getränkt impregnated
Getriebe 1. drive, drive unit/mechanism. 2. gear train, gear transmission
Getriebegehäuse gear housing
Getriebehohlwelle hollow drive shaft
Getriebekonstruktion drive design, drive unit design
Getriebemotor geared motor
getrieben foamed, expanded, blown
getrieben, chemisch chemically blown
getrieben, mechanisch mechanically blown *(foam)*
Getriebeöl gear oil
Getriebepumpe gear pump
Getriebetransmission gear train/transmission
Getriebeuntersetzung 1. gear reduction. 2. gear reduction unit
Getriebewelle drive shaft
Gew.-% percent by weight, % w/w
Gewächshaus greenhouse
Gewässerschutz protection of the waterways
Gewebe 1. cloth, fabric. 2. woven fabric *(where a distinction is made between* **Gewebe** *and* **Gewirke**, *q.v.)*
Gewebe, technische industrial fabrics
Gewebeabschnitte fabric offcuts
Gewebebeschichtung fabric coating
Gewebegrundierung fabric anchor coating *(of PVC paste)*
Gewebekunstleder (PVC) leathercloth, (PVC) coated fabric
Gewebelaminat fabric-based laminate
Gewebeprepreg fabric prepreg
Geweberand selvedge
Gewebereste fabric waste/scrap
Gewebesieb filter screen
gewebeverstärkt fabric reinforced
Gewerbeabfall industrial waste/refuse

gewerbehygienisch *relating to workshop or factory hygiene (for translation example see under* **arbeitshygienisch**)
Gewerbemüll industrial waste/refuse
gewerblich industrial
gewerbliche Mitarbeiter blue-collar workers
gewerbliche Arbeitsnehmer blue-collar/manual workers
Gewerkschaften *(trade)* unions
Gewicht weight
Gewicht, spezifisches specific gravity
Gewichtsabnahme weight loss, loss of weight
Gewichtsabnahmebestimmung weight loss determination
Gewichtsänderung change in weight
Gewichtsanteil weight content
gewichtsbelastet weight loaded
Gewichtsdosiereinrichtung weigh feeder
Gewichtsdosierung 1. weigh feeding. 2. weigh feeder
Gewichtseinsparungen weight savings
Gewichtsersparnis weight savings
gewichtsintensiv heavy
Gewichtskonstanz constant weight: **Die Gewichtskonstanz ist besser** a more constant weight is achieved
Gewichtsmittel weight average
gewichtsmittleres Molekulargewicht weight-average molecular weight
Gewichtspreis price per unit weight
Gewichtsprozent percent by weight, % w/w
Gewichtsreduzierung weight reduction
Gewichtsschwankungen weight fluctuations/variations
Gewichtsteile parts by weight
gewichtsträchtig heavy
Gewichtsveränderung change in weight
Gewichtsverlust weight loss, loss of weight
Gewichtsvorteil weight advantage
Gewichtszunahme weight increase, increase in weight
Gewinde thread
Gewindeanschluß threaded connection
Gewindebohrung threaded hole
Gewindebuchse threaded bush
Gewindedornausdrehvorrichtung threaded mandrel unscrewing device/mechanism
Gewindeeinsatz thread insert
Gewindeflanke, aktive thrust/front face of flight, leading edge of flight *(e)*
Gewindeflanke, passive rear face of flight, trailing edge of flight
gewindeformend 1. self-tapping *(screw)*. 2. thread-forming
Gewindekern threaded core
Gewindeloch threaded hole
Gewindepartie threaded part
gewindeprägend self-tapping *(screw)*
gewindeschneidend 1. self-tapping *(screw)*. 2. thread-cutting
Gewindespindel threaded spindle
Gewinn profit

Gewinn- und Verlustrechnung profit and loss account
Gewinnrückgang drop in profits
Gewirke knitted fabric
gezielt judicious, controlled, specific, selective, tailored, made-to-measure **gezielte Eigenschaften** specific properties; **gezielte Nukleierung** selective nucleation; **gezielter Einsatz von Stabilisatoren** judicious use of stabilisers
gezogen 1. drawn 2. pultruded
GF *abbr. of* **Glasfaser**, glass fibre: **PP-GF** glass fibre filled polypropylene. **PP-GF 30** polypropylene containing 30% glass fibre
GF-Gehalt glass fibre content
GFK *abbr. of* **glasfaserverstärkte Kunststoffe**, glass fibre reinforced plastics, GRP
GFK Wickelmaschine GRP filament winding machine
GFK-Blattfeder GRP leaf spring
GFK-Schrott GRP scrap
GF-Kunststoffe glass fibre reinforced plastics, GRP
GF-UP *abbr. of* **glasfaserverstärktes ungesättigtes Polyesterharz**, glass fibre reinforced unsaturated polyester resin, GRP
GF-UP-Mattenlaminat glass mat reinforced polyester laminate
GF-UP-Platte GRP sheet
GGVE *abbr. of* **Gefahrgutverordnung Eisenbahn**, German regulations covering the transport of dangerous goods by rail
GGVS *abbr. of* **Gefahrgutverordnung Straße**, German regulations covering the transport of dangerous goods by road
GHPD *abbr. of* gated high power decoupling
Gießanlage casting plant
gießbar pourable
Gießelastomer pourable elastomer
gießen 1. to cast *(resin)*. 2. to pour *(liquids, granules, powders)*
Gießereibindemittel foundry resin
Gießereiharz foundry resin
Gießereiindustrie foundry industry
Gießereikern foundry core
Gießereimodell foundry pattern
gießfähig pourable
gießfertig ready for pouring, ready to pour, pourable
Gießfolienanlage 1. cast film extrusion line. 2. film casting line *(for making, say cellulose acetate film from a solution)*
Gießform (casting) mould
Gießharz casting resin
Gießharzformstoff cured casting resin, cast resin, casting
Gießharzgehäuse cast (resin) housing
Gießharzhinterfütterung cast resin backing
Gießharzkörper synthetic resin casting
Gießharzmasse casting compound/mix
Gießharzmischung casting compound/mix
Gießharzsystem casting compound/mix

Gießharztechnik casting resin technology
Gießharzteil casting
Gießharzwandler cast resin transformer
Gießkörper casting, cast resin
Gießlegierung casting alloy
Gießling casting
Gießmaschine casting machine
Gießmasse casting compound
Gießmischung casting compound
Gießpaste (PVC) casting paste
Gießplatte cast sheet
Gießverfahren casting (process)
Gießvorgang casting operation/process
Gießwalze casting roll *(used in chill casting of film)*
Gießwalzeneinheit casting roll unit *(e)*
giftig poisonous, toxic
Giftigkeit toxicity
Giftmüll toxic waste
Giftstoff toxic substance
Gilbung yellowing
Gilbungsbeständigkeit resistance to yellowing
Gilbungsneigung tendency to become yellow
gilbungsresistent non-yellowig, resistant to yellowing
Gilbungsresistenz resistance to yellowing
gilbungsstabil non-yellowing, resistant to yellowing
Gips 1. gypsum 2. plaster (of Paris)
Gipsfaserplatte plasterboard
Gipsformling plaster moulding
Gipskarton plasterboard
Gipsmodell plaster model
Gipsplatte plasterboard
Gipsputz gypsum plaster
Gitter lattice
Gitterboxpalette cage pallet
Gitterschnittest cross-hatch adhesion test
Gitterschnittmethode cross-hatch adhesion test
Gitterschnittprobe cross-hatch adhesion test specimen
Gitterschnittversuch cross-hatch adhesion test
Gitterschnittwert cross-hatch adhesion
Gitterstruktur lattice structure
Glanz gloss
Glanzabbau reduction in gloss
Glanzbeständigkeit gloss retention
Glanzerhalt gloss retention
Glanzfarbe gloss paint
Glanzgebung imparting gloss
Glanzgrad (degree/amount of) gloss
Glanzgradverlust reduction in gloss, gloss reduction
Glanzhaltevermögen gloss retention
Glanzhaltung gloss retention
Glanzlack gloss paint/enamel
Glanzminderung reduction in gloss, gloss reduction
glanzsteigernd increasing/enhancing gloss
Glanzverlust reduction in gloss, gloss reduction
Glasanteil glass content

Glasarmierung glass reinforcement
glasartig glassy
Glasdoppelscheibe double glazing unit, sealed unit
gläsern glassy
Glasfalz rebate
Glasfaser glass fibre
Glasfaseranhäufungen glass fibre accumulations, local glass fibre concentrations
Glasfaseranteil glass fibre content: **Ohne Glasfaseranteil ist eine nur sehr geringe Verschleißwirkung vorhanden** only very little wear is apparent in the absence of glass fibres
Glasfaserausrichtung glass fibre orientation/alignment
glasfasergefüllt glass (fibre) filled
Glasfasergehalt glass fibre content
Glasfasergewebe glass (fibre) fabric, glass cloth
glasfaserhaltig glass fibre reinforced/filled
Glasfaserkabel fibre-optic cable, optic fibre cable
Glasfaserkonzentration glass fibre content
Glasfaserkunststoff glass fibre reinforced plastic, GRP
Glasfaserlaminat glass fibre laminate
Glasfasermatte glass fibre mat
Glasfasern, geschnittene chopped strands
Glasfaserorientierung glass fibre orientation
Glasfaserprepreg prepreg
Glasfaserroving roving
Glasfaserschutt glass fibre sweepings *(waste fibres less than 0.15 mm long)*
Glasfaserstrang (glass) strand
glasfaserverstärkt glass fibre reinforced
Glasfaserverstärkung glass fibre reinforcement
Glasfaserverteilung glass fibre distribution
Glasfaserverteilung, wirre randomly distributed glass fibres
Glasfaservliesstoff chopped strand mat
Glasfaserzusatz incorporation/addition of glass fibres
Glasfassadenbau curtain wall construction
Glasfilamentgewebe glass cloth, glass (fibre) fabric
Glasgehalt glass content
Glasgewebe glass cloth/fabric
Glasgewebeabdeckung glass cloth covering
Glasgewebeband woven glass tape, glass cloth tape
Glasgewebelaminat glass cloth laminate
Glasgewebeverstärkung glass cloth reinforcement
Glashartgewebe glass cloth laminate
Glashartgewebeplatte glass cloth laminate sheet
Glashartgeweberohr glass cloth laminate pipe
glasiert glazed
Glasinseln isolated glassy zones
glasklar crystal clear
Glasklarfolie crystal clear film
Glasklarheit crystal clarity

Glasklebung 1. bonding of glass. 2. bonded glass joint
Glaskugeln glass beads/spheres
Glaskurzfasern short/chopped (glass) strands
Glaslaminat glass fibre laminate
Glasleiste glazing bead
Glasmatte glass mat
glasmattenverstärkt glass mat reinforced
Glasmikrosphären glass mirobeads/microspheres
Glasnadelmatte needled glass mat
Glaspapier glass paper
Glasplatte glass plate
Glaspunkt glass transition temperature
Glasroving roving
Glasrovinggewebe roving cloth
Glasseide glass fibre
Glasseidenendlosmatte continuous strand (glass fibre) mat
Glasseidenfaser continuous glass fibre
Glasseidengewebe glass cloth/fabric
Glasseidenkurzfaser chopped glass fibre
Glasseidenmatte glass mat
Glasseidenroving roving
Glasseidenrovinggewebe woven roving
Glasseidenschnittmatte chopped strand (glass fibre) mat
Glasseidenstrang (glass) strand
glasseidenumsponnen glass braid sleeved
Glasseidenverstärkung glass fibre reinforcement
Glasstab glass rod
Glasstapelfaser staple (glass) fibre
Glasstapelfasergewebe woven staple fibre
Glasstopfen glass stopper
Glastemperatur glass transition temperature
Glasübergang glass transition *(i.e. the transition of a polymer from the glassy to the viscoelastic state)*
Glasübergangsbereich glass transition zone
Glasübergangsgebiet glass transition zone
Glasübergangspunkt glass transition temperature
Glasübergangstemperatur glass transition temperature
Glasumwandlung glass transition *(see explanatory note under* **Glasübergang***)*
Glasumwandlungsbereich glass transition zone
Glasumwandlungspunkt glass transition temperature
Glasumwandlungstemperatur glass transition temperature
Glasumwandlungstemperaturbereich glass transition temperature range
Glasur glaze
Glasverklebung 1. bonding of glass. 2. bonded glass joint
glasverstärkt glass reinforced
Glasverstärkung glass reinforcement
Glasvlies glass mat
Glaszustand glassy state

Glätt- und Abziehmaschine polishing and take-off unit
Glätteinheit polishing stack
Glättkalander polishing stack
Glättspalt polishing nip
Glättwalze polishing roll
Glättwalzen polishing stack
glattwandig smooth-bore *(pipe)*
Glättwerk polishing stack
gleichbleibend constant, consistent:
  gleichbleibend hohe Spritzteilqualität consistently high moulded-part quality
Gleichdralldoppelschnecke co-rotating twin screw *(e)*
Gleichdralldoppelschneckenkneter co-rotating twin screw compounder/plasticator
Gleichdrallmaschine co-rotating twin screw extruder
Gleichdrallschnecke co-rotating twin screw *(e)*
Gleichdrallschneckenextruder co-rotating twin screw extruder
Gleichdrallsystem co-rotating twin screws, co-rotating twin screw system
Gleichgewicht equilibrium
Gleichgewicht, thermisches thermal equilibrium
Gleichgewichtsabsorption equilibrium absorption
Gleichgewichtsdruck equilibrium pressure
Gleichgewichtskondensation equilibrium condensation
Gleichgewichtskurve equilibrium curve
Gleichgewichtsreaktion equilibrium reaction
Gleichgewichtssystem equilibrium system
Gleichgewichtswassergehalt equilibrium moisture content
Gleichgewichtszustand state of equilibrium
gleichlang equally long, of equal length
Gleichlauf synchronism
gleichlaufend co-rotating, rotating in the same direction *(twin screws) (e)*
gleichläufig co-rotating, rotating in the same direction *(twin screws) (e)*
Gleichrichter rectifier
gleichsinnig co-rotating, rotating in the same direction *(twin screws) (e)*
Gleichspannung d.c. voltage
Gleichspannungs-Drehankermagnetmotor d.c. permanent magnet motor
Gleichspannungssignal d.c. signal
Gleichstrom direct current, d.c.
Gleichstromantrieb d.c. drive
Gleichstrommotor d.c. motor
Gleichstromnebenschlußmotor d.c. shunt motor
Gleichstrompotential d.c. potential
Gleichstromquelle d.c. source
Gleichstromverstärker d.c. amplifier
Gleichung equation
Gleitadditiv lubricant
Gleitbeschichtung non-stick coating
Gleitbewegung sliding movement

Gleiteigenschaften 1. surface slip characteristics, anti-friction properties, frictional characteristics *(of materials).* **2.** lubricating properties *(of lubricants)*
gleitende Reibung sliding friction
gleitfähig having good surface slip
Gleitfähigkeit surface slip, anti-friction properties
Gleitfähigkeitsverhalten surface slip, anti-friction properties
Gleitfilm lubricating film
Gleitfläche sliding surface
Gleitgeschwindigkeit sliding speed/velocity
Gleitkalibrator low-friction calibrator *(e)*
Gleitlager slide bearing, friction bearing
Gleitmittel lubricant
Gleitmittelanteil lubricant content, amount of lubricant
Gleitmitteldosierung 1. addition of lubricant **2.** amount of lubricant (added)
Gleitmittelfilm lubricant film
gleitmittelfrei unlubricated, not containing lubricant
Gleitmittelgemisch lubricant blend
Gleitmittelkombination lubricant blend
Gleitmittelsystem lubricant blend
Gleitmittelverhalten lubricant performance, lubricating effect
Gleitmodul shear modulus
Gleitreibung sliding friction
Gleitreibungskoeffizient coefficient of sliding friction
Gleitreibungszahl coefficient of sliding friction
gleitsicher non-skid, skidproof
Gleitspiel sliding clearance
Gleitsubstanz lubricant
Gleitsystem lubricant blend
Gleitverhalten 1. surface slip behaviour anti-friction properties, frictional characteristics *(of materials).* **2.** lubricating properties *(of lubricants)*
Gleitvermögen slip
Gleitverschleiß wear due to sliding friction
Gleitvorgang slip/slipping/sliding (process)
Gleitwerkstoff anti-friction material
Gleitwert coefficient of sliding friction
Gleitwiderstand sliding resistance
Gleitwirkung lubricating effect
Gleitwirkung, äußere external lubrication
Gleitwirkung, innere internal lubrication
Gliederförderband articulated conveyor belt
glimmen to glow
Glimmentladung glow discharge
Glimmer mica
Glimmerheizbänder mica-insulated heater bands
Glimmerplättchen mica platelet
global 1. global, world-wide **2.** overall, total, general
Glockenmuffe socket
Glühbandabschneider incandescent wire cutter

Glühdornprüfung incandescent/glowing mandrel test
Glühdrahtfestigkeit incandescent/glowing wire resistance
Glühdrahtprüfung incandescent/glowing wire test
Glühdrahtschneiden incandescent wire cutting
Glühlampe light bulb, electric filament lamp
Glührückstand ash
Glühstabprüfung incandescent bar test
Glühverlust loss on ignition
Glutarsäure glutaric acid
Glutarsäureanhydrid glutaric anhydride
Glutbeständigkeit incandescence resistance
Glycerin glycerin
Glycerindistearat glycerin distearate
Glycerinmonooleat glycerin mono-oleate
Glycerinstearat glycerin stearate
Glycerintristearat glycerin tristearate
Glycerylphthalat glyceryl phthalate
Glycidol glycidol
Glycidylacrylat glycidyl acrylate
Glycidylester glycidyl ester
Glycidylether glycidyl ether
Glycidylgruppe glycidyl group
Glycidylmethacrylat glycidyl methacrylate
Glycidylrest glycidyl radical
Glycidylverbindung glycidyl compound
Glycin glycine
Glykol glycol
Glykoldiacetat glycol diacetate
Glykolether glycol ether
Glykoletheracetat glycol ether acetate
Glykolgruppe glycol group
Glykolsäurebutylester butyl glycolate
Glykolyse glycolysis
Glyptalharz glyptal resin, glycerin-phthalic acid resin
GM *abbr. of* **Glasmatte**, glass mat: **PP-GM** glass mat filled polypropylene
G-Modul torsion modulus
GMT *abbr. of* **glasmattenverstärkte Thermoplaste**, glass mat filled thermoplastics
GMT-Halbzeug glass mat filled thermoplastic material
GMT-UD *abbr. of* **unidirektional glasmattenverstärkte Thermoplaste** unidirectional glass mat reinforced thermoplastics
GMT-Verarbeitung processing of glass matt reinforced thermoplastics
goldbraun golden brown
GPC *abbr. of* **Gel-Permeations-Chromatographie**, gel permeation chromatography
GPF-Ruß general purpose furnace black
GPO *symbol for* propylene oxide rubber
gradkettig straight-chain
gradlinig straight-line *(curve)*
Grafik 1. graphics *(me)* **2.** diagram
Grafikarbeitsplatz graphics workstation *(me)*
Grafikbildschirm graphics display *(me)*

Grafikdatei graphics data file
Grafikdrucker graphics printer
grafikfähig with graphics capability *(me)*
Grafikfähigkeit graphics capability *(me)*
Grafikprozessor graphics processor *(me)*
Grafikroutine graphics routine/program *(me)*
Grafiksoftware graphics software *(me)*
Grafiksymbol graphic symbol
Grafiksystem graphics system *(me)*
Granulat 1. pellets, (polymer) granules. 2. moulding compound: *(The word granulate is hardly, if ever, used)*
Granulatbehälter feed/material hopper *(e,im)*
Granulatdosiergerät 1. granule metering unit *(if it actually measures out the material)*. 2. granule feed unit *(if no measuring out is involved)*
Granulatextrusion extrusion of granular compound *(as opposed to the extrusion of dry blends)*
Granulatform granular form
granulatförmig granular
Granulatgröße pellet/granule size
Granulatkörner pellets, granules
Granulatkühlvorrichtung pellet cooling unit
Granulator 1. granulator *(for cutting up scrap)*. 2. pelletiser *(for cutting up extruded strands to make pellets)*
Granulatteilchen polymer particles *(a free translation, but entirely acceptable) (sr)*
Granulattrichter (feed/material) hopper *(e,im)*
Granulattrocknungseinrichtung pellet drying unit
Granulatvorwärmer pellet/granule pre-heating unit
Granulieraggregat 1. granulator *(for cutting up scrap)* 2. pelletiser *(for cutting up extruded strands to make pellets) (sr)*
Granulieranlage 1. granulating line. 2. pelletising line *(sr)*: *(see explanatory note under* **granulieren***(*
Granulierdüse pelletising die *(sr)*
Granuliereinrichtung 1. granulator *(for cutting up scrap)*. 2. pelletiser *(for cutting up extruded strands to make pellets) (sr)*
granulieren 1. to granulate *(scrap)*. 2. to pelletise *(extruded strands) (sr)*
Granulierextruder reclaim extruder *(for compounding plastics scrap) (sr)*
Granulierflügel 1. granulator knife. 2. pelletiser knife *(sr)*
Granulierhaube cutting chamber hood *(sr)*
Granulierkopf pelletising die *(sr)*
Granulierlochplatte pelletising die *(sr)*
Granuliermaschine 1. granulator *(for cutting up scrap)*. 2. pelletiser *(for cutting up extruded strands to make pellets) (sr)*
Granuliermesserflügel 1. granulator knife. 2. pelletiser knife *(sr)*
Granulierplatte pelletising die *(sr)*
Granulierung 1. granulation *(of scrap)*. 2. pelletisation *(of extruded strands) (sr)*
Granuliervorrichtung 1. granulator *(for cutting up scrap)*. 2. pelletiser *(for cutting up extruded strands to make pellets) (sr)*
Granulierwasser water: *This is merely the water added to a pelletiser to carry off the pellets as they are produced, and should not be translated literally*
Granulierwerkzeug pelletising die *(sr)*
granulös granular
Graphik- see **Grafik-**
Graphit graphite
Graphitfaser graphite fibre
Graphitgewebe graphite fabric
Graphitierung graphitisation
Graphitpulver powdered graphite
Grasfangkorb grass box
Grat flash
gratfrei flash-free
Grathaut flash
Grauguß grey iron
Graumaßstab grey scale
grauschwarz charcoal grey
graviert engraved
gravimetrisch gravimetric
Gravitationsdosierung gravity feed
Greifer gripper
Greifereinheit gripper unit
Greifervorrichtung gripper arrangement
Gremium body *(of experts)*
Grenz- Certain words starting with **Grenz-** should normally be translated as limiting...; There are occasions, however, when it is equally correct to translate **Grenz-** as maximum, and the translator must decide from the context which is preferable.
Grenzbedingung boundary/limiting condition
Grenzbelastbarkeit critical/maximum load-bearing/-carrying capacity
Grenzbelastung critical/maximum load(ing)/stress
Grenzbiegespannung limiting flexural/bending stress
Grenzbruchdehnung limiting elongation at break
Grenzchubspannung limiting shear stress
Grenzdehnung limiting strain/elongation
Grenzdrehzahl 1. limiting speed. 2. limiting screw speed
Grenzdruck limiting pressure
Grenzfall borderline case
Grenzfläche 1. interface. 2. glueline
grenzflächenaktiv surface active
Grenzflächenebene 1. interfacial zone. 2. glueline
Grenzflächenenergie interfacial energy, surface tension
Grenzflächenkraft interfacial force
Grenzflächenspannung interfacial surface tension
Grenzfrequenz cut-off/limiting frequency
Grenzkontakt limit contact
Grenzkonzentration limiting concentration

Grenzkurve limiting curve
Grenzlast critical load/stress
Grenzlastspielzahl maximum number of stress/load cycles
Grenzmolekulargewicht limiting molecular weight
Grenzschicht boundary layer
Grenzspannung limiting stress
Grenztaster limit switch
Grenztemperatur limiting temperature
Grenzviskosität limiting viscosity number, intrinsic viscosity
Grenzwert limiting/threshold value, limit: **Sie strebt gegen einen maximalen Grenzwert** it tends towards a maximum value
Grenzwert, oberer upper limit
Grenzwert, unterer lower limit
Grenzwertgeber limit monitor
Grenzwertkontakt limit contact
Grenzwertmeldung limiting value signal
Grenzwertschalter limit switch
Grenzwertüberwachung limiting value control
Grenzzone boundary zone
Grenzzuhaltekraft limiting locking force *(im)*
Griff, trockener dry handle
griffest touch dry
Griffgefühl feel, handle
griffgünstig easily accessible
Grillgitter radiator grille
grob coarse
grob gemahlen coarsely ground
Grobbestandteile coarse particles
grobdispers coarse-particle
Grobeinstellung rough adjustment
Grobfraktion coarse fraction
Grobgewebe coarse fabric
Grobgrafik pixel graphics *(me)*
Grobgranulat coarse granules
Grobgut coarse material
grobkörnig coarse-particle
grobkristallin coarsely crystalline
grobmaschig coarse-mesh
grobteilig coarse-particle
Grobvakuum low vacuum
grobzellig coarse-cell
Großbauten large buildings
Großbehälter storage tank
Großbetrieb large company/concern/business
Großblasformanlage high-capacity blow moulding line/plant
Großblasformmaschine high-capacity blow moulding machine
Großcompoundieranlage high-capacity compounding line/plant
großdimensioniert large
große Produktionsserien long runs
Größenordnung order (of magnitude): **Die Durchmesser dieser Partikel liegen in der Größenordnung von 0.1 bis 1 μm** the particle size is of the order of 0.1-1 μm
Großextruder large extruder
Großflächenschmelzefilter large-area melt filter
großflächig large-area
großformatig large-size
Großformteil large moulding
Großgarage underground car park
Großhohlkörper large blow moulding
Großhohlkörperblasmaschine machine for making large blow mouldings
Großhohlkörperfertigung 1. production of large containers. 2. large-container production unit
Großmolekül macromolecule
Großpackung bulk container/pack
Großproduktion large-scale production
Großproduktionsmaschine large-scale production unit
Großraumschneidgranulator heavy-duty granulator *(sr)*
Großrechner mainframe computer *(me)*
Großrohr large-bore pipe
Großrohrkalibrierung 1. large-bore pipe calibration/sizing. 2. large-bore pipe calibration/sizing unit
Großrohrstraße 1. large-bore pipe extrusion line. 2. large-bore pipe production line *(general term)*
Großrohrwerkzeug large-bore pipe die *(e)*
Großschneidmühle heavy-duty granulator *(sr)*
Großserie large-scale production
Großserienanwendungen large-scale applications
Großserieneinsatz large-scale use
Großserienfertigung large-scale production
großserientauglich suitable for large-scale production
Großsilo large-capacity silo
Großsilolager large-capacity silo installation
großtechnisch large-scale, industrial-scale
Größtmaß maximum dimension
Größtwert maximum value
Großunternehmen large company/concern/business
Großverbraucher major/large-scale user
Großversuch production-scale trial, large-scale trial/test
Großvolumenschurre large-volume chute *(sr)*
großvolumig 1. bulky. 2. large-volume
Großwerkzeug large mould
großzügig dimensioniert generously dimensioned
Grund (screw) root surface *(e)*
Grundanforderung basic requirement
Grundanstrich primer coat
Grundausrüstung basic equipment
Grundausstattung basic equipment
Grundbaustein basic unit
Grundbauteile basic components
Grundbewegungen basic movements
Grunddaten basic data
Grundeigenschaften basic properties
Gründen, aus verfahrenstechnischen owing to the nature of the process
Grundentwicklung basic development
Grundfarbe primer

Grundflächenbedarf floor space requirements *(to install a machine)*
Grundforderung basic requirement
Grundgerät basic unit
Grundgerüst basic framework
Grundharz base resin
Grundieren application of primer
Grundierharz gelcoat resin
Grundierschicht primer coat
Grundierung primer
Grundierungsmittel primer
Grundierungspaste anchor coating paste
Grundkapital (share) capital, capital stock
Grundkennzeichen basic features
Grundkonstruktion basic design
Grundkonzept basic concept/idea
Grundkonzeption basic design
Grundlackierung primer coat
Grundlagenforschung fundamental/basic research
grundlegend basic, fundamental
Grundmaschine basic machine
Grundmaterial base material
Grundmerkmale basic features
Grundmolekül base molecule
Grundpaket basic package
Grundplatte base plate
Grundpolymer base polymer
Grundprinzip basic principle
Grundrahmen base frame
Grundregeln basic rules
Grundreihe basic range *(of machines, equipment etc.)*
Grundrezeptur basic formulation
Grundriß plan
Grundschnecke basic screw
Grundspiel flight clearance *(see explanatory note under* **Kopfspiel***, with which the word is synonymous)*
Grundstellung normal position
Grundstoff raw material
Grundstrich anchor coat *(for PVC paste)*
Grundsystem basic/standard system
Grundtyp 1. basic/standard grade *(of material).* 2. basic/standard model *(of machine)*
Grundüberlegungen basic considerations
Grundversion basic version
Grundvoraussetzung basic condition
Grundwasser ground water
Grundwasserschutz ground water protection
Grundwerkstoff base material
Grundzüge fundamentals
Grüner Punkt Green Dot *(this is a symbol printed on packs considered recoverable and recyclable through the* **Duales System***, q.v.)*
Grünling preform
Grünstandfestigkeit green strength
Gruppenterminal group terminal
GT *abbr. of* **Gewichtsteile**, parts by weight, p.b.w.
Gültigkeit validity
Gummi rubber *(see explanatory note under* **Kautschuk***)*
gummiähnlich rubbery, rubber-like
gummiartig rubbery, rubber-like
Gummibelag rubber covering
Gummidichtung rubber seal
Gummieigenschaften vulcanised rubber properties
gummielastisch rubbery-elastic
Gummielastizität rubber-like elasticity
gummiert 1. rubber covered *(roller etc.)* 2. rubberised
Gummifertigteil rubber article/component
Gummiformteil rubber moulding
Gummigegenwalze rubber-covered backing roll
gummihaltig rubber-based, containing rubber
Gummihandschuhe rubber gloves
Gummiindustrie rubber industry
Gummikleber rubber adhesive
Gummimehl powdered rubber
Gummipresseur rubber-covered back-up roll
Gummiprodukte rubber goods/products
Gummiproduzent rubber producer
Gummisack rubber bag
Gummisackverfahren rubber bag moulding
Gummischlauch rubber hose
Gummischneckenpresse rubber extruder
Gummispritzgießmaschine rubber injection moulding machine
Gummispritzgußteil injection moulded rubber article
Gummituchrakel knife-blanket coater, knife-over-blanket coater
gummiüberzogen rubber covered
Gummiüberzug rubber covering
Gummiverarbeiter rubber processor
Gummiwalze rubber covered (back-up) roll *(c)*
Gummiwaren rubber goods
GUS *abbr. of* **Gemeinschaft unabhängiger Staaten**, Commonwealth of Independent States
Gußasphalt poured asphalt
Gußform foundry mould
Gußloch pouring hole
Gußpolyamid cast polyamide/nylon
Gußteil casting
gut löslich readily soluble
Gutachten expert's report
Güte quality
Güteanforderungen quality requirements
Gütefaktor quality factor
Gütegrad quality
Gütekriterien quality criteria
Güterichtlinie quality guideline
Gütesicherung quality assurance
Güteüberwachung quality control
gutfließend free-flowing
Gutstoff usable material

H

Haarriß hairline crack
Haartrockner hair dryer
HAF-Ruß high abrasion furnace black
haftabweisend non-stick
Haftbrücke primer coat
Hafteigenschaften adhesive properties, adhesion
Haftelektrode contacting electrode
Haftetikett self-adhesive label
Haftfähigkeit adhesion
haftfest having good adhesion
Haftfestigkeit adhesive/bond strength, adhesion
Haftfestigkeitsmängel loss of adhesion, poor adhesion
Haftfläche adherend surface, surface to be bonded
Haft-Gleit-Effekt stick-slip effect
Haftgrund adherend surface
Haftgrundierung primer coat, tie coat
hafthindernd 1. non-stick, anti-adhesive. 2. interfering with adhesion
Haftklebeband adhesive tape
Haftklebeetikett self-adhesive label
Haftklebemittel pressure sensitive adhesive
Haftkleber pressure sensitive adhesive
Haftklebstoff pressure sensitive adhesive
Haftklebstoffdispersion pressure sensitive adhesive dispersion
Haftkraft adhesive force, adhesion
Haftmechanismus adhesion mechanism
Haftmittel 1. adhesive. 2. adhesion promoter
Haftprimer primer
Haftreibung static friction
Haftreibungskoeffizient coefficient of static friction
Haftreibungszahl coefficient of static friction
Haftschwierigkeiten adhesion problems
Haftstrich anchor coat *(for PVC paste)*
Haftsystem adhesive system
Haftung 1. adhesion. 2. responsibility, liability
Haftungsabweisung non-stick properties
Haftungseigenschaften adhesive properties, adhesion
Haftungseinbuße loss of adhesion
Haftungsfähigkeit adhesion, adhesive properties
Haftungsmängel poor/faulty adhesion
Haftungsmechanismus adhesion mechanism
haftungsmindernd adhesion-reducing
Haftungsprobleme adhesion problems
Haftungsschäden faulty/poor adhesion
Haftungsschwierigkeiten adhesion problems
Haftungssteigerung improved adhesion
haftungsverbessernd improving adhesion
Haftverbesserer coupling/bonding agent
Haftverlust loss of adhesion
haftvermittelnd coupling
Haftvermittler coupling/bonding agent
Haftvermittlung coupling

Haftvermögen adhesion
Haftwert adhesion: **hohe Haftwerte** excellent adhesion
Haftwirkung adhesive effect
Haftzugfestigkeit tensile bond strength
Haftzugversuch tensile bond test
Halbautomat semi-automatic machine
Halbautomatik semi-automatic system
halbautomatisch semi-automatic(ally)
Halbfabrikate semi-finished products
Halbfertigerzeugnisse semi-finished products
halbfest semi-solid
halbfett 50% oil length
halbflexibel semi-flexible
halbflüssig semi-liquid
halbhart 1. semi-rigid *(e.g. pipe, foam, sheet)*. 2. medium hard
Halbkreis semi-circle
halbkreisförmig semi-circular
Halbleiter semi-conductor
Halbleiteranordnung semi-conductor device/system
Halbleiterbauelement semi-conductor module
Halbleiterbaustein semi-conductor module
Halbleiterbauteil semi-conductor module
Halbleitereigenschaften semi-conductor properties/characteristics
Halbleiterleistungsschalter semi-conductor power switch
Halbleitermaterial semi-conductor material
Halbleiterschalter semi-conductor switch
Halbleiterspeicher semi-conductor memory *(me)*
Halbleitertechnik semi-conductor technology
halblogarithmisch semi-logarithmic
Halbmikromaßstab semi-microscale
halb-quantitativ semi-quantitative
Halbschale half-shell
halbschlagfest medium-impact
halbspröde semi-brittle
halbsteif semi-rigid
halbtechnisch semi-industrial: **halbtechnische Anlage** pilot plant; **halbtechnischer Maßstab** pilot plant scale
halbtransparent semi-transparent
halbverstärkend semi-reinforcing
Halbwertzeit half life period
Halbzeug 1. semi-finished products. 2. prepreg *(grp)*
Halbzeug, harzvorimprägniertes prepreg *(grp)*
Halbzeug, plattenförmiges sheets
Halbzeug, vorimprägniertes prepreg
Halbzeugprogramm range of semi-finished products
Halogen halogen
halogenfrei halogen-free
Halogengehalt halogen content
halogenhaltig halogen-containing, halogenated
Halogenhydrin halogen hydrin
halogeniert halogenated
Halogenkohlenwasserstoff halogenated hydrocarbon

Halogenradikal halogen radical
Halogenverbindung halogen compound
Halogenwasserstoff hydrogen halide
Halogenwasserstoffsäure halogen hydracid
Hals- und Bodenabfall neck and base flash *(bm)*
Halsabfälle neck flash *(bm)*
Halsbutzen neck flash *(bm)*
Halskalibrierung 1. neck calibration. 2. neck calibrating device *(bm)*
Halspartie neck section *(bm)*
Halsquetschkante neck pinch-off *(bm)*
Halsüberstände neck flash *(bm)*
Halswerkzeug neck mould *(bm)*
haltbar durable
Haltbarkeit durability
Haltebolzen retaining bolt
Haltedruck hold(ing)/dwell pressure *(im)*
Halterung mounting, support
Haltesteg spider leg *(e,bm)*
Hammermühle hammer mill *(sr)*
Hammerschlaglack hammer finish paint
Hand, aus einer from a single supplier, from one source *(i.e. several different units etc. all supplied by one and the same company)*
Handauflegen hand lay-up *(grp)*
Handauflegeverfahren hand lay-up (process) *(grp)*
handbedient hand/manually operated
Handbedienung manual operation
handbetätigt hand/manually operated
Handbetrieb manual operation
Handbetrieb, im manually
handbetrieben hand/manually operated
Handeinstellung manual setting/adjustment
Handelsbezeichnung trade name
Handelsmengen, in in commercial quantitites
Handelsname trade/brand name
Handelspartner trading partner
Handelsprodukt commercial product
Handelstyp commercial grade
handelsüblich commercial
handgefertigt hand-made
Handgetriebe manual drive
Handhabbarkeit ease of handling
Handhabung handling
Handhabungsautomat robot
handhabungsfreundlich easy to handle
Handhabungsgerät 1. handling device. 2. robot
Handhabungsrichtlinien handling instructions
handhabungssicher safe to handle
Handhabungssystem robotic system
Handlaminat hand lay-up laminate *(grp)*
Handlaminieren hand lay-up *(grp)*
handlaminiert made by hand lay-up *(grp)*
Handlaminierverfahren hand lay-up (process) *(grp)*
Handlaufprofil handrail profile
Handlingautomat robot
Handlingeigenschaften handling properties/characteristics
Handlinggerät 1. handling device/instrument. 2. robot
Handrad hand wheel
Handrührgerät hand mixer
Handschuhfach glove compartment
Handschuhkasten glove compartment
Handschuhkastenfach glove compartment
Handschweißgerät hand welding instrument
Handspritzen hand/manual spraying
Handstumpfschweißen manual butt welding
handtrocken touch-dry
Handventil manually operated valve
Handverfahren hand lay-up (process) *(grp)*
handwerklich manual *(e.g. referring to GRP fabrication techniques such as hand lay-up or spray lay-up. The versions of the word given in general dictionaries cannot be used here:)*
handwerklich gefertigt hand made
Hängeisolator suspended insulator
hantelförmig dumbbell-shaped
Hantelstab dumbbell-shaped test piece/specimen
Hantierungskapazität handling capacity
Haptik feel, handle *(see explanatory note under* **haptisch***)*
haptisch tactile. *The word haptic is not recommended since it is not common usage and few people - if any - would know what it meant*
Hardware hardware *(me)*
Hardwarebaustein hardware module *(me)*
Hardwareschnittstelle hardware interface *(me)*
harmlos, toxikologisch non-toxic
Harnstoff urea
Harnstoffabkömmling urea derivative
Harnstoffbindung urea linkage
Harnstoffbindungsanteil urea linkage content
Harnstoffbrücke urea bridge
Harnstoffderivat urea derivative
Harnstoff-Formaldehydharz urea-formaldehyde resin
Harnstoff-Formaldehydkondensat urea-formaldehyde condensate
Harnstoffharz urea resin, urea-formaldehyde resin
Harnstoffharzschaum urea-formaldehyde foam
Harnstoffharzschichtstoff urea laminate
Harnstoffpreßmasse urea moulding compound
hart eingestellt with a low plasticiser content *(PVC paste or compound)*
härtbar 1. thermosetting: **härtbare Harze** thermoset resins. 2. crosslinkable, curable, capable of being cured: **Das Harz ist drucklos härtbar** the resin can be cured without pressure
härtbar, nicht thermoplastic
Hartcompound unplasticised compound
Härte hardness
harte Einstellung low-plasticiser formulation *(PVC paste or compound)*
Härteabfall decrease in hardness
Härtegrad degree of hardness
hartelastisch energy-elastic

Hartelastizität energy elasticity
Härtemessung determination of hardness
härtend, reaktiv chemically curing
Härteprüfer hardness tester
Härteprüfgerät hardness tester
Härteprüfung hardness test
Härter hardener, catalyst
Härterei (steel) hardening shop
Härterkomponente hardener, catalyst
Härterpaste catalyst paste
Härterüberschuß excess hardener/catalyst
Härterunterschuß less hardener/catalyst: **bei einem 50-prozentigem Härterunterschuß** if there is 50% less hardener
Härteskala hardness scale
Härtetemperatur curing temperature
Hartextrusion extrusion of unplasticised PVC
Härtezeit curing time
Härtezyklus curing cycle
Hartfaserplatte hardboard
hartfließend with/having poor flow *(e.g. a moulding compound)*
Hartfolie rigid film
Hartfolienverarbeitung 1. fabrication of rigid film. 2. production of rigid film: **Emulsions-PVC für die Hartfolienverarbeitung** emulsion PVC for making rigid film
Hartgewebe fabric-based laminate
Hartgewebetafel fabric-based laminate sheet
Hartgranulat unplasticised compound
Hartgummi hard rubber, ebonite
Hartgußwalze case-hardened steel roll
Hartharz hard resin
Hartholzform hardwood mould *(t)*
Hartintegralschaumstoff rigid integral/structural foam
Hartkaolin non-plastic kaolin
hartkörnig hard-particle
hartmetallbestückt carbide tipped
Hartmetallwerkzeug carbide tipped tool
Hartpapier paper-based laminate
Hartpapierlaminat paper-based laminate
Hart-PVC rigid/unplasticised PVC, uPVC
Hart-PVC-Compound unplasticised PVC compound
Hart-PVC-Extrusion extrusion of unplasticised PVC
Hart-PVC-Folie rigid/unplasticised PVC film
Hart-PVC-Formmasse unplasticised/rigid PVC moulding compound
Hart-PVC-Formteil rigid PVC moulding
Hart-PVC-Platte rigid PVC sheet, uPVC sheet
Hart-PVC-Rohr rigid PVC pipe, uPVC pipe
Hart-PVC-Schaum rigid PVC foam
Hart-PVC-Spritzguß injection moulding of unplasticised PVC
Hart-PVC-Tafel rigid PVC sheet, uPVC sheet
Hartschaum rigid foam
Hartschaumisolierstoff rigid foam insulating material
Hartschaumkern rigid foam core
Hartschaumsandwichelement rigid foam composite element
Hartschaumstoff rigid foam
Hartschaumsystem rigid foam system
Hartsegment rigid segment
Hartsegmentgehalt rigid segment content
Hartspritzguß injection moulding of unplasticised PVC
Hartstahl hardened steel
Hartstahlmantel hardened steel (outer) layer
Härtung curing, crosslinkage
Härtungsablauf curing reaction
Härtungsbedingungen curing conditions
Härtungsbeschleuniger accelerator
Härtungsgeschwindigkeit curing speed/rate
Härtungsgrad degree of cure
Härtungskatalysator catalyst
Härtungsmechanismus curing mechanism
Härtungsmittel hardener, catalyst
Härtungsofen (curing) oven
Härtungsprogramm curing schedule
Härtungsreaktion curing reaction
Härtungsschrumpf curing shrinkage
Härtungsschwund curing shrinkage
Härtungsstufe curing stage
Härtungssystem curing system
Härtungstemperatur cure/curing temperature
Härtungsverhalten curing behaviour/characteristics
Härtungsverlauf curing pattern
Härtungszeit cure/curing time
Harturethanschaum rigid polyurethane foam
Hartverarbeitung processing without plasticiser, processing in unplasticised form *(PVC)*
Hartverbund rigid composite
Hartverchromen hard chroming
Harz resin
Harzansatz resin-catalyst mix
Harzanteil resin content
harzarm low-resin, with a low resin content
Harzaufnahme resin pick-up/take-up
Harzauftragsmenge amount of resin applied
harzbildend resin-forming
Harzdispersion resin dispersion
Harzdomäne resin domain
Harzfeinschicht gel coat *(grp)*
Harzfluß resin flow
Harzformstoff cured resin
harzgebunden resin bonded
Harzgehalt resin content
harzgetränkt resin impregnated
Harz-Härtermischung resin-catalyst mix
harzimprägniert resin impregnated
Harzkonzentration resin concentration
Harzlösung resin solution
Harzmatrix resin matrix
Harzmatte prepreg, sheet moulding compound, SMC *(grp)*
Harzmattenzuschnitte cut-to-size prepregs
Harznester resin accumulations
harzreich resin-rich, with a high resin content
Harzsäure rosin acid

Harzseife rosin soap
Harzträger filler
Harzüberschuß excess resin
Harzüberschußbehälter resin overflow tank
harzvorimprägniertes Halbzeug prepreg
Harzzufuhrleitung resin feed line
Harzzusatz addition of resin **bei weiteren Harzzusätzen...** if more resin is added...
Haube hood
häufigste Korngröße predominant particle size
Haufwerkschneider granulator *(sr)*
Hauptanschlußkabel main connecting cable
Hauptantrieb main drive (unit)
Hauptantriebswelle main drive shaft
Hauptanwendungsgebiet main field/area of use
Hauptaufgabe main task/function
Hauptbelastungsrichtung direction of main/principal stress
Hauptbewegungsrichtung main direction of movement
Hauptdispersionsgebiet primary glass transition range
Haupteinsatzgebiet main field/area of use
Hauptextruder main/principal extruder
Hauptforderung main/principal requirement
Hauptformulierungsbestandteile main/principal formulation components
Hauptkette main chain
Hauptkettenflüssigkristall linear liquid crystal
Hauptketten-LCP linear liquid crystal polymer
Hauptkettenpolymer linear polymer
Hauptlichtquelle main light source
Hauptmenü main menu *(me)*
Hauptmerkmale principal characteristics/features: **mechanische Hauptmerkmale** principal mechanical properties
Hauptmotor main motor
Hauptorientierungsrichtung main direction of orientation
Hauptschalter main switch
Hauptschnecke main screw
Hauptspannungsrichtung direction of greatest stress
Hauptspeicher main memory *(me)*
Hauptspindel main screw
Hauptsteuerwarte main control room
Hauptvalenz primary valency
Hauptvalenzbindung primary valency bond
Hauptversammlung general meeting *(of shareholders etc.)*
Hauptverteilerkanal main runner *(im)*
Hauptverteilersteg main runner *(im)*
Haus 1. firm, company: **von unserem Haus entwickelt** developed by us/ourselves, developed by our company. 2. house
Hausabflußrohr domestic waste pipe
hauseigen in-house
Hausgerät domestic/household appliance
Hausgeräteindustrie domestic/household appliance industry
Haushaltsabfall domestic/household refuse/rubbish
Haushaltsartikel household goods
Haushaltsgeräte household/domestic appliances
Haushaltsgerätelack household/domestic appliance paint/enamel
Haushaltsgeschäft hardware shop/store
Haushaltsmüll domestic/household refuse/rubbish
Haushaltswaren household goods
hausintern in-house
Hausmüll domestic/household refuse/rubbish
Hausmüllaufkommen creation/production of household refuse/rubbish
Hausmüllsammlung domestic/household refuse/rubbish collection
Hausmüllverbrennungsanlage domestic/household refuse/rubbish incinerator
Hausnorm internal standard (specification)
Hautbildung skinning
Hautbildungsresistenz skinning resistance
Hautkontakt skin contact
Hautpflegemittel skincare product
Hautreizungen skin irritations
Hautschutzcreme barrier cream
Hautschutzsalbe barrier cream
Hautverhinderungsmittel anti-skinning agent
Hautverhütungsmittel anti-skinning agent
Hautverträglichkeit skin tolerability *(see explanatory note under* **Verträglichkeit***)*
Hazenfarbskala Hazen colour scale
Hazenfarbzahl Hazen colour number/value
H-Brücke hydrogen bridge
HD *abbr. of* **Hochdruck,** high pressure
HDT-Wert heat distortion temperature
Hebeeinrichtung lifting device
Hebegeschwindigkeit lifting speed
Hebehub lifting stroke
Hebezeug lifting gear/tackle
Heckklappe rear hatch, tailgate, boot lid
Heckscheibe rear window
Heckschürze rear apron
Heckspoiler rear spoiler
Heckstoßfänger rear bumper
Heckteil rear end
Hecktür rear door
Heckverkleidung rear panel *(of a car)*
Heftpflaster sticking plaster
Heimwerker do-it-yourself worker
Heißabfüllung hot-filling *(of foodstuffs)*
Heißabschlag- die-face, hot-cut *(sr): This prefix is used in connection with pelletisers where strands are extruded and cut into pellets at the die face, whilst still hot*
Heißabschlageinrichtung die-face/hot-cut pelletiser *(sr)*
Heißabschlaggranulator die-face/hot-cut pelletiser *(sr)*
Heißabschlaggranuliereinrichtung die-face/hot-cut pelletiser *(sr)*
Heißabschlaggranulierung 1. die-face/hot-cut pelletisation. 2. die-face/hot-cut pelletiser *(sr)*

Heißabschlagvorrichtung die-face/hot-cut pelletiser *(sr)*
Heißabschlagwassergranulierung 1. hot-cut water-cooled pelletisation. 2. hot-cut water-cooled pelletiser
heißaushärtend heat-curing
Heißdampf superheated steam
Heißeinreißfestigkeit hot tear strength/resistance
Heißformschaumstoff hot-cured moulded foam
Heißgasschweißen hot air/gas welding
heißgewalzt hot-rolled
Heißgranulat hot-cut pellets
Heißgranuliermaschine 1. hot melt granulator. 2. die-face/hot-cut pelletiser *(sr)*: *(see explanatory note under* **Granulator**)
Heißgranulierung 1. hot melt granulation. 2. die-face/hot-cut pelletisation *(sr)*: *(see explanatory note under* **granulieren**)
Heißgranuliervorrichtung 1. hot melt granulator. 2. die-face/hot-cut pelletiser *(sr)*: *(see explanatory note under* **granulieren**)
heißhärtend heat curing
Heißhärter hot curing catalyst
Heißhärtung heat/hot curing
Heißkanal hot runner *(im)*
Heißkanalanguß hot runner feed system *(im)*
Heißkanalangußsystem hot runner feed system *(im)*
Heißkanalbauarten types of hot runner *(im)*: **Beispiele von Heißkanalbauarten** examples of hot runner systems
Heißkanalbeispiele examples of hot runner systems
Heißkanalblock hot runner unit, hot runner manifold block *(im)*
Heißkanalblockausführung 1. hot runner unit design. 2. hot runner unit construction *(im)*
Heißkanalbohrung hot runner *(im)*
Heißkanaldoppelwerkzeug hot runner two-cavity/-impression mould *(im)*
Heißkanaldüse hot runner nozzle *(im)*
Heißkanalelement hot runner unit *(im)*
Heißkanaletagenwerkzeug hot runner multi-daylight mould, hot runner stack mould *(im)*
Heißkanalform hot runner mould *(im)*
Heißkanalformenbau hot runner tooling *(im)*
Heißkanalinnendruck pressure inside the hot runner *(im)*
Heißkanalkonstruktion hot runner design *(im)*
Heißkanalmehrfachanguß hot runner multiple gating *(im)*
Heißkanalnadelventil hot runner needle valve *(im)*
Heißkanalnadelverschlußsystem hot runner needle shut-off mechanism *(im)*
Heißkanalnormalien standard hot runner mould units *(im)*
Heißkanalnormblock standard hot runner unit *(im)*
Heißkanalplatte hot runner manifold/plate *(im)*
Heißkanalregelgerät hot runner control unit *(im)*

Heißkanalrohrsystem hot runner system *(im)*
Heißkanalspritzgießwerkzeug hot runner injection mould *(im)*
Heißkanalsystem hot runner system *(im)*
Heißkanalverteiler hot runner *(im)*
Heißkanalverteileranguß hot runner feed system *(im)*
Heißkanalverteilerbalken hot runner unit, hot runner manifold block *(im)*
Heißkanalverteilerblock hot runner unit, hot runner manifold block *(im)*
Heißkanalverteilerplatte hot runner manifold/plate *(im)*
Heißkanalverteilersystem hot runner system *(im)*
Heißkanalwerkzeug hot runner mould *(im)*
Heißkaschierkleber hot laminating adhesive
Heißkaschierung heat lamination
Heißlauf- *see* **Heißkanal-**
Heißläufer hot runner *(im)*
Heißläuferbalken hot runner manifold bar *(im)*
Heißluft hot air
Heißluftalterung hot air ageing
Heißluftbeständigkeit hot air resistance
Heißluftdüsenkanal hot air tunnel
Heißluftkanal hot air tunnel
Heißluftlagerung hot air ageing
Heißluftofen hot air oven
Heißluftschrank hot air oven
Heißlufttrocknung 1. hot air drying 2. hot air drying unit
Heißmischung 1. dry blend. 2. hot mixing
Heißmischwerk hot mixing rolls
Heißprägen hot embossing
Heißprägeverfahren hot embossing (process)
Heißpressen hot press moulding, matched metal press moulding *(grp)*
Heißpreßverfahren hot press moulding, matched metal press moulding *(grp)*
Heißschaumstoff hot-cured foam
Heißschmelzextruder melt extruder, hot melt extruder, melt fed extruder
Heißschmelzkleber hot melt adhesive
Heißschmelzklebstoff hot melt adhesive
Heißschmelzmasse hot melt compound
Heißschmelzverklebung hot melt bonding
Heißsiegelautomat automatic heat sealing machine, automatic heat sealer
Heißsiegelbacken heat sealing bars
heißsiegelbar heat sealable
Heißsiegelbarkeit heat sealability
Heißsiegelbeschichtung heat sealable coating
heißsiegelfähig heat sealable
Heißsiegelgerät heat sealing instrument
Heißsiegellack heat sealing lacquer
Heißsiegelmasse heat sealing compound
Heißsiegelschicht heat sealable coating
Heißverklebung heat bonding
Heißverpressung hot pressing
Heißversiegelung heat sealing
heißvulkanisierend high temperature vulcanising/curing

Heißwasser hot water
Heißwasserbelastung 1. exposure to hot water. 2. immersion in hot water
heißwasserbeständig hot water resistant
Heißwasserheizanlage hot water heating unit
Heißwassertank hot water tank
Heißwasserumlaufheizung 1. circulating hot water heating. 2. circulating hot water heating system
Heißzerkleinern hot melt granulation *(sr)*
Heizbad heating bath
Heizball moulding bladder *(used in tyre manufacture)*
Heizband band heater, heater band
heizbar heatable
Heizblock hot runner unit, hot runner manifold block *(im)*
Heizbohrung heating channel
Heizelement 1. heating element. 2. heated tool/plate, hot plate *(w)*
Heizelementrohrschweißanlage heated tool pipe welding line
Heizelementschweißen heated tool welding
heizelementstumpfgeschweißt hot plate butt welded
Heizelementstumpfschweißen hot plate butt welding
Heizelementstumpfschweißverbindung hot plate butt weld(ed) joint
Heizenergie heating energy
Heizfläche heating surface/area
Heizkanal 1. heating channel *(im)*. 2. heating tunnel
Heizkeil heated tool
Heizkeilschweißen heated tool welding
Heizkeller boiler room
Heizkessel boiler
Heizkörper heater
Heizkörper, patronenförmiger cartridge heater
Heizkörperlack radiator paint
Heizkosten heating costs
Heizkreis heating circuit
Heizkreislauf heating circuit
Heiz-Kühleinrichtung heating-cooling unit
Heiz-Kühlmanschette heating-cooling collar/sleeve
Heiz-Kühlmischer heating-cooling mixer
Heiz-Kühlsystem heating-cooling system
Heiz-Kühlzone heating-cooling section/zone
Heizleistung heating unit consumption, heating capacity
Heizleistung, installierte installed/connected heating capacity
Heizleistung, zugeführte heat input
Heizleistungsaufnahme heat input
Heizmanschette heating sleeve/collar
Heizmantel heating jacket
Heizmedium heating medium
Heizmischer heating mixer
Heizöl fuel oil
Heizölauffangwanne fuel oil drip pan
heizölbeständig fuel oil resistant

Heizöllagertank fuel oil (storage) tank
Heizöltank fuel oil (storage) tank
Heizöltankblasmaschine machine for blow moulding fuel oil storage tanks
Heizöltankmaschine machine for blow moulding fuel oil storage tanks
Heizpatrone cartridge heater
Heizplatte 1. hotplate. 2. heating platen *(of press)*
Heizrohr heat pipe
Heizschlange heating coil
Heizschlauch moulding/curing bag *(used in tyre manufacture)*
Heizschrank 1. oven *(general term)*. 2. fusion oven *(as used in PVC paste processing)*
Heizspannung heater voltage
Heizspiegelschweißen hot plate welding
Heizspule heating coil
Heizstation heating station
Heizstrahler radiant heater
Heizstrecke 1. heating tunnel. 2. heating section
Heizstromkreis heating circuit
Heiztunnel heating tunnel
Heiz-und Kühlaggregat heating-cooling unit
Heizungsart type of heating
Heizungssteuerung heater controls
Heizwendel heating coil
Heizwert calorific value
heizwertarm with/having a low calorific value
heizwertreich with/having a high calorific value
Heizzone heating zone/section
Heizzylinder heating cylinder *(im)*
Helixströmung helical flow
hellfarbig light coloured
hellgrau light grey
Helligkeit brightness
Helligkeitsgrad brightness
Helmvisier safety helmet visor
Hemmschuh obstacle
Heptan heptane
herabgesetzt reduced
Herausforderung challenge
herausgearbeitet machined, cut *(e.g. test piece from press moulded sheet)*
herausnehmbar removable
hergestellt, elektroerosiv spark eroded
hergestellt, galvanisch made by electrodeposition
herkömmlich conventional
Herstell- see **Herstellungs-**
Hersteller manufacturer, producer
Herstellerangaben information supplied by the manufacturer
Herstellung 1. manufacture *(of finished products)*. 2. production *(of finished or semi-finished products)*: **Herstellung von Blaskörpern** manufacture/production of blow mouldings; **Herstellung von Vorformlingen** production of parisons
Herstellungprogramm 1. product range, range of products. 2. production schedule/plan/programme

Herstellungsabschnitt production stage
Herstellungsbedingungen production conditions
Herstellungsdatum manufacturing/production date
Herstellungshilfsmittel processing aid
Herstellungshinweise production guidelines
Herstellungskosten manufacturing/production costs
Herstellungspalette product range, range of products
Herstellungsparameter 1. production variable. 2. production conditions
herstellungstechnisch *relating to manufacture or production*: **herstellungstechnische Vorteile** production advantages, advantages from the production point of view
Herstellungstoleranz production tolerance
Herstellungsverfahren 1. manufacturing process. 2. production process *(see explanatory note under* **Herstellung***)*
Herstellungszeiten production times: **Die Herstellungszeiten der Werkzeuge können also erheblich verkürzt werden** moulds can be made much more quickly
hervorragend outstanding
Herz central/main part, heart; *(for translation example see under* **Herzstück***)*
Herzkurve heart-shaped groove *(bm,bfe)*
Herzkurvendorn mandrel with a heart-shaped groove *(bm,bfe)*
Herzkurveneinspeisung heart-shaped groove type of feed system *(bm,bfe)*
Herzkurvenkanal heart-shaped groove *(bm,bfe)*
Herzstück central/main part, heart: **Das Herzstück des Laminators ist die Kühlwalze** the most important part of the laminating line is the cooling roll
Heteroatom heteroatom
heterocyklisch heterocyclic
heterogen heterogeneous
Heterogenität heterogeneity
Heteropolymer heteropolymer
Hetsäure HET acid *(full chemical name: hexachloroendomethylene tetrahydrophthalic acid)*
Hexa *abbr. of* **Hexamethylentetramin**, hexamethylene tetramine
Hexachlor-p-xylol hexachloro-p-xylene
Hexadekanol hexadecanol
Hexafluorpropylen hexafluoropropylene
Hexahydrophthalsäure hexahydrophthalic acid
Hexahydrophthalsäureanhydrid hexahydrophthalic anhydride
Hexahydrophthalsäurediglycidylester diglycidyl hexahydrophthalate
Hexamethoxymethylmelamin hexamethoxymethyl melamine
Hexamethyldisilazan hexamethyl disilazane
Hexamethylenbrücke hexamethylene bridge
Hexamethylendiamin hexamethylene diamine

Hexamethylendiisocyanat hexamethylene diisocyanate
Hexamethylentetramin hexamethylene tetramine
Hexan hexane
Hexandioldiacrylat hexanediol diacrylate
Hexandioldiglycidylether hexanediol diglycidyl ether
hexanisch in hexane: **hexanische Polymerlösung** a solution of the polymer in hexane
Hexaphenylethan hexaphenyl ethane
HF *abbr. of* **Hochfrequenz**, high frequency
HFA *abbr. of* **Hydrofluoralkan**, hydrofluoroalkane
HFCKW *abbr. of* **Hydrofluorchlorkohlenwasserstoff**, hydrochlorofluorocarbon, HCFC
HF-Druckvorbehandlungsgerät instrument for the HF pretreatment of surfaces prior to printing
HFKW *abbr. of* **Hydrofluorkohlenwasserstoff**, hydrofluorocarbon, HFC
HF-Schaltung HF/high-frequency circuit
HF-Schweißanlage HF welding line
Hg-Säule mercury column
HHG *abbr. of* **Handhabungsgerät**, robot
HHPSA *abbr. of* **Hexahydrophthalsäureanhydrid**, hexahydrophthalic anhydride
hierarchisch hierarchal *(me)*
High-Solid-Lack high solids paint
Hilfestellung, verfahrenstechnische technical assistance
Hilfsabziehwerk auxiliary take-off (unit)
Hilfsaggregat ancillary unit
Hilfsantrieb auxiliary drive
Hilfseinrichtungen ancillary equipment
Hilfskern auxiliary core
Hilfslösemittel auxiliary solvent
Hilfsmenü auxiliary menu *(me)*
Hilfsmotor auxiliary motor
Hilfsschnecke ancillary/subsidiary screw
Hilfsstoff 1. auxiliary. 2. additive
Hilfsvorrichtungen ancillary equipment
Hilfswerkzeuge tooling aids
Hilfszylinder auxiliary cylinder
Himmel (car) roof liner
Hinderung, sterische steric hindrance
Hinterachse rear axle
hintereinandergeschaltet 1. arranged in tandem *(if there are only two units)*. 2. arranged in series *(if there are more than two units, e.g. a number of cooling units)*
Hintereinanderschaltung tandem arrangement
Hinterfüllung back filling
Hinterfüttern back filling
Hinterfütterungsmasse backing mix
Hintergrundprogramm background program *(me)*
Hinterschäumanlage foam backing plant
hinterschäumt foam backed

Hinterschliffwinkel clearance/relief angle
Hinterschneidungen undercuts
Hinterschnitte undercuts
hinterschnitten undercut
Hinterseite rear, back *(e.g. of a machine)*
Hinweise, sicherheitstechnische safety instructions
Hinweisleuchte warning light
Hinweistafel sign, notice
hinzufügen to add
Hitzealterung heat ageing
hitzebeansprucht subjected to high temperatures
hitzebelastbar heat resistant
Hitzebelastbarkeit heat resistance
Hitzebelastung exposure to heat, exposure to high temperatures
hitzebeständig heat resistant
Hitzebeständigkeit heat resistance
hitzeempfindlich heat sensitive, affected by heat
hitzefest heat resistant
hitzehärtbar heat/hot curing, heat setting
hitzehärtend heat/hot curing, heat setting
Hitzehärtung heat/hot curing, heat setting
Hitzelagerung heat ageing
Hitzereaktivierung heat activation *(of dry adhesive film)*
Hitzeresistenz heat resistance
Hitzeschild heat shield
Hitzeschockfestigkeit heat shock resistance
Hitzeschutzmittel heat stabiliser
hitzestabil thermally stable
Hitzestabilisator heat stabiliser
Hitzestabilisierung heat stabilisation
hitzestabilisiert heat stabilised
Hitzestabilität thermal stability
Hitzestau heat accumulation
hitzesterilisierbar heat sterilisable
hitzesterilisiert heat sterilised
hitzeunempfindlich unaffected by heat
hitzevernetzbar heat/hot curing
Hitzevulkanisation high temperature vulcanisation
HK 1. *abbr. of* **Heißkanal**, hot. 2. *abbr. of* **Herstellungskosten**, production/manufacturing costs
HKS *abbr. of* **Heißkanalsystem**, hot runner system
HKW *abbr. of* **Halogenkohlenwasserstoff**, halogenated hydrocarbon
HL-Schaumstoff high-load bearing foam
HM-Anlage HM-HDPE blown film plant
HM-Faser high modulus fibre
HM-Folie high molecular weight polyethylene film, paper-like polyethylene film
HMF-Ruß high modulus furnace black
HMMM-Harz hexamethoxymethyl melamine resin
HMS-Faser high modulus-strength fibre
Hobbock drum
Hobbybereich leisure sector

Hobeln planing
hochabriebfest highly abrasion resistant
hochabsorbierend highly absorbent
hochaggressiv highly corrosive/aggressive
hochaktiv highly reactive, high-reactivity
hochangereichert highly enriched
hocharomatisch highly aromatic
hochauflösend high-resolution *(me)*
Hochbau building *(as opposed to* **Tiefbau**, *q.v.)*
Hochbauten high rise buildings
hochbeanspruchbar heavy duty;
hochbeanspruchbare Thermoplaste engineering thermoplastics
hochbeansprucht highly stressed
hochbelastet highly stressed: **thermisch hochbelastet** subjected to high temperatures
hochbutadienhaltig with a high butadiene content
hochcarboxyliert highly carboxylated
hochchemikalienbeständig having excellent chemical resistance
hochdehnfähig highly elastic/extensible
hochdicht 1. high-density 2. very dense 3. very tight/watertight
hochdispers fine-particle
Hochdruck high pressure
Hochdruckdosieranlage high pressure metering/feed unit
Hochdruckflüssigkeitschromatographie high pressure liquid chromatography
Hochdruckgebläse high pressure blower
Hochdruckinjektion high pressure injection
Hochdruckinjektionsgerät high pressure injector
Hochdrucklaminat high pressure laminate
Hochdruckmischanlage high pressure mixing unit
Hochdruckplastifiziereinheit high pressure plasticising/plasticating unit
Hochdruckplastifizierung 1. high pressure plasticisation/plastication. 2. high pressure plasticising/plasticating unit, high pressure plasticator
Hochdruckpolyethylen low density polyethylene, LDPE
Hochdruckpolymerisation high pressure polymerisation
Hochdruckpumpe high pressure pump
Hochdruckpumpenaggregat high pressure pump unit
Hochdruckreaktor high pressure reactor
Hochdruckschlauch high pressure hose
Hochdruckspritzanlage high pressure spraying plant/unit
Hochdruckventil high pressure valve
Hochdruckverfahren high pressure process
Hochdruckwasserstrahl high pressure water jet
hochelastisch highly flexible
hochenergetisch high-energy
hochfest high-strength
Hochfestfaserverbund high strength fibre composite

hochfluorhaltig with a high fluorine content
hochfrequent high-frequency, HF
Hochfrequenzabschirmung high-frequency screening
Hochfrequenzfeld high frequency field
Hochfrequenzgenerator high frequency generator
Hochfrequenzisolation HF/high-frequency insulation
Hochfrequenzpotential high frequency potential
Hochfrequenzprägen high frequency embossing
Hochfrequenzschweißen high frequency welding
Hochfrequenzschweißmaschine high frequency welding machine
Hochfrequenzschweißverfahren high frequency welding (process)
Hochfrequenztechnik high frequency engineering
hochfrequenzverschweißbar capable of being high frequency welded **Die Folie ist hochfrequenzverschweißbar** the film can be high frequency welded
Hochfrequenzvibrator high frequency vibrator
hochfüllstoffhaltig highly filled
hochgasbeladen containing large amounts of gas
hochgefüllt highly filled
hochgenau high-precision
hochgeschmiert highly lubricated
Hochgeschwindigkeitsanlage high speed plant, high speed production line
Hochgeschwindigkeitsbeschichtung high speed spreading
Hochgeschwindigkeitsdispergierapparat high speed mixer
Hochgeschwindigkeitsextrusion high speed extrusion
Hochgeschwindigkeitskalibriersystem high speed calibrating/sizing system
Hochgeschwindigkeitsschneider 1. high speed cutter *(general term)*. 2. high speed slitter *(especially for film)*
Hochgeschwindigkeitsversuch high speed test
hochgiftig highly toxic
hochglänzend high gloss
Hochglanzfarbe gloss paint
Hochglanzlackierung high gloss finish
hochglanzpoliert highly polished
hochglanzverchromt high polish chromium plated
Hochglanzwalze highly polished roll
hochgleitfähig having very good surface slip, with very good surface slip
hochhitzebeständig high temperature resistant
hochintegriert highly integrated
hochklappbar capable of being tilted/swivelled upwards
hochkondensiert highly condensed
Hochkonjunktur economic boom
hochkonzentriert highly concentrated

hochkristallin highly crystalline
Hochlauf acceleration
Hochlauf, beim when running up to speed
hochlegiert high-alloy
Hochleistungs- 1. high capacity/performance. 2. high speed, fast cycling. 3. heavy duty: *Which of these alternatives is used will depend on the type of machine or process involved. What is correct in one case may not necessarily be so in another*
Hochleistungsanlage high capacity/performance plant
Hochleistungsblasextrusionsanlage high capacity/performance extrusion blow moulding plant
Hochleistungsblasfolienanlage high capacity/performance blown film plant
Hochleistungsblasformanlage high capacity/performance blow moulding line
Hochleistungsblasformautomat automatic high capacity/performance blow moulding machine
Hochleistungsdoppelschneckenextruder high speed twin screw extruder
Hochleistungeinschneckenextruder high speed single screw extruder
Hochleistungsextruder high speed extruder
Hochleistungsextrusion high speed extrusion
Hochleistungsextrusionsanlage high speed extrusion line
hochleistungsfähig heavy duty, high performance
Hochleistungsfaser high strength fibre
Hochleistungs-Faserverbundwerkstoff high performance fibre composite (material)
Hochleistungsform heavy duty mould
Hochleistungsgetriebe heavy duty gears
Hochleistungsgranulattrockner high performance pellet drier
Hochleistungsheizkörper heavy duty heater
Hochleistungsheizpatrone heavy duty cartridge heater
Hochleistungsinnenmischer heavy duty internal mixer
Hochleistungskabelummantelungsanlage high speed cable sheathing plant
Hochleistungskatalysator high performance catalyst
Hochleistungskneter heavy duty kneader/mixer
Hochleistungskunststoff engineering polymer, high performance plastic
Hochleistungsmischer heavy duty mixer
Hochleistungsplastifizieraggregat heavy duty plasticator
Hochleistungsrohrextrusionslinie high speed pipe extrusion line
Hochleistungsschlauchfolienanlage high capacity/performance blown film line
Hochleistungsschnecke heavy duty screw
Hochleistungsschneide heavy duty blade *(sr)*
Hochleistungsschneidmühle heavy duty granulator *(sr)*

Hochleistungsschnellmischer heavy duty high speed mixer
Hochleistungsseil high strength rope
Hochleistungsspritzeinheit high speed injection unit
Hochleistungsspritzgießen high speed injection moulding
Hochleistungsspritzgießmaschine fast cycling injection moulding machine
Hochleistungsthermoplast engineering thermoplastic
Hochleistungsverbundstoff advanced composite, high performance composite, high strength composite
Hochleistungsverpackungsmaschine high speed packaging machine: *This expression requires care in translation since it has also been encountered in connection with machines for making packaging containers (***Verpackungen***) in which case the word should be translated as* fast cycling machine for making packaging containers. *This, however, is a most unusual interpretation.*
Hochleistungswerkstoff high performance material
Hochleistungswerkzeug heavy duty mould
hochleitfähig highly conductive
Hochmodulfaser high modulus fibre
hochmodulig high modulus
hochmolekular high-molecular weight
hochorientiert highly oriented
hochpigmentiert highly pigmented
hochpolar highly polar
hochpoliert highly polished
hochpolymer high-polymeric
Hochpolymer high polymer
hochpräzis high-precision
hochqualitativ high quality
hochreaktiv highly reactive
hochreflektierend highly reflective
Hochregallager store/warehouse equipped with floor-to-ceiling shelves
hochrein very pure
hochrüsten to upgrade
hochschlagbeansprucht subjected to high impact (stress)
hochschlagfest high impact
hochschlagzäh high impact
hochschmelzend high-melting
Hochschule university
hochsensibel highly sensitive *(see explanatory note under* **sensibel***)*
hochshorig with a high Shore hardness
hochsiedend high-boiling
Hochsieder high-boiling solvent
Hochspannung high voltage/tension
Hochspannungsbelastung exposure to high voltage
Hochspannungsbereich high voltage range
Hochspannungsisolation high voltage insulation
Hochspannungsisolator high tension insulator

Hochspannungskabel high tension cable
Hochspannungskriechstromfestigkeit high voltage tracking resistance
Hochspannungslichtbogenfestigkeit high voltage arc resistance
Hochspannungsprüffeld high voltage test bed
Hochspannungsprüfung high voltage test
Hochspannungsschalter high voltage switch
Hochspannungstechnik high voltage/tension engineering
hochstabil extremely stable
höchstabriebfest having maximum abrasion resistance
Höchstdruck maximum pressure
Höchstdruckplastifizierung 1. ultra-high pressure plasticisation/plastication. 2. ultra-high pressure plasticator
Höchstdurchsätze maximum throughputs
Höchstkraft maximum force
höchstmolekular ultra-high molecular weight
hochstoßfest high impact
Hochstrukturruß high-structure carbon black
höchstschlagzäh ultra-high impact
Höchsttemperatur maximum temperature
Höchstwert maximum value
höchstzulässig maximum permissible
hochsubstituiert highly substituted
hochtechnologisch high-tech
Hochtemperaturanwendungen high temperature applications
hochtemperaturbelastbar high temperature resistant
Hochtemperaturbelastbarkeit high temperature resistance
Hochtemperaturbereich, im at high temperatures
hochtemperaturbeständig high temperature resistant
Hochtemperaturbeständigkeit high temperature resistance
Hochtemperatureigenschaften high temperature properties
Hochtemperaturen high temperatures
Hochtemperaturkabel high temperature cable
Hochtemperaturpackung high temperature packing
Hochtemperaturthermoplast high temperature thermoplastic
Hochtemperaturthermostat high temperature thermostat
Hochtemperaturverarbeitung high temperature processing
Hochtemperaturverfahren 1. high temperature process. 2. high temperature calendering (process)
Hochtemperaturverhalten high temperature behaviour/performance
Hochtemperaturversuch high temperature test
hochtourig high speed
hochtransparent completely transparent
hochungesättigt highly unsaturated
Hochvakuum high vacuum

Hochvakuumbedampfen high vacuum metallisation
Hochvakuumfett high vacuum grease
Hochvakuumkammer high vacuum chamber
Hochvakuummetallisierung high vacuum metallisation
hochverdichtet tightly compressed
hochverdünnt highly diluted
hochvernetzt highly crosslinked
hochverschleißfest extremely wear resistant, extremely hard wearing
hochverschleißwiderstandsfähig extremely wear resistant, extremely hard wearing
hochviskos high-viscosity
hochwärmebelastet subjected to high temperatures
hochwärmebeständig high temperature resistant
hochwärmeformbeständig high temperature resistant
Hochwärmeformstabiliserung 1. high temperature stabilisation. 2. high temperature stabiliser
hochwärmestabilisiert high temperature stabilised
hochwertig 1. high-quality/-grade. 2. high-valency
hochwertig, qualitativ high-quality
hochwetterecht highly weather resistant
hochwirksam highly effective
hochzäh very tough
Hochziehen lifting *(of paint film)*
hochzugfest high-tensile
höhenverstellbar vertically adjustable, adjustable in height, ...whose height can be adjusted
Höhenverstellung vertical adjustment
höhermolekular high-molecular weight
Hohlfaden hollow filament
Hohlform mould
Hohlformwerkzeug mould
hohlgebohrt hollow-bored
Hohlkammerplatte twin-wall sheet
Hohlkammerprofil window profile
Hohlkörper 1. blow moulding, blow moulded part. 2. blown container, bottle. 3. hollow article *(e.g. a PVC slush moulded part): The literal translation of this word, often encountered in translations, should never be used. Not only is* hollow body *ugly and bad English, but it is never used in this particular context by native English and American speakers*
Hohlkörper, blasgeformter 1. blow moulding, blow moulded part. 2. blown container, bottle
Hohlkörper, geblasener 1. blow moulding, blow moulded part. 2. blown container, bottle
Hohlkörper, technische industrial containers, industrial blow mouldings
Hohlkörper-Blas- und Füllautomat automatic blow moulding-filling machine
Hohlkörperblasanlage blow moulding line/plant

Hohlkörperblasautomat automatic blow moulding machine
Hohlkörperblasdüse parison die *(bm)*
Hohlkörperblasen blow moulding
Hohlkörperblasformmaschine blow moulding machine
Hohlkörperblasmaschine blow moulding machine
Hohlkörperblasprozeß blow moulding (process)
Hohlkörperblasverfahren blow moulding (process)
Hohlkörpercompound blow moulding compound
Hohlkörperfertigung blow moulding
Hohlkörperform blow(ing) mould
Hohlkörperhersteller 1. blow moulder. 2. bottle/container manufacturer
Hohlkörperherstellung blow moulding
Hohlkörperproduktionsanlage blow moulding line/plant
Hohlkörperprofil hollow section
Hohlkörperthermoplast blow moulding compound
Hohlkörperwerkzeug blow(ing) mould
Hohlnadel blowing/inflation needle *(bm)*
Hohlprofil hollow section
Hohlraum 1. (mould) cavity. 2. void
Hohlraumaufweitung (mould) cavity expansion
Hohlstab hollow rod
Hohlwelle hollow shaft
Hohlwellengetriebe hollow shaft drive
Hohlzylinder hollow cylinder
Holm tie bar/rod, column *(im)*
Holmabstand, lichter distance between tie bars *(im)*
holmenlos without tie bars *(im)*
Holmführung guide pillar system: **Spritzeinheit mit Holmführung** injection unit on guide pillars
Hologramm hologram
Holographie holography
holographisch holographic
Holsystem collecting system
Holzbearbeitungswerkzeuge woodworking tools
Holzfaserplatte (wood) chipboard
Holzform wooden mould
holzgemasert with a woodgrain finish
Holzkern wooden core
Holzkonstruktion wooden structure/construction
Holzlack wood lacquer
Holzlasur wood glaze
Holzleim wood glue/adhesive
Holzmehl wood flour
Holzmodell wooden model/pattern
Holzöl wood oil
Holzölalkydharz wood oil alkyd (resin)
Holzschutzlack wood lacquer
Holzschutzlasur wood glaze
Holzschutzlasurfarbe wood glaze
Holzschutzmittel wood preservative

Holzschutzzusatzmittel wood preservative
Holzspanplatte (wood) chipboard
Holzurmodell wooden master model/pattern
Holzversiegelung wood sealant
Holzwerkstoff wood-based material
Holzzellulose wood cellulose
homogen homogeneous
homogenisieren to homogenise
Homogenisierextruder compounding extuder
Homogenisierhilfe homogenising aid
Homogenisierleistung homogenising efficiency
Homogenisierung homogenisation
Homogenisierwirkung homogenising effect
Homogenisierzone transition section *(of a screw) (e)*
Homogenität homogeneity
Homogenität, thermische temperature uniformity: **um die thermische Homogenität der Schmelze zu verbessern** to make the melt temperature more uniform
Homolog homologue
homolytisch homolytic
Homopolyamid homopolyamide
Homopolykondensat homopolycondensate
Homopolymer homopolymer
homopolymer homopolymeric
Homopolymerfolie homopolymer-based film
Homopolymerisat homopolymer
homopolymerisiert homopolymerised
Honigwabe honeycomb
Honigwabenkern honeycomb core
Honigwabenstruktur honeycomb structure
Hookesch Hookean *(i.e obeying Hooke's Law)*
Hookesches Gesetz Hooke's Law
Horde tray
Horizontalausführung, in horizontally constructed: **Schneidmühle in Horizontalausführung** horizontal granulator
Horizontalbauweise, in horizontally constructed *(for translation example see under Horizontalausführung, in)*
Horizontalbeschickung 1. horizontal feeding. 2. horizontal feed system/mechanism
horizontales Ausschwimmen floating
horizontales Pigmentausschwimmen floating
Horizontalextruder horizontal extruder
Horizontalhub horizontal stroke
Horizontalschneidmühle horizontal granulator *(sr)*
Horizontalverschiebung horizontal displacement: **mit Horizontalverschiebung für einfache Entleerung** which can be moved horizontally to make emptying easier
Horizontalverstellung horizontal adjustment
Hostrechner host computer *(me)*
HPC-Ruß hard processing channel black
HPLC *abbr. of* high pressure liquid chromatography
HR 1. *abbr. of* high reactivity 2. *abbr. of* heat release 3. *abbr. of* high resilience
HR-Matrix high reactivity matrix
HRR *abbr. of* heat release rate

HR-Schaumstoff high resilience foam
HS *abbr. of* **Heizelementschweißen**, heated tool welding
HT-Faser high tensile fibre
HT-Matrix high temperature resistant matrix
HT-Polymer high temperature polymer
HTV 1. *abbr. of* **hoch-temperatur-vulkanisierend/vernetzend,** high temperature vulcanising. 2. *abbr. of* **Hochtemperaturverbrennung**, high temperature incineration
HT-Verfahren 1. high temperature process. 2. high temperature calendering (process)
HT-Werkstoff high temperature material
Hub stroke
Hubbegrenzung 1. stroke limitation. 2. stroke limit switch/mechanism
Hubeinstellung 1. stroke setting. 2. stroke adjusting mechanism
Hubkolbenpumpe reciprocating piston pump
Hubvolumen swept volume *(im)*
Hubwegmeßsystem stroke measuring system
Hubzahl number of strokes *(im)*
hubzahlabhängig depending on the number of strokes
Hubzähler stroke counter
Hubzählwerk stroke counter
Huckepackanordnung cascade arrangement
Hüftprothese artificial hip
Hüllkurve envelope
Hülsenausdrücksystem sleeve ejection system *(im)*
Hülsenauswerfer sleeve ejector *(im)*
Humanisierung humanisation: **ein Beitrag zur Humanisierung des Arbeitsplatzes** a contribution towards making work more pleasant
hüpfender Kitt bouncing putty
Hutablage rear parcel shelf *(in a car)*
Hüttenindustrie iron and steel industry
HWG *abbr. of* **Heißabschlag-Wasser-Granulierung:** 1. hot-cut water-cooled pelletisation. 2. hot-cut water-cooled pelletiser
HWZ *abbr. of* **Halbwertzeit,** half-life period
Hybridbindung hybrid linkage
Hybridgarn hybrid yarn
Hybridgewebe hybrid fabric
Hybridmikroelektronik hybrid microelectronics
Hybridpolymer hybrid polymer
Hybridschaltung hybrid integrated circuit
Hybridstruktur hybrid structure
Hybridverbundwerkstoff hybrid composite (material)
Hybridwerkstoff hybrid/composite material
Hydantoinesterimid hydantoin ester imide
Hydratation hydration
Hydrationswasser water of crystallisation
Hydratisation hydration
hydratisieren to hydrate
Hydratwasser water of crystallisation
Hydraulik hydraulics, hydraulic system
Hydraulikaggregat hydraulic unit

Hydraulikanlage hydraulics, hydraulic system
Hydraulikantrieb hydraulic drive
Hydraulikbehälter hydraulic accumulator
Hydraulikblock hydraulic unit
Hydraulikdruck hydraulic pressure
hydraulikdruckabhängig hydraulic pressure-dependent, depending on the hydraulic pressure
hydraulikdruckunabhängig hydraulic pressure-independent, irrespective of the hydraulic pressure
Hydraulikflüssigkeit hydraulic fluid
Hydraulikgerät hydraulic unit
Hydraulikkolben hydraulic ram
Hydraulikkreis hydraulic circuit
Hydraulikmotor hydraulic motor
Hydrauliköl hydraulic oil
Hydraulikölbedarf hydraulic oil requirements
Hydrauliköldruckanzeige hydraulic system pressure gauge
Hydraulikölstrom 1. amount of hydraulic oil. 2. hydraulic oil flow/stream
Hydrauliköltank hydraulic oil reservoir
Hydraulikpumpe hydraulic pump
Hydraulikspeicher hydraulic accumulator
Hydraulikspritzzylinder hydraulic injection cylinder
Hydraulikstellglied hydraulic control element *(plural:* hydraulic controls*)*
Hydrauliksystem hydraulic system, hydraulics
Hydrauliküberwachungspaket hydraulic monitoring package
Hydraulikventil hydraulic valve
Hydraulikzylinder hydraulic cylinder
hydraulisch hydraulic
hydraulisch abbindend hydraulic *(e.g. cement, mortar etc.)*
hydraulisches Bindemittel hydraulic adhesive/cement
Hydrazin hydrazine
Hydrazinderivat hydrazine derivative
Hydrazoverbindung hydrazo compound
Hydrazylradikal hydrazyl radical
hydrierbar capable of hydrogenation
hydriert hydrogenated
Hydrierung hydrogenation
Hydrobasislack water-based paint
Hydrochinon hydroquinone
hydrodynamisch hydrodynamic
Hydrofluoralkan hydrofluoroalkane
Hydrofluorchlorkohlenwasserstoff hydrochlorofluorocarbon, HCFC
Hydrofluorkohlenwasserstoff hydrofluorocarbon, HFC
Hydrolack water-based/waterborne paint
Hydrolyse hydrolysis
hydrolyseanfällig subject to hydrolysis
Hydrolysebeständigkeit hydrolysis resistance
hydrolyseinstabil not resistant to hydrolysis
Hydrolyseresistenz hydrolysis resistance
hydrolysierbar hydrolysable
hydrolysieren to hydrolyse

hydrolytisch hydrolytic
Hydromotor hydraulic motor
Hydroperoxid hydroperoxide
Hydroperoxidgruppe hydroperoxide group
Hydroperoxyradikal hydroperoxy radical
hydrophil hydrophilic
Hydrophilie hydrophilic properties/character
hydrophiliert made water miscible, made hydrophilic
hydrophob hydrophobic, water repellent
Hydrophobie water repellency, hydrophobic properties
hydrophobierend making water repellent
hydrophobiert water repellent, made water repellent
Hydrophobierung imparting water repellency, making water repellent, waterproofing
Hydrophobierungsmittel water repellent
Hydrophobierwirkung water repellent effect
Hydropumpe hydraulic pump
Hydrospeicher hydraulic accumulator
hydrostatisch hydrostatic
Hydroxidgrupe hydroxyl group
Hydroxidradikal hydroxyl group
Hydroxyalkylacrylester hydroxyalkyl acrylate
Hydroxyalkylester hydroxyalkyl ester
Hydroxyalkylvinylether hydroxyalkyl vinyl ether
Hydroxyäquivalent hydroxy equivalent
Hydroxycarbonsäure hydroxycarboxylic acid
Hydroxyester hydroxy ester
Hydroxyethylacrylat hydroxyethyl acrylate
Hydroxyethylcellulose hydroxyethyl cellulose
Hydroxyethylmethacrylat hydroxyethyl methacrylate
Hydroxyethylpropylester hydroxyethylpropyl ester
Hydroxyfunktionalität hydroxy functionality
hydroxyfunktionell hydroxyfunctional
Hydroxylacrylat hydroxyl acrylate
Hydroxylaminchlorhydrat hydroxylamine chlorohydrate
Hydroxylaminhydrochlorid hydroxylamine hydrochloride
Hydroxylaminsulfat hydroxylamine sulphate
hydroxylarm with a low hydroxyl content
Hydroxylendgruppe terminal hydroxyl group
Hydroxylgehalt hydroxyl content
Hydroxylgruppengehalt hydroxyl group content
hydroxylgruppenhaltig containing hydroxyl groups
hydroxyliert hydroxylated
hydroxylreich with a high hydroxyl group content
Hydroxylverbindung hydroxyl compound
Hydroxylwert hydroxyl value
Hydroxylzahl hydroxyl value
Hydroxymethylgruppe hydroxymethyl group
Hydroxymethylmethacrylat hydroxymethyl methacrylate
Hydroxypolyether hydroxypolyether
Hydroxypropylacrylat hydroxypropyl acrylate

Hydroxypropylester hydroxypropyl ester
Hydroxystearinsäure hydroxystearic acid
Hydrozyklon hydrocyclone
Hygroskopie hygroscopicity
hygroskopisch hygroscopic
Hygroskopizität hygroscopicity
Hyperbolfeder hyperbolic spring
Hysterese hysteresis
Hysteresekurve hysteresis curve
Hystereseschleife hysteresis loop

I

i.a. abbr. of **im Allgemeinen**, generally (speaking)
IBM-kompatibel IBM-compatible
IC abbr. of integrated circuit
ICP abbr. of intrinsically conductive polymer
IC-Schaltung integrated circuit
ideal-elastisch ideal-elastic
ideal-plastisch ideal-plastic
Identifizierung identification
Identifizierungskennzeichen identifying mark
Identitätsermittlung identification
i.d.R. abbr. of **in der Regel**, as a rule, generally (speaking)
Igelkopf knurled mixing section (of screw) (e)
IHS abbr. of **Integralhartschaum**, rigid integral/structural foam
ILS abbr. of **interlaminare Scherfestigkeit**, interlaminar shear strength
ILSS abbr. of interlaminar shear strength
ILSS-Versuch interlaminar shear test
IMC-Beschichtung in-mould coating
IMC-Verfahren in-mould coating
IMD abbr. of in-mould decoration
IMD-Verfahren in-mould decoration
IM-Faser intermediate-modulus fibre
Imidazol imidazole
Imidglied imide segment
Imidgruppe imide group
Imidhartschaumstoff rigid imide foam
imidisiert imidised
IML-Technik in-mould labelling
Immersionsversuch immersion test
Immission pollution; *translators must guard against rendering this as* immission, *which is never used.* **Immissionen** *are caused by* **Emissionen** - *in other words the word is synonymous with* **Umweltverschmutzung**
Immissionsschutzgesetzgebung anti-pollution legislation, pollution control legislation
immissionsschutzrechtliche Regelungen anti-pollution regulations, pollution control regulations
Immissionsschutzrechtsetzung anti-pollution legislation, pollution control legislation
Impedanz impedance
Implantat implant
implementieren to implement
Implosion implosion
Importabhängigkeit dependence on imports
imprägnieren to impregnate
Imprägnierharz impregnating resin
Imprägnierlasur (clear) impregnating varnish
Imprägniermaschine impregnating machine
Imprägniermittel impregnating agent
Imprägnierparameter impregnating conditions
Imprägniertiefe depth of impregnation
Imprägnierung impregnation
Impuls pulse, impulse
impulsartig pulse-like
Impulsfolge impulse sequence
Impulsschweißen impulse welding
Impulszähler impulse counter
IMR abbr. of in-mould release
IMR-Additiv in-mould release agent
IMR-System in-mould release agent
inaktiv inert
Inbetriebnahme putting into operation *(plant, machine etc.)*
Inbetriebsetzung starting-up
Inbrandsetzen ignition
Indan indane
Inden indene
indifferent inert
indifferent, physiologisch non-toxic, physiologically inert
Indifferenz inertness
Indikator indicator
Indiz indication
Indukionszeit induction period
Induktionsheizgerät induction heater, induction heating unit
Induktionsheizung induction heating
Induktionsperiode induction period
Induktionsphase induction phase
induktionsschleifengeführt induction loop controlled
Induktionsspule induction coil
Induktionszeit induction period
induktiv inductive
Industrie, kautschukverarbeitende rubber processing industry
Industrieabgase industrial waste gases
Industriebekleidung industrial clothing
Industriebodenbelag industrial floor screed
Industriedecklack industrial paint
Industrieeinbrennlack industrial stoving paint/enamel
Industriegesellschaft industrial society
Industriehalle factory building
Industriehandschuhe industrial gloves
Industriehygiene industrial hygiene
Industrieklima industrial atmosphere
Industrielack industrial paint
Industrieländer industrialised countries
industriell industrial

Industriemüll industrial refuse/waste
Industrienationen industrialised countries
Industrieplatten industrial sheets
Industrieregale industrial shelving
Industriereinigungsmittel industrial cleaner
Industrieroboter industrial robot
Industrierohr industrial pipe
Industriesack industrial bag/sack
Industrieschutzbekleidung industrial protective clothing
Industriesparte branch of industry
Industriestaaten industrialised countries
Industriezelle production unit
Industriezweig branch of industry
induziert induced
ineinandergreifend 1. intermeshing *(twin screws) (e)*. 2. linked, interconnected *(e.g. processes)*
ineinanderkämmend intermeshing *(twin scews)*
inert inert
Inertgas inert gas
Inertisierung neutralisation
Informationen information *(this word does NOT have a plural)*
Informationsausgabe 1. information output *(me)*. 2. feedback information
Informationsaustausch exchange of information
Informationsdichte information density *(me)*
Informationseingang information input
Informationsfluß information flow *(me)*
Informationsflut flood of information
Informationsgehalt information content
Informationsgehalt information content
Informationsquelle source of information
Informationssammelschiene data bus *(me)*
Informationsschleife data loop *(me)*
Informationsschnittstelle information interface *(me)*
Informationsschrift information sheet
Informationsspeicher databank, data storage device *(me)*
Informationstechnik information technology
Informationsträger data storage device
Informationsübermittlung transfer of information
Informationsübertragung information transfer, transfer of information
Informationsvorrat information store, database, databank
infrarot infra-red, IR
Infrarotbereich infra-red range
Infrarotbestrahlungslampe infra-red lamp
Infrarotdifferenzspektrometrie infra-red differential spectrometry
Infrarotlöten infra-red soldering
Infrarotmeßfühler infra-red sensor
Infrarotofen infra-red oven
infrarotspektrographisch infra-red spectrographic
Infrarotspektroskopie infra-red spectroscopy
Infrarotspektrum infra-red spectrum

Infrarotstrahler infra-red heater
Infrarottunnel infra-red heating tunnel
Infraschallbereich infrasonic range
Infrastruktur infrastructure
Ingenieur engineer
Ingenieurbau constructional/civil engineering
Ingenieurkunststoff engineering plastic
inhärent inherent
inhibierend inhibiting
Inhibierung inhibition
Inhibitor inhibitor
inhomogen inhomogeneous
Inhomogenitäten, thermische uneven temperature: **Schwankungen der Schneckendrehzahl führen zu thermischen Inhomogenitäten** screw speed fluctuations will result in uneven temperatures
Initiator initiator
Initiatordosierung 1. amount of initiator. 2. addition of initiator: **Wegen der Toxizität von Vinylchlorid ist man bestrebt das Öffnen der Reaktoren zur Initiatordosierung zu vermeiden** because of the toxicity of vinyl chloride, one should, if possible, avoid opening the reactors to add the initiator
Initiatorsuspension initiator suspension
initiieren to initiate
Initiierung initiation
Initiierungsmittel initiator
Injektionsblasnadel blowing pin, inflation needle *(bm)*
Injektionsdüse injection nozzle *(im)*
Injektionskolben injection plunger *(im)*
Injektionspunkt injection point
Injektionsspritze (hypodermic) syringe
injizieren to inject
inkompressibel incompressible
Inkompressibilität incompressibility
Inkrafttreten coming into force
Inkrement increment
inkrementell incremental
inländisch home, domestic
Inlandsbeteiligungen domestic holdings
Inlandsgeschäft home/domestic business
Inlandsumsatz domestic turnover
Inliner liner
Inmold-Coating-Beschichtung in-mould coating
Innenabmessungen internal dimensions
Innenauskleidung lining *(e.g. of a tank)*
Innenausstattung interior fitments/fittings/trim
innenbeheizt internally heated
Innenbeleuchtung interior lighting
Innenbereich inner zone
Innendruck 1. internal/inside pressure. 2. cavity pressure *(im)*
innendruckabhängig cavity pressure-dependent
Innendruckfestigkeit resistance to internal pressure
innendruckunabhängig cavity pressure-independent, independent of cavity pressure

Innendruckverformung deformation through internal pressure
Innendruckzeitstandmessung long-term failure test under internal hydrostatic pressure *(used for plastics pipes)*
Innendruckzeitstanduntersuchung long-term failure test under internal hydrostatic pressure *(used for plastics pipes)*
Innendruckzeitstandverhalten long-term resistance to internal hydrostatic pressure *(of plastics pipes)*
Innendruckzeitstandversuch long-term failure test under internal hydrostatic pressure *(used for plastics pipes)*
Innendruckzeitstandwert long-term resistance to internal hydrostatic pressure *(of plastics pipes)*
Innendurchmesser inside diameter
Inneneinsatz interior/inside use
Innenfarbe interior paint
Innenfiltration depth filtration
Innenflügel interior casement
innengekühlt cooled from the inside
innengepanzert internally hardened, hardened on the inside *(barrel) (e)*
Innengewinde internal thread
Innenhimmel (car) roof liner
Innenkalibrieren internal calibration/sizing
Innenkneter internal mixer
Innenkühlluft internal cooling air *(bfe)*
Innenkühlluftstrom 1. internal cooling air stream. 2. amount of internal cooling air
Innenkühlluftsystem internal air cooling system *(bfe)*
Innenkühlsystem internal cooling system
Innenkühlung 1. internal cooling. 2. internal cooling system
Innenlage core layer *(of a laminate)*
Innenluftdruck internal/inside air pressure *(bfe)*: Der Innenluftdruck der Blase the air pressure inside the film bubble
Innenluftkühlsystem 1. internal air cooling system. 2. internal bubble cooling system *(bfe)*
Innenluftkühlung 1. internal air cooling. 2. internal air cooling system
Innenluftstrom 1. internal air stream. 2. amount of air inside *(e.g. a film bubble)*
Innenmaße internal/inside dimensions
Innenmischer internal mixer
Innenpanzerung wear resistant liner *(of extruder barrel)*
Innenradius inside radius
Innenraum interior *(e.g. of building)*, vehicle interior
Innenraumeinsatz indoor use
Innenraumisolator indoor insulator
Innenwandfarbe interior paint
Innenzahnradpumpe internal gear pump
innerbetrieblich in-plant, in-house
innere Gleitwirkung internal lubrication
innere Reibung internal friction
innere Weichmachung internal plasticisation

inneres Gleitmittel internal lubricant
innerlich internal
innerlich weichgemacht internally plasticised
innermolekular intramolecular
Inreaktivität inertness
Insassensicherheit passenger safety
Inserttechnik insert moulding
instabil unstable
Instabilität instability
Installationsbereich sanitary sector
Installationskosten installation costs
installierte Leistung installed/connected load
Instandhaltung maintenance
Instanz authority
Instrumentenbrett instrument panel, dashboard
Instrumentenrahmen instrument surround
Instrumententräger instrument panel, dashboard
Instrumenttafel instrument panel, dashboard
integral integral, integrated
Integralhartschaumstoff rigid integral/structural foam
Integralschaum integral/sructural foam
Integralschaumstoff integral/structural foam
Integralschaumstoffkern integral/structural foam core
Integrationsbauweise integrated construction
integrationsfähig integrable
Integrationsfähigkeit integration capability
Integrationsgrad degree of integration
integrierbar integratable
integriert integrated
integrierter Schaltkreis integrated circuit
intelligentes Terminal intelligent terminal *(me)*
Intensivkneter intensive kneader
Intensivmischer intensive mixer
Intensivmischung intensive mixing
interatomar interatomic
interdisziplinär interdisciplinary
interessant, verfahrenstechnisch technically interesting, interesting from the processing point of view
Interferenz, elektromagnetische electromagnetic interference, EMI
Interferenzfarbe iridescent colour
Interferenzgrundmuster interference pattern
Interferenzlinien interference lines
Interferenzpigment iridescent pigment
Interferogramm interferogram
Interferometrie interferometry
interlaminar interlaminar
interlaminare Scherfestigkeit interlaminar shear strength
intermittierend intermittent
intermolekular intermolecular
interpenetrierend interpenetrating
Interpolation interpolation
interpoliert interpolated
intramolekular intramolecular
intrinsisch intrinsic
intrinsisch leitfähig intrinsically conductive
Intrusionsverfahren intrusion (process)

Intumeszenzmaterial intumescent compound
Investeinsparung (capital) investment savings
Investitionen (capital) investment
Investitionsaufwand capital outlay/investment, cost, expenditure *(for translation example see under* **Investitionskosten***)*
Investitionsausgaben capital expenditure/ investment
Investitionsbedarf need for capital investment
Investitionsbereitschaft readiness to invest
Investitionsgüter capital goods
Investitionskosten capital outlay/investment, cost, expenditure: **Obwohl die Investitionskosten nicht unbedingt ein entscheidendes Kriterium sind** although cost is not necessarily a major consideration
Investitionspreis capital outlay, cost
Investitionsprogramm investment programme
Investitionsrate rate of investment
Investitionsvolumen investment volume
Ion ion
Ionenabgabe ion release
ionenbildend ion-forming
Ionenbindung ionic bond
Ionenfänger ion interceptor
Ionengitter ionic lattice structure
Ionenkomplex ion complex
Ionenquelle ion source
Ionenstoß ion collision
Ionenverbindung ionic bond
Ionisationsbeständigkeit ionisation resistance
ionisch ionic
ionisierbar ionisable
ionisierend ionising
ionisiert ionised
Ionisierung ionisation
ionitrierbehandelt ionitrided
Ionitrieren ionitriding
ionitriert ionitrided
ionogen ionogenic
Ionogenität ionic character
ionomer ionomeric
Ionomerharz ionomer (resin)
i.o.Tiegel *abbr. of* **im offenen Tiegel**, open-cup: **Flammpunkt i.o. Tiegel** flash point, open-cup
IPN *abbr. of* interpenetrating network
IPN-Polymer interpenetrating network polymer, IPN polymer
IR *abbr. of* isoprene rubber
irreversibel irreversible
Irrigationsrohr irrigation pipe
IR-Spektroskopie infra-red spectroscopy
ISAF-Ruß intermediate super abrasion furnace black
isobar isobaric, at constant pressure
Isobare isobar
Isobornylacrylat isobornyl acrylate
Isobuten isobutene
Isobutylacrylat isobutyl acrylate
Isobutylen-Isoprenkautschuk isobutylene-isoprene rubber, butyl rubber
Isobutylmethacrylat iosobutyl methacrylate

Isobutylstearat isobutyl stearate
Isochinolin isoquinoline
isochor isochoric, at constant volume
Isochore isochore
Isochromatenaufnahme isochromatic photograph
isochron isochronous
Isocyanat isocyanate
isocyanatfrei isocyanate-free
Isocyanatgruppe isocyanate group
Isocyanatharz isocyanate resin
Isocyanatkleber isocyanate adhesive
Isocyanatkomponente isocyanate component
Isocyanatpräpolymer isocyanate prepolymer
Isocyanatverbindung isocyanate compound
Isocyanurat isocyanurate
Isododecan isododecane
Isolation insulation
Isolationseigenschaften insulating properties
Isolationsmasse 1. insulating compound 2. cable compound
Isolationsmischung 1. insulating compound 2. cable compound
Isolationsschicht insulating layer
Isolationsvermögen insulating properties
Isolationswerte insulating properties
Isolationswiderstand insulation resistance
Isolator insulator
Isolierbuchse insulating bush
Isoliereigenschaften insulating properties
Isoliergrund barrier coat
Isolierhülse insulating sleeve
Isolierkanal insulated runner *(im)*
Isolierkanalanguß insulated runner system, insulated runner feed system *(im)*
Isolierkanalangußplatte insulated runner unit/ plate *(im)*
Isolierkanalform insulated runner mould *(im)*
Isolierkanalsystem insulated runner system *(im)*
Isolierkanalwerkzeug insulated runner mould *(im)*
Isolierlack insulating varnish
Isoliermantel insulating jacket
Isoliermasse insulating compound
Isoliermedium insulating medium
Isolierplatte insulating plate
Isolierrohr insulating tube
Isolierschicht insulating layer
Isolierstoff insulating material
Isolierverglasung double glazing
Isoliervermögen insulating properties
Isolierverteiler insulated runner *(im)*
Isolierverteilerangießsystem insulated runner system *(im)*
Isolierverteilerkanal insulated runner *(im)*
Isolierwerkstoff insulating material
Isolierwerte insulating properties
Isolierwirkung insulating effect
Isomer isomer
isomer isomeric
Isomerie isomerism

isomerisiert isomerised
Isomerisierung isomerisation
Isooctanol iso-octanol
Isophoron isophorone
Isophorondiisocyanat isophorone diisocyanate
Isophthalat isophthalate
Isophthalsäure isophthalic acid
Isopren isoprene
Isoprenkautschuk isoprene rubber
Isopropanol isopropanol
Isopropanolamin isopropanolamine
isopropanolisch isopropanolic
Isopropenylnaphthalin isopropenyl naphthalene
Isopropylacetat isopropyl acetate
Isopropylidengruppe isopropylidene group
isosterisch isosteric
isotaktisch isotactic
Isotaxieindex isotactic index
isotherm isothermal
Isotherme isotherm
Isotopendickenmeßgerät isotope thickness gauge
isotrop isotropic
Isotropie isotropy
Isoverbindung iso-compound
Isoverglasung 1. double glazing. 2. sealed unit
ISO-Vorschrift ISO standard
Istgewicht actual weight
Istgröße actual value
Istkurve actual value curve
Istwert actual value
Istwertanzeige actual value display *(me)*
Istwertgeber actual value transducer *(me)*
Istwertkurve actual value curve
Istwertverarbeitung actual value processing *(me)*
Itaconsäure itaconic acid
IZ 1. *abbr. of* **Industriezelle**, production unit 2. *abbr. of* **Integralzelle**, integral unit
Izod-Kerbschlagähigkeit Izod notched impact strength

J

Jahresabschluß annual accounts
Jahresdurchschnitt yearly average
Jahresgewinn annual profit
Jahreshälfte half-year
Jahreskapazität annual capacity
Jahresproduktion annual production
Jahrestonnen tonnes p.a.
Jahresüberschuß annual surplus, surplus for the year
Jahresumsatz annual turnover
Jahreszeit time of the year
Jahreszeit, kalte winter months
jato *abbr. of* **Jahrestonnen**, tonnes p.a.
JFZ *abbr. of* **Jodfarbzahl**, iodine colour value
JIT *abbr. of* just-in-time
Jod iodine
Jodfarbskala iodine colour scale
Jodfarbzahl iodine colour value
jodometrisch iodometric
Joghurtbecher yogurt/yoghurt cup
justierbar adjustable
Justierbarkeit adjustability
justieren to adjust
Justierschraube adjusting screw
Just-in-time Lieferung just-in-time delivery
Just-in-time-Produktion just-in-time production/manufacturing

K

Kabel- und Drahtummantelungen cable sheathing and wire insulation
Kabelabfall cable scrap
Kabelaufbereitungsanlage cable reclaiming plant
Kabelei cable industry
Kabelgranulat cable (insulating) compound
Kabelhülle cable sheathing
Kabelindustrie cable industry
Kabelisolation cable insulation
Kabelisolationsmischung cable (insulating) compound
Kabelisoliermasse cabel (insulating) compound
Kabelisolierung cable insulation
Kabelkanal cable conduit
Kabelkopf cable sheathing die *(e)*
Kabelmantel cable sheathing
Kabelmasse cable (insulating) compound
Kabelmischung cable (insulating) compound
Kabel-PVC PVC cable compound
Kabelrezeptur cable compound formulation
Kabelschacht cable duct
Kabelschutzrohr cable conduit
Kabelsektor cable sector/industry
Kabelspritzkopf cable sheathing die *(e)*
Kabeltrommel cable drum
Kabeltype cable (insulating) compound
Kabelummantelung cable sheathing
Kabelummantelungsanlage cable sheathing line *(e)*
Kabelummantelungsdüse cable sheathing die *(e)*
Kabelummantelungsextruder cable sheathing extruder
Kabelummantelungswerkzeug cable sheathing die *(e)*
Kabelvergußmasse cable jointing compound

Kabelwerk cable factory/works
Kachel tile
Kaffeeautomat coffeemaker
Kaffeemaschine coffeemaker
Kalander calender
Kalanderabzugswalze calender take-off roll
Kalanderanlage calendering line
Kalanderbauform 1. calender design, type of calender: **Alle anderen Kalanderbauformen werden für spezielle Fälle benutzt** all other types of calender are used for special applications. 2. calender roll configuration
Kalanderbeheizung 1. calender heating. 2. calender heating system
Kalanderbeschichtung calender coating
Kalanderbeschickung 1. calender feeding: **Eine direkte Kalanderbeschickung ist möglich** the calender can be fed direct. 2. calender feed unit: **Der Extruder mit Breitschlitzdüse ist die ideale Kalanderbeschickung** the extruder fitted with a slit die is the ideal calender feed unit
Kalanderfolie calendered film/sheeting
Kalandergeschwindigkeit calender speed
Kalanderlinie calendering line
Kalandernachfolge calender downstream equipment
Kalanderspalt calender nip
Kalanderstraße calendering line
Kalanderströmung calender flow
Kalanderverfahren calendering
Kalanderwalze calender roll
Kalanderwalzenspalt calender nip
Kalandrierbarkeit ease of calendering: **Das Kriterium für die Kalandrierbarkeit einer Kautschukmischung** the criterion for the ease with which a rubber compound can be calendered
Kalandrieren calendering
Kalandriergeschwindigkeit calendering speed
Kalandrierprozeß calendering (process)
Kalandrierung calendering
Kalibrationskurve calibration curve
Kalibrator calibrating device/unit, sizing unit (e)
Kalibratorscheibe calibrating/sizing plate (e)
Kalibrierblasdorn calibrating blowing mandrel (bm): *The difference between a* **Kalibrierblasdorn** *and a* **Blasdorn** *(q.v.) is that the former shapes the neck of the bottle or container to the exact dimensions required, i.e. it calibrates or sizes it, whereas the latter inflates the parison without calibrating it*
Kalibrierblenden sizing/draw plates (e) *(used to calibrate very thin tubing)*
Kalibrierblendenpaket sizing/draw plate assembly (e)
Kalibrierbohrung calibrator bore (e)
Kalibrierbüchse calibrating/sizing sleeve (e)
Kalibrierdorn calibrating mandrel (e) *(not to be confused with* **Kalibrierblasdorn***, q.v.)*
Kalibrierdruck calibrating/sizing pressure (e)
Kalibrierdüse calibrating/sizing die (e)

Kalibriereinheit calibrating/sizing unit (e)
Kalibriereinrichtung calibrating device/unit, sizing unit (e)
kalibrieren to calibrate, to size (e)
Kalibrierhülse calibrating/sizing sleeve (e)
Kalibrierkonzept calibration unit design (e)
Kalibrierkorb calibrating/sizing basket, calibrating/sizing cage (bfe)
Kalibrierluft calibrating air (e)
Kalibrierplatte calibrating/sizing plate (e)
Kalibrierring calibrating/sizing ring (e)
Kalibrierrohr calibrating/sizing tube (e)
Kalibrierscheibe calibrating/sizing plate (e)
Kalibrierspalt calibrating/sizing nip (e)
Kalibrierstrecke calibrating/sizing section (e)
Kalibrierstrom calibrating current
Kalibriersystem calibrating/sizing system (e)
Kalibriertisch calibrating/sizing table (e)
Kalibrierung 1. calibration, sizing. 2. calibrating unit (e)
Kalibriervorrichtung calibrating/sizing device (e)
Kalibrierwerkzeug calibrating/sizing die (e)
Kaliglimmer potash mica, muscovite
Kalilauge caustic potash solution
Kaliumdichromat potassium dichromate
Kaliumhydroxyd potassium hydroxide
Kaliumpersulfat potassium persulphate
Kaliumtetrafluorborat potassium tetrafluoroborate
Kalkanstrich limewash, whitewash
Kalkfarbe limewash, whitewash
Kalkmilch milk of lime
Kalkputz lime plaster
Kalksandstein sand-lime brick
Kalkstein limestone
kalorisch caloric
kaltabbindend cold setting
Kaltarbeitsstahl cold-worked steel
Kaltbiegeeigenschaften cold bending properties
Kaltdach double-shell roof
kalte Jahreszeit winter months
Kältebeanspruchung low temperature exposure
Kältebelastung low-temperature exposure
Kältebeständigkeit low-temperature resistance
Kältebiegebeständigkeit cold bend resistance
Kältebiegeversuch cold bend test
Kältebruchfestigkeit 1. cold crack resistance (e.g. of PVC coated fabrics). 2. brittleness temperature
Kältebruchtemperatur 1. cold crack temperature (e.g. of PVC coated fabrics). 2. brittleness temperature
Kältebruchverhalten cold crack behaviour
Kältedämmstoff low-temperature insulating material
Kälteeigenschaften low-temperature properties
Kälteelastifizierungsvermögen low-temperature plasticiser efficiency
kälteelastisch *having low-temperature flexibility*

Kälteelastizität low-temperature flexibility
Kälteentgraten cryogenic deflashing
Kältefalzversuch low-temperature folding (endurance test)
kältefest low-temperature resistant
Kältefestigkeit low-temperature resistance
kälteflexibel *having low-temperature flexibility:* **Diese Typen sind kälteflexibel bis -55°C** these grades remain flexible down to -55°C
Kälteflexibilität low-temperature flexibility
Kälteisolierung insulation against cold
Kältekammer low-temperature test chamber
Kältekontraktion low-temperature shrinkage
Kältekreis cooling circuit
Kältekreislauf cooling circuit
Kältelagerung low-temperature ageing
Kälteleistung cooling capacity/efficiency *(for translation example see under* **Kühlleistung***)*
Kältemaschine refrigeration machine/unit
Kältemittel coolant, refrigerant
Kaltentgratung cryogenic deflashing
kalter Fluß cold flow
kalter Pfropfen cold slug *(im)*
Kälteschälfestigkeit low-temperature peel strength
Kälteschlagbeständigkeit low-temperature impact strength
kälteschlagfest low-temperature impact resistant
Kälteschlagfestigkeit low-temperature impact strength
Kälteschlagwert low-temperature impact strength
kälteschlagzäh low-temperature impact resistant
Kälteschlagzähigkeit low-temperature impact strength
Kälteschockfestigkeit low-temperature shock resistance
Kälteschrank low-temperature test chamber
Kälteprödigkeitspunkt low-temperature brittleness point, brittleness temperature
Kälteprödigkeitsversuch low-temperature brittleness test
Kältestandfestigkeit low-temperature resistance
Kältestauchfestigkeit low-temperature compressive strength
Kältethermostat low temperature thermostat
Kälteverhalten low-temperature performance/behaviour/properties
Kälteversprödung low-temperature embrittlement
Kälteversprödungsneigung tendency to become brittle at low temperatures
Kälteversprödungstemperatur low-temperature brittleness point, brittleness temperature
kältezäh low-temperature (impact) resistant
Kältezähigkeit 1. low temperature resistance 2. low temperature toughness, low temperature impact strength

Kältezähigkeit low-temperature (impact) strength
Kaltfluß cold flow
Kaltformen cold forming
Kaltformmethoden cold forming techniques/methods
Kaltformschaumstoff cold-cure moulded foam
Kaltfütterextruder cold feed extruder
kaltgefüttert cold-fed *(extruder)*
kaltgewalzt cold-rolled
Kaltgranulat cylindrical and/or diced pellets
Kaltgranulator 1. strand pelletiser. 2. dicer, strip pelletiser *(sr)*
Kaltgranuliermaschine 1. strand pelletiser. 2. dicer, strip pelletiser *(sr)*
kaltgranuliert 1. strand pelletised. 2. diced *(sr)*
Kaltgranulierverfahren strand and strip pelletisation *(sr)*
kalthärtend cold-curing
Kalthärter cold-curing catalyst
Kalthärtung cold-curing
Kalthärtungssystem cold-curing system
Kaltkanalangußsystem cold runner feed system *(im)*
Kaltkanalform cold runner mould
Kaltkanaltechnik cold runner injection moulding
Kaltkanalverteiler cold runner *(im)*
Kaltkanalwerkzeug cold runner mould *(im)*
Kaltkleben cold bonding
Kaltmischung cold mixing
Kaltpolymerisat cold rubber, cold polymerised rubber
Kaltpolymerisation cold polymerisation
Kaltprägen cold embossing
Kaltpressen cold press moulding *(grp)*
Kaltpreßmasse cold-curing moulding compound
Kaltpreßwerkzeug cold press moulding tool *(grp)*
Kaltschaumstoff cold-cure foam
kaltschlagzäh low-temperature impact resistant
Kaltschweißung cold weld *(this is a poor quality weld made at too low a temperature)*
Kaltsiegelkleber contact adhesive
Kalttauchverfahren cold dipping (process)
Kaltumformen cold forming
Kaltverformung cold forming
Kaltverstreckbarkeit cold stretching properties
Kaltverstreckung cold stretching
Kaltverteiler cold runner *(im)*
Kaltvulkanisation room temperature vulcanisation
kaltvulkanisierend room temperature vulcanising
Kaltwasser cold water
Kalzium- *see* **Calzium-**
Kamin vent
Kaminwirkung chimney effect
Kamm flight land *(e)*
Kämmbereich intermeshing zone *(e)* : **im Kämmbereich der Schnecken** where the screws intermesh
kämmend intermeshing *(screws)(e)*

kämmend, vollständig closely intermeshing *(screws)(e)*
Kammer 1. chamber *(e)* : This is a twin screw term denoting that part of the channel of one screw which is not filled by the flight of the other, i.e. the space which, in closely intermeshing screws, is filled with polymer melt. 2. compartment *(general term)*
Kammkante flight land edge *(e)*
Kämmspalt flight clearance *(see explanatory note under* **Kopfspiel***)*
Kämpfer transom *(of a window)*
Kämpferpfosten transom *(of a window)*
Kämpferprofil transom profile
Kanadabalsam Canada balsam
Kanal 1. channel, flow channel *(general term)*. 2. runner *(im)*
Kanalisation sewers, sewage system
Kanalrohr sewage pipe
Kanaltiefe (screw) channel depth *(e)* : This is the distance between the screw root surface and the barrel wall and should not be confused with **Gangtiefe** *(q.v.)* which is the distance between the root surface and the flight lands. **Kanaltiefe** in other words, includes the clearance between the flight lands and the barrel, **Gangtiefe** does not. See DIN 24450, Fig. 1.
Kanaltiefenverhältnis channel depth ratio *(e)*
Kanister jerrycan, canister
Kanisterblasmaschine canister blow moulding machine
Kantenabdeckung edge coverage *(of a paint)*
Kantenabtaster edge scanner
Kantenbedeckung edge coverage *(of a paint)*
Kantenbeflammung application of flame to the edge
Kantenbeschnitt 1. edge trim, edge trimmings. 2. edge trimmer, edge trimming unit. 3. trimming the edges *(of film or sheeting)*
Kantendeckkraft edge covering/hiding power
kantengerundet with rounded edges
Kantenkorrosion edge corrosion, corrosion near the edges
kantenorientiert edge-orientated
Kantenschneider edge trimmer
Kantenschnitt 1. edge trim, edge trimmings 2. edge trimmer, edge trimming unit 3. trimming the edges *(of film or sheeting)*
Kantenschutzecke corner protector
Kantenverhältnis aspect ratio
kantig angular
kanzerogen carcinogenic, cancer-causing
Kaolin kaolin
Kapazität 1. capacity. 2. capacitance *(electrical)*
Kapazitätsabbau capacity reduction
Kapazitätsaufstockung capacity increase/expansion, increase in capacity
Kapazitätsausbau capacity increase/expansion, increase in capacity
Kapazitätsauslastung capacity utilisation

Kapazitätsausweitung capacity increase/expansion, increase in capacity
kapazitätsidentisch with identical capacities
Kapazitätszuwachs capacity increase, increase in capacity
kapillar capillary
Kapillardüse capillary nozzle
Kapillare capillary
Kapillareffekt capillary effect
Kapillarenquerschnitt capillary diameter
Kapillarenwand capillary wall
Kapillarkräfte capillary forces
Kapillarrheometer capillary rheometer
Kapillarrheometrie capillary rheometry
kapillarrheometrisch capillary-rheometric
Kapillarröhre capillary
Kapillarrohrthermostat capillary thermostat
Kapillarviskosimeter capillary viscometer
Kapillarwirkung capillary action
Kapitalbeteiligung capital holding
Kapitalbindung tying up of capital
Kapitaleinsatz capital outlay/investment, cost
kapitalintensiv capital-intensive
Kapitalmarkt capital market
kapseln to encapsulate, to embed
Karbon- *see* **Carbon-**
Karbonatisation carbonation
karbonatisiert carbonated
Kardangelenk cardan joint
Kardanwelle cardan shaft
Kardanwellenabdeckung cardan shaft cover(ing)
Karkasse carcass
Karosse carbody
Karosserie carbody, vehicle body
Karosserielack carbody paint
kartesisch Cartesian
Kartusche cartridge
Karussellblasaggregat rotary/carousel-type blow moulding unit
Karussellspaltmaschine rotary/carousel-type cutter
Karzinogen carcinogen
karzinogen carcinogenic, cancer-causing
Karzinogenität carcinogenity
Kaschieranlage laminating plant
kaschierbar capable of being laminated
kaschieren 1. to laminate. 2. to conceal, to hide
Kaschierkleber laminating adhesive
Kaschierklebstoff laminating adhesive
Kaschiermaschine laminator, laminating machine
Kaschierpaste laminating paste
Kaschierstation laminating station/unit
Kaschierstrich laminating coating
Kaschierverfahren lamination, laminating *(process)*
Kaschiervorrichtung laminating unit
Kaschierwalze laminating roll
Kaschierwerk laminating unit
Kasein casein
Kaskade cascade

Kaskadenanordnung cascade arrangement
Kaskadenextruder cascade extruder
Kaskadenregelung 1. cascade control. 2. cascade control unit
Kaskadentemperaturregelung 1. cascade temperature control. 2. cascade temperature control unit
Kaskadenumlaufextruder cascade-type melt recirculation extruder
Kassette cassette *(me)*
Kassettenbandgerät cassette recorder
Kassettendecke coffered ceiling
Kassettengerät cassette recorder
Kassettenrecorder cassette recorder
Kassettensiebwechsler cassette screen changer
Kassettenspeicher cassette memory *(me)*
kastenförmig box-shaped
Katalysator catalyst
Katalysatorauswahl choice of catalyst
Katalysatorgift catalyst poison
katalysatorhaltig catalysed
Katalysatoroberfläche catalyst surface
Katalysatorpaste catalyst paste
Katalysatorreste catalyst residues
Katalysatorrückstände catalyst residues
Katalysatorvergiftung catalyst poisoning
Katalyse catalysis
katalysieren to catalyse
Katalysierung catalysation, mixing with catalyst
katalytisch catalytic
Kataphorese cataphoresis, electrophoresis
kataphoretisch cataphoretic, electrophoretic
katastrophaler Bruch catastrophic failure
katastrophales Versagen catastrophic failure
Kathode cathode
Kathodenstrahloszillograph cathode ray oscillograph
Kathodenstrahlröhre cathode ray tube
kathodisch cathodic
Kation cation
kationaktiv cationic
kationogen cationic
Kaufentscheidung decision to buy/purchase
Kaufkraft purchasing power
Kaufleute business executives
käufliches Programm ready-made program *(me)*
Kautschuk rubber: *Strictly speaking, the word refers to crude, i.e. unvulcanised rubber as opposed to* **Gummi** *(which is the word used to denote rubber in its vulcanised form). In English, both words are called rubber and no qualification (i.e. use of the words vulcanised or unvulcanised) is normally necessary. The word caoutchouc is no longer in use and should be avoided. Where a distinction is made between* **Gummi** *and* **Kautschuk**, *the translation should be along the following lines* **Durch Vulkanisation wird aus dem plastische Silikonkautschuk in den elastischen Silicongummi übergeführt** vulcanisation converts the silicone rubber from the plastic into the elastic state
Kautschukadditiv rubber additive
Kautschukanteil rubber content
Kautschukballen rubber bale, bale of rubber
Kautschukchemikalien rubber chemicals
Kautschukchemiker rubber chemist
Kautschukeigenschaften unvulcanised rubber properties
kautschukelastisch rubber-elastic
Kautschukelastizität rubber elasticity, rubber-like elasticity
Kautschukfüllstoff rubber filler
kautschukhaltig 1. rubber-containing *(general term)*. 2. rubber-modified *(if referring to PS, for example)*
Kautschukkomponente rubber component
Kautschuklatex rubber latex
Kautschukmischung rubber compound/mix
kautschukmodifiziert rubber modified
Kautschuknetzwerk rubber network
Kautschukphase rubber phase
Kautschukspritzguß injection moulding of rubber
Kautschuktechnologie rubber technology
kautschuktechnologisch *relating to rubber or rubber technology* **kautschuktechnologische Eigenschaften** rubber properties
Kautschukteilchen rubber particle
kautschuküberzogen rubber covered
kautschukverarbeitend rubber processing
Kautschukverarbeitungshilfe rubber processing aid
Kautschukverarbeitungsmaschine rubber processing machine
Kautschukverschnitt rubber blend
Kautschukverstärkung rubber reinforcement
Kavität (mould) cavity
Kavitäteninnendruck cavity pressure *(im)*
Kavitätenzahl number of cavities/impressions *(im)*
kavitieren to cavitate
Kegelanguß sprue gate *(im)*
Kegelanschnitt sprue gate *(im)*
kegelförmig conical, tapered
kegelig conical, tapered
Kegel-Platterheometer cone and plate rheometer
Kegel-Platteviskosimeter cone and plate viscometer
Kegelstumpf truncated cone *(as used in determining Shore hardness)*
Kegelventil cone valve
Kehlnaht internal angle weld, inside corner weld
keilförmig wedge-shaped
Keilriemen V-belt
Keilriemenantrieb V-belt drive
Keilriemenscheibe V-belt pulley
Keilriemenuntersetzung V-belt speed reducing drive
Keilvorschubverfahren advancing wedge method *(of determining film adhesion)*

Keilwinkel wedge angle, lip angle
Keimbildner nucleating agent
Keimbildung nucleation
Keimbildungsgeschwindigkeit nucleation rate
Kennbuchstabe code/identifying letter
Kenndaten characteristics, constants, properties
Kennfunktion characteristic function
Kenngröße characteristic (value), constant, property
Kennlinie curve, characteristic
Kennlinien-Linerarisierungsprogramm curve linerarisation program *(me)*
Kennwert characteristic (value), constant, property: **mechanische/elektrische/thermische Kennwerte** mechanical/electrical/thermal properties
Kennzahl characteristic (value), constant, property
Kennzeichen feature, characteristic
kennzeichnende Eigenschaften characteristic features
Kennzeichnung characterisation
kennzeichnungspflichtig requiring special markings/labelling
Keramikbänder ceramic band heaters, ceramic-insulated band heaters
Keramikheizbänder ceramic band heaters, ceramic-insulated band heaters
Keramikheizkörper ceramic band heater, ceramic-insulated band heater
Keramikheizung ceramic band heaters, ceramic-insulated band heaters
keramisch ceramic
Kerbe notch
Kerbeffekt notch effect
kerbempfindlich notch-sensitive
Kerbempfindlichkeit notch sensitivity
Kerbfestigkeit notch resistance
Kerbgrund notch base
Kerbradius notch radius
Kerbschlagbiegeversuch notched flexural impact test
Kerbschlagbiegezähigkeit notched flexural impact strength
Kerbschlagfestigkeit notched impact strength
Kerbschlagprobe notched impact test specimen
Kerbschlagtest notched impact test
Kerbschlagversuch notched impact test
Kerbschlagwert notched impact strength
kerbschlagzäh notch impact resistant
Kerbschlagzähigkeit notched impact test
Kerbschlagzugzähigkeit notched tensile impact strength
Kerbstellenwirkung notch effect
Kerbtiefe notch depth
Kerbunempfindlichkeit notch resistance
Kerbung notching
Kerbwirkung notch effect
kerbzäh notch resistant
Kerbzähigkeit notch resistance
Kern 1. root *(of a screw) (e)*. 2. mandrel *(part of a die) (e)*. 3. core, core plate *(im)*

Kernausschraubvorrichtung core unscrewing device/mechanism *(im)*
Kernbereich core region
Kernbüchse core box
Kerndrehvorrichtung core rotating device/mechanism *(im)*
Kerndurchmesser root diameter *(e)*
Kerne, eingelegte core inserts
Kerneigenschaften key properties, most important properties
Kerneinsatz core insert *(im)*
Kernenergie nuclear energy
Kernforschung nuclear research
Kernfrage crucial question
Kerngießen investment casting, lost wax casting
Kernkasten core box
Kernkraft nuclear power
Kernkraftwerk nuclear power station
Kernlage core layer
Kernmaterial core material
Kernmodell core pattern
Kernplatte core plate *(im)*
kernprogressiv with constantly increasing root diameter *(screw) (e)*
Kernprogressivität *term denoting that a screw is* **kernprogressiv** (q.v.)
Kernprogressivschnecke constant taper screw *(screw with constantly increasing root diameter)*
Kernreaktor nuclear reactor
Kernresonanz nuclear resonance
Kernresonanzspektroskopie nuclear resonance spectroscopy
Kernsand core sand
Kernsandbindemittel foundry resin
Kernschicht core
Kernseite moving mould half *(im)*
kernseitig on the moving mould half *(im)*: **kernseitige Werkzeughälfte** moving mould half
Kernspeicher internal memory *(of a process computer) (me)*
Kernstift core pin *(im)*
Kernstoffabfall nuclear waste
Kernstück central/main part *(for translation example see under* **Herzstück***)*
Kerntechnik nuclear engineering
kerntechnisch nuclear
Kernträger core plate *(im)*
Kernträgerplatte core plate *(im)*
Kernverbund sandwich structure/component
Kernverbundbau sandwich construction/structure
Kernverbundbauteil sandwich component
Kernverbundkonstruktion sandwich construction/structure
Kernversatz core misalignment/displacement : **um einen Kernversatz zu vermeiden....** to prevent the core getting out of alignment
Kernwandtemeratur core wall temperature *(im)*
Kernwerkstoff core material
Kernziehen core pulling *(im)*

Kernzug core puller *(im)*
Kernzugsteuerung 1. core puller control. 2. core puller control mechanism *(im)*
Kernzugvorrichtung core puller, core pulling mechanism *(im)*
Kerosin kerosene
Kerosinbrennertest kerosene burner test
Kerzenfilter cartridge/candle filter *(mf)*
Kessel reactor
Kesselwagen road tanker
Ketenacetal ketene acetal
Ketocarbonylgruppe keto-carbonyl group
Ketogruppe keto group
Keton ketone
Ketongruppe ketone group
Ketonharz ketone resin
Ketonhydroperoxid ketone hydroperoxide
Ketonperoxid ketone peroxide
Ketosäure keto acid
Kettbaum warp beam *(txt)*
Kettbaumfolie warp beam film
Kettdraht warp wire
Kette 1. warp *(txt)*. 2. chain
Kettenabbau chain degradation
kettenabbauend chain-degrading
Kettenabbruch chain termination
Kettenabschnitt chain segment
Kettenaktivierung chain activation
Kettenantrieb chain drive
kettenartig chain-like
Kettenaufbau chain structure
Kettenausrichtung chain alignment/orientation
Kettenbaum warp beam
Kettenbaustein chain segment
Kettenbeweglichkeit chain mobility
Kettenbruch chain fission/scission
Kettenbruchstück chain segment
Kettenende chain end
Kettenfaden warp thread
Kettenförderer chain conveyor
kettenförmig chain-like
Kettengebilde chain structure
Kettenglied chain segment
Kettenlänge chain length
Kettenmolekül chain molecule
Kettenorientierung chain orientation
Kettenradikal chain radical
Kettenreaktion chain reaction
Kettensegment chain segment
Kettenspaltung chain fission/scission
Kettenstart chain initiation
Kettenstarter chain initiator
kettensteif with rigid chains
Kettensteifigkeit chain stiffness
Kettenstruktur chain structure
Kettenstück chain segment
Kettenteil chain segment
Kettentrieb chain drive
Kettenübertragung chain transfer
Kettenübertragungsreaktion chain transfer reaction
Kettenverkürzung chain shortening

Kettenverlängerer chain extender, curing/crosslinking agent
Kettenverlängerung chain extension
Kettenverlängerungsreaktion chain extension reaction
Kettenverschlaufung chain entanglement
Kettenverzweigung chain branching
Kettenwachstum chain propagation
Kettfaden warp thread
Kettrichtung warp direction *(txt)*
Kettwirktechnik warp knitting *(txt)*
Kfz abbr. of **Kraftfahrzeug,** motor vehicle
Kfz-Anwendung automotive application
Kfz-Innenraum vehicle interior
Kfz-Rückleuchte rear light
Kfz-Sektor motor industry
Kfz-Stoßfänger (car) bumper
KI abbr. of **künstliche Intelligenz,** artificial intelligence
Kicker kicker
Kieselgel silica gel
Kieselgur kieselguhr, diatomaceous earth
Kieselsäure 1. silicic acid. 2. silicon dioxide, silica
Kieselsäuregel silica gel
Kieselsäuregerüst silica skeleton
kindersicher child-resistant, child-proof
Kinderspielwaren toys
kinematische Viskosität kinematic viscosity
Kinetik kinetics
kinetisch kinetic
kinetische Energie kinetic energy
Kippfenster bottom-hung window
Kippflügel tilting casement
Kippschalter tumbler switch
Kippvorrichtung tilting arrangement/mechanism/device
Kissenmaterial upholstery material
Kitt putty, mastic
Kitt, hüpfender bouncing putty
kittartig putty-like
Kittmasse putty, mastic
KKB abbr. of **Kunststoff-Kraftstoffbehälter,** plastics fuel tank
Klammerwert figure in brackets, bracketed figure
klappbar swing-back, swing-hinged, swivel-mounted, swivel-type, hinge-mounted, hinged *(machine unit)*
klappen to tilt
Klappfenster top-hung window
Klappflansch hinged flange
Klappflügel folding casement
Kläranlage sewage purification plant
Klarheit clarity
Klarlack clear lacquer/varnish
klarlöslich completely soluble (the **klar-** refers to the fact that a clear solution is obtained)
Klärschlamm (sewage) sludge
Klärschlammbecken (sewage) sludge purification basin
Klarschliffverbindung ground glass joint

Klarschrifttastatur character/typewriter/ QWERTY keyboard *(me)*
Klarsichtdeckel transparent cover
Klarsichtfolie transparent film
klarsichtig transparent
Klarsichtpackung see-through pack
Klartext plain language *(me) (i.e. plain English, German etc.)*
Klartextanzeige plain language display *(me)*
Klartextausdruck plain language print-out *(me)*
Klartextmeldung plain language signal *(me)*
Klartexttastatur character/typewriter/QWERTY keyboard *(me)*
Klassiereinheit grading unit *(for powders)*
klassieren to grade *(powders)*
Klassifizierung classification
klassisch classic
Klauenkupplung clutch coupling
Klebeband adhesive tape
Klebebeschichtung adhesive coating
Klebeeigenschaften adhesive properties
Klebefähigkeit adhesive properties
Klebefestigkeit bond/adhesive strength
Klebefilm film adhesive
Klebefläche adherend surface
Klebefolie 1. self-adhesive film 2. film adhesive 3. bonding sheet *(an alternative name for the prepregs used in the production of multi-layer laminates)*
Klebefolienbuchstabe self-adhesive letter
klebefreundlich easy to bond/stick
Klebefuge glueline
Klebefugenbruch glueline failure
Klebehaftung adhesion
Klebeharz adhesive resin
Klebeigenschaften adhesive properties
Klebekraft adhesive power/strength
Klebemasse adhesive
Klebemittel adhesive
Klebemörtel adhesive mortar
kleben to bond
klebend tacky, sticky
Kleber adhesive
Kleberlösung adhesive solution
Kleberreste adhesive residues
Kleberschicht adhesive coating/layer
Klebespalte glueline
Klebestreifen adhesive tape
Klebetechnik bonding technique
Klebeverbindung bonded joint
Klebevermögen adhesive power
Klebevorgang bonding process
Klebfestigkeit bond strength
Klebfläche adherend surface, surface to be bonded
Klebflächenvorbehandlung preparation of adherend surfaces
klebfrei tack-free
Klebfreiheit freedom from tack
Klebfuge glueline
Klebharz adhesive resin
Klebkraft adhesive power/strength

Klebnaht glueline
Klebneigung tendency to stick *(e.g. compound to metal rolls)*
klebrig tacky, sticky
Klebrigkeit tack, tackiness
Klebrigkeitsdauer open time
klebrigmachend tackifying
Klebrigmacher tackifier, tackifying agent
klebrigweich soft and tacky
Klebrohstoff adhesive resin
Klebschicht adhesive film/layer
Klebschichtstärke adhesive film thickness
Klebstoff adhesive
klebstoffabweisend non-stick, anti-adhesive
Klebstoffansatz adhesive mix: **Je größer der Klebstoffansatz, um so kürzer ist die Verarbeitungszeit** the greater the amount of adhesive mixed, the shorter will be the pot life
Klebstoffauftrag application of the adhesive
Klebstoffdämpfe adhesive fumes
Klebstoffharz adhesive resin
Klebstoffindustrie adhesives industry
Klebstoffmischung adhesive (mix)
Klebstoffolie film adhesive
Klebstoffformulierung adhesive formulation
Klebstoffpatrone adhesive cartridge
Klebstoffreste adhesive residues
Klebstoffrohstoff adhesive resin
Klebstoffschicht adhesive film/layer
Klebstoffsortiment range of adhesives
Klebstoffsystem adhesive (system)
Klebsystem adhesive (system)
klebtechnisch *relating to adhesives, bonding, etc.*: **klebtechnische Eigenschaften** adhesive properties
Klebung bonded joint
Klebverbindung bonded joint
Klebwerkstoff adhesive
Klebwert bond strength
Klebwirkung adhesive effect
Kleiderbügeldüse coathanger die *(e)*
Kleiderbügelkanal coathanger manifold *(e)*
Kleiderbügelverteilerkanal coathanger manifold *(e)*
Kleinbetrieb small company/business/concern/firm
Kleincharge small batch
kleine Produktionsserien short runs
kleinflächig small-area
Kleingerät small-scale unit
Kleinhohlkörper small blow moulding
Kleinrechner minicomputer *(me)*
Kleinrohr small-bore pipe
Kleinserien short runs
Kleinserienanwendungen small-scale applications
Kleinserienfertigung small-scale production
Kleinstmaß minimum dimension
Kleinstmenge extremely small amount/quantity
kleinstmöglich smallest possible
Kleinstserien extremely short runs
Kleinstwerkzeug extremely small mould

Kleinstwert miniumum value
kleintechnisch small-scale, pilot plant scale
Klemmabstand distance between clamps *(e.g. in tensile tests)*
Klemmbacken clamps, grippers
Klemme clamp
Klemmleiste terminal bar
Klemmpratze clamp
Klimaanlage air conditioning unit/equipment
Klimabedingungen climatic conditions
Klimagerät air conditioning unit
Klimakammer conditioning cabinet/chamber
Klimaraumlagerung ageing under standard climatic conditions
Klimaschrank conditioning cabinet
klimatisch climatic
klimatisiert air-conditioned, climatically controlled
Klimatisierung 1. air conditioning. 2. air conditioning unit
Klimawechselbeständigkeit resistance to changing climatic conditions
Klimazelle conditioning cabinet/chamber
Klimazelle conditioning cabinet
Klinkermauerwerk clinker walls
klümpchenfrei free from lumps
Klumpenbildung lump formation
klumpenfrei free from lumps
Kluppenrahmen tenter/stenter frame *(txt)*
Knallgasflamme oxyhydrogen flame
Knäuelung entanglement *(of molecules)*
Knautschzone crumple zone *(in a car)*
Knet bank *(c)*
Knetarm kneader blade
knetbar kneadable
Knetblock kneading block
Knetblockanordnung kneading block assembly
Knetelemente kneading discs
Kneter kneader, compounder
Knetergehäuse 1. kneader housing *(of a mixer or kneader)*. 2. barrel assembly *(of a compounding unit)*
Kneterschaufeln kneader blades
Knetflügel mixing/kneader blades
Knetgang mixing flight *(of a screw)*
Knetkammer mixing chamber/compartment
Knetraum mixing chamber/compartment
Knetschaufeln mixing/kneader blades
Knetscheiben kneading discs
Knetscheibenschneckenpresse screw compounder
Knetschnecke mixing/compounding/homogenising screw
Knetschneckeneinheit compounding screw unit
Knetschneckenwelle mixing/compounding/homogenising screw
Knetspalt mixing gap *(of a mixing screw)*
Knetsteg mixing flight *(of a screw)*
Knetwelle mixing/compounding/homogenising screw
Knetzahn kneading cog
Knick break *(in a curve)*

Knickfestigkeit 1. buckling resistance. 2. folding endurance *(e.g. of PVC leathercloth)*
Knicklast buckling stress
Knickpunkt kink *(in a curve)*
Knickung buckling, folding
Knickversuch folding test
Kniehebel toggle
Kniehebelformschließaggregat toggle mould clamping unit
Kniehebelgetriebe toggle gear
Kniehebelhub toggle stroke
Kniehebelmaschine toggle clamp machine
Kniehebelmechanismus toggle mechanism
Kniehebelpresse toggle press
Kniehebelschließsystem toggle clamp system
Kniehebelsystem toggle mechanism
Kniehebelverriegelung toggle lock mechanism/system
Kniehebelverschluß toggle clamp mechanism/system
Knistergeräusch crackling (noise)
knitterarm crush/crease resistant
Knotenpunkt point of intersection, nodal point
Koagulation coagulation
koagulieren to coagulate
Koagulierung coagulation
Koaleszenz coalescence
Koaxialkabel coaxial cable
Kobaltbeschleuniger cobalt accelerator
Kobaltblau cobalt blue
Kobalthärtung cobalt cure
Kobaltnaphthenat cobalt naphthenate
Kobaltoctoat cobalt octoate
Kobaltsikkativ cobalt drier
Kobalttrockenstoff cobalt drier
kochfest resistant to boiling water
Kochversuch boiling test
Kochwasser boiling water
kodiert coded *(me)*
kodiert, binär binary coded *(me)*
Koextrudieren coextrusion
Koextrusionsanlage coextrusion line
Koextrusionsblasanlage coextrusion blow moulding line/plant
Koextrusionsblasen coextrusion blow moulding
Koextrusionsblaskopf blown film coextrusion die
Koextrusionsblasversuch coextrusion blow moulding trial
Koextrusionsblaswerkzeug blown film coextrusion die
Koextrusionskopf coextrusion die head *(e)*
Koextrusionsverfahren coextrusion (process)
Koextrusionswerkzeug coextrusion die
Kofferraum boot
Kofferraumauskleidung boot liner
Kofferraumdeckel boot lid
Koflerbank Kofler hot bench
koflockuliert co-flocculated
kohärent coherent
Kohärenz coherence
Kohäsion cohesion

Kohäsionsbruch cohesive fracture/failure
Kohäsionseigenschaften cohesive properties
Kohäsionsenergie cohesive energy
Kohäsionsfähigkeit cohesion
Kohäsionsfestigkeit cohesive strength
Kohäsionskraft cohesive force
Kohäsionsriß cohesive failure
Kohäsionsversagen cohesive fracture/failure
kohäsiv cohesive
Kohäsivbruch cohesive fracture/failure
Kohäsivfestigkeit cohesive strength
Kohle 1. coal 2. charcoal 3. carbon
kohleartig carbon-like
Kohlefaser carbon fibre
kohlefaserverstärkt carbon fibre reinforced
Kohlelichtbogen carbon arc
Kohlelichtbogenstrahlung carbon arc radiation
Kohlelichtlampe carbon arc lamp
Kohlendioxid carbon dioxide
Kohlenfaserfüllung carbon fibre reinforcement
Kohlenfaserverstärkung carbon fibre reinforcement
Kohlenmonoxid carbon monoxide
Kohlensäure carbonic acid
kohlensäurefrei uncarbonated
kohlensäurehaltig carbonated *(beverage)*
Kohlenstoff carbon
kohlenstoffarm low-carbon *(e.g. steel)*
Kohlenstoffaser carbon fibre
Kohlenstoffasergewebe carbon fibre fabric
Kohlenstoffaserlaminat carbon fibre laminate
kohlenstoffaserverstärkt carbon fibre reinforced
Kohlenstoffaserverstärkung carbon fibre reinforcement
Kohlenstoffatom carbon atom
Kohlenstoffelektrode carbon electrode
Kohlenstoffkette carbon chain
Kohlenstoffkettenlänge carbon chain length
Kohlenstoff-Kohlenstoffbindung carbon-carbon bond
Kohlenstoffkurzfaser chopped carbon fibre
Kohlenstoffpigment carbon black
Kohlenstoffreste carbon residues
Kohlenstoffrückstände carbon residues
Kohlenstoffschaum carbon foam
Kohlenwasserstoff hydrocarbon
Kohlenwasserstoffgruppe hydrocarbon group
Kohlenwasserstoffharz hydrocarbon resin
Kohlenwasserstoffkautschuk hydrocarbon rubber
Kohlenwasserstoffkette hyrocarbon chain
Kohlenwasserstoffpolymer hydrocarbon polymer
Kohlenwasserstoffrest hydrocarbon radical
Kohlenwasserstoffverbindung hydrocarbon compound
Kohlenwasserstoffwachs hydrocarbon wax
Kohleteer coal tar
Kokille metal mould
Kokillenguß chill casting
Kokillenhartgußwalze chill-cast roll

Kolben 1. plunger, ram, piston. 2. flask
Kolbenakku ram accumulator *(e)*
Kolbenakkumulator ram accumulator *(e)*
Kolbenbewegung plunger/ram movement: **Hat der Dorn die Düse passiert, wird die Kolbenbewegung beendet** as soon as the mandrel has passed through the nozzle, the plunger stops moving
Kolbendichtung piston seal
Kolbeneinspritzsystem plunger injection system
Kolbenextruder ram extruder
kolbenfrei free from hoops *(see also explanatory note under* **Kolbenringe***)*
Kolbenhub piston/plunger/ram stroke
Kolbeninjektions-Spritzgießmaschine plunger injection moulding machine
Kolbeninjektor ram injector unit
Kolbenmaschine 1. plunger injection moulding machine. 2. ram extruder
Kolbenplastifizierung 1. plunger plasticisation. 2. plunger plasticising unit *(im)*
Kolbenplastifizierzylinder plunger plasticising cylinder *(im)*
Kolbenpumpe piston/reciprocating pump
Kolbenring piston ring
Kolbenringe hoops: *These are caused locally when film of uneven thickness is wound up tightly*
kolbenringfrei free from hoops *(see also explanatory note under* **Kolbenringe***)*
Kolbenspeicher ram accumulator *(e)*
Kolbenspritzeinheit plunger injection unit *(im)*
Kolbenspritzgießmaschine plunger injection moulding machine
Kolbenspritzsystem plunger injection mechanism *(im)*
Kolbenspritzzylinder plunger injection cylinder
Kolbenstange piston rod
Kolbenstopfaggregat ram feeder
Kolbenstrangpresse ram extruder
Kolbentriebwerk piston drive mechanism
Kolbenvorlaufgeschwindigkeit plunger advance speed
Kolbenvorlaufzeit plunger advance speed *(like many words ending in* **-zeit**, *this one, too, is synonymous with* **-geschwindigkeit***)*
kolbenwegabhängig plunger/ram stroke-dependent, depending on the plunger/ram stroke
kollabieren to collapse *(e.g. foam)*
Kollision collision
Kolloid colloid
kolloidal colloidal
Kolloidalbereich colloidal range
Kolloidmühle colloid mill
Kolonnenfüllkörper column packings
Kolophonium colophony, rosin
Kolophoniumester colophony/rosin ester
Kolorimeterrohr Nessler cylinder
Kombination combination

Kombinationsflachfolie composite/multi-layer flat film
Kombinationsgleitmittel lubricant blend
Kombinationsharz blending resin
Kombinationsmöglichkeit 1. *possibility of combining (e.g. different parts of a machine).* 2. compatibility: **Mit Nitrocellulose besteht nur eine sehr geringe Kombinationsmöglichkeit** its compatibility with nitrocellulose is very poor
Kombinationspartner blending component: **Ölfreie Polyesterharze sind in vieler Hinsicht die optimalen Kombinationspartner für Cellulosenitrat** Oil-free polyesters are, in many ways, the best resins for combining/blending with cellulose nitrate
Kombinationsverhältnis blending ratio
kombinierbar combinable
Kommando command, instruction *(me)*
Kommandosequenz command sequence *(me)*
Kommastelle decimal point
kommunal municipal
Kommunalmüll municipal waste/rubbish/refuse
Kommunikation communication
Kommunikationsarchitektur communication architecture *(me)*
Kommunikationsbus communication bus *(me)*
kommunikationsfähig with communications capability, communicating *(me)*
Kommunikationsnetz communication network *(me)*
Kommunikationsprozessor communication processor *(me)*
Kommunikationssystem communication system *(me)*
Kommunikationstechnik communication/ telecommunication (engineering)
Kommutatormotor commutator motor
kompakt compact, solid
Kompaktbauweise totally enclosed *(machine)*
Kompaktgerät compact unit/instrument
Kompaktheit compactness
Kompaktierung compaction, compression
Kompaktspritzen conventional injection moulding *(see explanatory note under* **Kompaktspritzguß***)*
Kompaktspritzgießmaschine conventional injection moulding machine *(see explanatory note under* **Kompaktspritzguß***)*
Kompaktspritzguß conventional injection moulding *(i.e. employing solid material - hence the word* **kompakt** *- as opposed to structural foam moulding)*
kompatibel compatible
kompatibilisieren to make compatible
Kompatibilität compatibility
Kompatibilitätsgrenze compatibility limit
Kompatibilitätsprobleme compatibility problems
komplementär complementary
Komplementärfarbe complementary colour

Komplettanlage complete plant
Komplex complex
Komplexbildner chelator, chelating agent
komplexiert doped
Komplexion complex ion
Komplexität complexity
kompliziert complex, complicated
Kompostieren composting
kompostierfähig compostable
Kompostierung composting
Kompoundier- *see* **Compoundier**
Kompressibilität compressibility
Kompression 1. compression. 2. compression ratio
Kompressionsbeanspruchung compressive stress
Kompressionsbereich compression/transition section *(of screw) (e)*
Kompressionsblasen compression blow moulding
Kompressionsblasformverfahren compression blow moulding (process)
Kompressionsdruck compression
Kompressionsentlastung decompression
Kompressionsformen compression moulding
Kompressionsmodul compressive modulus
Kompressionsphase packing phase *(im)*
Kompressionsschnecke compression screw *(e)*
Kompressionsverformung compressive deformation
Kompressionsverhältnis compression ratio *(e)*
Kompressionszeit packing time, packing phase *(im)*
Kompressionszone compression/transition section *(of a screw)(e)*
komprimierbar compressible
komprimieren to compact, to compress, to consolidate
Komprimierzone compression/transition section *(of a screw)(e)*
Kompromiß compromise
Kompromißlösung compromise solution
Kondensat condensate
Kondensation condensation
Kondensationsgrad degree of condensation
kondensationshärtend condensation-curing
Kondensationsharz condensation resin
Kondensationskunststoff condensation polymer
Kondensationspolymer condensation polymer
Kondensationspolymerisation condensation polymerisation
Kondensationsprodukt condensation product
Kondensationsreaktion condensation reaction
kondensationsvernetzend condensation-curing/-crosslinking
Kondensationsvernetzung condensation-curing/-crosslinkage
Kondensator 1. capacitor. 2. condenser: *(Use version 1. only in an electrical context)*
Kondensatorfolie capacitor film

Kondensatorplatten capacitor plates
kondensieren to condense
Kondenswasser condensation, condensed moisture
Kondenswasserschicht layer/film of condensation
konditioniert conditioned, annealed
Konditionierung conditioning, annealing
Konditionierungslagerung conditioning, annealing
konduktometrisch conductometric
konfektioniert compounded, formulated, made-up; *this word is often used for the sake of convenience rather than clarity, as in this example*: **konfektionierte Additive**: *from the context of the article this turned out to be nothing more than* predispersed additives. **In einigen Fällen empfiehlt sich der Einsatz von konfektionierten Pigmentpasten**, in some cases it is best to use ready-made pigment pastes. *See also example 24 in the appendix, and explanatory notes and translation examples under* **Konfektionierung**
Konfektionierung *This word was originally associated with the clothing industry, but nowadays is used very widely to express whatever an author wants it to mean. Translators must use their ingenuity in dealing with it. Here are some meanings*: manufacture, conversion, making-up *(various kinds of articles from film, coated fabric etc.)*, fabrication, compounding, formulation, solution. **Die Polymeren lassen sich gut lösen und aus einer Konfektionierung mit 50% Xylol heraus verarbeiten**. The polymers dissolve easily and can be processed in the form of a 50% solution in xylene. **Wegen ihres ausgeprägt hydrophoben Characters können die meisten Wachse nur schwer in eine wässrige Konfektionierung einbezogen werden**. Because of their pronounced hydrophobic character, most waxes are difficult to use in water-based formulations. *When the word does not seem to make sense in a given context, translators should use their own judgement and substitute whatever word they think fits the bill, even if it is not among those given above.*
Konfektionierungseigenschaften compounding properties
Konfektionierungsklebrigkeit building tack
...steigert die Konfektionierungsklebrigkeit von Gummimischungen ...acts as a tackifier for rubber compounds
Konfektionierungskosten fabricating costs
Konfektionierungsmaschine conversion equipment
Konfektionsklebrigkeit building tack, autohesion
Konfiguration configuration
konisch conical
Konizität draft, draw *(im)*

konjugiert conjugated *(e.g. double bond)*
Konjunktur economic situation/trend, market conditions
Konjunkturabhängigkeit dependence on the economy
konjunkturell economic
Konjunkturentwicklung economic development
Konjunkturerholung economic recovery
Konjunkturlage economic situation
Konjunkturpolitik economic policy
Konjunkturschwäche (economic) recession
Konjunkturverlauf economic trend
Konkurrent competitor
Konkurrenz competition
Konkurrenzfabrikat rival product
konkurrenzfähig competitive
Konkurrenzfähigkeit competitiveness
Konservierer preservative
Konservierung preservation
Konservierungsmittel preservative
Konservierungsstoff preservative
Konsistenz consistency
Konsistenzgeber thickener, thickening agent, consistency modifier
konsolidiert consolidated
konstant constant, even, uniform
Konstant, dielektrischer dielectric constant, permittivity
Konstanthaltung *keeping something constant*: **Geräte zur Konstanthaltung der Temperatur** devices used to maintain a constant temperature
Konstantklima constant test atmosphere
Konstantpumpe constant delivery pump, fixed displacement pump
Konstantspannungsquelle source of constant voltage
Konstantstrom constant current
Konstanz constancy : *Although this is literally correct, the word is not usually translated as such, e.g.* **größere Konstanz der Eigenschaften** *should be rendered as* more constant properties. *See also entries under* **Gewichtskonstanz** *and* **Temperaturkonstanz**
konstruieren 1. to design 2. to construct
Konstruieren, computergestütztes computer aided design, CAD
Konstruieren, rechnerunterstütztes computer aided design, CAD
konstruiert, richtig correctly designed
Konstrukteur designer
Konstruktion 1. design. 2. construction : *Version 1. will normally be the first choice, version 2. being rare since the word normally used to denote construction is* **Bau.** *The translator can only decide from the context when and whether to use version 2.*
Konstruktionsänderung design modification, change in design
Konstruktionsaufbau construction

Konstruktionsbüro drawing office, design studio
Konstruktionsdaten design data
Konstruktionsdetails 1. design details 2. structural details *(see explanatory note under* **Konstruktion***)*
Konstruktionsdichtungsmasse structural sealant
Konstruktionseinzelheiten 1. design details. 2. structural details *(see explanatory note under* **Konstruktion***)*
Konstruktionsform 1. design, type of design. 2. construction, type of construction *(see explanatory note under* **Konstruktion***)*
Konstruktionshinweise 1. design guidelines. 2. construction guidelines *(see explanatory note under* **Konstruktion***)*
Konstruktionsklebstoff structural adhesive
Konstruktionskonzept design concept
Konstruktionsleichtbau lightweight construction
Konstruktionsmaterial 1. engineering plastic/polymer 2. construction material
Konstruktionsmerkmale 1. design features. 2. structural features *(see explanatory note under* **Konstruktion***)*
Konstruktionsmöglichkeiten 1. design possibilities. 2. construction possibilities *(see explanatory note under* **Konstruktion***)*
konstruktionsorientiert design-orientated
Konstruktionsphase design stage
Konstruktionsprinzipien 1. design principles. 2. construction principles *(see explanatory note under* **Konstruktion***)*
Konstruktionsrichtlinien design guidelines
Konstruktionsthermoplast engineering thermoplastic
Konstruktions-TPE engineering thermoplatic elastomers
Konstruktionswerkstoff 1. construction material: **wird Stahl als Konstruktionswerkstoff für die Düse gewählt...** if the die is made of steel... 2. engineering polymer/plastic
Konstruktionswerkstoffe, thermoplastische engineering thermoplastics
Konstruktionszeichnung design drawing
konstruktiv 1. constructional, structural 2. *relating to design*: **konstruktive Änderungen** design modifications
konstruktiv aufwendig complex in design
Konsumartikel consumer goods
Konsumentengruppe consumer group
Konsumentenspritzguß injection moulding of consumer goods
Konsumgüter consumer goods
Konsumkunststoffe bulk plastics
Konsumspritzteile injection moulded consumer goods
Konsumwaren consumer goods
Kontaktbluten bleeding *(of pigments or dyes)*
Kontaktdruck contact pressure

Kontaktfläche 1. contact surface. 2. area of contact
Kontaktheizung 1. contact heating. 2. contact heating unit
Kontaktkleber contact adhesive
Kontaktklebstoff contact adhesive
Kontaktkühlung 1. contact cooling. 2. contact cooling unit
Kontaktkühlwalze contact cooling drum/roll
Kontaktleiste contact strip
kontaktlos 1. solid state, electronic. 2. non-contact, contactless
kontaktlose Steuerung solid-state controls
Kontaktpunkt contact point, point of contact
Kontaktstelle contact point
Kontaktthermometer contact thermometer
Kontaktverfahren contact moulding (process) *(grp)*
Kontaktwalze pressure roll
Kontaktwiderstand contact resistance
Kontaktwinkel contact angle
Kontamination contamination
kontaminiert contaminated
Kontermutter lock nut, check nut
Kontianlage continuous plant
kontinuierlich continuous(ly)
Kontinuumsmechanik continuum mechanics
Kontraextruder contra-extruder
Kontraschnecke contra-screw
Kontrastfarbe contrasting colour
kontrastreich high-contrast
Kontrollampe pilot light
Kontrollehre checking fixture
Kontrollgerät control instrument
kontrollieren to check
Kontrollmodell checking fixture
Kontrollprüfung quality control test
Kontrollrechner control computer *(me)*
Kontrollschrank control cabinet
Kontrollsignal control signal
Kontrollvariable control variable
Kontureinsatz mould insert *(im)*
Konturen contours
Konturentlüftung mould venting
Konturplatte cavity plate *(im)*
Konvektion convection
Konvektionsstrom convection current
Konvektionsverluste heat losses due to convection
konvergent convergent, converging
konvergierend converging
Konzentrat concentrate
Konzentratgranulat masterbatch
Konzentration concentration
Konzentrationsänderung change of concentration
Konzentrationsbereich concentration range
konzentrationsbezogen concentration-related
Konzentrationsgefälle concentration gradient
Konzentrationsgradient concentration gradient
Konzentrationssprung sudden increase in concentration

Konzentrationsstreuungen variations in concentration
Konzentrationsverhältnis concentration ratio
konzentriert concentrated
konzentrisch concentric
Konzentrizität concentricity
Konzept concept, idea, notion, design, scheme: *This word needs great care in translation when linked to another, e.g.* **Spritzblaskonzept,** *where it must be translated as unit or machine, or* **Battenfeld-Fischer Konzept VK 4-0.7,** *where the word has obviously been used instead of the less grand sounding but more correct word* **Maschine.** *Sometimes, indeed, the word should be omitted altogether, as in* **Streckblasanlagenkonzept CIB-24** *where* **Anlage** *and* **Konzept** *mean the same thing. Here are two more examples:* **Maschinenkonzepte in Bausatzform** modular machines; **Verfahrens- und Maschinenkonzepte für das Streckblasverfahren** stretch blow moulding processes and machines; **extremes Konzept** advanced design
Konzeptionsänderung design modification
Konzeptphase design stage: **bereits in der Konzeptphase** already at the design stage
Konzernplanung corporate/group planning
Konzernunternehmen subsidiary
konzipiert conceived, designed
Konzipierung design
Koordinatensystem coordinate system
Kopalharz copal
Köperbindung twill weave *(txt)*
Kopf 1. extruder head, die head *(e)*. 2. die *(e)*: *Version 2 should be used wherever* **Kopf** *is linked to words denoting manufactured products such as* **Schlauch, Rohr** *etc. See* **Rohrkopf, Schlauchkopf, Profilkopf, Schlauchfolienkopf.** *On the other hand,* **Speicherkopf** *is translated as* accumulator head
Kopfgranulator die-face/hot-cut pelletiser *(sr)*
Kopfgranulierung 1. die-face/hot-cut pelletisation. 2. die-face/hot-cut pelletiser *(sr)*
Kopfkonstruktion extruder head design, die head design *(e)*
Kopfspeicher melt accumulator *(bm)*
Kopfspeichermaschine melt accumulator machine *(bm)*
Kopfspeichersystem melt accumulator system *(bm)*
Kopfspiel flight clearance: *This is the clearance between the flight land of one screw and the root surface of the other in a twin screw assembly*
Kopfstütze head rest/restraint
Kopftemperatur die head temperature
Kopierfräsmaschine copy milling machine
Kopierfräsmodell copy-milling model
Kopiermodell copy-milling model
Kopoly-, kopoly- *see* **Copoly-** *and* **copoly-**
koppeln to couple, to connect
Kopplung coupling, linkage
Kopplungsmechanismus coupling mechanism
Kopplungsmöglichkeiten *possibilities of linking/connecting*: **Hydraulische Steuerungen mit einfachen Kopplungsmöglichkeiten mit elektronischen Steuerungen** hydraulic controls which can easily be linked to electronic controls
Kopplungssystem coupling system
Korb calibrating/sizing basket, calibrating/sizing cage *(bfe)*
Kord cord
Korn particle
*∗**kornartig** granular
Kornbeschaffenheit particle characteristics
Korndurchmesser particle size
Kornfeinheit particle size
Kornform particle shape
Korngröße particle size
Korngröße, häufigste predominant particle size
Korngrößenbereich particle size range
Korngrößenverteilung particle size distribution
Korngruppenverteilung particle size distribution
körnig granular
Kornklasse particle size range
Kornoberfläche particle surface
Kornpartikel particle
Kornporosität particle porosity
Kornspektrum particle size range
Kornstruktur particle structure
Körnung 1. particle size. 2. grit *(of sandpaper, emery paper etc.)*
Körnungsverteilung particle size distribution
Kornveränderung change(s) in particle characteristics
Kornverteilung particle size distribution
Kornzusammensetzung particle size distribution
Koronafestigkeit corona resistance
Koronavorbehandlung corona pretreatment
Körperflüssigkeiten body fluids
Körpergehalt solids content
körperreich high-solids
Körperschall solid-borne sound, sound transmitted by solids
Körperschallabsorption absorption of solid-borne sound
körperverträglich body-compatible
Korrekturfaktor correction factor
Korrekturwert correction factor
Korrelation correlation
Korrelationskoeffizient correlation coefficient
korrelieren to correlate
korrigieren to correct, to adjust

∗ *Words starting with* **Korn-**, *not found here, may be found under* **Partikel-** *or* **Teilchen-**

korrodierend corrosive
Korrosion corrosion
Korrosion, elektrochemische electrolytic corrosion
Korrosion, elektrolytische electrolytic corrosion
Korrosion, lochfraßartige pitting
korrosionsanfällig susceptible to corrosion
Korrosionsanfälligkeit susceptibility to corrosion
Korrosionsbeginn start of corrosion
Korrosionsbelastbarkeit corrosion resistance
Korrosionsbelastung exposure to corrosion
korrosionsbeständig corrosion resistant:
   korrosionsbeständiger Stahl stainless steel
Korrosionsbeständigkeit corrosion resistance
korrosionsempfindlich corrosion-sensitive, susceptible to corrosion
Korrosionserscheinungen signs of corrosion
korrosionsfest corrosion resistant
korrosionsfördernd corrosive
Korrosionsfreiheit freedom from corrosion
Korrosionsgefahr risk of corrosion
Korrosionsgeschützt protection against corrosion
Korrosionsgeschwindigkeit corrosion rate
korrosionshemmend corrosion-inhibiting
korrosionsinhibierend corrosion-inhibiting, anti-corrosive
Korrosionsinhibitor corrosion inhibitor
Korrosionsneigung tendency to corrode
Korrosionsreaktion corrosion reaction
Korrosionsschäden corrosion damage: *Note that the word damage is always used in its singular form in this context. The plural is something entirely different*
Korrosionsschutz corrosion protection, anti-corrosive effect
Korrosionsschutzanstrich anti-corrosive coating/finish
Korrosionsschutzeigenschaften anti-corrosive properties
korrosionsschützend anti-corrosive, protecting against corrosion
Korrosionsschutzfarbe anti-corrosive paint
Korrosionsschutzgrundfarbe anti-corrosive primer
Korrosionsschutzgrundierung anti-corrosive primer
Korrosionsschutzinhibitor *an illogical word, for* **schutz** *robs it of its meaning. Forget about protection and translate as* corrosion inhibitor
Korrosionsschutzlack anti-corrosive paint
Korrosionsschutzmittel anti-corrosive agent
Korrosionsschutzpigment anti-corrosive pigment
Korrosionsschutzprimer anti-corrosive primer
Korrosionsschutzsystem anti-corrosive paint
Korrosionsschutzwirkung anti-corrosive effect
Korrosionsspuren signs/traces of corrosion
korrosionstabil corrosion resistant
korrosionsverhindernd corrosion-preventing
korrosionsverursachend causing corrosion, corrosive
Korrosionsverursacher corrosive agent
Korrosionswirkung, elektrolytische electrolytic corrosion
korrosiv corrosive
Korrosivität corrosiveness
Korrosivwirkung corrosive effect
Korund corundum
Korundschlämme corundum slurry
Kosmetika cosmetics
Kosmetikindustrie cosmetics industry
Kostenanalyse cost analysis
Kostenaufwand cost, expense
kostenaufwendig costly, expensive
Kostenbewertung cost appraisal
kosteneffektiv cost-effective
Kosteneffizienz cost-effectiveness
Kosteneinsparung cost saving
Kostenersparnis cost saving
Kostenersparnisgründen, aus to save money
Kostenfaktor cost factor
Kosten-Festigkeitsverhältnis cost-strength ratio
Kostengegenüberstellung cost comparison
kostengerecht inexpensive, reasonably priced, economical, low-cost
Kostengründen, aus for financial/economic reasons
kostengünstig inexpensive, reasonably priced, economical, low-cost
kostenintensiv cost-intensive
Kosten-Leistungsverhältnis cost-performance ratio
Kostenniveau cost
Kosten-Nutzenrechnung cost-benefit calculation
Kosten-Nutzenüberlegungen cost-benefit considerations
Kosten-Nutzenverhältnis cost-benefit ratio
kostenoptimal cost saving
kostenoptimiert cost-effective
Kostenoptimierung achieving cost-effectiveness
Kostenplanung cost planning
Kostenrechnung cost calculation
Kostenreduzierung cost reduction
Kostensenkung cost reduction
kostensparend cost-saving
Kosten-Standzeitrelation cost-service life ratio
Kostensteigerung cost increase
Kostenstruktur cost structure
kostenträchtig expensive, costly
Kostenvergleich cost comparison
Kostenvorteil cost advantage
kostspielig costly, expensive
Kotflügel mudguard, fender
kovalent covalent
KPVK *abbr. of* **kritische Pigment-Volumen-Konzentration**, critical pigment volume concentration, c.p.v.c.
Kracken cracking

Kraft 1. force. 2. power. 3. energy. 4. stress: *In compound words containing* -kraft, *this can sometimes be translated as* tension, *e.g.* **Abzugskraft** *(q.v)*
Kraft, aufgelegte applied force
Kraftamplitude force amplitude
Kraftaufnehmer force transducer
Kraftaufwand amount of force needed ...**ohne großen Kraftaufwand** ...without having to use a lot of force
Kraftbedarf power requirement(s)
Kraft-Dehnungsdiagramm stress-strain diagram/curve
Kraft-Durchbiegungskurve force-deflection curve
Kräftegleichgewicht force equilibrium
Krafteinleitung applications of load(s)
Kraftfahrzeug motor vehicle
Kraftfahrzeugbau vehicle construction
Kraftfahrzeugdecklack automotive paint/enamel/finish
Kraftfahrzeugindustrie motor industry
Kraftfahrzeuginnenausstattung vehicle interior trim
Kraftfahrzeugtank fuel/petrol tank
Kraftfluß stress transmission/distribution, transmission/distribution of forces/stresses
kraftgeregelt power-controlled
Kraft-Längenänderungsdiagramm stress-strain diagram/curve
Kraftmaschine engine
Kraftmaximum maximum force
Kraftmeßdose force transducer
Kraftmeßeinrichtung force transducer
Kraftmeßplatte force transducer/sensor
Kraftmeßwaage force regulator
Kraftschluß *connection where power is transmitted by frictional contact*: **Die Kraft wird durch Kraftschluß übertragen** the power is transmitted by friction
kraftschlüssig non-positive *(connection or joint)*
Kraftstoff fuel, petrol
Kraftstoffbehälter fuel/petrol tank
kraftstoffbeständig fuel resistant
Kraftstoffdurchlässigkeit fuel permeability
Kraftstoffleitung fuel line
Kraftstoffpumpenbalg fuel pump bellows
Kraftstoffschlauch fuel hose/line
Kraftstofftank fuel/petrol tank
Kraftstoffverbrauch fuel consumption
krafttragend load-bearing
kraftübertragend load-/power-transmitting
Kraftübertragung transmission of power/force, power/force transmission
Kraft-Verformungsdiagramm 1. force-deformation diagram/curve *(general term)*. 2. stress-strain diagram/curve *(if referring to tensile stresses)*
Kraft-Verformungsverhalten stress-strain behaviour
Kraftwaage force regulator

Kraft-Wegdiagramm 1. force-deformation diagram *(general term)* 2. stress-strain diagram *(if referring to tensile stresses)*
Kraftwerk power station
Kragarm cantilever arm
Krankenhaus hospital
Krater crater
kraterähnlich crater-like
Kraterbildung cratering
Kraterfreiheit freedom from cratering
kratzbeständig scratch proof, mar resistant
kratzempfindlich easily scratched
Kratzer 1. scratch. 2. scraper
kratzfest scratch proof, mar resistant
Kratzfestbeschichtung scratch/mar resistant coating
Kratzfestigkeit scratch/mar resistance
kratzfestmachendes Additiv scratchproofing additive
Kratzfestmittel scratchproofing additive
Kratzputz stonechip-filled plaster
kratzunempfindlich mar resistant, scratchproof
Kräusellack wrinkle finish
krebserregend carcinogenic, cancer-causing
krebserzeugend carcinogenic, cancer-causing
Kreditkarte credit card
Kreide chalk, whiting
kreiden to chalk
Kreidung chalking
Kreidungsbeständigkeit resistance to chalking
Kreidungseffekt chalking (effect)
Kreidungsresistenz resistance to chalking
Kreisbogenschablone arc-shaped jig
Kreisel impeller
Kreiselpumpe rotary pump
Kreisfläche circular surface/area
kreisförmig circular
Kreiskolbenpumpe annular piston pump, orbital pump
Kreislauf circuit: **wird im Kreislauf gefördert** is circulated
Kreislaufpumpe circulating/centrifugal pump
Kreislaufverfahren circulation
Kreislaufwasser circulating water
Kreismesser 1. rotor knife *(sr)*. 2. rotary knife *(general term)*
Kreismesserwelle 1. rotor *(sr)*. 2. knife rotor *(general term) (see also explanatory note under* **Messerwelle**)
Kreisplatte circular plate
Kreisprofildüse round profile die *(e)*
Kreisprozeß cyclic process
kreisrund round, circular
Kreissäge circular saw
Kreisschneidevorrichtung circular cutting device
Kresol cresol
Kresol-Formaldehydkondensat cresol-formaldehyde condensate
Kresolharz cresol resin
Kresolnovolak cresol novolak
Kresolresol cresol resol

Kresylglycidylether cresyl glycidyl ether
Kreuzgelenk universal joint/coupling
Kreuzgelenkkupplung universal joint/coupling
Kreuzköperbindung cross twill weave *(txt)*
Kreuzschienenverteiler crossbar distributor
Kreuzspule cheese, cross-wound bobbin *(txt)*
Kreuzungspunkt point of intersection
Kriechbeanspruchung creep stress
Kriechbelastung creep stress
Kriechdehnung creep strain/elongation
Kriechen creep
Kriechfaktor creep factor
Kriechfestigkeit creep strength
Kriechgeschwindigkeit creep rate
Kriechkurve creep curve
Kriechmodul creep modulus
Kriechmodulkurve creep modulus curve
Kriechmodullinie creep modulus curve
Kriechneigung tendency to creep
Kriechrate creep rate
Kriechspannung creep stress
Kriechspur carbonised track, leakage path *(caused by leakage current:)* **Veränderungen durch Kriechspuren und Erosion** changes due to tracking and erosion
Kriechstrecke tracking, leakage current: **Da unter bestimmten Voraussetzungen bei Kriechstrecken Flammen auftreten können...** since, under certain conditions, tracking can cause flames to be produced...
Kriechstrom leakage current, tracking
kriechstromfest tracking resistant
Kriechstromfestigkeit tracking resistance
Kriechstromprüfung tracking resistance test
Kriechstromsicherheit tracking resistance
Kriechstromzeitbeständigkeit long-term tracking resistance
Kriechverformung creep deformation
Kriechverhalten creep behaviour
Kriechversuch creep test
Kriechwegbildung tracking
Kriechwiderstand creep resistance
Kristall crystal
kristallartig crystal-like
Kristallgitter crystal lattice
kristallin crystalline
Kristallinität crystallinity
Kristallinitätsgrad degree of crystallinity
Kristallisation crystallisation
Kristallisationsbedingungen crystallising conditions
kristallisationsbeständig non-crystallising
kristallisationsfähig crystallisable
Kristallisationsfähigkeit crystallisability
kristallisationsfrei non-crystallising
Kristallisationsgeschwindigkeit rate of crystallisation
Kristallisationsgrad degree of crystallinity
Kristallisationsneigung tendency to crystallise
kristallisationsstabil non-crystallising
Kristallisationstemperatur crystallisation temperature
Kristallisationstendenz tendency to crystallise
Kristallisationszustand crystalline state
kristallisierbar crystallisable
kristallisieren to crystallise
Kristallit crystallite
Kristallitorientierung crystallite orientation
Kristallitphase crystallite phase
Kristallitschmelztemperatur crystalline melting point
Kristallöl white spirit
Kristallstruktur crystal structure
Kristallwachstum crystal growth
Kristallwachstumsgeschwindigkeit rate of crystal growth
Kristallwasser water of crystallisation
Kriterium criterion *(plural:* criteria*)*
kritisch critical
Kronenkork crown cork
Krümel crumbs
Krümelform crumb form *(e.g. a rubber)*
krümelig crumbly, in crumb form
Krümmer elbow
Krümmerkopf side-fed die *(bm)*
Krümmung curvature
Krümmungsradius radius of curvature
Krümmungswinkel angle of curvature
kryogen cryogenic
Kryostat cryostat, low temperature thermostat
KSP *abbr. of* **Kunststoffpyrolyse**, pyrolysis of plastics waste
KSP-Anlage plant for the pyrolysis of plastics waste
KT *abbr. of* **Kunststoff-Temperatur**, melt temperature
KTL *abbr. of* **kathodische Elektrotauchlackierung**, cathodic electrophoretic/electrodeposition painting
kubisch cubic, cubical
kubischer Ausdehnungskoeffizient coefficient of cubical/volume expansion
Kuchen filter cake
Küchengerät 1. food processor. 2. kitchen appliance
Küchengeschirr kitchenware
Küchenmaschine food processor
kugelähnlich spherical
Kugelblase spherical bubble
Kugeldruckhärte ball indentation hardness
Kugeldruckprüfung ball indentation test
Kugeleindruckhärte ball indentation hardness
Kugeleindruckprüfung ball indentation test
Kugeleindruckverfahren ball indentation test
Kugelelektrode spherical electrode
Kugelfallversuch falling ball test
Kugelfallviskosimeter falling sphere viscometer
kugelförmig spherical
Kugelgraphit spherical-particle graphite
kugelig spherical
Kugelkorn spherical particle
Kugelkühler spherical condenser
Kugellager ball bearing
Kugellagerkäfig ball bearing cage

Kugelmühle ball/pebble mill
Kugelschaum foam consisting of spherical cells
Kugelstruktur spherical shape
Kugeltank spherical tank
Kugelventil ball valve
Kühlabschnitt cooling section/zone
Kühlabzugsanlage cooling take-off unit
Kühlanlage cooling plant
Kühlbad 1. quench bath *(for cooling extruded pipe or film)*. 2. cooling bath *(general term)*
kühlbar coolable
Kühlbohrung cooling channel
Kühldorn cooling mandrel *(bfe)*
Kühldüse cooled die *(e) (see explanatory note under **Kühldüsenverfahren**)*
Kühldüsenverfahren extrusion through a cooled die: *This is a special process for producing void-free solid profiles from partly crystalline thermoplastics, described in Kunststoffe 67(1977)10, p.594*
Kühldüsenwerkzeug cooled die *(e) (see explanatory note under **Kühldüsenverfahren**)*
Kühleffekt cooling effect
Kühlemulsion soluble oil
Kühlergrill radiator grille
Kühlermaske radiator grille
Kühlerschlauch radiator hose
Kühlerschutzgitter radiator grille
Kühlerverkleidung radiator grille
Kühlerzarge radiator surround
Kühlfahrzeug refrigerated truck
Kühlflüssigkeit coolant
Kühlgerät cooling unit
Kühlgleichmäßigkeit even cooling
Kühlhaus cold store
Kühlindustrie refrigeration industry
Kühlintensität cooling intensity
Kühlkanal 1. cooling channel *(im)*. 2. cooling section
Kühlkanalabstand distance between the cooling channels
Kühlkanalanordnung cooling channel layout
Kühlkanalkreis cooling circuit
Kühlkanalwandungen cooling channel walls
Kühlkapazität cooling capacity/efficiency *(for translation example see under **Kühlleistung**)*
Kühlkette refrigerated distribution chain
Kühlkreis cooling circuit
Kühlkreislauf cooling circuit
Kühlkreislauf, geschlossener closed-circuit cooling system
Kühlleistung cooling capacity/efficiency: **Hier spielt besonders die Kühlleistung eine Rolle** here, efficient cooling is particularly important
Kühlluft cooling air
Kühlluftanblaswinkel cooling air impingement angle *(bfe)*
Kühlluftaustrittsgeschwindigkeit cooling air delivery rate *(bfe)*
Kühlluftgebläse cooling air blower
Kühlluftmenge amount of cooling air
Kühlluftring cooling/air ring *(bfe)*
Kühlluftstrom 1. amount of cooling air: **der verfügbare Kühlluftstrom** the amount of cooling air available. 2. cooling air stream/flow: **Die Aufteilung des Kühlluftstroms in mehrere Teilströme hat folgende Vorteile** dividing the cooling air stream into several smaller ones has the following advantages
Kühlluftströmung cooling air stream/flow *(bfe)*
Kühlluftsystem air cooling system
Kühlmantel cooling jacket
Kühlmedium cooling medium, coolant
Kühlmischer cooling mixer
Kühlmittel coolant
Kühlmittelschlauch coolant hose
Kühlmöbelschaumstoff foam for freezers and refrigerators
Kühlpartie cooling section/zone
Kühlphase cooling phase
Kühlring cooling/air ring *(bfe)*
Kühlringlippen cooling ring lips *(bfe)*
Kühlringlippenspalt cooling ring orifice *(bfe)*
Kühlrippen cooling fins
Kühlschlange cooling coil
Kühlschrank refrigerator
Kühlschrankgehäuse refrigerator housing
Kühlschrankinnengehäuse refrigerator liner
Kühlschrankinnenverkleidung refrigerator liner
Kühlstation cooling station
Kühlstift cooling pin *(im)*
Kühlstrecke cooling section/zone
Kühlsystem cooling system
Kühltechnik refrigeration engineering
Kühltrog cooling trough, quench tank
Kühltrommel cooling drum
Kühltunnel cooling tunnel
Kühlturm cooling tower
Kühlung 1. cooling. 2. cooling system/unit
Kühlungsverhältnisse cooling conditions
Kühlventilator cooling fan
Kühlwalze 1. cooling roll. 2. chill roll *(e)*
Kühlwalzenverfahren cast film extrusion, chill roll casting/extrusion
Kühlwanne cooling trough, quench tank
Kühlwasser cooling water
Kühlwasserablauf cooling water outlet
Kühlwasserablauftemperatur cooling water outlet temperature
Kühlwasseranschluß cooling water connection
Kühlwasserausgleichsbehälter radiator tank
Kühlwasserdurchflußmenge cooling water flow rate, cooling water throughput
Kühlwasserkasten radiator tank
Kühlwasserkreislauf cooling water circuit
Kühlwasserleitung cooling water line
Kühlwassernetz cooling water supply
Kühlwasserregulierventil cooling water control valve
Kühlwasserrückgewinnungsanlage cooling water recovery/recycling unit
Kühlwasserverbrauch cooling water consumption

Kühlwasserverbraucher cooling water consumer
Kühlwasserverteiler cooling water manifold
Kühlwasserzufluß 1. cooling water inlet. 2. cooling water supply: **bei verstopfter Düse erfolgt die Unterbrechung des Kühlwasserzuflusses** if the nozzle becomes blocked, the cooling water supply is interrupted
Kühlwasserzufuhr 1. cooling water inlet. 2. cooling water supply *(for translation example see under* **Kühlwasserzufluß***)*
Kühlwasserzulauf 1. cooling water inlet. 2. cooling water supply *(for translation example see under* **Kühlwasserzufluß***)*
Kühlwasserzulauftemperatur cooling water inlet temperature
Kühlwendel cooling coil
Kühlzeit cooling time, setting time *(im)*
Kühlzone cooling section/zone
Kulminationspunkt culminating point
Kumaronharz coumarone resin
Kumulation accumulation, cumulation
kumulativ cumulative
Kundenbedürfnisse customers' requirements
Kundendienst after-sales service
kundenspezifisch *relating to or concerned with customers' requirements*: **Je nach kundenspezifischen Anforderungen** depending on customers' requirements
Kundenvorlage customer's sample
Kundenzeichnung customer's drawing *(e.g. of a moulded part)*
Kunstgewerbe arts and crafts
kunstgewerbliche Artikel craft objects/articles
Kunstharz synthetic resin
Kunstharzbeton polymer concrete
Kunstharzbindemittel synthetic resin binder
kunstharzgebunden synthetic resin bound
Kunstharzklebstoff synthetic resin adhesive
Kunstharzlack synthetic resin-based paint
Kunstharzpulver powdered synthetic resin
kunstharzvergütet synthetic resin modified
Kunstkautschuk synthetic rubber
kunstkautschukbeschichtet synthetic rubber coated/covered *(e.g. rolls)*
Kunstleder (PVC) leathercloth, (PVC) coated fabric
Künstlerfarben artists' colours *(it is interesting to note that, although* **Farbe** *is always paint,* **Farben** *are invariably referred to as colours when used by artists)*
künstlich artificial, synthetic
Kunststoff plastic, polymer: *The word is sometimes used instead of* **Kunstharz***: e.g.* **Polyesterkunststoff***, which should be translated as* polyester resin
Kunststoff, technischer engineering plastic *(see explanatory note under* **technisch***)*
Kunststoff, thermoplastischer thermoplastic (polymer)
Kunststoffabbau polymer degradation

Kunststoffabfall 1. plastics waste *(in a post-consumer context)*. 2. plastics scrap *(in a plastics processing context)*
Kunststoffabfälle, vermischte mixed plastics waste
Kunststoffabfallgemisch mixed plastics waste
Kunststoffabfallmischung mixed plastics waste
Kunststoffadditiv plastics additive
Kunststoffaufbereitung plastics compounding
Kunststoffaufbereitungsmaschine plastics compounder
Kunststoffausschuß plastics scrap
Kunststoffbearbeitung plastics machining
Kunststoffbeschichtungsanlage plastics coating plant/line
Kunststoffblend polymer blend/alloy, polyblend
Kunststoffbranche plastics industry
Kunststoffdispersion synthetic resin dispersion, polymer dispersion
Kunststoffdispersionskleber dispersion/emulsion/latex adhesive
Kunststoffe, selbstverstärkende self-reinforcing polymers, liquid crystal polymers
Kunststoffeinsatz plastics usage
Kunststoffemballage plastics pack
Kunststoffentsorgung plastics waste disposal
Kunststofferzeuger plastics producer/manufacturer
Kunststoffgemisch mixed plastics waste
kunststoffgerecht plastics-oriented
Kunststoffgießharzkörper synthetic resin casting
Kunststoffgranulat plastics pellets/granules
Kunststoffhersteller polymer producer/manufacturer
Kunststoffhohlkörper 1. blow moulding, blow moulded part. 2. blown container, bottle. 3. hollow item *There is no need to translate the first part, since* **Kunststoff** *will invariably be obvious from the context. See also explanatory note under* **Hohlkörper**
Kunststoffiltration polymer melt filtration
Kunststoffisolator synthetic resin insulator
Kunststoffkarte plastics card
Kunststoffkraftstoffbehälter plastics fuel tank: **Kunststoffkraftstoffbehälter aus PE** polyethylene fuel tank
Kunststofflack 1. paint for plastics 2. polymer-based paint
Kunststofflaminat synthetic resin laminate
Kunststofflatex synthetic resin latex
Kunststofflegierung polymer blend, polyblend, polymer alloy
Kunststofflinse plastics lens
Kunststofflösung polymer solution
Kunststoffmaschine plastics processing machine
Kunststoffmasse 1. polymer melt 2. moulding compound

Kunststoffmatrix 1. polymer matrix *(if referring to thermoplastics)*. 2. resin matrix *(if referring to UP, EP and similar resins)*
Kunststoffmischabfall mixed plastics waste
Kunststoffmischung polymer compound
kunststoffmodifiziert synthetic resin modified
Kunststoffmolekül polymer molecule
Kunststoffmüll plastics waste
Kunststoffneuware virgin polymer
Kunststoffolie plastics film/sheeting
Kunststofformmasse plastics moulding compound
Kunststoffphase polymer phase
Kunststoffprepreg prepreg
Kunststoffpreßteil (plastics) moulding
Kunststoffprofil-Ablängeautomat automatic device for cutting plastics profiles into lengths
Kunststoffqualität melt/polymer quality
Kunststoffrecyclat recycled plastics material/waste/scrap
Kunststoffrecycling recycling of plastics waste
Kunststoffrecyclingbetrieb plastics recycling establishment/firm
Kunststoffrecyclingprodukte recycled plastics products
Kunststoffregenerat plastics reclaim/regrind
Kunststoffregenerierung plastics recycling
kunststoffreich with/having a high plastics content *(e.g. household rubbish)*
Kunststoffrohr plastics pipe
Kunststoffrohstoff plastics/polymer raw material
Kunststoffrückstände 1. plastics waste. 2. polymer residues
Kunststoffrückstände, vermischte mixed plastics waste
Kunststoffschleudermaschine centrifugal casting machine
Kunststoffschmelze polymer melt
Kunststoffschneckenpresse plastics extruder
kunststoffspezifisch plastics-related
Kunststoffspritzerei plastics injection moulding shop
Kunststofftank plastics fuel tank
kunststofftechnisch *relating to plastics*:
kunststofftechnische Eigenschaften plastics properties
Kunststofftechnologie plastics/polymer technology
Kunststoffteil plastics moulding
Kunststoffteilchen polymer particle
Kunststofftemperatur melt temperature
Kunststoffüberzug plastics coating
kunststoffumhüllt 1. synthetic resin potted 2. plastics sheathed
Kunststoffumhüllung 1. synthetic resin potting 2. plastics sheathing
kunststoffummantelt plastic sheathed
Kunststoffummantelung plastics sheathing
Kunststoffverarbeitung plastics processing
kunststoffvergütet synthetic resin modified

Kunststoffverklebung 1. bonding of plastics. 2. bonded plastics joint
Kunststoffverwertung plastics waste recycling
Kunststoffwerkstoff plastics material
Kunststoffwirtschaft plastics industry
Kupfer copper
Kupferbelag copper cladding
Kupfer-Beryllium copper-beryllium
Kupferfolie copper foil
Kupfergeflecht copper braid, copper wire mesh
kupferkaschiert copper-clad
Kupferkaschierung copper cladding
kupferlaminiert copper-clad
Kupferlegierung copper alloy
Kupfernaphthenat copper naphthenate
Kupferrohrheizschlange copper heating coil
Kupfertiefdruck rotogravure (printing)
Kupferumflechtung copper braid
Kupplung coupling
Kupplungsbelag clutch lining
Kupplungsflüssigkeit clutch fluid
Kupplungsmittel coupling agent
Kupplungspedal clutch pedal
Kupplungsreaktion coupling reaction
Kupplungsschicht primer/anchor coat
Kurbelwelle crankshaft
Kurbelwellendichtung crankshaft seal
Kurbelwellenlagerdeckel camshaft bearing cap
Kurve curve
Kurvenast part of the curve
Kurvenschar group/family of curves
Kurventeil part of the curve
Kurvenzug curve
kurzbauend short, squat
Kurzbeschreibung brief/outline description
Kurzbewitterung accelerated weathering
Kurzbewitterungsgerät accelerated weathering instrument
Kurzbewitterungsversuch accelerated weathering test
Kürzel abbreviation
Kurzfasergelege chopped strand non-woven/bonded fabric
Kurzfasern 1. chopped strands *(if referring to glass fibres)*. 2. chopped fibres *(if referring to fibres other than those made of glass)*
Kurzfaserprepreg chopped strand prepreg
Kurzfaservlies chopped strand (glass) mat
kurzfristig at short notice, quickly, immediately, short-term, for a short time
kurzgeschnitten short, chopped *(e.g. fibres)*
Kurzglasfasern short/chopped strands
kurzglasfaserverstärkt chopped strand reinforced
kurzglasverstärkt chopped strand reinforced
Kurzhonen honing, superfinishing
Kurzhub short stroke
Kurzhubhonen honing, superfinishing
kurzhubig short-stroke
Kurzkettenverzweigung short-chain branching
kurzkettig short-chain
Kurzkohlenstoffasern short/chopped carbon

fibres
Kurzkompressionsschnecke short-compression zone screw
kurzlebig short-lived
Kurzölalkydharz short-oil alkyd (resin)
Kurzölalkydharzlack short-oil alkyd paint
kurzölig short-oil
Kurzprüfung accelerated test
Kurzschluß short circuit
Kurzschlußankermotor squirrel cage motor
Kurzschneckenextruder short-screw extruder
Kurzschreibweise abbreviated form, abbreviation
Kurztest accelerated test
kurzwellig short-wave
Kurzzeitbeanspruchung short-term stress
Kurzzeitberstversuch accelerated bursting test
Kurzzeitbewitterungsverhalten short-term weathering resistance
Kurzzeitbewitterungsversuch accelerated weathering test
Kurzzeitbruchlast short-term breaking stress
Kurzzeitdurchschlagfestigkeit short-term dielectric strength
Kurzzeiteigenschaften short-term properties
kurzzeithärtend fast curing
kurzzeitig briefly, short-term, accelerated
Kurzzeitkriechversuch accelerated creep test
Kurzzeitlagerung 1. short-term ageing. 2. short-term immersion
Kurzzeitlängung short-term elongation
Kurzzeitprüfung 1. accelerated test. 2. short-term test
Kurzzeitverhalten short-term behaviour/performance
Kurzzeitversuch 1. accelerated test. 2. short-term test
Kurzzeitwerte 1. accelerated test results. 2. short-term test results
Kurzzeitzugbeanspruchung short-term tensile stress
Kurzzeitzugversuch accelerated tensile test
KW *abbr. of* **Kohlenwasserstoff,** hydrocarbon
KW-Harz hydrocarbon resin
K-Wert 1. K-value. 2. heat transfer coefficient, coefficient of heat transmission

L

labil unstable *(e.g. a chemical compound)*
Labilität instability *(of a chemical compound)*
Labor laboratory
Laboransatz laboratory formulation
Laborbedingungen laboratory conditions
Laborbeschichtungsanlage laboratory coating machine/line

Laboreinwellenextruder laboratory single-screw extruder
Laborextruder laboratory extruder
Laborfolienblasanlage laboratory film blowing line
Laborkalander laboratory calender
Labormaßstab laboratory scale
Labormeßextruder laboratory extruder
Labormischwalzwerk laboratory mixing rolls
Laborprobe laboratory sample/specimen
Laborprüfung laboratory test/trial
Laborraum laboratory
Laborversuch laboratory test/trial
Laborziehgerät laboratory thermoforming machine
Laborzweiwellenkneter laboratory twin-screw mixer/kneader
Lack 1. paint. 2. lacquer, varnish. 3. enamel
Lack, lösemittelarmer high-solids paint
Lack- und Farbenindustrie paint/coatings industry
Lackadditiv paint additive
Lackansatz paint formulation
Lackaufbau coating composition
Lackbenzin white spirit
Lackbindemittel paint resin, surface coating resin
Lackbranche coatings/paint industry
Lackchemiker paint chemist
Lackdraht lacquered/varnished wire
Lackeigenschaften paint properties
Lackfachleute paint experts
Lackfachmann paint expert
Lackfarbe paint
Lackfarbenbestandteil paint constituent/component
Lackfehler paint defect
Lackfestkörper paint solids
Lackfeststoffgehalt paint solids content
Lackfilm 1. paint film. 2. lacquer film/coating
Lackfilmaufbau paint film structure
Lackgewebe varnished fabric
Lackglasgewebe varnished glass cloth
Lackhaftung paint adhesion
Lackharz paint resin, surface coating resin
Lackhilfsmittel paint additive
Lackierbarkeit paintability
Lackiereinbrenntemperatur stoving/baking temperature
lackierfähig paintable
Lackierofen paint drying/stoving oven
lackiert painted
Lackierung paint film, finish
Lackierungsdefekt paint film defect
Lackindustrie paint/coatings industry
Lackkonsistenz paint consistency
Lackkunstharz synthetic paint resin
Lacklaborant paint chemist
Lackleinöl refined/varnish linseed oil
Lacklösemittel paint solvent
Lackoberfläche paint film surface
Lackpapier varnished paper

Lackpolyester polyester surface coating resin
Lackpolymer paint resin, surface coating resin
Lackreste paint residues
Lackrezeptierung formulating a paint, developing a paint formulation
Lackrohstoff paint resin, surface coating resin
Lackschicht paint film
Lackschichtablösung detachment/lifting of the paint film
Lackstörung paint defect
Lacksystem surface coating system, paint system
Lacktechnik paint technology
Lacktechniker paint technologist
lacktechnisch *relating to paints*:
  lacktechnische Eigenschaften paint properties
Lacktechnologie paint technology
lacktechnologisch *relating to paints*:
  lacktechnologische Eigenschaften paint properties
Lackzusatz paint additive
Lactam lactam
Lactamschmelze lactam melt
Lacton lactone
Ladenbau shopfitting
Ladendiebstahl shoplifting
Ladung 1. charge. 2. load, cargo
Ladungsdichte charge density
Ladungsmonitor charge monitor
Ladungssignal charge signal
Ladungsträger charge carrier
Ladungsverstärker charge amplifier
Ladungsverteilung charge distribution
Lage 1. position 2. layer
lagegeregelt position-controlled
Lagenbindung interlaminar adhesion
Lagenzahl number of layers
Lager 1. bearing. 2. warehouse, storeroom
Lagerbedingungen storage conditions
Lagerbehälter storage tank
Lagerbestand stock in hand
lagerbeständig having a good shelf life
Lagerbeständigkeit shelf/storage life, storage stability
Lagerbuchse bearing bush
Lagerdauer 1. ageing period. 2. time of immersion, immersion period *(see explanatory note under* **Lagerung**). 3. storage period
Lagerdruck bearing pressure
Lageregelkreis position control circuit
Lagereigenschaften shelf/storage life, storage stability
lagerfähig *related to* **Lagerfähigkeit**: Die Harzlösung ist mindestens 1 Jahr
lagerfähig the resin solution has a shelf life of at least one year OR: can be kept for at least one year
Lagerfähigkeit shelf/storage life, storage stability
Lagerfähigkeitsgarantie guaranteed shelf life
Lagerfläche storage area

Lagergebäude warehouse
Lagergehäuse bearing housing
Lagerhalle warehouse
Lagerhaltung storage, warehousing
Lagerkäfig bearing cage
Lagerkapazität storage capacity
Lagerkonservierungsmittel preservative
Lagerkosten warehousing/storage costs
Lagerkraft bearing force
Lagermodul storage modulus
Lagerplatz storage space
Lagerraum storeroom
Lagerreibung bearing friction
Lagerschale bearing bush
Lagersilo (bulk storage) silo
lagerstabil 1. having a long shelf life. 2. stable
Lagerstabilität shelf/storage life, storage stability: **hohe Lagerstabilität** long shelf life
Lagerstabilitätsschwierigkeiten shelf life problems
Lagerstelle bearing seat
Lagertank storage tank
Lagerückmeldung position feedback *(im)*
Lagerung 1. immersion *(of a test specimen in a liquid)*. 2. ageing *(of a test specimen in a medium other than a liquid, e.g. hot air. This is the alternative word to use since it is not good English to talk of immersing something in air)*. 3. storage
Lagerungsbedingungen 1. ageing conditions. 2. storage conditions *(see explanatory note under* **Lagerung**)
Lagerungsbehälter storage tank
Lagerungsdauer 1. ageing period. 2. time of immersion, immersion period. 3. storage period *(see explanatory note under* **Lagerung**)
Lagerungseigenschaften shelf/storage life, storage stability
Lagerungstemperatur 1. ageing temperature. 2. immersion temperature. 3. storage temperature *(see explanatory note under* **Lagerung**)
Lagerungsversuch 1. ageing test. 2. immersion test *(see explanatory note under* **Lagerung**)
Lagerungszeit 1. ageing period. 2. time of immersion, immersion period. 3. shelf/storage life *see explanatory note under* **Lagerung**)
Lagervorschriften storage regulations
Lagerzapfen journal
Lagerzeit 1. shelf/storage life. 2. storage period: **nach längerer Lagerzeit...** when it has been stored for some time...
Lambda-Wert, λ-Wert thermal conductivity
lamellar lamellar
Lamelle lamella *(plural:* lamellae)
lamellenartig lamellar
Lamellenstruktur lamellar structure
laminar laminar
laminare Schichtenströmung laminar flow
laminare Strömung laminar flow
laminares Fließen laminar flow

Laminarfluß laminar flow
Laminarflußverteiler laminar flow distributor
Laminat laminate
Laminataufbau laminate structure
Laminatband laminate web *(grp)*
Laminatdicke laminate thickness
Laminateigenschaften laminate properties
Laminatoberfläche laminate surface
Laminator laminating unit
Laminatpapier laminating paper
Laminatschale laminate shell
Laminatschicht laminate layer
Laminatstruktur laminate structure
Laminieranlage laminating plant
Laminierdruck laminating pressure
laminieren to laminate
Laminierform laminating mould *(grp)*
Laminierharz laminating resin
Laminierkalander laminating calender
Laminierkleber laminating adhesive
Laminiermaschine laminator, laminating machine
Laminierpaste laminating paste
Laminierpresse laminating press
Laminiersystem laminating system
Laminierverfahren laminating process *(grp)*
Laminiervorgang laminating process
Lammfellscheibe lambswool polishing wheel
Lammfellwalze lambswool roller
Lampe 1. lamp. 2. (light) bulb. 3. light
Lampenfassung lampholder
Lampenschirm lampshade
Lampensockel lamp socket
LAN *abbr. of* local area network *(me)*
Länder Federal States
Länder, alte the former W. Germany
Länder, neue the former E. Germany
Landmaschinen agricultural machinery
Landmaschinenlack agricultural machinery/equipment paint
Landwirtschaft agriculture
landwirtschaftlich agricultural
Landwirtschaftsfolie agricultural film/sheeting
langandauernd long-lasting
langanhaltend long lasting
langdauernd long lasting
Längenänderung change in length
Längenausdehnung linear expansion, coefficient of linear expansion
Längenausdehnungskoeffizient, thermischer coefficient of linear expansion
Längenkontraktion longitudinal contraction
Längenschrumpf longitudinal shrinkage
Längenschwindung longitudinal shrinkage
Längenschwund longitudinal shrinkage
langfaserverstärkt continuous strand-reinforced *(grp)*
Langfeldleuchte strip light
langfristig long-term
Langfristprognose long-term forecast
Langglasfasern long glass fibres
langglasverstärkt reinforced with long glass strands *(grp)*
Langhalsrundkolben long-neck round-bottom flask
Langhalsstehkolben long-neck flat-bottom flask
langhubig long-stroke
Langkettenverzweigung long-chain branching
langkettig long-chain
Langkettigkeit long-chain character
Langkompressionsschnecke long-compression zone screw *(e)*
langlebig long-lasting
Langlebigkeit longevity, long (service/working) life
Langölalkydharz long-oil alkyd (resin)
Langölalkydharzlack long-oil alkyd paint
langölig long-oil
längs longitudinal(ly)
Längsachse longitudinal axis
langsamlaufend slow speed
Langsamläufer 1. slow speed machine. 2. slow speed mixer
Langsamrührer slow-speed stirrer/mixer
Längsausdehnung longitudinal expansion
Längsbewegung longitudinal/axial movement
Längsdehnung longitudinal expansion
längsgenutet longitudinally grooved
längsgerichtet longitudinally oriented
Längskontraktion longitudinal shrinkage
Längskraft longitudinal force
Längsmarkierungen 1. longitudinal marks *(general term)*. 2. spider lines *(e,bm)*
Längsmesser longitudinal knife/knives *(sr)*
Längsmischung longitudinal/axial mixing
Längsmischvermögen longitudinal/axial mixing efficiency
Längsnuten longitudinal grooves
längsorientiert longitudinally oriented
Längsorientierung orientation in machine direction
Längsrichtung machine direction, longitudinal direction
Längsscherung longitudinal shear
Längsschlitzentgasungsgehäuse barrel vented through longitudinal slits
Längsschneideeinrichtung longitudinal cutter
Längsschneidevorrichtung longitudinal cutter
Längsschnitt 1. longitudinal cut. 2. longitudinal section
Längsschrumpf longitudinal shrinkage
Längsschrumpfung longitudinal shrinkage
Längsschwindung longitudinal shrinkage
Längsspannung longitudinal stress
Längsspritzkopf straight-through die, in-line die *(e)*
Längsspritzwerkzeug straight-through/in-line die *(e)*
Längsstreckmaschine longitudinal stretching unit
Längsstreckwerk longitudinal stretching unit
Längsströmung longitudinal flow
Längstrenneinrichtung longitudinal cutter
längsverschiebbar longitudinally adjustable

längsverstreckt longitudinally oriented *(film)*
Längsverstreckung longitudinal stretching
Längswellmaschine longitudinal corrugating machine
Längswellvorrichtung longitudinal corrugating device
langwellig long wave
Langzeitalterungsbeständigkeit long-term ageing resistance
Langzeitbeanspruchung long-term stress
Langzeitbelastung long-term stress
Langzeitbelastungseigenschaften long-term load bearing properties
Langzeitbestrahlung long-term irradiation
Langzeitbetrieb long-term operation
Langzeitdruckbeanspruchung long-term compressive stress
Langzeitdurchschlagfestigkeit long-term dielectric strength
Langzeiteffekt long-term effect
Langzeiteigenschaften long-term properties
Langzeiteinsatz long-term use
Langzeiterfahrung long-term experience
Langzeitfestigkeit fatigue strength, endurance limit
Langzeitfilter long-life filter
langzeitig long-term
Langzeitkontakt long-term contact
Langzeitkriechverhalten creep behaviour *(the first part need not be translated because creep is a long-term phenomenon anyway)*
Langzeitlagerung 1. long-term ageing. 2. prolonged immersion *(of test piece in a liquid)*
Langzeitprüfung 1. creep test. 2. long-term test
Langzeitqualität long-term performance
Langzeitschmierung permanent lubrication
Langzeitschutz long-term protection
Langzeitschweißfaktor long-term welding factor
Langzeitstabilisator long-term stabiliser
Langzeitstabilität long-term stability
Langzeittemperaturbeständigkeit long-term heat resistance
Langzeittest 1. creep test. 2. long-term test
Langzeituntersuchung 1. creep test. 2. long-term test
Langzeitverhalten long-term performance
Langzeitversuch 1. creep test. 2. long-term test
Langzeitwasserlagerung prolonged immersion in water
Langzeitwirkung long-term effect
Lappen rag
***Lärm** noise
lärmarm quiet in operation, quiet-running, low-noise, silent *(machine, pump, etc.)*
Lärmdämmung sound insulation
Lärmemission sound/noise emission
Lärmemissionswert 1. sound emission value *(in a scientific context).* 2. noise level *(in a non-scientific context, e.g. when discussing noisy machines such as granulators)*

lärmgedämpft soundproofed
Lärmminderung noise reduction
Lärmminderungsmaßnahmen noise reduction measures
Lärmschutzwand sound barrier
Lasche, doppelte double strap
Lasche, einfache single strap
Laserabtastgerät laser scanning analyser
Laseranlage laser equipment
Laserbeschrifter laser marker, laser marking/etching device
Laserbeschriftung laser marking/etching
Laserbeschriftungsprogramm laser marking/etching program *(me)*
Laserbeschriftungssoftware laser marking/etching software *(me)*
Laserdruck laser printing
Laserdrucker laser printer
Laserenergie laser energy
lasergesteuert laser controlled
Laserhärtung laser (beam) curing
Laserleistung laser output
Laserlicht laser beam
Laserpyrolyse laser pyrolysis
Laserschneidverfahren laser beam cutting (process)
lasersensibel laser sensitive
Laserstrahl laser beam
Laserstrahlhärtung laser (beam) curing
Laserstrahlschneiden laser (beam) cutting
lasieren to glaze
Last stress, load
Last, aufgelegte applied stress/load
Lastamplitude stress amplitude
Lastanlegung application of stress
Lastaufbringung application of stress
Lastaufnahmefähigkeit load-bearing/-carrying capacity
Lastaufnahmevermögen load-bearing/-carrying capacity
lastbeaufschlagt under stress
Lastdruck load pressure
Lastenheft list of requirements, (technical) specification
Lastfrequenz load frequency
lastkompensiert load-compensated
Lastspiel stress/load cycle
Lastspielfrequenz stress/load cycle frequency
Lastspielzahl number of stress/load cycles
Laststandzeit time under stress, stress duration
Laststeigerung load increment *(in tensile testing)*
Lasttragevermögen load-bearing capacity
Lasttrennschalter isolating switch
Lastübertragung stress/load transfer
Last-Verformungsdiagramm 1. force-deformation diagram *(general term)* 2. stress-strain diagram *(if referring to tensile stresses)*
Lastverteilung stress/load distribution
Lastwechsel stress/load cycle

---

* Words starting with **Lärm-**, *not found here, may be found under* **Geräusch-** *or* **Schall-**

Lastwechselzahl number of stress/load cycles
Lastzyklus stress/load cycle
Lasur glaze, clear lacquer/varnish *(usually for wood)*
Latex latex, emulsion, dispersion
Latexpolymer polymer latex
Latexteilchen latex particle
Lauf, geräuscharmer quiet-running
Laufeigenschaften, ruhige smooth passage through the machine
laufend, unrund eccentric
Läufer run *(of paint film)*
Lauffläche 1. tyre tread. 2. wearing surface *(e.g. of a screw)*
Laufflächenmischung tyre tread compound
Laufflächenvulkanisat vulcanised tyre tread compound
Laufflächenwerkstoff tyre tread compound
Laufmeterzahl number of running meters
Laufmittel eluant
Laufrad impeller
Laufrichtung, in in machine direction
Laufruhe quietness in operation
Laufruhe, hohe very quiet in operation, very quiet-running
laufruhig quiet in operation, quiet-running
Laufspiel running clearance *(of bearings)*
Laufstreifen tyre tread
Lauge alkali solution
Laugenbeständigkeit alkali resistance
Laurinlactam laurinlactam
Laurinsäure lauric acid
Lauroylperoxid lauroyl peroxide
LCD-Display liquid crystal display, LCD
LCD-Flüssigkristallanzeige liquid crystal display, LCD
LC-Polymer liquid crystal polymer
L/D-Verhältnis L/D ratio
Lebensdauer (working) life: **vom Werkzeug wird elne lange Lebensdauer erwartet** moulds are expected to last a long time
Lebensdauerprofil, thermisches thermal endurance profile
Lebensdauerprognose estimated (working) life
Lebensmittel foodstuffs
Lebensmittelbedarfsgegenstände consumer goods for food contact applications
Lebensmittelbehälter food pack/container
Lebensmittelbereich food sector
Lebensmittelbranche food industry/sector
lebensmittelecht suitable for food contact applications
Lebensmittelgesetz food regulations: **im Sinne des Lebensmittelgesetzes** complying/in accordance with the food regulations
Lebensmittelindustrie food industry/sector
lebensmittelnah in contact with foodstuffs **Anwendungen im lebensmittelnahen Bereich** food contact applications
lebensmittelrechtlich *relating to food regulations:* **lebensmittelrechtliche Bestimmungen/Anforderungen/**
Empfehlungen food regulations
lebensmittelrechtliche Unbedenklichkeit suitability for food contact aplications **Die Folie zeichnet sich durch lebensmittelrechtliche Unbedenklichkeit aus** the film can be used OR is suitable for applications involving direct contact with foodstuffs
Lebensmittelverpackung 1. food packaging. 2. food pack
Lebensmittelverpackungsfolie food packaging film
Lebensraum biotope
Lebenszyklus life cycle
Leckage leak, leakage
leckagefrei leak-proof, non-leaking
Leckprüfgrät leak tester
Leckströmung leakage flow *(e)*
Ledernarbung leather grain effect
LED-Leuchtanzeige LED display
Leerfahren purging *(im) (emptying the machine by operating it without adding further material)*
Leergut empties
Leerhub dry cycle *(im)*
Leerlaufleistung no-load power
Leerlaufzeit downtime, shut-down period
Leerlaufzyklen, Anzahl der number of dry cycles *(usually per minute) (im)*
Leervolumen void volume
Legieren alloying, blending
Legierung alloy
Legierungskomponente alloying component
Legierungspartner alloying component
Lehne 1. back rest 2. arm rest
Lehre gauge, jig, fixture
Lehrenbohrwerk drill jig
Lehrling apprentice
leicht applizierbar easy to apply
leicht drapierbar easily draped
leicht pigmentierbar easily pigmented
leichtablösend easy-release
Leichtbau lightweight construction
Leichtbaukonstruktion lightweight construction
Leichtbaustruktur lightweight structure
Leichtbauweise lightweight construction: **Ein Extruder in Leichtbauweise** a lightweight extruder
Leichtbauwerkstoff lightweight material
Leichtbeton lightweight concrete
leichtbrennbar highly flammable
leichtentflammbar highly flammable
leichtentformbar easily demoulded
leichtentzündlich highly flammable
leichtfließend free-flowing, easy-flow
leichtflüchtig readily volatile
Leichtfraktion light fraction
Leichtfüllstoff lightweight filler
leichtgewichtig lightweight
leichtlaufend easy-process *(e.g. PVC)*
Leichtmetallgehäuse light metal housing
Leichtölfraktion light oil fraction
Leichtspat gypsum

leichttrennend easy-release
Leichtwerkstoff lightweight material
Leim glue
Leimauftragungsgerät glue applicator
Leimfarbe distemper
Leimharz adhesive resin
Leinöl linseed oil
Leinölalkyd linseed oil alkyd
Leinölfettsäure linseed oil fatty acid
Leinwandbindung plain weave *(txt)*
Leistung 1. performance, efficiency, capacity *(e.g. of a machine)*. 2. output *(of power)* 3. input *(of power)* 4. rating *(electrical)* 5. energy, power. 6. service *(provided by a company)*
Leistung, abgegebene power output
Leistung, aufgenommene power/energy input, power/energy used
leistung, bemessene rating *(of an engine)*
Leistung, elektrische power
Leistung, installierte installed/connected load
Leistung, mechanische 1. productive capacity/output *(of a machine)*. 2. power, horsepower *(of an engine)*
Leistung, thermische thermal efficiency
Leistung, zugeführte power input
Leistungsabfall 1. drop in power. 2. decline in performance
Leistungsabgabe power/energy output
leistungsabhängig output-dependent, depending on the output
Leistungsangaben 1. performance data. 2. output data *(e.g. of an engine)*
Leistungsangebot services offered *(by a company)*
Leistungsaufnahme 1. energy used/consumed: **durch die höhere Leistungsaufnahme** because more energy is used. 2. power input
Leistungsbedarf power/energy consumption, power/energy requirements
Leistungsbewertung assessment of performance
Leistungsbilanz energy balance
Leistungsdaten performance data *(of a machine)*
Leistungsdichte power density
Leistungserhöhung increased outputs/efficiency **bei der weiteren Leistungserhöhung von Profilanlagen...** as outputs of profile extrusion lines increase...
leistungsfähig powerful, efficient, capable, high-output, high-performance
Leistungsfähigkeit efficiency, output, performance *(e.g. of a machine or material)*
Leistungsfahrbreite 1. range of outputs. 2. operating range
Leistungsgarantie guaranteed output
Leistungsgrenze performance limit
Leistungskraft efficiency
Leistungsmerkmale performance features
Leistungsminderung reduced outputs/efficiency: **man muß bei diesen Schnecken eine Leistungsminderung von 15-20% in Kauf nehmen** with these screws, one must accept output reductions of 15-20%
Leistungsniveau performance
Leistungsprofil overall performance
Leistungsreserven power/energy reserves
Leistungsschalter power switch
leistungsschwach inefficient
leistungssparend power/energy saving
Leistungsspritzguß high speed injection moulding
leistungsstark powerful, efficient, high-output, high performance
Leistungssteigerungen increased outputs/efficiency
Leistungsteil power unit
Leistungstransformator power transformer
Leistungsvermögen performance
Leitblech metal guide plate *(sr)*
Leitbleche 1. bubble guides *(bfe)*. 2. collapsing boards *(bfe)*
leitend conductive
Leiter 1. conductor *(of electricity)*. 2. head *(of a department)*. 3. ladder
Leiterbild printed circuit pattern
Leiterisolierung wire insulation
Leiterplatte printed circuit board
Leiterpolymer ladder polymer
Leitertemperatur conductor temperature
leitfähig conductive
leitfähig, intrinsisch intrinsically conductive
Leitfähigkeit conductivity
Leitfähigkeitsruß conductive carbon black
Leitlinie guideline
Leitplatten 1. bubble guides *(bfe)*. 2. collapsing boards *(bfe)*
Leitrechner central/master computer *(me)*
Leitrechnerschnittstelle central computer interface *(me)*
Leitschicht conductive layer/coating
Leitstandterminal master terminal *(me)*
Leitstangen guide bars, bubble guides *(bfe)*
Leitungswasser tap water
Leitungswassernetz water mains/supply
Leitwalze guide roll *(bfe)*
Lenkrad steering wheel
Lenkradkranz steering wheel rim
Lenkrolle guide roll
Lenksäule steering column
Leseeinrichtung reader *(me)*
Lesestift code pen *(me)*
Leuchtanzeige light signal
Leuchtanzeigetableau light signal indicator panel
Leuchtdiode light-emitting diode, LED *(me)*
Leuchtdiodenadapter LED adaptor *(me)*
Leuchtdiodenmatrix LED matrix *(me)*
Leuchte lamp, lighting fixture, light fitting, light
Leuchtenabdeckung lamp cover
Leuchtenfassung lamp holder
Leuchtengehäuse lamp housing
Leuchtenklemme lamp mounting
Leuchtenreflektor light reflector

Leuchtkörper light fitting
Leuchtröhre fluorescent lamp/tube
Leuchtröhrenhalterung fluorescent lamp holder
Leuchtschaltbild illuminated circuit diagram
Leuchtstofflampe fluorescent lamp/tube
Leuchtstoffröhrenhalter fluorescent lamp holder
Leuchtziffernanzeige LED display
Lewisbase Lewis base
Lewissäure Lewis acid
L-Form four-roll L-type calender
lichstabil lightfast
Lichtabsorption light absorption
Lichtabsorptionskoeffizient light absorption coefficient
Lichtalterung light ageing
Lichtausbeute luminous efficiency
lichtbeständig light resistant
Lichtbeständigkeit light resistance
Lichtbogen arc
lichtbogenbeständig arc resistant
Lichtbogenbeständigkeit arc resistance
lichtbogenfest arc resistant
Lichtbogenfestigkeit arc resistance
lichtbogenlöschend arc-suppressing
Lichtbogenschweißen arc welding
Lichtbrechungsindex refractive index
Lichtdecke illuminated ceiling
Lichtdrücker push-button light switch
Lichtdurchgang light transmission
lichtdurchlässig translucent
Lichtdurchlässigkeit light transmission
Lichtdurchlässigkeitszahl percentage light transmission
lichte Weite zwischen den Säulen distance between tie bars *(im)*
lichtecht lightfast
Lichtechtheit lightfastness
Lichteigenschaften light resistance
Lichteinwirkung action/effect of light, exposure to light: **Sie sind gegen Lichteinwirkung stabilisiert** they are light stabilised
lichtempfindlich light sensitive
Lichtempfindlichkeit light sensitivity
Lichtenergie light energy
lichter Abstand zwischen den Säulen distance between tie bars *(im)*
lichter Holmabstand distance between tie bars *(im)*
lichter Säulenabstand distance betwee tie bars *(im)*
Lichtgitter light diffusing grid
lichthärtend light curing
lichtinduziert light-induced
Lichtintensität light intensity
Lichtkuppel dome light
Lichtleiter optic(al) fibre
Lichtleitfaser optical fibre
Lichtleitkabel optic(al) cable
Lichtmikroskop optical microscope
Lichtmikroskopie optical microscopy
lichtmikroskopische Aufnahme photomicrograph
Lichtplatte light diffusing panel
Lichtpunktlinienschreiber light spot line recorder
Lichtquelle light source
lichtreflektierend light-reflecting
Lichtreflexion light reflection
Lichtschranke light beam guard
Lichtschutz light stabilisation
Lichtschutzausrüstung light stabilisation: **Wir haben Produkte für normale und hohe Lichtschutzausrüstung** we can offer products suitable for normal and increased light stability; *(see also translation examples under* **Ausrüstung***)*
Lichtschutzeffekt light stabilising effect
Lichtschutzmittel light stabiliser
Lichtschutzsystem light stabilising system
Lichtschutzzusatz light stabiliser
Lichtstabilisatorwirkung light stabilising effect
lichtstabilisierend light stabilising
lichtstabilisiert light stabilised
Lichtstabilisierung light stabilisation
Lichtstabilisierwirkung light stabilising effect
Lichtstabilität light stability
Lichtstrahl light beam
Lichtstrahloszillograph light beam oscillograph
Lichtstreukoeffizient light scattering coefficient
Lichtstreuung light scattering
Lichtstreuwirkung light scattering effect
Lichttechnik lighting technology
Lichttransmission light transmission
Lichttransmissionsgrad degree of light transmission
lichtundurchlässig opaque, impermeable to light
Lichtundurchlässigkeit opacity, impermeability to light
Lichtvorhang light beam guard
Lichtwand translucent wall
Lichtweite distance between tie bars *(im)*
Lichtwellenleiter optic(al) cable
Lichtwellenleiterkabel optical cable
Lichtwerbung illuminated advertising sign
Lieferant supplier
Lieferantennachweis list of suppliers
Lieferantenverzeichnis list of suppliers
Lieferantenwahl choice of supplier
Lieferantenwechsel change of supplier
lieferbar available **lieferbar 50% in Butyldiglykol** supplied as a 50% solution in butyl diglycol
Lieferbedingungen delivery conditions, terms of delivery
Lieferform form in which supplied, physical form as supplied: **in Lieferform** as supplied; **Lieferform 45% in Butylacetat** supplied as a 45% solution in butyl acetate
Lieferprogramm range *(of machines, products, etc.)*
Lieferschein delivery note
Lieferschwierigkeiten delivery problems

Lieferspezifikation delivery specification
Lieferung 1. delivery. 2. consignment
Lieferzeit delivery date
Liegezeit cream time *(PU foaming)*
Liganden ligands
Lignin lignin
LIM *abbr. of* **Lichtmikroskop** optical microscope
LIM-Verfahren liquid injection moulding (process)
Lineal ruler
linear linear
linear-elastisch linear-elastic
linearer Ausdehnungskoeffizient coefficient of linear expansion
linearer Wärmeausdehnungskoeffizient coefficient of linear expansion
Linearisierung linearisation
Linearität linearity
Linearmotor linear motor
Linearphthalat linear phthalate
Linearverstärker linear amplifier
linienförmig linear
Linienmarkierungen 1. longitudinal marks *(general term)*. 2. spider lines *(e,bm)*
Linienschreiber chart recorder
Linksgewinde left-hand thread
Linolensäure linolenic acid
Linse lens
Linsengranulat pellets
Lippe, biegsame adjustable/flexible lip *(e)*
Lippe, einstellbare adjustable/flexible lip *(e)*
Lippe, feste fixed lip *(e)*
Lippenbalken restrictor/choke bar *(e)*
Lippenelemente (die) lip elements *(e)*
Lippenpaar die lips *(e)*
Lippenparallelführung die land *(e)*
Lippenpartie die lips *(e)*
Lippenspalt 1. die orifice/gap *(e)*. 2. cooling ring orifice *(bfe)*
Lippenverstellmöglichkeit 1. possibility of adjusting the die lips. 2. die lip adjusting mechanism
Lippenverstellung 1. die lip adjustment. 2. die lip adjusting mechanism *(e)*
Literaturhinweise bibliography
Lithiumcarbonat lithium carbonate
Lithiumhydroxid lithium hydroxide
lithiumorganisch organo-lithium
Lithiumoxid lithium oxide
lizensiert licensed
Lizenzgeber licensor
Lizenzgebühr lincense fee
Lizenznehmer licensee
L-Kalander four-roll L-type calender
LkW *abbr. of* **Lastkraftwagen** lorry, truck
Lkw-Lauffläche lorry tyre tread
Lochbild mould fixing diagram, mould fixing details, standard platen details *(diagram showing pattern of holes in platen for mould mounting) (im)*
Lochblech perforated metal sheet, perforated sheet metal

Lochdornhalter breaker plate-type mandrel support *(e,bm)*
Lochdornhalterkopf die with breaker plate-type mandrel support *(e,bm)*
Lochdornhalterwerkzeug die with breaker plate-type mandrel support *(e,bm)*
Lochdüse 1. perforated die. 2. pelletising die *(sr)*
Lochen perforation
Lochfraß pitting
lochfraßartige Korrosion pitting
Lochfraßkorrosion pitting
Lochkarte punched card *(me)*
lochkartengesteuert controlled by punched cards
Lochkartenleser punched card reader *(me)*
Lochkartensteuerung 1. punched card control. 2. punched card control system
Lochplatte 1. pelletising die. 2. breaker plate *(e)*. 3. perforated disc: **Der Vakuumtopf enthält eine Lochplatte mit großer Bohrungszahl** the vacuum pot contains a perforated disc with many holes
Lochring breaker plate *(e)*
Lochscheibe breaker plate *(e)*
Lochscheibendornhalter breaker plate-type mandrel support *(e)*
Lochstanzwerk blanking tool
Lochstempel punch
Lochstreifen punched tape, paper tape *(me)*
Lochstreifenleser punched tape reader *(me)*
Lochstreifenstanzer paper tape punch *(me)*
Lochsuchgerät pore detector
Lochtragring breaker plate-type mandrel support *(e)*
locker loose
Lockflamme pilot flame
logarithmisch logarithmic
Logarithmus logarithm
Logik logic *(me)*
Logikausgang logic output *(me)*
Logikbaustein logic module *(me)*
Logikelement logic element *(me)*
Logikfunktion logic function *(me)*
Logikschaltkreis logic circuit *(me)*
Logiksignal logic signal *(me)*
Logistik logistics
logistisch logistic
Lohn wage
Lohn- *This prefix, linked to the word for a process, signifies that that process is carried out by an outside firm against payment* (**Lohn**), *e.g.* **Lohnspritzgießen, Lohnvermahlen, Lohnblasformen** *etc. Such terms should be translated by putting* contract *in front of the English word for the process, e.g.* contract blow moulding, contract injection moulding *etc.*
lohnintensiv wage-intensive
Lohnkosten labour costs
lokal local(ised)
Lokalisierung localisation

Lokalversprödung local(ised) embrittlement
Longitudinalwelle longitudinal wave
lösbare Verbindung temporary joint
löschbar erasable *(me)*
Löscheinheit erase head *(me)*
löschen 1. to erase, to clear *(me)* 2. to cancel
Löschen erasure *(me)*
Löseeigenschaften 1. dissolving power *(of a solvent)*. 2. solvating power/capacity *(of a plasticiser for PVC)*
Lösefähigkeit 1. dissolving power *(of a solvent)*. 2. solvating power/capacity *(of a plasticiser for PVC)*
Lösemittel solvent
Lösemittelabdunstung solvent evaporation
Lösemittelabgabe solvent emission/ evaporation/release
Lösemittelanteil solvent content
lösemittelarm low-solvent: **lösemittelarmer Lack** high solids paint
Lösemittelbeständigkeit solvent resistance
Lösemitteldämpfe solvent vapours
Lösemittelechtheit solvent resistance
Lösemittelemission solvent emission/release/ evaporation
lösemittelempfindlich affected by solvents
Lösemittelempfindlichkeit susceptibility to solvent attack
lösemittelfest solvent resistant, resistant to solvents
lösemittelfrei solvent-free
lösemittelfreundlich solvent-compatible
Lösemittelgehalt solvent content
Lösemittelgemisch solvent mixture/blend
lösemittelhaltig solvent-based/-containing
Lösemittelkleber solvent-based adhesive
Lösemittelklebstoff solvent-based adhesive
Lösemittelkombination solvent blend
Lösemittellack solvent-based paint
lösemittellöslich solvent-soluble, soluble in solvents
Lösemittellöslichkeit solubility in solvents
Lösemittelphase solvent phase
Lösemittelreaktivierung solvent activation *(of a dry adhesive film)*
lösemittelreduziert with reduced solvent content
lösemittelreich with a high solvent content
Lösemittelresistenz solvent resistance
Lösemittelrest residual solvent
Lösemittelrestgehalt residual solvent content
Lösemittelretention solvent retention
Lösemittelrückgewinnung 1. solvent recovery. 2. solvent recovery unit
Lösemitteltyp type of solvent
Lösemittelverdunstung solvent evaporation
Lösemittelverlust solvent loss
Löser solvent
Lösetemperatur 1. solvation temperature *(of plasticiser for PVC)*. 2. dissolving temperature *(of a solvent)*
Löseverhalten solubility characteristics

Lösevermögen 1. solvation power/capacity *(of plasticiser for PVC)*. 2. dissolving power *(of a solvent)*
Losgröße batch size
löslich soluble
löslich, begrenzt sparingly soluble
löslich, beschränkt sparingly soluble
löslich, gut readily soluble
löslich, schlecht having poor solubility
löslich, schwer difficult to dissolve
löslich, weitgehend largely soluble
Löslichkeit solubility
Löslichkeitsbereich solubility range
Löslichkeitseigenschaften solubility characteristics
Löslichkeitsverhalten solubility characteristics
Losnummer batch number
Lösung solution
Lösung, ethanolische ethanol solution
Lösung, methanolische methanol solution
Lösungseffekt dissolving effect
Lösungsmittel-, lösungsmittel- *see* **Lösemittel-, lösemittel-**
Lösungspolymerisat solution polymer
Lösungspolymerisation solution polymerisation
lösungspolymerisiert solution polymerised
Lösungsverhalten dissolving characteristics
Lösungsvermittler solution aid
Lösungsviskosität solution viscosity
Lösungsvorgang dissolving process
Lösungsvorschlag suggested solution
Loswechsel change of batch
Lötbad solder bath
Lötbadbeständigkeit solder bath resistance
Lötbadfestigkeit solder bath resistance
Lötbadlagerung solder bath immersion
löten to solder
Lötwelle soldering iron
LP *abbr. of low profile, an English expression widely used in German texts*
LS *abbr. of low-shrinkage*
LT-Verfahren 1. low temperature process. 2. low temperature calendering (process)
lückenlos complete, total
Luftansaugung air intake
Luftausschluß exclusion of air
Luftaustritt air outlet
Luftauswerfer pneumatic ejector *(im)*
Luftballon balloon
Luftbefeuchter air humidifier
Luftbeladung 1. introduction of air 2. air content *(e.g. of foam)*
Luftbelastung air/atmospheric pollution
Luftblase air bubble
luftblasenfrei free from air bubbles
Luftbürste air brush
luftdicht airtight
Luftdruck 1. air pressure. 2. atmospheric pressure
Luftdurchsatz air throughput
Luftdurchwirbelung forced air circulation
Luftdusche air cooling device

Lufteinschlüsse entrapped air, air entrapments
Lufteintrag introduction of air
Lufteintritt air inlet
Luftemission atmospheric emission
Lüften (mould) breathing
Lüfter fan
Lüfterflügel fan blade
Lüfterzarge fan housing/shroud
Luftfahrt 1. air travel. 2. aircraft industry
Luftfahrtbehörden aviation authorities
Luftfahrtindustrie aircraft industry
Luftfahrtwerkstoff aviation/aircraft material
Luftfeuchte 1. relative humidity. 2. atmospheric humidity/moisture, moisture in the atmosphere
Luftfeuchte, relative relative humidity
Luftfeuchteempfindlichkeit sensitivity to atmospheric humidity
Luftfeuchtegehalt atmospheric moisture content
Luftfeuchtigkeit 1. relative humidity. 2. atmospheric humidity
Luftfilter air filter
luftforciert trocknend force drying
luftfrei free from air
Luftführungskanal air duct
luftgekühlt air cooled
luftgetrocknet air dried
lufthärtend air curing
Luftkanal air duct
Luftkissen air cushion
Luftkissenboot hovercraft
Luftkompressor air compressor
Luftkraft aerodynamic force
Luftkühlaggregat air cooling unit
Luftkühlgerät air cooling unit
Luftkühlring cooling/air ring *(bfe)*
Luftkühlsystem air cooling system
Luftkühlung 1. air cooling. 2. air cooling unit/system
luftlos airless
luftloses Sprühverfahren airless spraying
Luftmesser air knife
Luftofen circulating air (drying) oven
Luftpolster air cushion
Luftpolsterfolie bubble film
Luftrakel air knife
Luftraum air space
Luftreifen pneumatic tyre
Luftsauerstoff atmospheric oxygen
Luftschall air-borne sound
Luftschwert air knife
Luftspalte air gap
Luftsteuerventil pneumatic control valve
Luftstrom 1. air stream. 2. amount of air **große Luftströme** large amounts of air
Luftströmung air stream/flow
Luftströmungsverhältnisse air flow conditions
lufttrocknend air drying
Lufttrocknung air drying
Luftturbulenz atmospheric turbulence
Luftumwälzofen air circulating oven

Luftumwälzung 1. air circulation. 2. air circulating system: **Zum Erzielen einer gleichmäßigen Temperaturverteilung muß eine Luftumwälzung vorgesehen werden** to achieve uniform temperature distribution, the air should be made to circulate freely OR: an air circulating system should be provided
Luftumwälzungssystem air circulating system
Luftumwälzvorrichtung air circulation unit
Lüftungsbaustein ventilation brick
Lüftungskanal ventilation shaft
Lüftungsschlitz ventilation slit
Luftverschmutzung air/atmospheric pollution
Luftverteilungselemente air distributing elements *(bfe)*
Luftverunreinigung air/atmospheric pollution
Luftwendestangen air deflector bars *(bfe)*
Luftwiderstand (aerodynamic) drag
Luftwiderstandsbeiwert drag coefficient/factor
Luftzirkulation air circulation
Luftzirkulationsrate air circulation rate
Luftzufuhr 1. air supply. 2. air inlet
Luftzuführungselemente air supply elements *(bfe)*
Luftzuführungsschlauch air supply line
Luftzutritt access of air
Lunker void
lunkerfrei void-free
LVN limiting viscosity number, intrinsic viscosity
LWL abbr. of **Lichtwellenleiter**, optic(al) cable
lyophil lyophilic
lyophob lyophobic
lyotrop lyotropic

M

machbar feasible
machbar, technisch technically feasible
machbar, wirtschaftlich economically viable
Machbarkeit feasibility
Machbarkeitstudie feasibility study
mager short-oil
Magnesit magnesite
Magnesiumlegierung magnesium alloy
Magnesiumoxid magnesium oxide
Magnesiumsilikat magnesium silicate
Magnesiumstearat magnesium stearate
Magnet 1. magnet. 2. solenoid
Magnetanker magnetic armature
Magnetband magnetic tape *(me)*
Magnetbandbeschichtung magnetic tape coating
Magnetbandgerät magnetic tape unit *(me)*
Magnetbandkassette magnetic tape cassette *(me)*

Magnetbandkassettengerät cassette recorder *(me)*
Magnetbeleg magnetic ink document *(me)*
Magnetfeld magnetic field
magnetisches Feld magnetic field
magnetisierbar magnetisable
Magnetitpigment magnetite pigment
Magnetkarte magnetic card *(me)*
Magnetkartenleser magnetic card reader *(me)*
Magnetplatte magnetic disk *(me)*
Magnetplattenlaufwerk magnetic disk drive *(me)*
Magnetplattenspeicher magnetic disk, disk storage device *(me)*: *There is no need to translate the last part of this word since a magnetic disk is itself a memory*
Magnetstreifen magnetic strip
Magnetträger 1. magnetic tape. 2. magnetic disk *(me)*
Magnetventil solenoid valve
Magnetventiltreiber solenoid (operated) valve driver
Mahlaggregat grinder, grinding unit
Mahlbahn grinding face *(sr)*
Mahldauer grinding time
Mahlglasfasern milled glass fibres
Mahlgut 1. material being ground, material to be ground, millbase. 2. regrind
Mahlgutansatz millbase formulation
Mahlguteinlauf feed opening/port/throat *(e) There is no need here for a literal translation since the meaning will be obvious from the context*
Mahlgutformulierung millbase formulation
Mahlgutzusammensetzung millbase composition
Mahlkammer 1. grinding compartment. 2. cutting chamber *(sr)*
Mahlkörper grinding medium
Mahlmedium grinding medium
Mahlpaste millbase
Mahlprozeß grinding process/operation
Mahlraum 1. grinding compartment. 2. cutting chamber *(sr)*
Mahlwalzwerk grinding rolls
Maisstärke maize starch
MAK *abbr. of* **maximale Arbeitsplatzkonzentration** maximum allowable concentration, MAC
Makroalkylradikal macroalkyl radical
Makroaufbau macrostructure
Makrobefehl macro-instruction/-command *(me)*
Makrobereich macro-range/-region
makro-Brownsche Bewegung macro-Brownian movement
Makrobruch macrofracture
Makromolekül macromolecule
makromolekular macromolecular
Makromolekülkette macromolecule chain
Makromonomer macromonomer
Makroradikal macroradical
Makroriß macrocrack

Makroschaum macrofoam
makroskopisch macroscopic
Makrostruktur macrostructure
MAK-Wert MAC, maximum allowable concentration
Maleinat maleinate
Maleinatharz maleic resin
maleiniert maleinised
Maleinimid maleinimide
maleinisiert maleinised
Maleinsäure maleic acid
Maleinsäureanhydrid maleic anhydride
Maleinsäureester maleinate
Malerbürste paint brush
Malerlack decorator's paint
Malerpinsel paint brush
Manganviolett manganese violet
Mängelkorrektur fault correction
Mannichbase Mannich base
Manometer manometer
Manschette sleeve
Mantelmasse cable (sheathing) compound
Mantelmaterial cable (sheathing) compound
Mantelmischung cable (sheathing) compound
Mantelthermoelement jacketed thermocouple
manuell manual
MAP *abbr. of* manufacturing automation protocol
marginal marginal
maritim maritime
Marke *a blanket expression which can mean anything the writer can't or won't express more explicitly, e.g.* type, grade, compound, resin, dispersion, machine *etc., depending on the context.*
Markenname trade/brand name
Markenzeichen trade mark
Marketingstrategie marketing policy
Markierungsbeleg mark-sense card *(me)*
Markise awning
Markstein milestone *(in a company's history)*
Marktakzeptanz market acceptance
Marktanteil market share
Marktaufteilung market breakdown
Marktausweitung market expansion
Marktchancen market opportunities
Marktdurchdringung market penetration, penetration of the market
Marktführer market leader
marktgängig currently available, (available) on the market
Marktgliederung market breakdown
Marktlücke market gap
marktorientiert market-orientated
Marktpotential market potential
Marktprognose market forecast
Marktsättigung market saturation
Marktstudie market study
marktüblich standard, ordinary: **auf einer marktüblichen Blasformmaschine** on an ordinary blow moulding machine
Marktverhältnisse market conditions
Marktwirtschaft market economy

Marmor marble
Marmorieren marbling
marmoriert marbled
Martens, Wärmeformbeständigkeit nach
 Martens heat distortion temperature
Martensgrad Martens temperature
Martens-Wärmeformbeständigkeit Martens heat distortion temperature
Martenswert Martens temperature
Martenszahl Martens temperature
Maschengewebepaket screen pack *(mf)*
Maschenware knitted fabrics/materials
Maschenweite mesh width *(mf)*
Maschenzahl mesh count
Maschine machine
maschinell mechanical: **maschinelle Einrichtungen** equipment
Maschinen- und Apparatebau machine and equipment manufacture
Maschinenabschaltung machine cut-out
Maschinenaufspannplatte platen *(im)*
Maschinenausfall machine breakdown: **...kann einen Maschinenausfall bewirken** ...can cause the machine to break down
Maschinenausfallzeit machine downtime
Maschinenausfuhr machine exports
Maschinenauslastung *using the machine to capacity*: **Nicht nur ist die Maschinenauslastung geringer...** not only is the machine not run to full capacity...
Maschinenausnutzungsrate machine utilisation (rate)
Maschinenausrüstung equipment
Maschinenbau mechanical engineering
maschinenbaulich *relating to machinery or equipment. This word should be treated like* **maschinenbautechnisch** *and* **maschinentechnisch**
maschinenbautechnisch *relating to machinery or equipment*: **Der maschinenbautechnisch ideale Schneckenextruder ist der Einschneckenextruder** from the technical point of view, the single-screw extruder is the ideal machine
Maschinenbediener machine operator
Maschinenbedienpersonal machine operators
maschinenbedingt due to the machine
Maschinenbedingungen machine parameters
Maschinenbeschädigungen damage to a machine *(for translation example see under* **Bedienungsfehler***)*
Maschinenbetrieb machine operation: **Ein besonders wirtschaftlicher Maschinenbetrieb wird gewährleistet** the machine is specially economic to run
Maschinenbett 1. machine base. 2. machine frame
Maschinenbewegungen machine movements
Maschinencode machine code
Maschinendaten machine data
Maschinendatenerfassung machine data collection/acquisition *(me)*

Maschinendüse machine/injection nozzle *(im)*
Maschineneffizienz machine efficiency
Maschineneinflußgrößen factors influencing machine performance
Maschineneinlauf feed opening/port/throat *(e,im)*
Maschineneinrichtung equipment
Maschineneinstelldaten machine settings
Maschineneinsteller machine setter
Maschineneinstellgröße machine setting
Maschineneinstellparameter machine setting
Maschineneinstellprogramm machine setting program *(me)*
Maschineneinstellungen machine settings
Maschineneinstellwerte machine settings
Maschinenelemente machine components
Maschinenerstausrüstung equipment included when the machine was originally supplied: **Die Wahlausrüstungen müssen bei der Maschinenerstausrüstung festgelegt werden** optional equipment must be selected at the time the machine is ordered
Maschinenführer machine operator
Maschinenführung machine control
Maschinenfunktionen machine functions
Maschinengängigkeit smooth passage through the machine: **Durch das Antiblockmittel wird nicht nur die Blockneigung der Folien verringert, sondern auch deren Maschinengängigkeit bei der Weiterverarbeitung verbessert** The antiblocking agent not only reduces the tendency of the film to block, but also enables it to pass more smoothly through the machine during subsequent conversion into finished products.
Maschinengeschwindigkeit machine speed
Maschinengestell machine frame
Maschinenhersteller machine manufacturer
Maschinenhydraulik machine's hydraulic system
Maschinenkonzept machine *(see explanatory note under* **Konzept***)*
Maschinenkonzeption 1. machine. 2. machine design
Maschinenkosten machine costs
Maschinenlagerung machine support
Maschinenlänge 1. machine length. 2. extruder length
Maschinenlängsachse longitudinal machine axis
Maschinenlauf machine operation: **geräuscharmer Maschinenlauf** the machine is quiet-running OR: quiet in operation
Maschinenlebensdauer machine life: **während der Maschinenlebensdauer** during the machine's lifetime
Maschinenleistung machine output/performance/capacity
maschinenlesbar machine-readable *(me)*
Maschinenlieferant machine supplier
maschinennah machine-oriented/-sensible *(me)*

Maschinennutzung machine utilisation
Maschinennutzungsgrad machine productivity
Maschinenoberteil top part of the machine
Maschinenparameter machine parameter/variable
Maschinenpark 1. available machines/equipment: **Die Vorplastifizierung ist vom Produktionsprogramm und vom Maschinenpark abhängig** pre-plasticisation depends on the range of products being made and on the equipment available. 2. machines **In den letzten Jahren wurden erhebliche Mittel für einen auf die PP-Verarbeitung ausgerichteten Maschinenpark investiert** in recent years, considerable amounts have been invested in machines for processing polypropylene
Maschinenpersonal machine operators
Maschinenprogramm 1. range of machines. 2. machine operating program *(me)*
Maschinenpult control desk/panel/console
Maschinenrahmen machine frame
Maschinenreihe range of machines
Maschinenrückseite back of the machine
maschinenseitig 1. on the machine: **maschinenseitig angebracht** attached to the machine. 2. relating to the machine: **maschinenseitige Stellgrößen** machine settings; *Sometimes the word can be omitted, e.g.:* **Maschinenseitig bietet sich eine hydraulische Ausstoßvorrichtung als ideale Lösung an;** *Here, the word has obviously been used as padding, so that the translation should read* the ideal solution is to use an hydraulic ejector
Maschinensprache machine language *(me)*
Maschinenständer machine frame
Maschinenstatus machine status
Maschinenstellgröße machine variable
Maschinenstellwert machine variable
Maschinensteuerprogramm machine control program *(me)*
Maschinensteuerung machine control system, machine controls
Maschinensteuerungsdatei machine control data file *(me)*
Maschinenstillstand 1. machine stoppage. 2. stopping of the machine **bei Maschinenstillstand** if the machine stops
Maschinenstillstandzeit machine downtime
Maschinenstopp stopping of the machine **bei einem unvorhergesehenem Maschinenstopp** if the machine stops suddenly
Maschinenstörung machine malfunction
Maschinenstumpfschweißung machine butt welding
Maschinenstundensatz machine hour rate
Maschinentakt machine cycle
Maschinentaktsteuerung machine cycle control

maschinentechnisch *relating to machinery or equipment:* **Die verfahrenstechnischen Weiterentwicklungen des klassischen Spritzgießverfahrens in letzter Zeit sind direkte Folgen der maschinentechnischen Entwicklungen** recent technical developments in the classic injection moulding process are the direct result of developments in the machine sector; **maschinentechnische Voraussetzungen** machine requirements; *Like so many words of this kind, this one, too, is often used as padding, as in this example, where it can be omitted.* **Diese Funktionen werden durch die nachfolgend beschriebene maschinentechnische Ausrüstung realisiert** these functions are carried out by the equipment described below; **Deshalb empfiehlt sich eine maschinentechnische Trennung** it is therefore advisable to use separate machines
Maschinentrichter (feed/material) hopper *(e,im)*
Maschinenüberwachung machine control
Maschinenumbaukosten machine rebuilding costs
Maschinenumbausatz machine conversion unit
Maschinenverhalten machine performance
Maschinenverschleiß machine wear: **um den Maschinenverschleiß in Grenzen zu halten** to keep wear of the machine within limits
Maschinenvorderseite front of the machine
Maschinenwartung machine maintenance
maschinenwaschbar machine washable
Maschinenzuhaltekraft locking force *(im)*
Maschinenzuhaltung 1. locking force *(im)*. 2. locking mechanism *(im)*
Maschinenzustand machine status *(me)*
Maschinenzyklus machine cycle
Maschinenzylinder injection cylinder *(im)*
Maske mask *(me)*
Maß, nach 1. custom-built *(machine)*. 2. tailor-made *(plastic)*
Maßabweichungen dimensional deviations
Maßänderungen dimensional changes
Maßbeständigkeit dimensional stability
***Masse** 1. material *(general term)*. 2. melt. 3. mass, weight: *Version 2. describes the compound which has been heated to a plastic condition in an extruder barrel or the cylinder of an injection moulding machine. The term* stock *is sometimes used in this connection although this, strictly speaking, refers to uncured rubber compounds*
Maße, nicht werkzeuggebundene mould-independent dimensions: *These are dimensions formed across the mould parting line, in the direction of moulding*
Maße, werkzeuggebundene mould-dependent dimensions: *These are dimensions formed entirely in one mould part, i.e. dimensions not affected by the mating parts of the mould*

---

* *Words starting with* **Masse-**, *not found here, may be found under* **Material-** *or* **Schmelze-**

Masseanhäufungen material accumulations
Masseaustritt 1. melt delivery. 2. *denoting the place where the melt leaves the machine*: **Mit der Zylindertemperierung der ABC Extruder kann von der Einzugszone bis zum Masseaustritt ein optimales Temperaturprofil gefahren werden** thanks to the barrel temperature control system of ABC extruders, an optimum temperature profile can be achieved from the feed zone right through to the die. 3. escape of material: **Ein Masseaustritt an der Entgasungsöffnung kann nur durch Unterdosierung verhindert werden** material can only be prevented from escaping through the vent by underfeeding
Masseaustrittstemperatur melt exit temperature: **Die Masseaustrittstemperaturen liegen unter 250°C** the temperature of the melt as it leaves the die is below 250°C
Massebehälter (feed/material) hopper *(e,im)*
Massebohrung melt flow-way, runner *(im)*
Massedruck melt pressure
Massedruckanzeige melt pressure indicator
Massedruckaufnehmer melt pressure sensor/transducer
Massedruckgeber melt pressure gauge/indicator
Massedruckverlauf melt pressure profile
Massedurchsatz 1. melt throughput. 2.material throughput *(if the product is obviously not in melt form)*. 3. output rate, mass flow rate *(through an extruder or die)*, output
Masseeinzug material feed
Massefluß 1. melt flow. 2. material flow
Massehomogenität melt homogeneity
Massekanal 1. melt flow-way *(general term)*. 2. runner *(im)*
Masseknet bank *(c)*
Massen-% % w/w, percent by weight
Massenfertigung mass production
Massengüter mass-produced goods
Massenkunststoffe bulk plastics
Massenprodukte mass-produced goods/articles
Massenproduktion mass production
Massenspeicher mass/bulk storage unit *(me)*
Massenspektrometer mass spectrometer
Massenspektrometrie mass spectrometry
massenspektrometrisch mass-spectrometric
Massenspektroskopie mass spectroscopy
Massenspektrum mass spectrum
Massenteile parts by weight, p.b.w.
Massenträgheit moment of inertia
Massenträgheitskraft force of inertia
Massenträgheitsmoment moment of inertia
Massepolster melt cushion
Massepolymerisat bulk polymer
Massepolymerisation bulk polymerisation
Massepolymerisationsanlage bulk polymerisation plant
Massepolymerisationsverfahren bulk polymerisation (process)

Masse-PVC bulk PVC
Masse-PVC-Anlage bulk PVC polymerisation plant
Massespeicher melt accumulator *(bm)*
Massespeichersystem melt accumulator system *(bm)*
Massestrang melt strand, extrudate
Massestrom 1. melt stream, melt. 2. melt throughput. 3. material throughput *(if the product is obviously not in melt form)*. 4. output rate, mass flow rate *(through an extruder or die)*, output: *This word requires great care in translation, as shown by the following examples. Here, the context in which the word is used is extremely important*; **Die erforderliche Wärme wird durch ein Heizelement erzeugt, welches im Massestrom liegt** the necessary heat is produced by a heating element immersed in the melt stream; **bei gleichbleibendem Massestrom** for a constant melt throughput; **Hier wird vorausgesetzt, daß der Massestrom trotz ansteigender Fließwiderstände in gleicher Höhe bleibt, daß also die Maschine keinen abnehmenden Massestrom liefert. Es ist ebenfalls vorausgesetzt daß die Abmessungen des Fließquerschnitts über den Fließweg konstant bleiben und nicht eine Verteilung des Massestroms auf einen größer werdenden Umfang stattfindet.** Here, it has been assumed that the melt throughput remains the same, despite increasing flow resistance, so that the machine supplies a constant amount of melt. It has also been assumed that the runner diameter remains constant along the whole distance so that the melt does not have to be distributed over a constantly increasing area
Masseteilströme separate melt streams *(e) (melt divided by spider)*
Massetemperatur 1. melt temperature *(e,im)*. 2. material temperature *(general term, where* **Masse** *is obviously not melt, e.g. in calendering)*
Massetemperaturanzeige melt temperature indicator
Massetemperaturfühler melt thermocouple
Massetemperaturmessungen melt temperature determinations: **Wegen technischer Schwierigkeiten, die Massetemperaturmessungen im Zylinder bereiten...** since it is technically difficult to measure melt temperatures inside the barrel...
Massetemperaturverteilung melt temperature distribution
Massethermoelement melt thermocouple
Massetrichter (feed/material) hopper *(e,im)*
Masseveränderung change in weight
Masseverteilerkanal runner *(im)*

Masseweg 1. (melt) flow path: 2. (melt) flowway. 3. runner *(im) (for translation example see under* **Fließweg***)*
maßgenau dimensionally accurate
Maßgenauigkeit dimensional accuracy
maßgerecht dimensionally accurate
maßgeschneidert 1. custom-built *(machine)*. 2. tailor-made *(plastic)*
maßhaltig dimensionally stable
Maßhaltigkeit dimensional stability
massiv solid
Massivbauweise solid construction
Massivgummi solid rubber
Massivgummidichtung solid rubber seal
Massivguß solid casting
Massivkörper solid body
Massivplatte solid sheet
Massivprofil solid profile
Maßkonstanz dimensional accuracy
Maßnahmen, verfahrenstechnische technical measures
Maßnahmenkatalog list of measures to be taken
Maßschwankungen dimensional variations/fluctuations
Maßstab scale
Maßstab, großtechnischer on a large scale, on an industrial scale
Maßstab, kleintechnischer on a small scale, on a pilot plant scale
Maßstab, halbtechnischer pilot plant scale
Maßstabilät dimensional stability
Maßstabsvergrößerung scaling-up, scale-up
Maßstreuungen dimensional variations
Maßtoleranzen dimensional tolerances
Maßtreue dimensional accuracy
Maßungenauigkeiten dimensional inaccuracies
Maßzahl measure
Mastikation mastication
Mastiziereffekt masticating effect
mastizieren to masticate
***Materialanhäufungen** material accumulations
Materialansammlungen material accumulations
Materialaufgabeschacht feed opening/port/throat *(e,im)*
Materialaufwand amount of material needed/used
Materialausschußquote amount of waste/scrap produced: **Eine Reduktion der Materialausschußquote ist gewährleistet** less scrap is produced
Materialaustrag product discharge
Materialaustrageschacht delivery chute
Materialaustritt 1. melt delivery. 2. *denoting the place where the melt leaves the machine.* 3. escape of material *(for translation examples see under* **Masseaustritt***)*
Materialauswahl choice of material
materialbedingt material-related
Materialbeschaffenheit material properties/characteristics
materialbezogen material-related
Materialbruch material failure
Materialdosierung 1. material feed. 2. material feed system. 3. material metering system
Materialdurchsatz 1. melt throughput. 2. material throughput *(if the product is obviously not in melt form)*. 3. output rate, mass flow rate *(through an extruder or die)*, output
Materialeckdaten key material constants/properties
materialeigen material-specific
Materialeinfüllöffnung feed opening/port/throat *(e,im)*
Materialeinfülltrichter (feed/material) hopper *(e,im)*
Materialeinlaßventil material inlet valve
Materialeinlauf (feed/material) hopper *(e,im)*
Materialeinlaufstück feed section/zone *(e)*
Materialeinsatz use of material: **Verringerung des Materialeinsatzes** reducing the amount of material used
Materialeinsparungen material savings: **Durch diese Maßnahmen sind Materialeinsparungen möglich** these measures enable material to be saved OR: enable material savings to be effected
Materialeinzug material feed
Materialeinzugszone feed section/zone *(e)*
Materialermüdung material fatigue
Materialersparnis material saving
Materialfluß 1. material flow. 2. melt flow
Materialfülltrichter (feed/material) hopper *(e,im)*
materialgerecht suitable or correct for a given material: **bei materialgerechter Verarbeitung** if the material is processed correctly (**nicht materialgerecht** *would obviously be* incorrectly); *For other translation examples see under* **werkstoffgerecht***. See also explanatory notes under* - **gerecht**
materialimmanent material-inherent
Materialkenngröße material constant
Materialkennwert material constant/property
Materialkuchen 1. (piece of) pre-compressed moulding compound. 2. pre-compressed material *(cm)*
Materialmangel lack of material
Materialmenge amount of material
Materialmenge, vordosierte pre-weighed amount of material
Materialprüfgerät material testing instrument
Materialprüfinstrument material testing instrument
Materialprüftechnik material testing
Materialrecycling material recycling/recovery
Materialrückfluß reverse flow, backflow **Um einen Materialrückfluß zu verhindern** to prevent the melt flowing back
Materialschlauch film bubble *(bfe)*
Materialschmelze polymer melt

---

* *Words starting with* **Material-***, not found here, may be found under* **Masse-** *or* **Schmelze-**

materialschonend without damaging the material
materialsparend material-saving
materialspezifisch material-related, specific to the material
Materialstrahl jet of material *(im) (see explanatory note and translation example under* **Freistrahl***)*
Materialstrang extrudate
Materialstrom 1. melt stream, melt. 2. melt throughput. 3. material throughput *(if the product is obviously not in melt form)*. 4. output rate, mass flow rate *(through an extruder or die)*, output *(for translation examples see under* **Massestrom***)*
Materialtemperatur 1. material temperature 2. melt temperature
Materialtrichter (feed/material) hopper *(e,im)*
Materialüberschuß excess material
Materialverhalten material performance, behaviour of the material
Materialverwertung material recovery/recycling
Materialwulst 1. bead (of material). 2. bank *(c)*
Materialzufuhr 1. material feed. 2. material feed system: **...die eine gleichmäßige Materialzufuhr sicherstellt** which ensures that the material is fed evenly *(i.e. to the machine)*
Materialzufuhr, automatische 1. automatic material feed. 2. automatic material feed system
Materialzuführung 1. material feed. 2. material feed system
Materialzufuhrvorrichtung material feed system
materielles Recycling material recycling
Matrixdrucker dot-matrix printer *(me)*
Matrixharz matrix resin
Matrixwerkstoff matrix material
Matrize 1. female mould *(t)*. 2. cavity plate *(im)*
Matrizenbelüftung forcing air into the mould cavity *(to help in ejection of the moulded part) (im)*
matrizenseitig on the fixed mould half, on the cavity half *(im)*
Matrizenwerkzeug mould
matt matt
Mattenlaminat glass mat laminate *(grp)*
Mattenverstärkung glass mat reinforcement
mattiert mat
Mattierungseffekt flatting effect
Mattierungsmittel flatting agent
Mattlack 1. matt-finish paint. 2. matt-finish lacquer
Mattlackierung matt finish
Mauerfläche masonry surface
Mauerwerk masonry, brickwork
Maus mouse *(me)*
Maximaldrehzahl maximum speed
Maximaldruck maximum pressure
Maximaldrucküberwachung maximum pressure control

maximale Arbeitsplatzkonzentration maximum allowable concentration, MAC
Maximalgrenzwert maximum value
Maximalkonzentration maximum concentration
Maximalkraft maximum force
Maximalschrumpfung maximum shrinkage
Maximaltemperatur maximum temperature
Maximalverformung maximum deformation
Maximalviskosität maximum viscosity
Maximalwert maximum value
MCAE *abbr. of* mechanical computer aided engineering
MDE *abbr. of* **Maschinendatenerfassung**, machine data acquisition/capture/ collection
MDI *abbr. of* **Diphenylmethandiisocyanat**, diphenylmethane di-isocyanate, MDI
Mechanik 1. mechanism, moving parts *(of a machine)*. 2. mechanical properties
mechanisch mechanical
mechanisch getrieben mechanically blown *(foam)*
mechanische Beanspruchung mechanical loading
mechanische Bearbeitung machining
mechanische Festigkeit mechanical strength
mechanische Leistung 1. productivity/output capacity, capacity *(of a machine)* 2. power, horsepower *(of an engine)*
mechanische Verschäumung mechanical foaming
mechanische Werkstatt engineering workshop
mechanische Werte mechanical properties
mechanischer Schaum mechanically blown foam
mechanischer Verlustfaktor loss factor
mechanisches Verhalten mechanical performance
Mechanisierungsgrad degree of mechanisation
Medianwert mean value
Medienbelastung exposure to chemicals, exposure to chemical agents
Medienbeständigkeit chemical resistance
Medium medium
medizinisch medical **medizinische Artikel** medical equipment
Medizintechnik medical engineering
Meerwasser seawater
mehrachsig multi-axial
mehradrig multi-conductor, multi-wire
Mehraufwand increased/extra/added expense/ effort/trouble/difficulty: **technischer Mehraufwand** technically more difficult *(see also translation examples under* **Aufwand***)*
mehraxial multi-axial
mehrbasisch polybasic
Mehrbitprozessor multi-bit processor *(me)*
mehrdirektional multi-directional
Mehretagenspritzgießwerkzeug multi-daylight injection mould
Mehretagentechnik stack mould technology *(im)*

Mehretagenwerkzeug stack mould, multi-daylight mould
mehretagig multi-daylight *(im,tm)*
mehrfach angebunden multiple-gated *(im)*
Mehrfachabzug multiple take-off (unit)
Mehrfachaddition polyaddition
Mehrfachanbindung multi-point gating *(im)*
Mehrfachanguß multi-point gating *(im)*
Mehrfachanschnitt multi-point gating *(im)*
Mehrfachanspritzung multi-point gating *(im)*
Mehrfachbandanschnitt multiple film gate *(im)*
Mehrfachblasdüse blown film coextrusion die
Mehrfachdosierung 1. multiple metering. 2. multiple feed. 3. multiple metering system. 4. multiple feed system
Mehrfachdüse 1. multiple die *(e)*. 2. multiple nozzle *(im)*
Mehrfachentgasung multiple venting
Mehrfachextrusion repeated extrusion
Mehrfachextrusionskopf multiple die head *(e)*
Mehrfachform multi-cavity mould, multi-impression mould
Mehrfachformwerkzeug multi-cavity mould, multi-impression mould
Mehrfachgewinde multiple thread
Mehrfachheißkanalform multi-cavity/-impression hot runner mould *(im)*
Mehrfachhub multi-stroke
Mehrfachkopf multiple die head *(e)*
Mehrfachpapiersack multi-ply paper sack
Mehrfachpunktanschnitt multi-point pin gate, multiple pin gate *(im)*
Mehrfachraupenabzug multiple caterpillar take-off (unit)
Mehrfachregler multiple controller, multiple control unit
Mehrfachschlauchkopf multi-parison die *(bm)*
Mehrfachschnecke multiple screw *(e)*
Mehrfachspritzgießen multi-cavity injection moulding
Mehrfachspritzgußwerkzeug multi-cavity/-impression injection mould
Mehrfachtunnelanschnitt multiple tunnel gate *(im)*
Mehrfachverarbeitung repeated processing
Mehrfachwerkzeug 1. multi-cavity/-impression mould *(im)*. 2. multi-orifice die *(e)*
Mehrfarbendruckwerk multi-colour printing unit
Mehrfarbengrafik multi-colour graphics
Mehrfarbenplotter multi-colour plotter *(me)*
Mehrfarbenspritzgießen multi-colour injection moulding
Mehrfarbenspritzgießmaschine multi-colour injection moulding machine
mehrfunktionell multi-functional
mehrgängig multiple-flighted, multi-start *(screw) (e)*
Mehrgängigkeit multi-start construction *(of screw) (e)*
Mehrkammerprofil multi-compartment profile
Mehrkanalbreitschlitzwerkzeug multiple-manifold slit die *(e)*

Mehrkanaldüse multiple-manifold die *(e)*
Mehrkanallinienschreiber multi-channel line recorder
Mehrkanalwerkzeug multiple-manifold die *(e)*
Mehrkernphenol multi-ring phenol
Mehrkomponentendosieranlage multi-component metering/feed equipment
Mehrkomponentenlegierung polymer blend/alloy, polyblend
Mehrkomponenten-Reaktionsharzsystem resin-catalyst system. *Since a* **Reaktionsharzsystem** *invariably consists of several components, i.e. resin and catalyst, there is no need to translate the first part of the term*
Mehrkomponenten-schaumspritzgießverfahren sandwich moulding *(im)*
Mehrkomponentenspritzaggregat multi-component injection unit
Mehrkomponentenspritzgießen sandwich moulding *(im)*
Mehrkomponentenspritzgießformteil sandwich moulding, sandwich moulded part
Mehrkomponentenspritzgießmaschine sandwich moulding machine *(im)*
Mehrkomponenten-TSG-Verfahren multi-component structural foam moulding (process)
Mehrkosten extra/additional costs
Mehrkreisgerät multi-circuit unit
Mehrkreiskühlsystem multi-circuit cooling system
Mehrkreisregelsystem multi-circuit control system
Mehrkreisüberwachungssystem multi-circuit monitoring system
Mehrlagenleiterplatte multi-layer printed circuit board
Mehrlagenschaltung multi-layer circuit, multi-layer printed circuit board
Mehrlagenverbund multi-layer composite
mehrlagig multi-layer
Mehrleistung productivity increase, greater/increased output
Mehrlochkopf multi-strand die *(e)*
Mehrlochspritzdüse multi-bore injection nozzle *(im)*
Mehrlochwerkzeug multi-strand die *(e)*
Mehrnutzenschneidevorrichtung multi-reel slitter
Mehrphasenkunststoff multi-phase polymer
Mehrphasenpolymer polymer blend/alloy, polyblend
Mehrphasenstruktur multi-phase structure
Mehrphasensystem multi-phase system
mehrphasig multi-phase
Mehrphasigkeit multi-phase character
Mehrplattenwerkzeug multi-plate mould, multi-part mould, multi-daylight mould *(im,tm)*
Mehrpreis extra/added cost
Mehrprozessorsystem multi-processor system *(me)*

Mehrschichtbetrieb 1. multi-shift operation. 2. multi-shift factory
Mehrschichtblascoextrusion blown film coextrusion
Mehrschichtblasfolie multi-layer blown film
Mehrschichtblasfolienanlage multi-layer blown film line
Mehrschichtblaskopf multi-layer blown film die
Mehrschichtbreitfolie extruded multi-layer film/sheeting, coextruded film/sheeting
Mehrschichtdüse 1. coextrusion die *(general term)*. 2. flat film coextrusion die. 3. blown film coextrusion die
Mehrschichtenaufbau 1. multi-layer coating/covering. 2. multi-layer screed
Mehrschichtenbelag 1. multi-layer coating/covering. 2. multi-layer screed
Mehrschichtenlackierung multi-coat finish
Mehrschichtenverbundstoff multi-layer composite (material)
Mehrschichtextrusionsanlage coextrusion line/plant
Mehrschichtextrusionsdüse 1. coextrusion die *(general term)*. 2. flat film coextrusion die *(e)*. 3. blown film coextrusion die *(bfe)*
Mehrschichtextrusionsverfahren coextrusion (process), multi-layer extrusion (process)
Mehrschichtflachdüse coextrusion slit die *(e)*
Mehrschichtflachfolienanlage flat film coextrusion line
Mehrschichtfolie composite film, multi-layer film, coextruded film
Mehrschichtfolienblasanlage multi-layer blown film line
Mehrschichtfolienblaskopf multi-layer blown film die
Mehrschichthohlkörper multi-layer blow moulding
mehrschichtig multi-layer
Mehrschichtkopf 1. coextrusion die *(general term)*. 2. flat film coextrusion die. 3. blown film coextrusion die
Mehrschichtplattendüse sheet coextrusion die, multi-layer sheet die *(e)*
Mehrschichtschlauchfolie multi-layer blown film
Mehrschichtschlauchkopf multi-layer parison die *(bm)*
Mehrschichtsystem multi-coat system
Mehrschichttechnik co-extrusion, multi-layer extrusion
Mehrschichtverbund multi-layer composite
Mehrschichtwerkzeug coextrusion die
Mehrschneckenextruder multi-screw extruder
Mehrschneckenmaschine 1. multi-screw extruder. 2. multi-screw compounder
Mehrschneckensystem multi-screw system
Mehrstationenanlage multi-station plant
Mehrstationenmaschine multi-station machine
Mehrstellendosierung 1. multi-point dispensing. 2. multi-point dispensing unit
Mehrstoffsystem multi-component system

Mehrstrangextrusion multi-strand extrusion
Mehrstufenprozeß multi-stage process
Mehrstufenpumpe multi-stage pump
mehrstufig multi-stage
mehrteilig multi-part
Mehrverbrauch increase in consumption
Mehrverteilerwerkzeug multi-manifold die *(e)*
Mehrwalzenabzug multi-roll take-off (unit)
Mehrwalzenabzugsvorrichtung multi-roll take-off (unit)
Mehrwalzenauftragswerk multi-roll applicator
Mehrwalzenstuhl multiple roll mill
Mehrwegbehälter multi-trip/returnable container
Mehrwegflasche multi-trip/returnable bottle
Mehrweggebinde multi-trip/returnable container
Mehrweggetränkeflasche multi-trip/returnable beverage bottle
Mehrwegglasflasche multi-trip/returnable glass bottle
Mehrwegpackmittel multi-trip/returnable packaging
Mehrwegverpackung multi-trip/returnable packaging
Mehrwellenanordnung multi-screw arrangement *(e)*
Mehrwellenextruder multi-screw extruder
Mehrwellensystem multi-screw system
mehrwertig 1. polyvalent. 2. polyhydric *(if an alcohol)*
Mehrzahl majority, greater number, most of
Mehrzonenschnecke multi-section screw *(e)*
Mehrzweckanlage multi-purpose plant
Mehrzweckharz general purpose resin
Mehrzweckklebstoff general purpose adhesive
Mehrzweckkopf multi-purpose extrusion die
Mehrzweckpolyesterharz general purpose polyester resin
Mehrzweck-TPE general purpose thermoplastic elastomers
MEK 1. *abbr. of* **Methylethylketon,** methyl ethyl ketone. 2. *abbr. of* **maximale Emissions-Konzentration,** maximum emission concentration
MEK-Wert maximum emission concentration
Melamin melamine
Melamin-Formaldehydharz melamine resin, melamine-formaldehyde resin
Melaminharz melamine resin, melamine-formaldehyde resin
Melaminharzglashartgewebe melamine glass fibre laminate
Melaminharzpreßmasse melamine moulding compound
Melaminharzschaumstoff melamine foam
Melaminharzschichtstoff melamine laminate
Melaminsalz melamine salt
Membran(e) 1. membrane. 2. diaphragm
Membranpumpe diaphragm pump
Membranschalter touch-sensitive key, membrane switch
Membrantastatur touch-sensitive/membrane keyboard

Mengendurchflußregler flow controller, flow control instrument
Mengenleistung 1. output, output rate. 2. throughput *(mostly of liquids through a pipe for example). Version 1 is preferable in connection with polymer processing*
mengenmäßig in terms of quantity, quantitative, quantity-wise: **Die mengenmäßige Zugabe liegt zwischen 1 und 2%** The amount added is between 1 and 2%
Mengenproduktivität productivity in terms of quantity
Mengenregelung 1. volume/flow control. 2. volume/flow control unit/device
Mengenregelventil flow control valve
Mengenregler volume/flow control unit
Mengensteuerung 1. volume/flow control. 2. volume/flow control unit/device
Mengenstrom flow rate
Mengenstromregelung 1. volume/flow control. 2. volume/flow control device
Mengenventil volume/flow control valve
Mengenzuwachs quantity increase
Mennige red lead
Mennigeprimer red lead primer
Mensur graduated flask/cylinder
Menü menu *(me)*
Menüanweisung menu instruction *(me)*
menügeführt menu-driven *(me)*
menügesteuert menu-driven *(me)*
Menümaske menu mask *(me)*
menüorientiert menu-oriented *(me)*
Menüsteuerung menu control *(me)*
Menütablett menu tablet *(me)*
Mercaptanendgruppe terminal mercaptan group
Mercaptangruppe mercaptan group
Mercaptanhärter mercaptan hardener
Mercaptid mercaptide
Mercaptobenzimidazol mercaptobenzimidazole
Mercaptobenzothiazol mercaptobenzothiazole
Mercapto-Butylzinnstabilisator mercaptobutyl-tin stabiliser
Mercaptoether mercaptoether
Mercaptogruppe mercapto group
mercaptomodifiziert mercapto-modified
Mercapto-Octylzinnstabilisator mercapto-octyl-tin stabiliser
Mercaptopropylsilan mercaptopropyl silane
Mercaptosilan mercaptosilane
mercaptosilanbehandelt mercaptosilane-treated
Mercaptoverbindung mercapto compound
Merkblatt (technical) data/information sheet
Merkblattwerte figures given in technical data sheets
Merkmal feature, characteristic
Merkmale, konstruktive 1. design features. 2. structural features *(see explanatory note under Konstruktion)*
Merkmale, verfahrenstechnische technical features

mesomomorph mesomorphous
Meß-, Steuer- und Regeltechnik measuring and control technology
Meß- und Regeltechnik measuring and control technology
Meßaufnehmer sensor, transducer, probe
Meßausgang measuring output
meßbar measurable
Meßbaustein measuring module
Meßbedingungen test conditions
Meßbereich measuring range
Meßblende measuring diaphragm
Meßbrücke bridge
Meßbuchse measuring socket
Meßdaten test data
Meßdatenauswertung data interpretation *(me)*
Meßdatenerfassungssystem data acquisition system *(me)*
Meßdatenflut flood of test data
Meßdose sensor, transducer, probe
Meßdüse experimental die *(e)*
Meßeinrichtung measuring equipment
Meßelektrode measuring electrode
Meßelement measuring element
Messer knife
Messerabstand distance between knives *(sr)*
Messerbalken 1. slitter bar *(used to slit film into tape)*. 2. knife block *(of a granulator)(sr)*
Messereinstellvorrichtung knife setting mechanism *(sr)*
Meßergebnis test result
Messerkreis cutting circle *(sr)*
Messermühle granulator *(sr)*
Messerreihe row of knives *(sr)*
Messerrotor rotor *(sr)(see explanatory note under Schneidrotor)*
Messerspalt knife clearance *(sr)*
Messerstandzeit (granulator) knife life *(sr)*
Messerwalze rotor *(sr)(see explanatory note under Schneidrotor)*
Messerwalzengranulator 1. granulator *(sr)*. 2. pelletiser *(sr)*: *Since* **Messerwalzen** *are invariably used in scrap granulators and strand pelletisers, there is no need to translate the first part of this term*
Messerwalzengranulierung 1. granulation. 2. pelletisation. 3. granulator. 4. pelletiser *(see explanatory note under* **Messerwalzengranulator***)*
Messerwechsel replacement/change of knives *(sr)*: **bei einem notwendig werdendem Messerwechsel** *when the knives need changing*; **leichter Messerwechsel** *knives are easily replaced*
Messerwelle 1. rotor *(sr) (the rotors of granulators always carry knives, so that this need not be emphasised)*. 2. knife rotor *(general term)*
Messerzahl number of knives *(sr)*
Meßextruder laboratory extruder
Meßfehler measuring error
Meßform experimental mould

Meßfrequenz test frequency
Meßfühler sensor, transducer, probe
Meßgeber sensor, transducer, probe
Meßgefäß measuring vessel
Meßgenauigkeit measuring accuracy: **Es erlaubt das Messen der Schlauchtemperaturen mit einer Meßgenauigkeit von etwa 5%** it enables the film bubble temperature to be measured with an accuracy of about 5%
Meßgerät measuring instrument
Meßgeräteausstattung measuring equipment
Meßgröße quantity *(to be or being measured)*, variable
Messingkalibrierung brass calibrator
Meßinstrument measuring instrument
Meßkneter laboratory compounder
Meßkolben graduated flask
Meßkonzept measuring system *(see also explanatory note under* **Konzept***)*
Meßkopf measuring head
Meßlänge measured distance/length, gauge length
Meßlasche sensor, transducer, probe
Meßmarke bench/gauge mark
Meßmethode method of determination
Meßort measuring point
Meßprotokoll test record
Meßpunkt measuring point
Meßreihe test series
Meßresultat test result
Meßschieber vernier
Meßschrieb test record
Meßsensor sensor, transducer, probe
Meßsignal test signal
Meßsonde sensor, transducer, probe
Meßspanne measuring range
Meßspannung test voltage
Meßstelle measuring point
Meßstift sensor pin
Meßstrecke measured distance, distance being OR to be measured
Meßsubstanz substance under test
Meßtechnik measuring technology
meßtechnisch *relating to measurement*: **Der Einfluß der Feuchtigkeit wird meßtechnisch erfaßt** the effect of moisture is measured
Meßuhr dial gauge
Meßumformer transducer
Messung measurement, determination
Meßverfahren test method, method of determination
Meßverstärker transducer
Meßvorrichtung measuring device/instrument
Meßvorschrift test specification
Meßwalzwerk laboratory roll mill
Meßwandler transducer
Meßwert 1. test data. 2. measured value *(as opposed to* **Rechenwert***, q.v.)*. 3. measured variable
Meßwertauflösung data resolution *(me)*

Meßwertaufnehmer sensor, probe, transducer
Meßwertausgabe data output *(me)*
Meßwertauswertung data evaluation
Meßwertdrucker data printer *(me)*
Meßwerteingang data input *(me)*
Meßwerterfassung data acquisition/gathering *(me)*
Meßwerterfassungsgerät data collector
Meßwertgeber sensor, transducer, probe: **Für die Druckmessung wurde ein handelsüblicher Meßwertgeber eingesetzt** the pressure was measured with an ordinary pressure transducer
Meßwertgerät sensor, transducer, probe
Meßwertspeicher data bank/memory *(me)*
Meßwertspeicherinhalt data bank contents *(me)*
Meßwertstreuungen data scattering
Meßwertüberwachung 1. data monitoring/supervision. 2. data monitor
Meßwertverarbeitung data processing
Meßzelle measuring cell
Meßzylinder measuring cylinder
Metallabrieb metal abrasion
Metallabscheidegerät metal separator
Metallalkoxid metal alkoxide
metallbedampft metallised
Metallcarboxylat metal carboxylate
Metallchlorid metal chloride
Metalldithiocarbamat metal dithiocarbamate
Metalldithiophosphat metal dithiophosphate
Metalldrahtgewebe wire gauze, wire cloth *(mf)*
Metalldruckgußform metal diecasting mould
Metalleffektlack metallic finish
Metalleinlage metal insert
Metalleinlegeteil metal insert
Metalleinsatz metal insert
Metallelastizität metal elasticity
Metallelemente, eingebaute metal inserts
metallfrei metal-free
Metallgewebefilter wire mesh screen/filter
Metallgitter expanded metal
metallhaltig metal-containing
Metallicbasislack metallic paint
Metallicdecklack metallic finish
metallisch metallic
metallisiert metallised
Metallisierung metallisation
Metallisocarboxylat metal isocarboxylate
Metallkatalysator metal catalyst
Metallkern metal core
Metallklebung 1. bonding of metals. 2. bonded metal joint
Metallkomplex metal complex
Metallkonstruktion metal structure/construction
Metallmercaptid metal mercaptide
metallorganisch organo-metallic
Metalloxid metal oxide
Metallpulver metal powder
Metallseife metal soap
Metallseifenstabilisator metal soap stabiliser
Metallseparator metal separator

Metallstearat metal stearate
Metallsuchanlage metal detector
Metallsuchbrücke metal detector
Metallsuchgerät metal detector
Metallteilchen metal particle
metallumflochten metal braided
Metalluntergrund metal substrate
Metallvergiftung metal poisoning *(of catalyst)*
Metallverklebung 1. bonding of metals. 2. bonded metal joint
Metallwinkel metal angle piece
metaständig in the meta position
Metastellung meta-position
metasubstituiert meta-substituted
Metathese double decomposition, metathesis
Metatheseabbau double decomposition, metathesis
Meteringschnecke metering-type screw *(e)*
Meteringzone metering section/zone *(e)*
Methacrolein methacrolein
Methacrylat methacrylate
Methacrylatgruppe methacrylate group
Methacrylatharz acrylic resin
Methacrylatoligomer methacrylate oligomer
Methacrylatpolymer polymethacrylate
Methacrylbindemittel acrylic binder
Methacrylfestharz solid acrylic resin
Methacrylharz acrylic resin
Methacrylimid methacrylimide
Methacrylnitril methacrylonitrile
Methacrylsäure methacrylic acid
Methacrylsäureallylester allyl methacrylate
Methacrylsäureester methacrylate
Methacrylsäureester-Butadien-Styrolcopolymerisat methyl methacrylate-butadiene-styrene copolymer, MBS copolymer
Methacrylsäure-Methacrylnitrilcopolymerisat methacrylic acid-methacrylonitrile copolymer
Methacrylsäuremethylester methyl methcrylate
Methan methane
Methandiphenyldiisocyanat diphenylmethane diisocyanate, MDI
Methanol methanol
methanolisch methanolic
methanolische Lösung methanol solution
methanollöslich methanol-soluble
methanolverethert methanol etherified
Methodik method
methoxyfunktionell methoxyfunctional
Methoxy(l)gehalt methoxy/methoxyl content
Methoxy(l)gruppe methoxy/methoxyl group
Methoxypropylacetat methoxypropyl acetate
Methylacetat methyl acetate
Methylacrylat methyl acrylate
Methylalkohol methyl alcohol
Methylalkylsilikonharz methyl alkyl silicone resin
Methylbenzylalkohol methyl benzyl alcohol
Methylcellulose methyl cellulose
Methylcyanacrylat methyl cyanoacrylate
Methyldiethanolamin methyl diethanolamine

Methylenblau methylene blue
Methylenbrücke methylene bridge
Methylenchlorid methylene chloride
Methylengruppe methylene group
Methylether methyl ether
Methylethylketon methyl ethyl ketone, MEK
Methylethylketonperoxid methyl ethyl ketone peroxide
Methylglykol methyl glycol
Methylglykolacetat methyl glycol acetate
Methylgruppe methyl group/radical
methyliert methylated
Methylierung methylation
Methylinden methyl indene
Methylmethacrylat methyl methacrylate
Methylmethacrylat-Butadien-Styrol methyl methacrylate-butadiene-styrene, MBS
Methylnaphthalin methyl naphthalene
Methylolgruppe methylol group
Methylolharnstoff methylol urea
Methylolmelamin methylol melamine
Methylolphenolharz methylol phenol resin
Methylphenylpolysiloxan methylphenyl polysiloxane
Methylphenylsilikonharz methylphenyl silicone resin
Methylpolysiloxan methyl polysiloxane
Methylrest methyl radical/group
Methylrot methyl red
Methylseitengruppe methyl side group
Methylsilikon methyl silicone
Methylsilikonfestharz solid methyl silicone resin
Methylsilikonharz methyl silicone resin
Methylsilikonöl methyl silicone fluid
Methylstyrol methyl styrene
Methyltetrahydrophthalsäureanhydrid methyl tetrahydrophthalic anhydride
Methylzinnmercaptid methyl-tin mercaptide
Methylzinnstabilisator methyl-tin stabiliser
Methylzinnverbindung methyl-tin compound
metrisch metric
MFI *abbr. of* melt flow index
MFI-Wert melt flow index
MF-Schaumkunststoff melamine foam
MFT *abbr. of* **Mindestfilmbildungstemperatur**, minimum film-forming temperature
MG *abbr. of* **Molekulargewicht**, molecular weight
MG-Verteilung molecular weight distribution
Migration migration *(usually of plasticiser)*
migrationsarm low-migration
migrationsbeständig migration resistant
Migrationsbeständigkeit migration resistance
migrationsecht migration resistant
Migrationsechtheit migration resistance
migrationsfrei non-migrating
Migrationsneigung migration tendency
Migrationsrate migration rate
Migrationsverhalten migration behaviour
Migrieren migration
mikroanalytisch microanalytical

Mikrobenbefall microbial attack
Mikrobereich micro-range/-region
mikrobiell microbial
mikrobiologisch microbiological
mikro-Brownsche Bewegung micro-Brownian movement
Mikrobruch microfracture
Mikrocomputer microcomputer *(me)*
mikrocomputergesteuert microcomputer-controlled
Mikrocomputersteuerung microcomputer control *(me)*
Mikrodosierer micro-metering unit
Mikroelektronik microelectronics
Mikrogefüge microstructure
Mikrogel microgel
mikroglaskugelhaltig containing glass microbeads
Mikroglaskugeln glass microbeads
Mikrogranulat microgranular compound
Mikrogravimetrie microgravimetry
mikrogravimetrisch microgravimetric
Mikrohärte microhardness
Mikrohohlglaskugeln hollow glass microbeads/microspheres
Mikrohohlkugeln hollow microbeads/microspheres
Mikrohohlraum microcavity
mikrokristallin microcrystalline
Mikrokugeln microbeads
Mikroluftbläschen microscopic air bubbles
Mikromaßstab microscale
mikromechanisch micromechanical
Mikrometer micrometer
mikronisiert micronised
Mikroorganismus microorganism
Mikroperlpolymerisation microbead/microsuspension polymerisation
Mikrophase microphase
Mikroporen pinholes
mikroporös 1. microcellular *(foam)*. 2. microporous *(general term)*
Mikroprozessor microprocessor *(me)*
mikroprozessorgesteuert microprocessor controlled *(me)*
mikroprozessorgestützt microprocessor controlled
Mikroprozessorregler microprocessor control unit
Mikroprozessorsteuerung microprocessor control
Mikroprozessorsystem microprocessor system
Mikroprozessortechnik microprocessor technology *(me)*: **mit Mikroprozessortechnik gesteuert** microprocessor controlled
Mikroquerriß transverse microcrack
Mikrorechner microcomputer *(me)*
Mikrorechnerbaustein microcomputer module *(me)*
Mikrorechnereinsatz use of a microcomputer
mikrorechnergeregelt microcomputer controlled
mikrorechnergesteuert microcomputer controlled
mikrorechnergestützt microcomputer aided
Mikrorechnerperipherie microcomputer peripheral *(me)*
Mikroriß microcrack
Mikroschalter microswitch
Mikroschaum microfoam
Mikroschnitt microtome section
mikroskopisch microscopic
Mikrosonde microprobe
Mikrospalte microcrack
Mikrospeicher micro-store, micro-memory *(me)*
Mikrosphären microbeads, microspheres
Mikro-S-PVC microsuspension/microbead PVC
Mikrostruktur microstructure
mikrostrukturell microstructural
Mikrosuspensionspolymerisation microsuspension/microbead polymerisation
Mikrosuspensions-PVC microsuspension/microbead PVC
Mikrotom microtome
Mikrotomschnitt microtome section
Mikrotomschnittaufnahme microtome section photomicrograph
Mikrowaage microbalance
Mikrowellen microwaves
Mikrowellenbereich microwave range
Mikrowellendurchlässigkeit microwave permeability
Mikrowellenheizung microwave heating
Mikrowellenherd microwave oven
Mikrowellenofen microwave oven
Mikrowellenplasmaanlage microwave plasma plant
mikrozellulär microcellular
Milchflasche milk bottle
Milchprodukte dairy products
Milchsäure lactic acid
Milchverpackung 1. milk carton. 2. milk bottle
Milchwirtschaft dairy industry
Minderung reduction
Mindestanforderungen minimum requirements
Mindestbaugröße minumum size *(of a machine)*
Mindestbelastbarkeit minimum load-bearing/-carrying capacity
Mindestdicke minimum thickness
Mindestdruck minimum pressure
Mindestdurchsatz 1. minimum output rate. 2. minumum throughput *(see also explanatory note under* **Durchsatz***)*
Mindestfilmbildungstemperatur minimum film forming temperature
Mindesthärtetemperatur minimum curing temperature
Mindestlosgröße minimum batch size
Mindestmolmasse minimum molecular weight
Mindestschußgewicht minimum shot weight *(im)*
Mindestverfilmungstemperatur minimum film-forming temperature
Mindestwanddicke minimum wall thickness

Mindestwert minimum figure/value
Mindestzeitstandwert minimum time-to-failure
Mineralfarbe silicate paint
Mineralfüllstoff mineral filler
mineralgefüllt mineral filled
mineralisch mineral
Mineralöl petroleum, mineral oil
Mineralsäure mineral acid
mineralverstärkt mineral-filled
Mineralwolle rock/mineral wool
Miniaturaufnehmer miniature transducer, mini-transducer
Miniaturausführung, in miniature:
  Kraftaufnehmer in Miniaturausführung miniature force transducer, mini-force transducer
miniaturisiert miniaturised
Miniaturisierung miniaturisation
Minidrucker mini-printer *(me)*
Minikassette mini-cassette
Minimalwert minimum value
Minimalzykluszeit minimum cycle time
minimieren to minimise
Minimierung minimisation
Minimumgrenzwert minimum value
Minimumkühlzeit minimum cooling time
Minustemperaturen sub-zero temperatures
Mipo abbr. of **Mischpolymerisat,** copolymer
Misch- und Knetextruder mixing and compounding extruder
Mischaggregat mixer, mixing unit
Mischanlage mixing equipment/unit
Mischaufwand amount of mixing required
mischbar miscible
Mischbarkeit miscibility, compatibility
Mischbedingungen mixing conditions
Mischbehälter mixing vessel
Mischbereich mixing section/zone *(e)*
Mischeffekt mixing effect/performance, homogenising effect/performance
Mischeinrichtung mixing equipment/unit
Mischelement mixing element
Mischemulgator mixed emulsifier, emulsifier blend
Mischer mixer
Mischerinhalt mixer contents
Mischerschaufel mixer blade
Mischextruder compounding extruder
Mischflügel rotor blade
Mischgarn hybrid/mixed yarn
Mischgefäß mixing vessel
Mischgerät mixer
Mischgrad degree of mixing, mixing efficiency:
  Je höher der Mischgrad, desto größer ist die Homogenität des Extrudates the more efficient the mixing, the more uniform will be the extrudate
Mischgut 1. mixer contents. 2. materials being mixed. 3. materials to be mixed
Mischhärter hardener blend
Mischkammer mixing chamber/compartment
Mischkatalysator mixed catalyst, catalyst blend

Mischkessel mixing vessel
Mischkesselinhalt 1. mixing vessel contents. 2. mixing vessel capacity
Mischkondensat co-condensate
Mischkopf 1. mixing head. 2. mixing torpedo *(e,im)*
Mischkristall mixed crystal
Mischkunststoffverwertung recycling/reclamation of mixed plastics waste
Mischleistung mixing efficiency
Mischlinie mixing line
Mischnoppen mixing pins *(on screw)*
Mischorgan mixing rotor
Mischperoxid peroxide blend
Mischphase mixed phase
Mischpolymer copolymer
Mischpolymerisat copolymer
Mischpolymerisation copolymerisation
Mischprozeß mixing process/operation
Mischring mixing element
Mischscheibe mixing element
Mischschnecke mixing/compounding/homogenising screw
Mischsteg mixing flight
Mischstift mixing stud/pin *(on a screw)*
Mischteil 1. mixing/homogenising section *(of screw)*. 2. torpedo *(of screw)*. Use version 2. if the **Mischteil** is at the end of the screw, otherwise use version 1. 3. mixing element
Mischteilschnecke 1. screw equipped with a mixing section. 2. screw equipped with a torpedo *(see explanatory note under* **Mischteil***)*
Mischtrommel mixing drum
Mischung 1. mix, mixture. 2. compound *(e.g. based on PVC)*
Mischungsansatz batch *(e.g. of resin-catalyst mix)*
Mischungsaufbau mix/compound formulation
Mischungsaufbereitung compounding
Mischungsbestandteil mix component, compound ingredient
Mischungsgrundkomponente basic mix/compound ingredient
Mischungsgüte mixing efficiency
Mischungsherstellung compounding
Mischungspartner mix component
Mischungsverhältnis mixing ratio
Mischverfahrensparameter mixing conditions
Mischverhalten mixing performance
Mischversuch mixing trial
Mischvorgang mixing process/operation/procedure
Mischwalzwerk mixing rolls
Mischwelle mixing/compounding/homogenising screw
Mischwerk mixer
Mischwerkskopf mixer head
Mischwerkzeug mixing rotor
Mischwirkung mixing/effect/performance, homogenising effect/performance
Mischzeit mixing time/period

Mischzone mixing section/zone *(e)*
Mischzyklus mixing cycle
Mißbrauch misuse
Mitarbeiter employee
Mitarbeiter, gewerbliche blue-collar workers
Mitarbeiterschulung staff training
Mitarbeiterstab staff, employees
Mitläuferfolie release film
Mitläuferpapier release paper
Mitteilungsverfahren notification procedure
mittel 1. mean, average 2. medium
Mittel, flüssige liquid assets
Mittelbeanspruchung mean load
Mittelbetrieb medium-sized company/business/concern/firm
Mittelbettmörtel medium-bed mortar
Mitteldehnung mean strain
mitteldispers medium-particle
Mitteleinsatz use of funds/money
mittelfett medium-oil
mittelflexibel moderately flexible
mittelfristig medium-term
mittelhart 1. medium hard. 2. semi-rigid *(e.g. pipe, foam, sheet etc.)*
mittelkettig medium-chain
Mittellinie centre line
mittelmolekular medium molecular weight
Mittelöl medium-heavy oil
Mittelölalkydharz medium-oil alkyd (resin)
Mittelölalkydharzlack medium-oil alkyd paint
mittelölig medium-oil
mittelreaktiv moderately reactive
mittelschlagfest medium impact
mittelschlagzäh medium impact
Mittelschnecke main screw
mittelsiedend medium-boiling
Mittelsieder medium-boiling solvent
Mittelspannung 1. medium voltage/tension. 2. mean stress
Mittelspannungsbereich medium voltage range
Mittelspannungskabel medium-voltage cable
mittelständisch medium-size *(business, firm etc.)*
mittelstoßfest medium-impact
Mitteltemperatur mean/average temperature
mittelviskos medium-viscosity
Mittelwalze centre roll
mittelwellig medium wave
Mittelwert mean/average value
Mittelwert, arithmetischer arithmetic mean
mittelzäh 1. medium-viscosity. 2. medium impact
Mittenabstand centre distance
Mittenauswerfer central ejector *(im)*
mittlere Teilchengröße mean/average particle size
mittlerer Teilchendurchmesser mean/average particle size
mittleres Molekulargewicht average molecular weight
Mitverwendung, anteilige incorporation of small amounts

Mizelle micelle
MMI *abbr. of* **Mensch-Maschine-Interface**, man-machine interface
Möbelausrüstungsteile furniture fittings
Möbelindustrie furniture industry
Möbellack furniture lacquer/varnish
Möbelstück piece of furniture
Mobilität mobility
Modell model, pattern
Modellbau modelmaking
Modellduplikat duplicate model
Modellführungen pattern guides
Modellgesetz model law
Modellgröße model quantity
Modellharz model-making resin
Modellieren modelling
Modellplatte pattern plate
Modelluntersuchung model test
Modellvorstellung model concept/representation
Modellwerkstatt pattern-making shop
Modellwerkstoff model-/pattern-making material
Modernisierung modernisation
Modifier modifier, modifying agent
Modifikator modifier, modifying agent
Modifikatorpartikel modifier particle
Modifikatorteilchen modifier particle
Modifizierharz modifying resin
Modifiziermittel modifier, modifying agent
modifiziert modified
modifiziert, organisch organically modified
Modifizierung modification
Modifizierungsgruppe modifying group
Modifizierungsmittel modifier, modifying agent
Modul modulus
Modularität modular character
modulartig modular
Modulwert modulus
Modus mode
Möglichkeit 1. possibility. 2. option. 3. facility: ...**mit Möglichkeit zum Einlegen von Sieben** ...with facilities for inserting screens
Möglichkeiten, verfahrenstechnische processing possibilities
molare Masse molecular weight
Molekül molecule
Molekülabstand distance between the molecules
Molekularaufbau molecular structure
Molekularbewegung molecular movement
Molekulargewicht molecular weight
Molekulargewicht, gewichtsmittleres weight-average molecular weight
Molekulargewicht, mittleres average molecular weight
Molekulargewichtmittelwert average molecular weight
molekulargewichtsabhängig depending on the molecular weight
Molekulargewichtsänderung change in molecular weight

Molekulargewichtsverteilung molecular weight distribution
Molekulargewichtsverteilungskurve molecular weight distribution curve
Molekulargewichtzahlenmittel number-average molecular weight
Molekularmasse molecular weight
Molekularorientierung molecular orientation
Molekularstruktur molecular structure
Molekularzustand molecular state
Molekülaufbau molecular structure
Molekülbaustein molecule unit
Molekülbeweglichkeit molecule mobility
Molekülbewegung molecule movement
Molekülfragment molecule segment
Molekülgebilde molecular structure
Molekülgitter molecular lattice structure
Molekülgröße molecule size
Molekülkette molecule chain
Molekülkettenlänge molecule chain length
Molekülkettenorientierung molecule chain orientation/alignment
Moleküllage molecule layer
Moleküllänge molecule length
Molekülmasse molecular weight
Molekülorientierung molecular orientation
Molekülschicht molecular layer, layer of molecules
Molekülsegment molecule segment
Molekülstruktur molecular structure
Molekülvergrößerung molecule enlargement
Molenbruch molar fraction
Molgewicht molecular weight
Molkerei dairy
Molkereiprodukte dairy products
Molkereisektor dairy industry
Molmasse molecular weight
Molmassenverteilung molecular weight distribution
Molverhältnis molar ratio
Molwärme molar heat
Molybdändisulfid molybdenum disulphide
Molybdäntrioxid molybdenum trioxide
Molybdat molybdate
Molybdatorange molybdenum orange
Molybdatrot molybdenum red
Momentanwert instantaneous value
Monatsproduktion monthly production/output
Monatstonnen tonnes per month
Monoalkylzinnchlorid mono-alkyl tin chloride
Monoalkylzinnstabilisator mono-alkyl tin stabiliser
Monoalkylzinnthioglykolsäureester mono-alkyl tin thioglycolate
Monoalkylzinnverbindung mono-alkyl tin compound
monoaxial uniaxial
Monoaxialreckanlage uniaxial stretching unit/equipment
Monocarbonsäure monocarboxylic acid
Monocarbonsäureester monocarboxylate
Monocarbonsäurevinylester vinyl monocarboxylate
Monochlorphenol monochlorophenol
Monofil monofilament
Monofilament monofilament
Monofilamenttyp monofilament grade *(of compound)*
Monofilanlage monofilament extrusion line
Monofilwerkzeug monofilament die *(e)*
monofunktionell monofunctional
Monolage monolayer
monolithisch monolithic
monomer monomeric
Monomer monomer
monomerarm with a low monomer content
Monomerbaustein monomer unit
Monomereinheit monomer unit
Monomerenverhältnis monomer ratio
monomeres Vinylchlorid vinyl chloride monomer
Monomergemisch monomer blend
monomerhaltig monomer-containing: Stark monomerhaltiges Polyamid 6 wird dadurch leichter erkannt als monomerarmes Nylon 6 with a high monomer content is thus more easily identified than nylon 6 with a low one
Monomerkonzentration monomer concentration
Monomerkugel monomer storage tank
monomerlöslich monomer soluble
Monomermolekül monomer molecule
Monomerradikal monomer radical
Monomerschmelze monomer melt
Monomertröpfchen monomer droplet
Monomertropfen monomer droplet
Monomerweichmacher monomeric plasticiser
Monomethylolharnstoff monomethylol urea
monomolekular monomolecular
monoolefinisch mono-olefinic
Monoorganosilan mono-organosilane
Monoorganozinnverbindung mono organotin compound
Monosäure monoacid
Monostyrol styrene monomer
monosubstituiert mono-substituted
Monosulfid monosulphide
Montage 1. assembly, mounting, fitting, installation. 2. assembly shop
Montageband assembly line
montagefreundlich easy to fit/install
Montageklebstoff assembly adhesive
Montagelehre assembly fixture
montageleicht easy to fit/install
Montageleim assembly glue
Montagestation assembly station
Montansäure montanic acid
Montansäureester montanate, montanic acid ester
Montanwachs montan wax
montieren to assemble, to mount, to fit, to install
Mooney-Anvulkanisationszeit Mooney scorch value
Mooney-Plastizität Mooney plasticity

Mooney-Viskosität Mooney viscosity
Moosgummi foam rubber
Moosgummidichtung foam ruber seal
Moosgummimischung foam rubber mix
Moosgummiprofil foam rubber profile
Morpholin morpholine
Morphologie morphology
morphologisch morphological
Mörtel mortar
Mörtelbelag mortar screed
Mörtelfuge mortar joint
Mörtelmischung mortar mix
motorbetrieben motor driven
Motorblock engine block
Motordrehzahl motor speed
Motorenleistung, gesamte installierte total installed/connected motor power
Motorgeräusch engine noise
Motorhaube bonnet
motorisch angetrieben motor driven
motorischer Antrieb motor drive
Motorleistung (electric) motor power
Motoröl motor oil
Motorölwanne oil sump
Motorradhelm crash helmet
Motorradschutzhelm crash helmet
Motorraum engine compartment
Motorschutzschalter motor safety switch
Motorträger engine mount
Motorwirkleistung effective motor power
MPC-Ruß medium processing channel black
MSR *abbr. of* **Messen, Steuern, Regeln**: MSR-System measuring and control system
MSR-Elektronik electronic measuring and control system
MSR-Geräte measuring and control instruments
MT-Ruß medium thermal black
Muffenkonstruktion socket(-type) construction
Muffenschweißung socket fusion welding
Muffenschweißverbindung welded socket joint
Muffenverbindung socket joint
Mühle 1. mill. 2. granulator *(sr)*. 3. pelletiser *(sr)* *(see explanatory note under* **Schneidgranulator***)*
Mühlenbaureihe 1. granulator range *(sr)*. 2. pelletiser range
Müll refuse, waste, rubbish, garbage
Müllaufbereitungsanlage refuse sorting plant
Müllaufkommen creation/production of waste/refuse
Mülldeponie waste/rubbish tip, waste/rubbish dump, landfill site/tip
Müllentsorgung waste/refuse disposal
Müllentsorgungsfirma waste/refuse disposal firm/contractor
Müllhalde waste/rubbish tip, waste/rubbish dump, landfill site/tip
Müllheizkraftwerk waste/refuse/garbage fuelled power plant
Müllheizwerk waste/refuse/garbage fuelled power plant
Müllkraftwerk waste/refuse/garbage fuelled power plant
Müllproblem waste disposal problem
Müllsack rubbish bag, dustbin liner
Müllsortieranlage waste/refuse sorting plant
Müllsortierung sorting of waste
Mülltonne dustbin, rubbish bin
Müllverbrennung waste/refuse incineration
Müllverbrennungsanlage waste/refuse incinerator
Müllverwertung waste recycling **thermische Müllverwertung** obtaining energy through waste incineration
multiaxial multiaxial
Multiblockcopolymer multi-block copolymer
multidirektional multi-directional
Multifunktionalität multi-functionality
multifunktionell multifunctional
Multilayerschaltung 1. multi-layer printed circuit board. 2. multi-layer circuit
Multiplexer multiplexor *(me)*
multivalent polyvalent
Mundstück (outer) die ring *(e)*
Mundstückring (outer) die ring *(e)*
Mundstückringspalt die gap/orifice *(e)*
Mundstückspalt die gap/orifice *(e)*
Mund-zu-Mund-Propaganda word-of-mouth publicity/recommendation
Muscovit potash mica, muscovite
Muscovitglimmer muscovite, potash/white mica
Musikplatte compact disc, CD
Musterabspritzungen moulding trials
Musterherstellung preparation of samples
Mustermenge sample quantity
Musterplättchen test piece/specimen
Mutagenität mutagenity
Mutter nut
Muttergesellschaft parent company
Muttermodell master pattern
MVI *abbr. of* melt volume index

N

NA *abbr. of* **Nitrosamin**, nitrosamine
Nabe hub
Nacharbeitung 1. secondary finishing *(e.g. of a moulding)*. 2. re-machining *(e.g. of a worn machine part)*
Nacharbeitungskosten secondary finishing costs
Nachbaratome adjacent atoms
Nachbarländer neighbouring countries
Nachbarmoleküle adjacent molecules
Nachbearbeitung 1. secondary finishing *(e.g. a moulding)*. 2. re-machining *(e.g. a worn machine part)*

nachbearbeitungsfrei *not requiring any secondary finishing* **nachbearbeitungsfreie Formteile** mouldings which require no subsequent finishing
Nachbehandlung after-/post-treatment
Nachbildung simulation
Nachblähen post-expansion *(of foam)*
Nachbrenndauer burning time/duration
Nachbrenner afterburner
Nachbrennzeit burning time/duration
nachchloriert post-chlorinated
nachdosieren to add more **Es empfiehlt sich daher, 1-2% Alterungsschutzmittel nachzudosieren** it is therefore advisable to add another 1-2% antioxidant
Nachdruck hold(ing)/follow-up pressure *(im)*
Nachdruckdauer hold(ing) pressure time *(im)*
Nachdruckhöhe hold(ing)/follow-up pressure *(im)*: The **höhe** *part of this word should be left untranslated since it adds nothing to its meaning, as shown by this example*: **Nachdruckhöhe und Nachdruckzeit sind so zu wählen, daß an den Formteilen keine Einfallstellen zu erkennen sind** the holding pressure and holding pressure time should be such that no sink marks are produced
Nachdruckniveau hold(ing)/follow-up pressure *(see explanatory note under **Nachdruckhöhe** which applies here, too)*
Nachdruckphase hold(ing) pressure phase
Nachdruckprogramm hold(ing) pressure programme *(im)*
Nachdruckpulsationen hold(ing) pressure fluctuations *(im)*
Nachdruckstufen hold(ing) pressure stages *(im)*
Nachdruckumschaltung change-over to hold(ing) pressure *(im)*
Nachdruckverlauf hold(ing) pressure profile *(im)*
Nachdruckzeit hold(ing) pressure time *(im)*
Nacheilung, mit slower, more slowly: **Walze A läuft gegenüber Walze B mit Nacheilung** roll A rotates more slowly than roll B
Nachfolge downstream unit/equipment
Nachfolgeaggregat downstream unit/equipment
Nachfolgeeinheit downstream unit/equipment
Nachfolgemaschine downstream machine
Nachfolgemodell subsequent/later model: **Alle Maschinen der Baureihe FA (Nachfolgemodell der Baureihe BKM)...** all machines in the FA range (successors to the BKM range)...
nachfolgend 1. downstream, adjacent to, followed by, following. 2. subsequent *(for translation examples see under **nachgeschaltet**)*
Nachfolgeprozeß 1. downstream process/operation. 2. subsequent process/operation
Nachfolgevorrichtung downstream unit/equipment
Nachfrage 1. inquiry. 2. demand: **Angebot und Nachfrage** supply and demand

Nachfragebelebung increased demand
Nachfragerückgang drop in demand
nachgeschaltet 1. downstream, adjacent to, followed by, following: **Ein Metallsuchgerät wurde dem Zerkleinerer nachgeschaltet** a metal detector was placed downstream from the granulator; **Die Blasstation mit der nachgeschalteten Flaschenentnahme** the blowing station with the adjacent bottle collecting unit; **...mit nachgeschalteten Quetschwalzen** ...followed by pinch rolls. 2. subsequent: **...die in einem nachgeschalteten Prozeß spanend bearbeitet werden** ...which are subsequently finished by machining
nachgiebig yielding
Nachgiebigkeit yieldingness
Nachglühdauer afterglow time/duration
Nachglühzeit afterglow time/duration
Nachhärtung post-curing
Nachhärtungsbedingungen post-curing conditions
Nachhärtungstemperatur post-curing temperature
nachjustieren to readjust
Nachkommastelle place after the decimal point; **drei Nachkommastellen** three digits after the decimal point
Nachkondensation post-condensation
Nachkristallisation post-crystallisation
Nachkühlstrecke post-cooling section
Nachpolymerisation post-polymerisation
Nachrichtensatellit communications satellite
Nachrichtentechnik telecommunication (engineering)
nachrüstbar *that which can be retrofitted, i.e. added later:* **Erforderlich ist ein fünfter Heizkreis, der nachrüstbar ist** a fifth heating circuit is necessary, which can be retrofitted
Nachschwindung post-/after-shrinkage
Nachschwund post-/after-shrinkage
nachspannen to tighten
nachstellen to (re)adjust
Nachteile, verfahrenstechnische technical limitations, processing disadvantages
nachtempern to anneal
Nachtropfen drooling *(im)*
Nachtschicht night shift
Nachverbrennung 1. afterburning. 2. afterburner
nachvernetzend post-curing/-crosslinking
Nachvernetzung post-curing
Nachvulkanisation post-vulcanisation, post-curing
Nachvulkanisationsstabilisator post-vulcanisation stabiliser
Nachweis 1. proof. 2. detection
nachweisempfindlich sensitive *(e.g. method of determination, chemical test etc.)*
Nachweisgrenze detection limit
Nackenstütze neck support, head rest/restraint

Nadelausreißfestigkeit needle extraction resistance
Nadelausreißkraft needle extraction force
Nadelausreißversuch needle extraction test
Nadelausreißwiderstand needle extraction resistance
Nadeldrucker dot-matrix printer
Nadeldüse needle valve nozzle *(im)*
Nadelfilzteppich needlefelt carpet
nadelförmig needle-shaped
Nadellager pin bearing
Nadeln needling
Nadelstichbildung pinholing *(of a paint film)*
Nadelstiche pinholes
Nadelventil needle valve *(im)*
Nadelventilangußsystem valve gating system *(im)*
Nadelventilverschluß needle shut-off mechanism *(im)*
Nadelventilwerkzeug valve gated mould *(im)*
Nadelverschluß needle valve *(im)*
Nadelverschlußdüse needle valve nozzle *(im)*
Nadelverschlußheißkanalsystem needle valve hot runner system
Nadelverschlußspritzdüse needle valve nozzle *(im)*
Nadelverschlußsystem needle shut-off mechanism *(im)*
Nadelwalze pin roller
Nadelwalzenfibrillator pin roller fibrillator
Nadelwalzenverfahren pin roller process *(for making fibrillated film)*
Näherung approximation
Näherungsschalter proximity switch
Nahost Middle East
Nahrungsmittel foodstuffs
Nahrungsmittelindustrie food industry
Nahtgüte weld quality
nahtlos seamless, jointless
nahtlos seamless
Nahtqualität weld quality
Nahtverbindung seam
Naphtha naphtha
Naphthalin naphthalene
2,6-Naphthalin-Dicarboxylsäure 2,6-naphthalene dicarboxylic acid
Naphthalinsulfonsäure naphthalene sulphonic acid
Naphthen naphthene
Naphthenat naphthenate
naphthenisch naphthenic
Naphthensäure naphthenic acid
Naphthylisocyanat naphthyl isocyanate
Naphthylrest naphthyl radical
Narbenbeständigkeit grain retention *(e.g. of grained or embossed sheet during thermoforming)*
Narbeneffektlackierung grain finish
Narbung grain
narkotisch narcotic
narrensicher foolproof
Naßabriebfestigkeit wet scrub resistance

Naßankermagnet wet-armature solenoid
Naß-auf-Naß-Auftrag wet-on-wet application
Naßbüchse water cooled feed section/zone *(e)*
naßchemische Bestimmung wet analysis
Naßdampf water vapour
Naßfestigkeit 1. wet strength. 2. water resistance
Naßfestigkeitswert wet strength (value)
Naßfestmittel wet strength agent
Naßfilm wet film
Naßfilmdicke wet film thickness
Naßfilmstärke wet film thickness
Naßfilmviskosität wet film viscosity
Naßgemisch wet mix
naßgepreßt press moulded with/using liquid resin *(grp)*
Naßgleitverschleiß wear due to wet sliding friction
Naßhaftung wet adhesion
Naß-in-Naß-Lackierung wet-in-wet paint application
Naß-in-Naß-Verfahren wet-in-wet method
Naßklebstoff wet adhesive
Naßlaminat laminate made by wet lay-up *(grp)*
Naßlaminierverfahren wet lay-up (process) *(grp)*
Naßmahlen wet grinding
Naßmörtel wet mortar
Naßpolyester polyester dough moulding compound, polyester DMC
Naßpreßverfahren liquid resin press moulding *(grp)*
Naßraum wet-process room
Naßschichtdicke wet film/coating thickness
Naßschliff wet grinding
Naßspinnverfahren wet spinning (process)
Naßtrennanlage wet separation unit
Naßtrennen wet separation
Naßverklebung wet bonding
Naßwickelverfahren filament winding (process)
Naßzelle bathroom
Naßzerkleinern wet shredding
Naßzustand wet state
Natriumbenzoat sodium benzoate
Natriumbisulfit sodium bisulphite
Natriumcarbonat sodium carbonate
Natriumdampfhochdrucklampe high pressure sodium vapour lamp
Natriumdampfniederdrucklampe low pressure sodium vapour lamp
Natriumdichromat sodium dichromate
Natriumhexametaphosphat sodium hexametaphosphate
Natriumhydroxid sodium hydroxide, caustic soda
Natriumlaurat sodium laurate
Natriumpolyacrylat sodium polyacrylate
Natriumsulfit sodium sulphite
Natriumtetrafluorborat sodium tetrafluoroborate
Natronlauge caustic soda solution
natur natural

Naturbewitterung natural/outdoor weathering
naturfarben natural
Naturgummi natural rubber
Naturharz natural resin
Naturkautschuk natural rubber
Naturkautschukvulkanisat natural rubber vulcanisate, vulcanised natural rubber
Naturkreide natural chalk
Naturprodukt natural product
Naturstein natural stone
NBR *symbol for* nitrile rubber
NC 1. *abbr. of* nitrocellulose. 2. *abbr. of* numeric control *(me)*
NCO-Gehalt isocyanate content
NC-Programm numeric control program, NC program *(me)*
NC-Programmierung numeric control programming, NC programming *(me)*
NC-Steuerung numeric control *(me)*
NC-Technik numeric control technology/ engineering
ND *abbr. of* **Niederdruck,** low pressure
NE *abbr. of* **nicht-eisern,** non-ferrous
Nebelleuchte fog lamp
Nebenaggregat ancillary unit
Nebenbestandteil by-product
Nebendispersionsgebiet secondary glass transition range
Nebeneffekt side effect
Nebeneinrichtungen ancillary equipment
Nebenextruder ancillary/subsidiary extruder
Nebenfunktion ancillary/subsidiary/secondary function
Nebenkanal secondary/branch runner *(im)*
Nebenkosten extra/additional costs
Nebenprodukt by-product
nebenproduktfrei free from by-products
Nebenreaktion side/secondary reaction
Nebenschluß-Gleichstromgetriebemotor shunt-wound geared DC motor
Nebenschlußmotor shunt motor
Nebenschnecke ancillary/subsidiary screw
Nebenvalenz secondary valency
Nebenvalenzkräfte secondary valency forces
Nebenwirkungen side effects
Nebenzeit downtime, shut-down period, non-productive time
Neck-in neck-in *(term applied to transverse shrinkage of film between the die lips and the chill roll)*
negativ negative
Negativabguß negative impression
Negativform female tool *(t)*
Negativformen female forming *(t)*
Negativverfahren female forming *(t)*
Negativwerkzeug female tool *(t)*
Neigung slope *(of a curve)*
Neigungswinkel angle of slope
nematisch nematic
NE-Metall non-ferrous metal
Nenndehnung 1. nominal strain. 2. nominal elongation/extension

Nenndrehmoment nominal torque
Nenndruck nominal pressure
Nenndurchmesser nominal diameter
Nenngröße nominal size
Nenninhalt nominal capacity
Nennkapazität nominal capacitance
Nennleistung rating *(e.g. of an engine)*
Nennmaße nominal dimensions
Nennspannung 1. nominal voltage 2. nominal stress
Nennstrom nominal current
Nennviskosität nominal viscosity
Nennvolumen nominal volume
Nennweite nominal width
Neopentyldiglycidylether neopentyl diglycidyl ether
Neopentylglykol neopentyl glycol
Neoprenkautschuk neoprene rubber
Nest (mould) cavity
Nesteinsatz cavity insert
Nestwandtemperatur cavity wall temperature *(im)*
Nettogewicht net weight
Nettozugang net increase
Netz mains (system)
Netzanschluß power supply
Netzausfall breakdown of the power supply
Netzdichte crosslink density
Netzdruck mains pressure
Netzfehler wetting fault/defect
Netzkabel mains cable
Netzmittel wetting agent
Netzmittelkonzentration wetting agent concentration
Netzmittellösung wetting agent solution
Netzraster grid
Netzspannung mains voltage
Netzstromversorgung, mit mains powered
Netzwasser mains water
Netzwerk network
Netzwerk, durchdringendes interpenetrating network
Netzwerkstruktur network structure
Neuanstrich 1. new coat of paint. 2. re-painting
Neuauslegung re-designing: **Neuauslegung der Strömungskanäle** re-designing the flow channels
Neubauten new buildings
Neuberechnung re-calculation
neuentwickelt newly/recently developed, new
Neuentwicklung recent/new development
Neugranulat virgin compound
Neuinvestitionen new investments
Neukonstruktion new/recent design: **eine weitere Neukonstruktion** another recent design
neukonzipiert re-designed
Neumaterial virgin material
Neuprogrammierung re-programming *(me)*
neurezeptiert reformulated
neutral neutral
neutral, chemisch chemically inert

Neutralisationsmittel neutralising agent
Neutralisieren neutralisation
neutralisiert neutralised
Neutralisierung neutralisation
Neutralisierungsmittel neutralising agent
Neutron neutron
Neutronenaktivierung neutron activation
Neuware virgin material
Neuwarenmaterial virgin material
Newtonsch Newtonian
Newtonsche Flüssigkeit Newtonian liquid
Newtonsches Fließen Newtonian flow
NF *abbr.* of **Niederfrequenz**, low-frequency
Nfz *abbr.* of **Nutzfahrzeug**, commercial vehicle
nichtarmiert unreinforced
Nichtbeachtung non-observance
nichtbeschichtet uncoated
nichtbeständig not resistant, attacked *(material in presence of chemicals)*
nichtbewittert unweathered
nichtbrennbar incombustible, non-flammable
Nichtbrennbarkeit non-flammability, flame/fire resistance
nichtdirektional non-directional, random *(grp)*
Nichteisenmetall non-ferrous metal
nichtelastisch non-elastic, plastic
nichtentflammbar non-flammable
Nichtentflammbarkeit non-flammability
nichtextrahierbar non-extractable
Nichtextrahierbarkeit non-extractability
nichtfasrig non-fibrous
nichtflüchtig non-volatile
Nichtflüchtigkeit non-volatility
nichtgelatinisierend non-solvating *(plasticiser)*
nichtgelierend non-solvating *(plasticiser)*
nichtgewebt non-woven
Nichtgleichgewichtszustand non-equilibrium state
nichthaftend non-stick
nichthärtbar thermoplastic
nicht-Hookesch non-Hookean *(i.e. not obeying Hooke's Law)*
nichtineinandergreifend non-intermeshing *(screws) (e)*
nichtionisch non-ionic
nichtionogen non-ionic
nichtisotherm non-isothermal
nichtkämmend non-intermeshing *(screws) (e)*
nichtkorrosiv non-corrosive
nichtkristallin non-crystalline, amorphous
nichtkristallisierend non-crystallising
nichtleitend non-conductive
Nichtleiter non-conductor
nichtlinear non-linear
Nichtlinearität non-linearity
Nichtlösemittel non-solvent
Nichtlöser non-solvent
nichtmagnetisch non-magnetic
nichtmetallisch non-metallic
nichtmodifiziert unmodified
Nicht-Nahrungsmittelverpackung non-food packaging

nichtnewtonsch non-Newtonian, pseudoplastic
nichtnewtonsche Flüssigkeit non-Newtonian liquid
nichtnukleiert non-nucleated
nichtoxidiert unoxidised
nichtplastifiziert unplasticised
nichtpolar non-polar
nichtpolymerisiert unpolymerised
nichtradikalisch non-radical
nichtreagiert unreacted
nichtreaktiv inert
nichtreflektierend non-reflecting
Nichtreifenanwendungen non-tyre applications
nichtrostender Stahl stainless steel
nichtrutschend non-slip
nichtsaugend non-absorbent
nichtsaugfähig non-abosrbent
nichtschmelzend infusible
nichtschrumpfend non-shrink(ing)
nichtsolvatierend non-solvating
nichtstabilisiert unstabilised
nichttoxisch non-toxic
nichttransparent non-transparent, opaque
nichttrocknend non-drying
nichttropfend non-drip
nichtumgesetzt unreacted
nichtverfärbend non-staining
nichtvergilbend non-yellowing
nichtvernetzt un-cured/-crosslinked/-vulcanised
nichtverpastbar non-paste making
nichtverstärkend non-reinforcing
nichtverstreckt unstretched
Nichtvorhandensein absence:
  Nichtvorhandensein von Abquetschmarkierungen absence of pinch-off welds
nichtwäßrig non-aqueous
nichtweichgemacht unplasticised
nichtzellig non-cellular, solid
nichtzellular non-cellular, solid
nichtzerstörend non-destructive *(test)*
nichtzinnstabilisiert non-tin stabilised
Nickeltitangelb nickel titanium yellow
niederdicht low-density
Niederdruckabscheider low pressure separator
Niederdruckabtastung low pressure scanning device
Niederdruckformmasse low pressure moulding compound
Niederdruckplasma low-pressure plasma
Niederdruckpolyethylen high density polyethylene, HDPE
Niederdruckpolymerisation low pressure polymerisation
Niederdruckreaktor low pressure reactor
Niederdruckspritzgießmaschine low pressure injection moulding machine
Niederdruckverfahren low pressure process
Niederdruckwerkzeugschutz low pressure mould safety device
niederenergetisch low-energy
niederfrequent low-frequency

Niederfrequenztechnik low-frequency engineering
niedergespannt having low surface tension
Niederhaltedruck clamping pressure *(t)*
Niederhaltekraft clamping force *(t)*
Niederhalter clamping device/mechanism *(t)*
Niederlassung branch (office)
niedermolekular low-molecular weight
Niederpolymer low molecular weight polymer
Niederschlag 1. rain(fall) 2. precipitate
Niederschlagsfeuchtigkeit condensed moisture
niedershorig with a low Shore hardness
Niederspannung low voltage/tension
Niederspannungsbereich low voltage range
Niederspannungskabel low voltage cable
Niederspannungstransformator low voltage transformer
niedertourig slow speed
niederviskos low-viscosity
niedrigaktiv low-reactivity
niedrigbauend low
Niedrigbauweise low construction
niedrigdicht low-density
niedrigenergetisch low-energy
niedrigflüchtig low-volatility
niedriggelierend 1. with a low fusion point *(PVC)*. 2. with a low gel point *(EP, UP resins etc.)*
Niedrigkompressionsschnecke low-compression screw *(e)*
niedrigmodulig low-modulus
niedrigmolekular low-molecular weight
niedrigpigmentiert with a low pigment content
Niedrigpreisimporten low-priced imports
niedrigreaktiv low-reactivity
niedrigschmelzend low-melting
niedrigsiedend low-boiling
Niedrigsieder low-boiling solvent
Niedrigstrukturruß low-structure carbon black
Niedrigtemperaturflexibilität low-temperature flexibility
niedrigviskos low-viscosity
Nieten riveting
Nietverbindung riveted joint
n.i.o. *abbr. of* **nicht in Ordnung**, unsatisfactory, inadequate, out of order
nitriergehärtet nitrided
Nitrierschicht nitrided layer: *Since nitriding is not a coating process but consists of heating steel in ammonia or a liquid salt bath to harden the surface,* **Schicht** *cannot be translated as* film *or* coating
Nitrierstahl nitrided steel
Nitrierstahlgehäuse nitrided steel barrel *(e)*
Nitrilgehalt nitrile content
Nitrilgruppe nitrile group
Nitrilkautschuk nitrile rubber
Nitrillatex nitrile rubber latex
Nitrilseitengruppe nitrile side group
Nitrocellulose nitrocellulose
Nitrocelluloselack nitrocellulose lacquer
Nitrolack nitrocellulose lacquer

Nitrosamin nitrosamine
Nitrosoperfluorbuttersäure nitrosoperfluorobutyric acid
Nitrosoverbindung nitroso compound
Niveau level
Niveau, mechanisches mechanical properties
Niveaukontrolle 1. level control. 2. level controller, level monitoring device
Niveaumelder level indicator
Niveauregelung 1. level control. 2. level controller, level monitoring device
Niveauüberwachung 1. level control. 2. level controller, level monitoring device
Niveauwächter level controller, level monitoring device
Nivelliermase levelling compound
nivelliert levelled
NMR *abbr. of* nuclear magnetic resonance
NMR-Spektrum nuclear magnetic resonance spectrum, NMR spectrum
Nocken cam
Nockenmischteil knurled mixing section *(of screw) (e)*
Nockenrotor cam-type rotor
Nockenscheibe cam disc
Nockensystem cam system
Nockenwelle camshaft
Nockenwellendichtung camshaft seal
Nomenklatur nomenclature
Noniusskala vernier
Norm standard specification
normalaktiv normally reactive
Normalausführung standard design/model
Normalausrüstung standard equipment
Normalbedingungen normal conditions
Normalbenzin standard grade petrol
Normalbetrieb normal operation
normalbrennbar flammable
normalentflammbar flammable
normalhärtend normal-curing
Normalien 1. standard units. 2. standard mould units: **Normalien für den Werkzeugaufbau** standard mould units
Normalienbauelemente 1. standard units. 2. standard mould units
normalisiert standard, standardised
Normalklima standard conditioning atmosphere *(see explanatory note under* **Normklima***)*
Normalklimahärtung room temperature curing
normalkonisch conventional conical *(screw) (e)*
Normalplastifiziereinheit standard plasticising unit
Normalpolystyrol standard polystyrene
normalschlagfest normal impact
normalschlagzäh normal impact
Normalschnecke conventional/standard screw
Normalschurre ordinary chute
Normalspritzguß conventional injection moulding
normalstoßfest normal impact
Normaltemperatur room temperature

Normaltemperaturaushärtung room temperature curing
normalviskos normal-viscosity
normalzäh normal impact
Normbaugruppe standard unit
Normbaustein standard unit
Normbedingungen standard conditions
Normdicke standard thickness
Normengremium standards committee/association
normgerecht standard
normiert standardised
Normierung standardisation
Normkleinstab small standard test piece/specimen, small standard specimen
Normklima standard conditioning atmosphere: *This is usually followed by figures such as 23/50, which means 23°C and 50% relative humidity*
Normprobe standard test piece/specimen
Normprobekörper standard test piece/specimen
Normprogramm standard range
Normschulterprobe standard dumbbell test piece/specimen
Normschulterstab standard dumbbell test piece/specimen
Normstab standard test piece
Normstandardheißkanalblock standard hot runner unit
Normstrecke standard length/distance
Normteil standard unit/component
Normung standardisation
Normvorschrift standard specification
Normzugstab standard tensile test piece
Notauseinrichtung emergency cut-out (switch/device)
Notausschalter emergency cut-out (switch/device), emergency stop button
Notkühlsystem emergency cooling system
Notkühlung emergency cooling
Notleuchte emergency lamp/light
Notprogramm emergency program *(me)*
Notstromaggregat emergency power supply unit
Novelle amendment *(to an existing law or regulation)*
novellieren to amend *(an existing law or regulation)*
Novolakharz novolak resin
Nuance shade, colour
Nuancieren tinting
Nuancierpaste tinting paste
nukleierend nucleating
nukleiert nucleated
Nukleierung nucleation
Nukleierungshilfsmittel nucleating agent
Nukleierungsmittel nucleating agent
Nukleierungsvorgang nucleating process
Nukleierungswirkung nucleating effect
nukleophil nucleophilic

Null ausziehbar, auf can be smoothed to a feather edge
Nullabgleich zero balance/adjustment
Nulleinstellung zero adjustment
Nullprobe reference sample
Nullpunkt zero point
Nullpunkteinstellung zero adjustment/setting
Nullpunktkompensation zero compensation
Nullpunktkorrektur zero correction
Nullpunktüberwachung zero point monitor
Nullpunktverschiebung zero shift
Nullserie trial run, pilot plant run, pre-production trial
Nullstellung zero position/setting/adjustment
Nullviskosität viscosity at zero shear rate
Nullwert zero value
numerisch numeric
numerische Steuerung numeric control, NC
Nummernschild (car) number plate
Nur-Lesespeicher read only memory *(me)*
Nutbuchse grooved bush
Nutbuchsenextruder grooved-barrel extruder
Nutbuchsenlaborextruder grooved-barrel laboratory extruder
Nutenbreite groove width
Nutenbuchse grooved bushing *(e)*
Nuteneinzugszone grooved feed section/zone *(e)*
Nutenextruder grooved-barrel extruder
Nutenscherteil grooved shear section, grooved smear head
Nutentorpedo grooved smear head *(e)*
Nutentorpedoschnecke grooved torpedo screw
Nutenzahl number of grooves
Nutenzylinder grooved barrel *(e)*
nutzbares Volumen effective volume
Nutzen 1. benefit, use. 2. reel
Nutzen-Aufwandbetrachtung cost-benefit considerations
Nutzfahrzeug commercial vehicle
Nutzinhalt effective capacity
Nutzleistung effective power
Nutzsignal useful signal
Nutzung 1. utilisation, exploitation. 2. recycling, recovery
Nutzung, thermische energy recovery/recycling
Nutzungsdauer useful/service life
Nutzungsgrad productivity
Nutzvolumen effective volume

O

o.a. *abbr. of* **oben angegeben**, mentioned above, above mentioned
Oberbegriff general term
Oberfläche surface
Oberfläche, orangenschalenartige orange peel effect
Oberfläche, spezifische specific surface area
Oberflächenabrieb surface abrasion
Oberflächenabtrag surface wear
Oberflächenadditiv wetting agent, surface active agent
oberflächenaktiv surface active
oberflächenaktivierend surface activating
Oberflächenaktivierung surface activation
Oberflächenaktivität surface acitivty
Oberflächenangriff surface attack
Oberflächenausrüstung 1. surface finish 2. surface finishing
Oberflächenbeeinträchtigungen surface defects/blemishes
oberflächenbehandelt 1. surface treated. 2. coated *(e.g. filler particles)*
Oberflächenbehandlung coating, surface treatment
Oberflächenbehandlungssystem surface coating system
Oberflächenbelegung surface deposit
Oberflächenbeschaffenheit surface finish
oberflächenbeschichtet coated
Oberflächenbeschichtung surface coating
Oberflächenbeständigkeit surface durability
Oberflächenbrillanz (surface) sparkle
oberflächenbündig flush with the surface
Oberflächendefekt surface mark/blemish/defect
Oberflächendelamination surface delamination
Oberflächendiffusion surface diffusion
Oberflächeneigenschaften surface properties
Oberflächenenergie surface energy
Oberflächenerosion surface erosion
Oberflächenfehler surface mark/blemish/defect
Oberflächenfeuchte surface moisture
oberflächengehärtet surface hardened *(see explanatory note under* **Oberflächenvergütung**)
Oberflächengewichtsverhältnis surface area-weight ratio
Oberflächenglanz 1. surface polish. 2. surface gloss *(of a plastics article)*
Oberflächenglätte surface smoothness
Oberflächengleitmittel release agent, surface lubricant
Oberflächengriff handle: **mit trockenem Oberflächengriff** with a dry handle, dry to the touch
Oberflächengüte surface finish
Oberflächenhärte surface hardness
Oberflächenharz gelcoat resin
Oberflächenharzschicht gelcoat
Oberflächenhaut 1. outer skin *(e.g. of a PVC particle)*. 2. surface skin
Oberflächenklebrigkeit surface tack
Oberflächenkonturen surface contours
Oberflächenkorrosion surface corrosion
Oberflächenkratzfestigkeit scratch resistance
Oberflächenladung surface charge
Oberflächenleitfähigkeit surface conductivity
Oberflächenmarkierungen 1. surface marks/blemishes/defects. 2. bank marks *(c)*
Oberflächenmatte surfacing mat *(grp)*
Oberflächenmittel wetting agent, surface active agent
Oberflächenmodifizierung surface modification
Oberflächenmorphologie surface morphology
oberflächennah near the surface
Oberflächenneubildung surface renewal
Oberflächenoptik surface appearance
oberflächenoptimiert with a high quality surface finish
oberflächenorientiert surface-orientated
Oberflächenoxidation surface oxidation
Oberflächenpanzerung wear resistant coating, hard face coating
oberflächenpoliert with a highly polished surface
Oberflächenqualität surface finish
Oberflächenrauhigkeit surface roughness
Oberflächenreibung surface friction
Oberflächenrisse surface cracks
Oberflächenschäden, thermische surface charring
Oberflächenschädigung surface damage
Oberflächenschicht 1. surface layer. 2. gel coat *(grp)*
Oberflächenschlieren surface streaks
Oberflächenschutz surface protection/coating
Oberflächenspannung surface tension
Oberflächenspannungsgradient surface tension gradient
Oberflächenstörungen surface defects
Oberflächenstruktur surface texture
Oberflächentechnik surface finishing, painting
Oberflächentemperatur surface temperature
Oberflächentemperaturfühler surface thermocouple
oberflächenunbehandelt 1. not surface treated 2. un-coated *(e.g. filler particles)*
Oberflächenunebenheit surface irregularity
Oberflächenunruhe surface irregularities
Oberflächenveränderung surface changes
Oberflächenveredlung 1. surface treatment. 2. painting
oberflächenvergütet surface hardened *(see explanatory note under* **Oberflächenvergütung**)

Oberflächenvergütung surface hardening (of metal parts to increase wear resistance)
Oberflächenverletzung surface damage
Oberflächenversprödung surface embrittlement
Oberflächenverunreinigungen surface contaminants, surface dirt
Oberflächenvlies overlay/surfacing mat (grp)
Oberflächenvorbehandlung surface preparation
Oberflächenvorbereitung surface preparation
Oberflächenwickler surface-drive winder
Oberflächenwiderstand surface resistance
Oberflächenwiderstand, spezifischer surface resistivity
oberflächlich superficial(ly), on the surface
Obergrenze upper/top limit
Oberheizkörper top heater
oberirdisch above ground
Oberkolbenpreßautomat automatic downstroke press
Oberkolbenpresse downstroke press
Oberkolbenspritzpreßautomat automatic downstroke transfer moulding press
Oberlippe, flexible adjustable/flexible lip (e)
Oberputz finishing coat (of plaster)
Oberteil top part
Oberwalze top roll
Oberwerkzeug top (mould) force (cm)
objektiv objective
obligatorisch obligatory
Ochsenjochprofil ox-bow profile (c)
Octadecylalkohol octadecyl alcohol
Octylacrylat octyl acrylate
Octylfettsäureester octyl fatty acid ester
Octylzinncarboxylat octyl tin carboxylate
Octylzinnmercaptid octyl tin mercaptide
Octylzinnstabilisator octyl tin stabiliser
Octylzinnverbindung octyl tin compound
ODR abbr. of oscillating disc rheometer
ofenhärtend heat/oven curing
Ofenruß furnace black
ofentrocknend oven drying
offene Zeit open assembly time
offener Becher open cup (in flash point determination)
offener Steuerkreis open-loop control circuit
Offenlegungsschrift patent application filed but not yet accepted
offenporig open-cell (foam)
Offenporigkeit open-cell character
offenzellig open-cell (foam)
Offenzelligkeit open-cell character: **Folgende Rezeptur liefert eine hohe Offenzelligkeit des Schaumes** the following formulation produces a largely open-cell foam OR: produces a foam with a high proportion of open cells
Offline off-line (me)
Öffnung opening, aperture
Öffnungsbewegung 1. opening movement. 2. mould opening movement

Öffnungsgeschwindigkeit 1. opening speed. 2. mould opening speed
Öffnungshub 1. opening stroke. 2. mould opening stroke
Öffnungskraft 1. opening force. 2. mould opening force
Öffnungsraum space pressurised during opening movement (of toggle clamp unit)
Öffnungsstellung open position (im)
Öffnungsvorgang 1. opening movement. 2. mould opening movement
Öffnungsweg 1. opening stroke. 2. mould opening stroke
Öffnungswegbegrenzung 1. opening stroke limiting device. 2. mould opening stroke limiting device
Öffnungsweite 1. opening stroke. 2. mould opening stroke
Offsetdruckfarbe offset printing ink
Offsetfarbe offset printing ink
o.g. abbr. of **oben genannt**, above mentioned
OH-Funktionalität hydroxyfunctionality
Ohmsches Gesetz Ohm's Law
OHZ abbr. of **OH-Zahl**, hydroxyl value
OH-Zahl hydroxyl value
Ökobilanz 1. environmental audit. 2. ecological balance
Ökodaten ecodata
Ökokreislauf ecocycle
Ökologie ecology
ökologisch ecological
ökologisch unbedenklich ecologically safe
ökonomisch economic, economical
ökonomisch sinnvoll making economic sense
ökopolitisch ecopolitical
Ökoprofil ecoprofile
Ökosystem ecosystem
Oktyl- see **Octyl-**
Ölabfluß oil outlet
Ölabsorptionszahl oil absorption value
ölabweisend oil repellent
Ölaufnahme oil absorption
Ölaufnahmezahl oil absorption value
Ölbad oil bath
Ölbedarf oil requirements: **geringer Ölbedarf** low oil requirements
Ölbehälter oil reservoir
ölbeheizt oil heated
ölbeständig oil resistant
Ölbeständigkeit oil resistance
Öldiffusionspumpe oil diffusion pump
Oleat oleate
Olefin olefin
Olefincopolymer olefin copolymer
olefinisch olefinic
Olefinpolymer polyolefin
Ölentspannung oil decompression
Oleum oleum, fuming sulphuric acid
ölfest oil resistant
Ölfilterung 1. oil filtration. 2. oil filtration unit
ölfrei oil-free
Ölfüllung, Gewicht ohne weight without oil

Ölgehalt oil content
ölgestreckt oil-extended
Ölharz oleo-resin
Ölhydraulik oil-hydraulic system
ölhydraulische Presse hydraulic press
Oligoalkylenglykol oligoalkylene glycol
Oligoalkylenterephthalat oligoalkylene terephthalate
Oligobutylenglykolether oligobutylene glycol ether
Oligobutylenterephthalat oligobutylene terephthalate
Oligoester oligoester
Oligoesteracrylat oligoester acrylate
Oligoether oligoether
Oligoglykol oligoglycol
Oligomer oligomer
oligomer oligomeric
Oligomerisationsgrad degree of oligomerisation
Oligomerkette oligomer chain
Oligopolmarkt oligopolistic market
Oligourethan oligourethane
Ölinnentemperierung internal oil heating system
Öl-in-Wasser-Emulsion oil-in-water emulsion, O/W emulsion
Ölkautschuk oil masterbatch
Ölkompression oil compression
Ölkreislauf oil circuit
Ölkrise oil crisis
Ölkühler oil cooling unit
Ölkühlung 1. oil cooling. 2. oil cooling system
Ölländer oil producing countries
Öllänge oil length
Ölmengenbedarf oil requirement, amount of oil needed/required
ölmodifiziert oil modified
Ölmotor hydraulic motor
Ölniveau oil level
Ölniveauwächter oil level monitor
Ölphase oil phase
Ölpreiskrise oil crisis
Ölpreisschock oil crisis
Ölpumpe hydraulic pump
ölreich long-oil
Ölreste oil residues
Ölrückstände oil residues
Ölsäureester oleate, oleic acid ester
Ölspiegel oil level
Ölstand oil level
Ölstandanzeiger oil level indicator
Ölstandkontrolle 1. oil level control. 2. oil level control unit
Ölstandüberwachung 1. oil level monitoring. 2. oil level monitor
Ölstrom 1. amount of oil: **ein Teil des zur Verfügung stehenden ölstroms** some of the oil available. 2. oil flow
Ölstromregler oil flow regulator
Ölsumpf oil sump
Öltank hydraulic oil reservoir

Öltankbelüftung oil tank ventilation
Öltankfüllung 1. oil reservoir capacity. 2. amount of oil inside the reservoir: **Ölstandskontrolle mit Schauglas zum Beobachten der Öltankfüllung** oil level control unit with sight glass to observe the oil level inside the reservoir
Öltemperaturwächter oil temperature monitor
Öltemperiergerät oil heating unit
öltemperiert oil heated
Öltemperierung 1. oil temperature control. 2. oil temperature control system. 3. oil heating system
Ölthermostat oil-filled thermostat
Ölumlaufheizung circulating oil heating system
Ölumlaufschmierung 1. circulating oil lubrication. 2. circulating oil lubricating system
Ölumlauftemperierung 1. circulating oil temperature control. 2. circulating oil temperature control system
ölverstreckt oil-extended
Ölverteuerung oil price increase
Ölwanne oil sump
Ölzahl oil absorption value
Ölzufluß oil inlet
Online on-line *(me)*
Onlinebetrieb on-line operation
Onlinelackierbarkeit on-line paintability
Onlinelackierung on-line painting
Onlinemessung on-line determination/measurement
OPA-Folie oriented nylon film
opak opaque
opaleszierend opalescent
OPP-Maschine oriented polypropylene blow moulding machine
OPP-Streckblasautomat automatic oriented PP stretch blow moulding machine
OPP-Verfahren blow moulding of oriented polypropylene
Optik 1. optics 2. appearance
optimal optimum, best (possible), most suitable. *This has become a buzz word and cannot always be taken at face value. In the following examples it would be a nonsense to translate it literally*: **Die Einhaltung einer optimalen Aufheizgeschwindigkeit ist erforderlich, da bei zu schneller oder zu langsamer Aufheizung...** it is important to raise the temperature gradually, since excessively fast or excessively slow heating will... **...sodaß die Folienschnitzel optimal gereinigt werden können** ... so that the cut-up film scrap can be thoroughly cleaned. *Translators should use their own judgement when dealing with this word.*
Optimalbedingungen ideal/optimum conditions
optimieren to perfect, to upgrade, to improve
Optimierung upgrading, improvement
Optimum, optimum 1. optimum. 2. maximum: *The context will determine when to use version 2. Whereas it is in order to refer to*

optimum properties and optimum quality, there are occasions when version 2. is preferable. See explanatory note under **Schlagzähigkeitsoptimum**
optisch 1. optical. 2. visual *(e.g. a signal)*.
optisches Aussehen appearance
optische Datenspeicherplatte optic(al) disc *(me)*
optische Speicherplatte optic(al) disk *(me)*
optischer Fehler surface defect
optischer Plattenspeicher optic(al) disk *(me)*
optischer Speicher optic(al) disk *(me)*
optoelektronisch optoelectronic
Orangenhauteffekt orange peel effect
orangenschalenartige Oberfläche orange peel effect
ordentliches Vorstandsmitglied full member of the Board of Directors
Ordinate ordinate
Ordnungszahl atomic number
Ordnungszustand state of order
Organisationsstruktur corporate structure
organisatorisch organisational
organisch organic
organisch modifiziert organically modified
Organoalkoxysilan organoalkoxysilane
Organochloralkoxysilan organochloroalkoxysilane
organofunktionell organo-functional
Organogruppe organic group/radical
Organokunststoff organic polymer
organometallisch organo-metallic
organomodifiziert organo-modified
Organooligosiloxan organo-oligosiloxane
organophil organophilic
Organophosphat organophosphate
Organophosphit organo-phosphite
Organopolysiloxan organo-polysiloxane
Organorest organic radical/group
Organosilan organo-silane
Organosiliciumverbindung organo-silicon compound
Organosiloxan organosiloxane
Organosol organosol
Organotitanat organo-titanate
Organozinnchlorid organo-tin chloride
Organozinnmercaptid organo-tin mercaptide
Organozinnmercaptocarbonsäureester organo-tin mercaptocarboxylate
Organozinnpolymer organotin polymer
Organozinnstabilisator organo-tin stabiliser
Organozinnverbindung organo-tin compound
orientiert, biaxial biaxially oriented
Orientierung orientation
Orientierungserscheinungen signs of orientation
orientierungsfrei non-oriented, free from orientation
Orientierungsgrad degree of orientation
Orientierungshilfe guide
Orientierungsspannungen frozen-in stresses
Orientierungspunkt point of reference

Orientierungsrichtung direction of orientation
Orientierungsverteilung orientation distribution
Orientierungszustand state of orientation
Orientierungszuwachs increase in orientation
Originalfüllung original contents
Originalgebinde original container
originalgetreu faithful (to the original)
Originalmaterial virgin material
originaltreu faithful (to the original)
Originalverpackung original packaging/container
Originalviskosität original/initial viscosity
OROM *abbr. of* optical read only memory *(me)*
orthogonal orthogonal
Orthophthalsäure ortho-phthalic acid, o-phthalic acid
orthoständig in the ortho position
Orthostellung ortho-position
orthosubstutuiert ortho-substituted
orthotrop orthotropic
ortsfest fixed, stationary *(machine)*
Ortsschaum in-situ foam
Ortsschaumstoff in-situ foam
OSI *abbr. of* open systems interconnection *(me)*
OSM *abbr. of* **Ozonschutzmittel**, anti-ozonant
Osmiumtrioxid osmium trioxide
Osmometrie osmometry
Osmose osmosis
osmotisch osmotic
östliche Staatshandelsländer Eastern bloc countries
OSW *abbr. of* **Ozonschutzwachs**, anti-ozonant wax
Oszillation oscillation, vibration
Oszillationsversuch oscillation/vibration test
oszillierend oscillating
Oszillograph oscillograph
Ottokraftstoff Otto fuel
Ottomotor Otto engine
Outserttechnik outsert moulding
Ovalleuchte bulkhead lamp
oxalkyliert oxalkylated
oxethyliert ethoxylated
Oxialkylengruppe oxyalkylene group
Oxialkylgruppe oxyalkyl group
Oxidation oxidation
Oxidationsanfälligkeit oxidation sensitivity
Oxidationsbeständigkeit oxidation resistance, resistance to oxidation
oxidationsempfindlich oxidation-sensitive, susceptible to oxidation
Oxidationsempfindlichkeit oxidation sensitivity
oxidationsfähig oxidation-sensitive, susceptible to oxidation
Oxidationsgeschwindigkeit oxidation rate, rate of oxidation
Oxidationsgrad degree of oxidation
Oxidationskatalysator oxidation catalyst
Oxidationsmittel oxidising agent
Oxidationsprodukt oxidation product
Oxidationsschutzmittel anti-oxidant
Oxidationsstabilisator anti-oxidant

Oxidationsstabilität oxidation resistance
Oxidationsstufe oxidation number
oxidationsunempfindlich not sensitive/ susceptible to oxidation
oxidativ oxidative
oxidierbar oxidisable
oxidierend oxidising
Oxidschicht oxide film
Oxiethylengruppe oxyethylene group
Oximethylenformaldehyd oxymethylene formaldehyde
Oximethylengruppe oxymethylene group
Oximverbindung oxime compound
Oxipropylengruppe oxypropylene group
Oxy-, oxy- see **Oxi-, oxi-**
Ozon ozone
Ozonalterung ozone ageing
Ozonalterungsgerät ozone ageing apparatus
Ozonalterungsschrank ozone ageing cabinet
Ozonalterungsversuch ozone ageing test
Ozonangriff ozone attack
Ozonbeständigkeit ozone resistance
Ozonbewitterung ozone weathering
Ozoneinfluß effect of ozone
Ozonfestigkeit ozone resistance
Ozongefährdung risk to the ozone layer
ozongeschützt ozone-stabilised
ozoninduziert ozone-induced
ozonisiert ozonised
Ozonisierung ozonisation
Ozonkonzentration ozone concentration
Ozonloch ozone hole, hole in the ozone layer
Ozonprüfkammer ozone test cabinet
Ozonriß ozone crack
Ozonrißbildung ozone cracking
ozonschädigend ozone-depleting, ozone-damaging
Ozonschädigung depletion of the ozone layer, damage to the ozone layer, ozone depletion
Ozonschicht ozone layer
ozonschichtschädigend ozone-depleting, ozone-damaging
ozonschichtunschädlich harmless for the ozone layer
Ozonschutz protection against ozone
Ozonschutzmittel anti-ozonant
Ozonschutzsystem anti-ozonant system
Ozonschutzwachs anti-ozonant wax
Ozonschutzwirkung anti-ozonant effect

P

Packgut product *(to be packaged or already packaged. This translation will be perfectly correct in context)*
Packmittel (packaging) container, pack
Packmittelabfall packaging waste
Packphase packing phase *(im)*
Packstoff packaging material
Packung 1. pack, package. 2. packing *(i.e. a seal)*
Packungsdichte packing density
Packungshohlkörper (packaging) container
Packvorgang packaging operation
Paddel paddle
Paddelmischer paddle mixer
PAEK abbr. of polyarylether ketone
Paketlieferung package deal
Palette 1. pallet. 2. range; *When the word is used as a suffix it usually means* range of..., *e.g.***Weichmacherpalette** range of plasticisers. **Holzpalette,** *on the other hand, would be a* wooden pallet
Palmitat palmitate
Palmitinsäure palmitic acid
PAN abbr. of phenyl α-naphthylamine
Pannenbehebung elimination of faults *(e.g. in a moulding process) (for translation example see under)* **Störungsbehebung**)
Pannensuche fault location *(e.g. in a moulding process; for translation example see under* **Störungslokalisierung**)
panzern to harden, to hard-face
Panzerschicht wear resistant coating, hard face coating
Panzerung 1. hardening, hard-facing. 2. wear resistant coating, hard face coating
Papierbahn paper web
Papierbeschichtung paper coating
Papierchromatographie paper chromatography
papierchromatographisch paper chromatographic
Papierfabrik paper mill
Papierfolie paper-like film
Papierimprägniermittel paper impregnating agent
Papierkleber paper adhesive
Papierklebstoff paper adhesive
Papierlack paper varnish
Papierlaminat paper-based laminate
Papiersack paper sack
Papierschichtstoff paper-based laminate
Pappdose (small) cardboard container
Pappkarton cardboard box
Papptrommel cardboard drum
Parabelantenne parabolic aerial, satellite dish
Parabelblattfeder parabolic leaf spring
Parabelfeder parabolic spring
parabolisch parabolic
Parabolschüssel parabolic/satellite dish
Paradichlorbenzol para-dichlorobenzene
paraffinisch paraffinic
Paraffinkette paraffin chain
Paraffinkohlenwasserstoff aliphatic hydrocarbon
Paraffinwachs paraffin wax
Parallelbus parallel bus *(me)*
Parallelerscheinungen side effects

Parallelführung 1. die land *(e)*. 2. parallel travel: **Die Parallelführung der beweglichen Aufspannplatten ist dann gewährleistet...** parallel travel of the moving platens is assured if...
parallelgeschaltet arranged in parallel
Parallelhub parallel stroke
Parallelität parallelism
parallelliegend lying parallel
Parallelschaltung parallel arrangement
Parallelschnittstelle parallel interface *(me)*
Parallelströmung parallel flow
Parallelteil die land *(e)*
Parallelverteiler parallel runners *(im)*
Parallelzone die land *(e)*
Parameter parameter, factor, quantity, characteristic feature: **Das Ergebnis wird durch folgende Parameter bestimmt** the result is determined by the following factors; **Der Parameter der die Thermoplaste characterisiert ist...** the characteristic feature of thermoplastics is...; **Die Zähigkeit des strömenden Mediums ist ein wichtiger Parameter** the viscosity of the flowing medium is an important factor; *The word must sometimes be translated very freely in order to make sense, as in this example;* **Große Anstrengungen werden unternommen, um gleichbleibende Parameter während des Betriebs innezuhalten** great efforts are made to maintain constant processing conditions, **verfahrenstechnisch relevante Parameter** technically important parameters; **prozeßführende Parameter** processing conditions
Paramethylstyrol p-methyl styrene
Parametriersoftware programming software *(me)*
Parametrierung programming *(me)*
Paraphenylendiamin p-phenylene diamine
paraständig in the para position
Parastellung para-position
parasubstituiert para-substituted
Paratoluolsulfonsäure p-toluenesulphonic acid
Paraxylol p-xylene
Park 1. available machines/equipment. 2. machines *(for translation examples see under* **Machinenpark**)
Parkbank park bench
Parkdeck parking area
Parkettversiegelung sealing varnish for parquet floors
Parkettversiegelung parquet floor sealant (composition)
Parkhaus multi-storey car park
Parkplatz car park
Partialdruck partial pressure
Partialester partial ester
Partialgasdruck partial gas pressure
Partialströme separate melt streams

Partiekontrolle batch control
partiel partial
partiellmethyliert partly methylated
partiellverethert partly etherified
Partieprüfung batch control
***Partikel** particle
Partikelagglomeration particle agglomeration
Partikelbrücke particle bridge *(mf)*
Partikelform particle shape
Partikelgrenzfläche particle interface
Partikelgröße particle size
Partikeloberfläche particle surface
Partikelverschweißung fusion of the particles
partikulär particulate
PAS *abbr. of* plastics additive system
Passagierflugzeug passenger aircraft
Paßbohrung accurately bored/drilled hole
Passergenauigkeit exact/accurate fit
Paßflächen mating surfaces
Paßgenauigkeit exact/accurate fit
passiv pasive
Passivation passivation
passivierend passivating
Passivität passivity
Paßstück adaptor
Passung eine enge Passung a tight fit
Paßwort password *(me)*
Paste paste
Pastellnuance pastel colour/shade
Pastellton pastel shade
Pastenansatz paste formulation
Pastenbereitung paste making/preparation
pastenbildend paste-making *(properties of PVC)*
Pastenentlüftung paste deaeration
Pastenextrusion paste extrusion *(ptfe)*
Pastenharz 1. (PVC) paste resin 2. pigment paste resin, resin suitable for pigment pastes
Pastenharzbindemittel pigment paste binder resin
Pastenherstellung paste preparation
Pastenpolymerisat paste polymer, paste-making polymer
Pasten-PVC PVC paste resin, paste-making PVC
Pastenrezeptur paste formulation
Pastentauchverfahren paste dipping (process)
Pastentyp paste-making grade *(of PVC)*
Pastenverarbeitung paste making: **PVC für die Pastenverarbeitung** paste-making PVC; **Dibasische Bleiphosphite werden häufig bei der Pastenverarbeitung verwendet** dibasic lead phosphites are widely used in pastes
Pastenverschnittharz paste extender resin
Pastenviskosität paste viscosity
Pastenware PVC paste resin, paste-making PVC
pastös paste-like
Patentanmeldung patent application

---

* *Words starting with* **Partikel**-, *not found here, may be found under* **Korn**- *or* **Teilchen**-

Patentanspruch patent claim
Patentanwalt patent agent
Patentblatt patent journal
Patenterstellung granting of a patent
Patenterteilung granting of a patent
patentfähig patentable
patentiert patented
Patentinhaber patentee
Patentlösung panacea
Patentrechte patent rights
patentrechtlich geschützt patented
Patentschrift patent specification
Paternosteranlage paternoster-type conveyor, bucket conveyor
Patrize 1. male mould *(t)*. 2. core plate *(im)*
Patrone cartridge
patronenförmiger Heizkörper cartridge heater
Patronenheizkörper cartridge heater
Pausenzeit 1. interval. 2. change-over time *(im)*
Pausenzeituhr interval timing mechanism
Pausfolie tracing film
Pb-frei lead-free
PBN *abbr. of* phenyl ß-naphthylamine
PC 1. *abbr. of* polycarbonate 2. *abbr. of* personal computer
PC-Steuerung stored-program control *(me)*
PDE *abbr. of* **Prozeßdatenerfassung**, process data collection/acquisition/ capture
PDE/PDV-Anlage process data acquisition and processing system *(me)*
PDV *abbr. of* **Prozeßdatenverarbeitung**, process data processing *(me)*
PE hart high density polyethylene, HDPE
PE weich low density polyethylene, LDPE
Peaktemperatur peak/maximum temperature
PEBA *abbr. of* polyether block amide
PE-Beutel polyethylene/polythene bag
PEC *abbr. of* **Polyethylen chloriert,** chlorinated polyethylene
PE-Farbkonzentrat polyethylene/PE masterbatch
PE-HD *equivalent to* HDPE, high density polyethylene
PE-HD-HMW *equivalent to* HMW-HDPE, high molecular weight high density polyethylene
PE-HD-UHMW *equivalent to* UHMW-HDPW, ultra-high molecular weight high density polyethylene
PE-Innenhülle PE/polyethylene/polythene liner
PE-Innensack PE/polyethylene/polythene liner
PEK *abbr. of* polyether ketone
PEKEKK *abbr. of* polyether ketone ether ketone ketone
PE-LD *equivalent to* LDPE, low density polyethylene
PE-LLD *equivalent to* LLDPE, linear low density polyethylene
Pelletieren pelleting
PE-MD *equivalent to* MDPE, medium density polyethylene
PE-ND *equivalent to* LDPE, low density polyethylene

Pendelhammer pendulum
Pendelhammergerät pendulum impact tester
Pendelhammergeschwindigkeit pendulum impact speed
Pendelhärte pendulum hardness
pendelnd freely suspended
Pendelrollenlager self-aligning roller bearing
Pendelsäge floating saw
Pendelschlagversuch pendulum impact test
Pendelschlagwerk pendulum impact tester
Pendeltür swingdoor
Pendelwalze floating roller
Penetrationsgrad degree of penetration
Penetrationsvermögen penetrating power
penetrierfähig capable of good penetration *(e.g. primers, impregnating agents etc.)*
Pentabrombenzylacrylat pentabromobenzyl acrylate
Pentachlorphenol pentachlorophenol
Pentaerythritester pentaerythritol ester
Pentaerythrit(ol) pentaerythritol
Pentaerythrittriacrylat pentaerythritol triacrylate
Pentan pentane
Per perchloroethylene, tetrachloroethylene
Peradipinsäure peradipic acid
Perameisensäure performic acid
Perbenzoesäure perbenzoic acid
Perbernsteinsäure persuccinic acid
Perbuttersäure perbutyric acid
Percarbonat percarbonate
Perchlorethylen perchloroethylene, tetrachloroethylene
perchloriert perchlorinated
Perdampf tetrachloroethylene vapour/fumes
Peressigsäure peracetic acid
Perester per-ester
Perfluoralkoxy perfluoroalkoxy, PFA
perfluoralkylmodifiziert perfluoroalkyl modified
Perfluorcarbon perfluorocarbon
Perfluorethylenpropylen perfluoroethylene propylene, FEP
perfluoriert perfluorinated
Perforationsfestigkeit piercing resistance
perforiert perforated
periodisch periodic
periodisch arbeitend operating in cycles
peripher peripheral
Peripherie peripheral, peripheral unit *(me)*
Peripheriebaugruppe peripheral (unit) *(me)*
Peripheriebaustein peripheral, peripheral unit *(me)*
Peripheriebauteil peripheral, peripheral unit *(me)*
Peripheriegerät 1. peripheral equipment *(me)*. 2. ancillary equipment
Peripherieleitung peripheral cable
Peripherieprozessor peripheral processor *(me)*
Perketal perketal
Perlen beads
Perlglanzeffekt pearlescent effect
Perlglanzpigment pearlescent pigment
Perlmühle pearl mill

Perlpolymerisation bead/suspension polymerisation
Permanentmagnet permanent magnet
Permanenz permanence
Permeabilität permeability
Permeabilitätsdaten permeability data
Permeabilitätskoeffizient permeability coefficient
Permeabilitätsverhalten permeability behaviour
Permeabilitätswert permeability coefficient
Permeant permeant
Permeation permeation
Permeationsgeschwindigkeit permeation rate
Permeationskoeffizient permeability coefficient
Permeationsrate permeation rate
Permeationsverhalten permeation behaviour: **günstiges Permeationsverhalten** impermeability
Permeationswert permeability coefficient
permeiert permeated
Peroxid peroxide
Peroxidbedarf peroxide requirement
Peroxiddosierung amount of peroxide
peroxidfrei peroxide-free
Peroxidgruppe peroxide group
peroxidhärtend peroxide-curing/-crosslinking
peroxidisch aushärtbar peroxide-curing/-crosslinking
peroxidische Vernetzung peroxide crossinkage/cure
peroxidisch-vernetzend peroxide-crosslinking/-curing
Peroxidkonzentration peroxide concentration
Peroxidmenge amount of peroxide
Peroxidmolekül peroxide molecule
Peroxidradikal peroxide radical
Peroxidsuspension peroxide suspension
peroxidvernetzbar peroxide-curing/-crosslinking
Peroxidvernetzer peroxide curing/crosslinking agent
peroxidvernetzt peroxide-crosslinked/-cured
Peroxidvernetzung peroxide cure/crosslinkage
Peroxidvulkanisat peroxide-vulcanised/-cured rubber
Peroxidvulkanisation peroxide vulcanisation/cure
Peroxidzerfall peroxide decomposition
Peroxidzersetzung peroxide decomposition
Peroxydicarbonat peroxydicarbonate
Peroxyradikal peroxide radical
Peroxyverbindung peroxy compound
Perpropionsäure perpropionic acid
Persäure per-acid
Persenning tarpaulin
Personal staff, employees
personalarm intended to convey the idea of few workers/employees/ operators, etc.: **personalarme Fertigung** production with a skeleton staff
Personalaufwand 1. personnel expenditure, wages and salaries. 2. number of personnel
personalaufwendig labour-intensive
Personalcomputer personal computer, PC
personalintensiv labour-intensive
Personalkosten labour costs, wages and salaries
Personalmangel shortage of staff, staff shortage
Personalschwierigkeiten staff problems
Personenautomobil (motor) car
Personenkraftwagen (motor) car
Personenwagen (motor) car
Perspektiven possibilities, prospects
Persulfat persulphate
Perverbindung per-compound, peroxide
PES-Laminat polyethersulphone laminate
PETP-Flasche PET bottle *Although it would be theoretically correct to write* PETP bottle, *this is not correct usage. All British and American plastics and packaging journals* ALWAYS *use the abbreviated form given above. Quite often, in fact, lower case is used, e.g.* pet bottle, *despite the obvious ambiguity thus introduced.*
PETP-Hohlkörper PET container/bottle *(see explanatory note under* **PETP-Flasche)**
Petrochemie petrochemistry
petrochemisch petrochemical
Petrolether petroleum ether
PE-UHMW *equivalent to* UHMW-PE, ultra-high molecular weight polyethylene
PE-VLD *equivalent to* VLDPE, very low density polyethylene
PE-X *abbr. of* crosslinked polyethylene
Pflanzentopf flower pot
pflanzlich vegetable *(e.g. oils or fats)*
Pflanztrog plant trough
pflegefrei maintenance-free
pflegeleicht easy-care
Pflegeleichtausrüstung easy-care finish
Pflegeleichtigkeit easy-care properties
Pflegemittel *This word has no exact English equivalent.* **Pflegemittel** *can be creams and polishes for furniture, leather etc., as well as creams, lotions etc. for the hair. The context will usually provide the clue as to the best way of dealing with the word but, if in doubt, one can always use the all-embracing* creams and polishes
Pflichtenheft specification
Pfosten 1. post 2. mullion (of a window)
Pfropfcopolymer graft copolymer
Pfropfcopolymerisation graft copolymerisation
Pfropfen, kalter cold slug *(im)*
Pfropfenhalterung cold slug retainer *(im)*
Pfropfenströmung plug flow
Pfropfpolymer(isat) 1. graft polymer. 2. graft copolymer: *Version 1 is acceptable, although version 2 is more correct chemically*
Pfropfpolymerisation 1. graft polymerisation. 2. graft copolymerisation *(see explanatory note under* **Pfropfpolymer)**
pfropfpolymerisiert 1. graft polymerised. 2. graft copolymerised *(see explanatory note under* **Pfropfpolymer)**

Pfropfung grafting
Pfropfungsgrad degree of grafting
Pfropfverfahren graft polymerisation/ copolymerisation *(see explanatory note under* **Pfropfpolymer***)*
PGC *abbr. of* **Pyrolyse-Gas-Chromatographie**, pyrolysis gas chromatography
Pharmaindustrie pharmaceutical industry
Pharmazeutika pharmaceutical products/ preparations
Phase, dispergierte disperse phase
Phase, disperse disperse phase
Phasenanschnittsteuerung phase interface control *(me)*
Phasengleichgewicht phase equilibrium
Phasengrenze phase boundary
Phasengrenzfläche phase boundary
Phasenlänge die land *(e)*: *This rather unusual word was said by the author of the source article to be the same as* **Parallelführung** - *hence the unexpected English equivalent*
Phasenmorphologie phase morphology
Phasenstabilität phase stability
Phasenstruktur phase structure
Phasentrennung phase separation
Phasenübergang phase transition
Phasenumkehr phase inversion
Phasenumschlag phase reversal
Phasenumschlagpunkt phase reversal point
Phasenumwandlung phase reversal
Phasenversatz phase displacement
Phasenverschiebung phase displacement
Phasenverteilung phase distribution
phasenverträglich phase-compatible
Phasenwinkel phase angle
PHB *abbr. of* polyhydroxy butyrate
Phenol phenol
phenolarm low-phenol
Phenolflüssigharz liquid phenolic resin
Phenol-Formaldehydharz phenol-formaldehyde resin
Phenol-Formaldehydkondensat phenol-formaldehyde condensate
Phenolharz phenolic resin
Phenolharz, technisches phenolic engineering resin
Phenolharzformmasse phenolic moulding compound
Phenolharzhartpapier phenolic paper laminate
Phenolharzkleber phenolic adhesive
Phenolharzlaminat phenolic laminate
Phenolharzmatte phenolic prepreg
Phenolharzpapierlaminat phenolic paper laminate
Phenolharzpreßmasse phenolic moulding compound
Phenolharzschaumstoff phenolic foam
Phenolharzschichtstoff phenolic laminate
Phenolharzwerkstoff phenolic material
phenolisch phenolic
phenolmodifiziert phenol-modified
Phenolnovolak phenol novolak

Phenolnovolakglycidyletherharz phenol novolak glycidyl ether resin
Phenolphthalein phenolphthalein
Phenolpulverharz powdered phenolic resin
Phenolresol phenol resol
Phenolschaum phenolic foam
Phenoplaste phenolics: *Whereas it is correct to call* **Aminoplaste** *aminoplastics, the expression phenoplastics is not accepted usage, the materials always being referred to as* phenolics
Phenoplastformmasse phenolic moulding compound
Phenoplastformteil phenolic moulding
Phenoplastharz phenolic resin
Phenoplastpreßmasse phenolic moulding compound
Phenoplastpreßstoff moulded phenolic material: **Gleitlager aus Phenoplastpreßstoff** phenolic sliding bearings
Phenoxyradikal phenoxy radical
Phenoxyrest phenoxy radical/group
Phenylendiamin phenylene diamine
Phenylethylrest phenylethyl group
Phenylglycidylether phenylglycidyl ether
Phenylgruppe phenyl group/radical
Phenylharnstoff phenyl urea
Phenylisopropylrest phenylisopropyl group
Phenylmethylsilicon phenylmethyl silicone
Phenylmethylsiliconharz phenylmethyl silicone resin
Phenylmethylsiliconöl phenylmethyl silicone fluid
Phenylmethylvinylpolysiloxan phenylmethyl vinyl polysiloxane
Phenylnaphthylamin phenyl naphthylamine
Phenylpropylsiloxan phenylpropyl siloxane
Phenylrest phenyl radical
Phenylring phenyl ring
Phenylseitengruppe phenyl side group
Phenylsiliconharz phenyl silicone resin
Phenylurazol phenyl urazole
pH-Gerät pH meter
phlegmatisiert retarded
Phlegmatisierungsmittel retarding agent
Phonogerät stereo equipment
Phonoindustrie recording industry
Phosgen phosgene
Phosphat phosphate
phosphatiert phosphated
Phosphatierung phosphating
Phosphatweichmacher phosphate plasticiser
Phosphor phosphorus
Phosphoreszenz phosphorescence
Phosphor-Halogenverbindung phosphorus-halogen compound
phosphorhaltig phosphorus-containing, containing phosphorus
Phosphorigsäureester phosphite, phosphorous acid ester
Phosphorsäure phosphoric acid

Phosphorsäureester phosphate, phosphoric acid ester
Phosphorverbindung phosphorus compound
Photoabbau photodegradation
photoabbaubar photodegradable
Photoätzen photoetching, photoengraving, process engraving
photochemisch photochemical:
   photochemisch abbaubar photodegradable
photochrom photochromic, photosensitive
Photodegradation photodegradation
Photoelektron photoelectron
photoempfindlich photosensitive
Photoempfindlichkeit photosensitivity
photogenarbt photoetched, photoengraved
photoinduziert photo-induced
Photoinitiator photo-initiator
photokatalytisch photocatalytic
Photolack resist, photoresist (material)
photolytisch photolytic
Photon photon
Photooxidation photo-oxidation
Photopolymer photoreactive polymer
photopolymerisierbar photopolymerisable
photoreaktiv photoreactive
Photoresistentwickler photoresist developer
Photosensibilisator photosensitiser
photothermisch photothermal
Photozelle photoelectric cell
pH-Regler pH regulator
Phthalat phthalate
Phthalatharz phthalate resin
Phthalatweichmacher phthalate plasticiser
Phthalocyaninblau phthalocyanine blue
Phthalocyaninfarbstoff phthalocyanine pigment
Phthalocyaningrün phthalocyanine green
Phthalsäure phthalic acid
Phthalsäureanhydrid phthalic anhydride
Phthalsäurediallylester diallyl phthalate
Phthalsäuredibutylester dibutyl phthalate, DBP
Phthalsäurediethylhexylester diethylhexyl/dioctyl phthalate, DOP
Phthalsäurediglycidylester diglycidyl phthalate
Phthalsäureester phthalate
PHV abbr. of polyhydroxy valerate
pH-Wert pH (value)
physikalisch physical
physikalisch trocknend physically drying
physikalische Trocknung physical drying
physikochemisch physico-chemical
physiologisch einwandfrei non-toxic
physiologisch indifferent non-toxic, physiologically inert
physiologisch nicht einwandfrei toxic
physiologisch unbedenklich non-toxic
physiologische Unbedenklichkeit non-toxicity
physiologisches Verhalten toxicological properties
Physisorption physisorption
PID-Verhalten PID characteristics
Piezoeffekt piezo-electric effect
piezoelektrisch piezoelectrical

Piezoelektrizität piezoelectricity
piezoresistiv piezoresistive
Pigment pigment
Pigmentagglomerat pigment agglomerate
Pigmentanreibung 1. pigment grinding. 2. pigment paste
Pigmentanreicherung pigment enrichment
Pigmentanteil pigment content: *When the plural version is used, the -anteile part can often be omitted in translations, e.g.* **Sie können Pigment- und Füllstoffanteile enthalten** they can contain pigments and fillers
Pigmentaufnahmevermögen pigment wetting power/properties
Pigmentaufschlämmung pigment slurry
Pigmentausschwimmen, horizontales floating
Pigmentausschwimmen, vertikales flooding
pigmentbenetzend pigment wetting
Pigmentbenetzung 1. pigment wetting 2. pigment wetting properties
Pigmentbenetzungseigenschaften pigment wetting properties
Pigmentbindevermögen pigment binding power
Pigmentdispersion pigment dispersion/paste
Pigmentdispersionsadditiv pigment dispersing agent
Pigmentdosierung 1. addition of pigment 2. amount of pigment (used or added)
Pigmentfarbstoff pigment
pigmentfrei unpigmented
Pigmentgehalt pigment content
Pigmentgranulat pigment granules
Pigmenthärte pigment hardness
pigmentierbar, leicht easily pigmented
pigmentiert pigmented
Pigmentierung 1. pigmentation. 2. pigment content
Pigmentierungsart type of pigment
Pigmentierungsgrad pigment content
Pigmentierungshöhe pigment content
Pigmentkonzentration pigment concentration
Pigmentkonzentratpaste pigment masterbatch
Pigmentkorn pigment particle
Pigment-Kunststoffkonzentrat pigment masterbatch
Pigmentmassenkonzentration pigment mass concentration, p.m.c.
Pigmentnester (local) pigment accumulations
Pigmentnetzer pigment wetting agent
Pigmentoberfläche pigment surface
Pigmentpartikel pigment particle
Pigmentpartikeldurchmesserverteilung pigment particle size distribution
Pigmentpartikelgrößenverteilung pigment particle size distribution
Pigmentpartikelverteilung pigment particle size distribution
Pigmentpaste pigment paste
Pigmentpräparation pigment paste
Pigmentpulver powdered pigment
Pigmentruß carbon black

Pigmentsedimentation pigment sedimentation, settling out of pigment
Pigmentstabilisator pigment stabiliser
Pigmentstruktur pigment structure
Pigmentteilchen pigment particle
Pigmenttragevermögen pigment wetting power/properties
Pigmentvermahlen pigment grinding
Pigmentverteiler pigment dispersing agent
Pigmentverteilung pigment dispersion
Pigmentverträglichkeit pigment compatibility
Pigmentvolumenkonzentration pigment volume concentration, p.v.c.
PI-Laminat polyimide laminate
Pilotanlage pilot plant
Pilotenkanzel cockpit
Pilotmaschine pilot plant-scale machine
Pilotproduktionsanlage pilot production plant
Pilotprojekt pilot project
Pilotserie pilot run
Pilz mould
Pilzanguß diaphragm gate *(im)*
Pilzbefall mould attack
Pilzbefall attack by mould: **resistent gegen Pilzbefall** mould resistant
pilzhemmend fungicidal
Pinholebildung pinholing
Pinole mandrel *(e,bm)*
Pinolenblaskopf side-fed blown film die *(bfe)*
Pinolenkopf side-fed die *(e,bm)*
Pinolenkopf, seitlich angeströmter side-fed die *(e,bm)*
Pinolenkörper mandrel *(e,bm)*
Pinolenschlauchkopf side-fed die *(e,bm)*
Pinolenspritzkopf side-fed die *(e,bm)*
Pinolenwerkzeug side-fed die *(e,bm)*
Pinsel brush
Pinselapplikation application by brush
Pinselreiniger brush cleaner
Pixelgrafik pixel graphics *(me)*
PkW abbr. of **Personenkraftwagen**, (motor) car
Pkw-Lauffläche car tyre tread
Pkw-Radblende hub cap
plan flat, even, smooth
Plane tarpaulin
Planen, rechnerunterstütztes computer aided planning, CAP
Planenstoff tarpaulin material
Planet planetary screw/spindle/gear *(e)*
Planetengetriebe planetary gear
Planetenmischer planetary mixer
Planetenmischkneter planetary mixer
Planetenrührwerk planetary mixer
Planetenspindel planetary screw/spindle/gear *(e)*
Planetschnecke planetary screw/spindle/gear *(e)*
Planetwalze planetary screw/spindle/gear *(e)*: *The word **Walze** is used because the action of these spindles is that of a roller: the melt is rolled out into a thin layer, which is mixed, rolled out again and so on. The word roller should never be used in this context*
Planetwalzenextruder planetary-gear extruder
Planetwalzenteil planetary-gear section *(e)*
planparallel plane parallel
Planparallelität parallelism: **Gute Planparallelität ist bei diesen Platten wichtige Voraussetzung** it is important for these platens to be parallel
planpoliert plane polished
Planungsaufwand amount of planning (necessary)
Planungsschritte planning stages
Plasma plasma
Plasmabehandlung plasma treatment
Plasmaentladung plasma discharge
plasmageschnitten cut with a plasma arc
Plasmamodifizierung plasma treatment
Plasmapolymerisation plasma polymerisation
plasmapolymerisiert plasma polymerised
Plasmaschicht plasma film/coating
Plasmaveredelungsverfahren plasma treatment
Plasmaverfahren plasma method/technique
Plasmavorbehandlung plasma pretreatment
Plastbeton polymer concrete
Plaste plastics *(this term is encountered in literature originating in the former East Germany)*
Plastifikat (polymer) melt
Plastifikatgüte melt quality
Plastifikator plasticator
Plastifizieraggregat plasticising/plasticating unit, plasticator
Plastifizierbeginn, bei 1. at the start of plasticisation. 2. when the polymer starts to melt
Plastifizierbereich transition section/zone *(of screw)*
Plastifizierdruck melt pressure: *The word does not, as might be supposed, denote the pressure which plasticises the compound, but the pressure of the compound during plasticisation - in other words the melt pressure*
Plastifiziereinheit plasticising/plasticating unit, plasticator
Plastifiziereinrichtung plasticising/plasticating unit, plasticator
Plastifiziereinschneckenextruder single screw compounding/plasticating extruder
plastifizieren 1. to plasticise *(i.e. to soften by adding plasticiser, e.g. to PVC)*. 2. to plasticise, to plasticate *(i.e. to soften by heating, e.g. in an extruder)*. 3. to soften
Plastifizierende end of plasticisation, **bei Plastifizierende** when plasticisation has been completed
Plastifizierextruder compounding/plasticating extruder
Plastifiziergrad degree of plasticisation/plastication
Plastifizierhilfe plasticising/plasticating aid

Plastifizierkammer plasticising chamber/ compartment, plasticating chamber/ compartment
Plastifizierkanäle plasticising grooves
Plastifizierkapazität plasticising capacity
Plastifizierleistung plasticising capacity
Plastifiziermaschine plasticator
Plastifiziermenge plasticising capacity
Plastifizierorgan plasticising/plasticating element
Plastifizierphase plasticising time *(im)*
Plastifizierschaft plasticating shaft
Plastifizierschnecke plasticising/plasticating screw
Plastifizierspindel plasticising/plasticating screw
Plastifizierstrom plasticising rate: *On no account should* **-strom** *here be rendered as* flow *or* current *as suggested in Euromap Recommendations No. 5 and No. 10.* **Plastifizerstrom** *differs from* **Plastifizierleistung** *in that the former is the actual amount of material plasticised in a given time, expressed in g/s, the latter the maximum amount of material a machine can plasticise in a given time, expressed in* kg/h
Plastifizierstrombedarf required plasticising rate
Plastifiziersystem plasticising/plasticating system
Plastifizierteil compounding unit/section: **zweiwelliger Plastifizierteil** twin screw compounding unit
Plastifizierung 1. plasticisation, plastication. 2. plasticising/plasticating unit
Plastifizierungsmittel plasticiser
Plastifizierungstempertur 1. plastication temperature. 2. softening point *e.g. in welding*
Plastifizierungsverfahren plasticisation, plastication, plasticising/plasticating process
Plastifizierverhalten plasticising performance
Plastifiziervermögen plasticising performance: *this general term covers* **Plastifizierstrom** *(q.v.) as well as* **Plastifizierleistung** *(q.v.)*
Plastifiziervolumen plasticising capacity/rate
Plastifiziervorgang plasticisation, plastication, plasticising/plasticating process
Plastifizierzeit plasticising time *(im)*
Plastifizierzone 1. transition section *(of a screw)* 2. plasticising section/unit *(of a machine)*
Plastifizierzylinder 1. plasticising/plasticating cylinder *(im)*. 2. barrel *(e)*
Plastikabfälle plastics waste
Plastikakkumulator melt accumulator *(bm)*
Plastikfolie plastics film/sheeting
Plastilin plasticine
plastisch plastic, soft: **plastischer Kunststoff** polymer melt
plastische Seele fluid centre
Plastisol plastisol, (PVC) paste
Plastisolschicht plastisol/paste coating
Plastizier- *see* **Plastifizier-**

Plastizität plasticity
Plastomer thermoplastic (material)
Plasttechnik plastics technology *(see explanatory note under Plaste)*
Plateausohle platform sole
Platine blank
Platinkatalysator platinum catalyst
platinkatalysiert platinum catalysed
Plättchen platelet
plättchenförmig platelet-like *(e.g. filler particles)*
Platte sheet
Platten, größte lichte Weite zwischen den maximum daylight between platens *(im)*
Plattenablage sheet stacking unit
Plattenabzug sheet take-off (unit)
Plattenanlage sheet extrusion line
plattenartig sheet-like
Plattenauswerfer ejector plate *(im)*
Plattenbahn sheet web
Plattenbreitschlitzdüse sheet die *(e)*
Plattendruck platen pressure *(cm)*
Plattendurchbiegung platen deflection *(im)*
Plattendüse sheet die *(e)*
Plattenelektrode plate electrode
Plattenextrusion sheet extrusion
Plattenextrusionsanlage sheet extrusion line
Plattenextrusionslinie sheet extrusion line
plattenförmiges Halbzeug sheets
Plattenformmaschine sheet thermoforming line
Plattenglättanlage sheet polishing unit
Plattengroßanlage wide sheet extrusion line
Plattengröße platen size *(im)*
Plattenheizung platen heating system, platen heaters
Plattenkopf sheet extrusion head, sheet die
Plattenlaufwerk disk drive *(me)*
Plattenmaterial sheets
Plattenpaket mould plate assembly *(im)*
Plattenpresse platen press *(cm)*
Plattenspeicher magnetic disk, disk storage device *(me) (see explanatory note under* **Magnetplattenspeicher***)*
Plattenspeicher, optischer optic(al) disk *(me)*
Plattenstraße sheet extrusion line
Plattentemperatur platen temperature *(cm)*
Plattenware sheets
Plattenwärmeaustauscher plate heat exchanger
Plattenwerkzeug 1. sheet die *(e)*. 2. multi-plate/-part mould, multi-daylight mould *(im)*
Plattenziehen sheet calendering
Plattenzuschnitt cut-to-size piece of sheet, cut blank *(t)*
platzaufwendig requiring a lot of space
Platzbedarf 1. (amount of) space required. 2. floor space *(required to install a machine)*: **geringer Platzbedarf** small space requirements
platzintensiv requiring a lot of space
platzsparend requiring little space, space saving
Plausibilitätskontrolle plausibility check
Pleuel connecting rod

Pleuelstange connecting rod
plissiert pleated *(mf)*
Plissierung pleating *(mf)*
Plotter plotter *(me)*
Plotterausdruck plotter printout *(me)*
PMK *abbr. of* **Pigmentmassenkonzentration**, pigment mass concentration, p.m.c.
Pneumatikzylinder pneumatic cylinder
pneumatisch pneumatic
pockennarbig pockmarked
POF *abbr. of* **Polycarbonat optische Faser**, PC/polycarbonate optic(al) fibre
Poisson-Zahl Poisson's ratio
polar polar
Polarisationsmikroskop polarising microscope
Polarisator polariser
polarisierbar polarisable
Polarisierbarkeit polarisability
polarisiert polarised
Polarisierung polarisation
Polarität polarity
polieren to polish
polierfähig capable of being polished: **Kunststoffe sind polierfähig** plastics can be polished
Polierscheibe polishing wheel
Polster melt cushion
Polstermaterial 1. upholstery material. 2. cushioning material *(e.g. foam for packaging)*
Polstermöbel upholstered furniture
Polsterregelung cushion control *(im)*
Polsterverhalten cushioning properties
Polteppich pile carpet
polumschaltbar pole-reversing/-changing
Polyacetal polyacetal, POM
Polyacetylen polyacetylene
Polyacrylamid polyacrylamide
Polyacrylat polyacrylate
Polyacrylatbasis, auf polyacrylate-based
Polyacrylatharz acrylic resin
Polyacrylester polyacrylate
Polyacrylnitril polyacrylonitrile
Polyacrylsäure polyacrylic acid
Polyacrylsäureester polyacrylate
Polyaddition polyaddition
Polyadditionsharz polyaddition resin
Polyadditionsprodukt polyaddition product
Polyadditionsreaktion polyaddition reaction
Polyaddukt polyadduct
Polyadipat polyadipate
Polyadipinsäureester polyadipate
Polyalkohol polyalcohol
Polyalkylenoxidmethacrylat polyalkylene oxide methacrylate
Polyalkylenterephthalat polyalkylene terephthalate
Polyalkyloxygruppe polyalkyloxy group
Polyamid polyamide, nylon, PA
Polyamidblock polyamide block
Polyamidfasergewebe nylon fabric
Polyamidharz polyamide resin
Polyamidimid polyamide imide

Polyamidkette polyamide chain
Polyamidkord nylon cord
Polyamidwiederholungseinheit polyamide repeat unit
Polyamin polyamine
Polyaminhärter polyamine hardener/catalyst
Polyaminoamid polyaminoamide
Polyaminoamidhärter polyaminoamide hardener
Polyanilin polyaniline
Polyaramid polyaramide
Polyarylamid polyarylamide
Polyarylat polyarylate
Polyarylether polyaryl ether
Polyaryletherketon polyaryl ether ketone, PAEK
Polyarylsulfon polyaryl sulphone
Polyäth- *see* **Polyeth-**
Polyazomethin polyazomethine
Polybenzimidazol polybenzimide azol
Polybeton polymer concrete
Polybismaleinimidharz poly-bis-maleinimide resin
polybromiert polybrominated
Polybutadien polybutadiene
Polybutadien-Acrylnitrilcopolymerisat butadiene-acrylonitrile copolymer/rubber
Polybutadienkautschuk polybutadiene rubber
Polybutadienöl polybutadiene oil
Polybuten polybutene
Polybutylenterephthalat polybutylene terephthalate, PBTP
Polybutylthiophen polybutyl thiophen
Polybutyltitanat polybutyl titanate
Polycaprolactam polycaprolactam
Polycarbonat polycarbonate, PC
Polycarbonsäure polycarboxylic acid
Polycarbonsäureanhydrid polycarboxylic (acid) anhydride
Polychlorbutadien polychlorobutadiene, polychloroprene
Polychlorbutadienkautschuk polychlorobutadiene rubber
polychloriert polychlorinated
Polychloropren polychloroprene
Polychloroprenkautschuk polychloroprene rubber
Polychlortrifluorethylen polychlorotrifluoroethylene
polycyklisch polycyclic
Polydialkylsiloxankette polydialkyl siloxane chain
Polydiallylphthalat polydiallyl phthalate
Polydicyclopentadien polydicyclopentadiene
Polydien polydiene
Polydimethylsiloxan polydimethyl siloxane
Polydiorganosiloxan polydiorganosiloxane
polydispers polydisperse
Polyederschaum macrofoam, foam consisting of polyhedral cells
polyedrisch polyhedral
Polyelektrolyt polyelectrolyte

Polyen polyene
Polyepoxid polyepoxide
Polyester polyester
Polyesteracrylat polyester acrylate
Polyesteramid polyester amide
Polyesteransatz polyester mix/formulation
Polyesterarmierung polyester reinforcement
Polyesterbasis, auf polyester-based
Polyesterbaustein polyester unit
Polyesterbeton polyester concrete
Polyestercarbonat polyester carbonate
Polyesterfasergewebe polyester fabric
Polyesterfaservlies non-woven polyester fabric
Polyestergarn polyester yarn
Polyestergewebe polyster fabric
Polyesterglykol polyester glycol
Polyesterharz polyester resin
Polyesterharzform polyester mould
Polyesterharzformmasse polyester moulding compound
Polyesterharzformstoff cured polyester resin
Polyesterharzlack polyester lacquer/varnish
Polyesterharzmatte polyester prepreg, polyester sheet moulding compound
Polyesterharzpreßmasse polyester moulding compound
Polyesterimid polyester imide
Polyesterimidharz polyester imide resin
Polyesterketon polyester ketone
Polyesterkord polyester cord
Polyesterkunststoff polyester resin
Polyesterpolyol polyester polyol
Polyesterpreßmasse polyester moulding compound
Polyesterreaktionsharz polyester resin, UP resin
Polyesterrezeptur polyester formulation
Polyesterurethankautschuk polyester urethane rubber
Polyether polyether
Polyetheracrylat polyether acrylate
Polyetheramid polyether amide
Polyetheramin polyether amine
Polyetherblock polyether block
Polyetherblockamid polyether block amide
Polyetherester polyether ester
Polyetheretherketon polyetherether ketone, PEEK
Polyetheretherketonketon polyether ether ketone ketone, PEEKK
Polyetherglykol polyether glycol
Polyetherimid polyether imide, PEI
Polyetherketon polyether ketone, PEK
Polyetherketonetherketonketon polyether ketone ether ketone ketone, PEKEKK
polyethermodifiziert polyether modified
Polyetherpolycarbonat polyether polycarbonate
Polyetherpolyol polyether polyol
Polyethersegment polyether segment
Polyethersiloxan polyether siloxane
Polyethersulfon polyether sulphone, PES

Polyethertriacrylat polyether triacrylate
Polyetherurethan polyether urethane
Polyetherurethankautschuk polyether urethane rubber
polyethoxyliert polyethoxylated
Polyethylen hart high density polyethylene, HDPE
Polyethylen hoher Dichte high density polyethylene, HDPE
Polyethylen niedriger Dichte low density polyethylene, LDPE
Polyethylen weich low density polyethylene, LDPE
Polyethylene polyethylene, polythene, PE *(the second version is not used in the USA)*
Polyethylenformmasse polyethylene moulding compound
Polyethylenglykol polyethylene glycol
Polyethylenkette polyethylene chain
Polyethylenoxid polyethylene oxide
Polyethylenschaumstoff polyethylene foam
Polyethylenschlauchfolie blown polyethylene film
Polyethylenterephthalat poleythylene terephthalate, PETP, PET
Polyethylenwachs polyethylene wax
Polyfluoracrylat polyfluoroacrylate
Polyfluoralkylen polyfluoroalkylene
Polyfluorethylenpropylen polyfluoroethylene propylene
Polyfluorsiloxan polyfluorosiloxane
polyfunktionell polyfunctional
Polyglycerylcitrat polyglyceryl citrate
Polyglyceryltartrat polyglyceryl tartrate
Polyglycidylurethan polyglycidyl urethane
Polyglykolfettsäureester polyglycol fatty acid ester
Polyglykolterephthalsäureester polyglycol terephthalate
Polyharnstoff polyurea
Polyharnstoffamid polyurea amide
Polyhexafluorpropylen polyhexafluoropropylene
Polyhydrazid polyhydrazide
Polyhydroxybenzoat polyhydroxybenzoate
Polyhydroxybutyrat polyhydroxybutyrate
Polyhydroxypolyether polyhydroxypolyether
Polyhydroxystearinsäure polyhydroxystearic acid
Polyhydroxyvalerat polyhydroxyvalerate
Polyhydroxyverbindung polyhydroxyl compound
Polyimid polyimide, PI
Polyisobutylen polyisobutylene
Polyisocyanat polyisocyanate
Polyisocyanatharz polyisocyanate resin
polyisocyanatvernetzbar polyisocyanate-curing
polyisocyanatvernetzt polyisocyanate-cured
Polyisocyanurat polyisocyanurate
Polyisopren polyisoprene
Polykarbamid polyurea
Polykation polycation

Polykondensat polycondensate
Polykondensatharz polycondensation resin
Polykondensation polycondensation
Polykondensationsharz polycondensation resin
Polykondensationsprodukt polycondensation product
polykonjugiert polyconjugated
polykristallin polycrystalline
Polylacton polylactone
Polylaurinlactam polylaurin lactam
Polymaleinat polymaleinate
Polymer polymer
polymer polymeric
Polymer, flüssigkristallines liquid crystal polymer
Polymerabbau polymer degradation
Polymerakku all-plastics (car) battery
Polymerakkumulator all-plastics (car) battery
Polymeranalytik polymer analysis
Polymeranalytiker polymer analyst
polymeranalytisch *relating to polymer analysis*: **ein weiteres polymeranalytisches Werkzeug...** another tool/technique for analysing polymers..
Polymerbatterie all-plastics (car) battery
polymerbeschichtet polymer coated
Polymerbeton polymer concrete
Polymerbildungsreaktion polymerisation reaction
Polymerdispersion polymer dispersion
Polymerenschädigung polymer damage
Polymerforschung polymer research
Polymergemisch polymer blend/alloy, polyblend
Polymergerüst polymer framework
Polymergruppe polymer group, group of polymers
Polymerharz (polymer) resin
Polymerhauptkette main polymer chain
Polymerisat polymer
Polymerisatdispersion polymer dispersion
Polymerisatgemisch polymer blend/alloy, polyblend
Polymerisation polymerisation
Polymerisationsabbruch polymerisation termination
Polymerisationsanlage polymerisation plant
Polymerisationsbedingungen polymerising conditions
polymerisationsfähig polymerisable
Polymerisationsgeschwindigkeit polymerisation rate, rate of polymerisation
Polymerisationsgrad degree of polymerisation; **...ist ein Niederdruckpolyethylen mit sehr hohem Polymerisationsgrad** ...is a high density polyethylene with a very high molecular weight *(the higher the degree of polymerisation, the higher the molecular weight - hence the apparent departure from the original in this particlar case)*
polymerisationshemmend polymerisation-inhibiting
Polymerisationshilfsmittel polymerisation aid

Polymerisationsinitiator polymerisation initiator
Polymerisationskessel polymerising vat
Polymerisationsreaktion polymerisation reaction
Polymerisationsstarter polymerisation initiator
Polymerisationstechnik method of polymerisation
Polymerisationstemperatur polymerisation temperature
Polymerisationsumsatz percentage polymerisation
Polymerisationsverfahren polymerisation (process)
Polymerisationsverlauf polymerisation process: **Abbildung 2 zeigt den Einfluß der Temperatur auf den Polymerisationsverlauf** Fig. 2 shows the effect of temperature on the polymerisation process
polymerisationsverzögernd polymerisation-retarding
Polymerisationswärme heat of polymerisation
Polymerisatlösung polymer solution
Polymerisatplatte plastics sheet
Polymerisatschmelze polymer melt
polymerisierbar polymerisable
polymerisieren to polymerise
Polymer-Keramikhybrid(e) polymer-ceramic hybrid
Polymerkette polymer chain
Polymerklasse polymer group, group of polymers
Polymerknäuel polymer tangle
Polymerkonzentration polymer concentration
Polymerlatex polymer dispersion
Polymerlegierung polymer blend/alloy, polyblend
Polymerlösung polymer solution
Polymermatrix polymer matrix
Polymermischung polymer blend/alloy, polyblend
polymermodifiziert polymer modified
Polymermolekül polymer molecule
Polymermörtel polymer mortar
Polymernetzwerk polymer network
polymeroptische Faser optic(al) fibre
Polymerphase polymer phase
Polymerpulver polymer powder
Polymerradikal polymer radical
Polymerschmelze polymer melt
Polymerstruktur polymer structure
Polymertechnik polymer technology
Polymerteilchen polymer particle
Polymerüberzug polymer coating
polymerveredelt polymer/resin coated
Polymerverschnitt polymer blend/alloy, polyblend
Polymerweichmacher polymeric plasticiser
Polymerwerkstoff polymer material/substance
Polymethacrylamid polymethacrylamide
Polymethacrylat polymethacrylate
Polymethacrylatharz acrylic/polymethacrylate resin

Polymethacrylimid polymethacrylimide, PMI
Polymethacrylnitril polymethacrylonitrile
Polymethacrylssäureester polymethacrylate
Polymethylenbrücke polymethylene bridge
Polymethylenharnstoff polymethylene urea
Polymethylenkette polymethylene chain
Polymethylmethacrylat polymethyl methacrylate, PMMA
Polymethylstyrol polymethyl styrene
Polymethyltrifluorpropylsiloxan polymethyl trifluoropropyl siloxane
Polymilchsäure polylactic acid
Polynomansatz polynomial theorem
Polynorbon polynorbone
polynuklear polynuclear
Polyoctenylkautschuk polyoctenyl rubber
Polyol polyol
Polyolefin polyolefin
Polyolefinformmasse polyolefin moulding compound
Polyolefinkautschuk polyolefin rubber
Polyolefinleichtschaum lightweight polyolefin foam
polyolgehärtet polyol-cured
Polyolhärter polyol hardener/catalyst
Polyolkombination polyol blend
Polyoxyalkylendiglycidylether polyoxyalkylene diglycidyl ether
Polyoxyalkylengruppe polyoxyalkylene group
Polyoxyalkylenmethylsiloxan-Copolymer polyoxyalkylene methyl siloxane copolymer
Polyoxyalkylensiloxan-Copolymer polyoxyalkylene siloxane copolymer
Polyoxymethylen polyoxymethylene, polyacetal, POM
Polyoxymethylmelamin polyoxymethyl melamine
Polyphenylenether polyphenylene ether, PPE
Polyphenyletherharz polyphenylene ether resin
Polyphenylenoxid polyphenylene oxide, PPO
Polyphenylensulfid polyphenylene sulphide, PPS
Polyphenylensulfidsulfon polyphenylene sulphide sulphone, PPSS
Poly-1,4-phenylenvinylen poly-1,4-phenylene vinylene
Polyphenylsiloxan polyphenyl siloxane
Polyphosphorsäure polyphosphoric acid
Polyphthalamid polyphthalamide
Polyphthalocyanin polyphthalocyanine
Poly-p-phenylen poly-p-phenylene
Polypropylen polypropylene, PP
Polypropylenglykol polypropylene glycol
Polypropylenoxid polypropylene oxide
Polypropylenwachs polypropylene wax
Polypyrrol polypyrrole
Polysaccharid polysaccharide
Polysilanverbindung polysilane compound
Polysiloxan polysiloxane
polysiloxanmodifiziert polysiloxane modified
Polystyrol polystyrene, PS

Polystyrol, dehnbares expandable polystyrene
Polystyrolformmasse polystyrene moulding compound
Polystyrolgranulat polystyrene granules
Polystyrolhartschaum expanded polystyrene: *Since this material is never anything but hard there is no need to translate the term literally*
Polystyrolschaum expanded polystyrene
Polystyrolschaumanlage expanded polystyrene production line
Polystyrolschaumkugeln expanded polystyrene beads
Polystyrolschaumplatte expanded polystyrene sheet
Polystyrolschaumstoff expanded polystyrene
Polystyrolschaumstoffverpackung expanded polystyrene pack
Polysulfid polysulphide
Polysulfidbindung polysulphide linkage
Polysulfidharz polysulphide resin
polysulfidisch polysulphide
Polysulfidkautschuk polysulphide rubber
Polysulfon polysulphone, PSU
Polyterephthalat polyterephthalate
Polytetrafluorethylen polytetrafluoroethylene, PTFE
Polytetrahydrofuran polytetrahydrofuran
Polytetramethylenglykol polytetramethylene glycol
Polytitansäurebutylester polybutyl titanate
Polytrifluorchlorethylen polytrifluorochloroethylene
Polytrifluorethylen polytrifluoroethylene
Polytrimethylsilylacetylen polytrimethylsilyl acetylene
Polyurethan polyurethane, PU
Polyurethanhartschaum rigid polyurethane foam
Polyurethanhartschaumstoff rigid polyurethane foam
Polyurethanharz polyurethane resin
Polyurethanintegralschaum polyurethane integral/structural foam
Polyurethanlack polyurethane paint
Polyurethanortsschaum in-situ polyurethane foam
Polyurethanschaumstoff polyurethane foam
Polyurethanstrukturschaumstoff polyurethane structural/integral foam
Polyurethanweichschaum flexible polyurethane foam
Polyvinylacetal polyvinyl acetal
Polyvinylacetat polyvinyl acetate
Polyvinylalkohol polyvinyl alcohol
Polyvinylbutyral polyvinyl butyral
Polyvinylbutyrat polyvinyl butyrate
Polyvinylcarbazol polyvinyl carbazole
Polyvinylchlorid polyvinyl chloride, PVC
Polyvinylchlorid hart rigid/unplasticised polyvinyl chloride, rigid/unplasticised PVC, uPVC

Polyvinylchlorid weich plasticised polyvinyl chloride, plasticised PVC
Polyvinylchloridacetat vinyl chloride-vinyl acetate copolymer
Polyvinylether polyvinyl ether
Polyvinylfluorid polyvinyl fluoride
Polyvinylformal polyvinyl formal
Polyvinylidenchlorid polyvinylidene chloride
Polyvinylidenfluorid polyvinylidene fluoride, PVDF
Polyvinylisobutylether polyvinyl isobutyl ether
Polyvinylstearylcarbamat polyvinyl stearyl carbamate
Polyvinyltoluol polyvinyl toluene
Pore 1. cell *(of foam)*. 2. pore *(general term)*
Porenbeton expanded/aerated/foamed concrete
Porenbild cell structure
Porendichte pore density
porenfrei pore-free
Porenfreiheit absence of pores
Porenfüller primer sealer, grain filler
Porengröße cell size
Porengrößenverteilung cell size distribution
Poreninhalt cell content
Porenstruktur cell structure
Porensuchgerät pore detector
porös porous
Porosität porosity
Portalbauweise, in gantry-type *(machine)*
Porzellan porcelain
Porzellankugelmühle porcelain ball mill
Positioniereinrichtung positioning device
Positioniergenauigkeit accurate positioning: **eine recht große Positioniergenauigkeit des Aggregats kann so erreicht werden** the unit can thus be positioned very accurately
Positionierung positioning
Positionierungsbaugruppe positioning module *(me)*
Positionierungslöcher locating/guide holes
positionsgenau accurately positioned
positiv positive
Positivabguß positive impression
Positivschablone positive template
Positivwerkzeug male tool *(t)*
Positron positron
Potentialdifferenz potential difference
potentialfrei zero-potential
Potentiometer potentiometer
Potentiometrisch potentiometric
Potenzansatz power law
Potenzfließgesetz power law
Potenzgesetz power law
Potenzgesetzexponent power law index
Potenzgesetzflüssigkeit power law fluid
Potenzgesetzschmelze power law melt
Potenzgesetzstoff power law fluid *(i.e. a fluid exhibiting power law flow)*
Potenzgesetzverhalten 1. power law flow *(if referring to a polymer melt)*. 2. power law behaviour/characteristics *(general term)*
Potenzzahl power index

Pourpoint pour point
PP-Co *abbr. of* polypropylene copolymer
PPD *abbr. of* **Paraphenylendiamin**, p-phenylene diamine
PP-EM *abbr. of* **elastomermodifiziertes Polypropylen**, elastomer-modified polypropylene
PP-Recyclat polypropylene reclaim, recycled polypropylene
PPS 1. *abbr. of* **Polyphenylensulfid**, polyphenylene sulphide 2. *abbr. of* **Produktions-Planung und Steuerung**, production planning and control
PPS-System production planning and control system
Prägeanlage embossing unit
Prägedruck embossing pressure
Prägeeinrichtung embossing unit
Prägekalander embossing calender
Prägemaschine embossing machine
Prägen 1. embossing, stamping. 2. compression *(if the word is used to describe the second stage of* **Spritzprägen** *(q.v.) as in Kunststoffe 69,(1979)5, p.254, fig.22)*
Prägepresse embossing press
Prägespalt embossing nip
Prägespritzen injection-compression moulding
Prägevorrichtung embossing unit/device
Prägewalze embossing roller
Prägewerk embossing unit
Prägewerkzeug embossing tool
Pralinen chocolates
Pralldämpfer bumper
Prallmahlanlage impact grinder
Prallmühle impact mill/grinder *(sr)*
Prallplatte baffle plate
Prallscheibenmühle impact mill grinder *(sr)*
Prandtl-Zahl Prandtl number
Präpolymer prepolymer
Praxisbedingungen practical conditions
praxisbewährt proven
Praxiserfahrung practical experience
Praxisergebnisse practical results
praxisgerecht practice-oriented, meeting practical requirements/conditions
praxisnah *resembling actual practice*: **...sind erst nach Prüfung unter praxisnahen Bedingungen möglich** ...are possible only after testing under simulated service/operating conditions
praxisrelevant important for practical use: **praxisrelevante Eigenschaften** properties of practical importance/signifance
Praxisrelevanz practical importance
Praxistest practical test
praxisüblich *as in actual practice* **unter praxisüblichem Verarbeitungsbedingungen** under normal processing conditions
Praxisverhalten behaviour/performance under practical conditions, behaviour in practice: **Die Faktoren die das Praxisverhalten der Formteile bestimmen...** the factors which

determine how the mouldings will behave in practice OR: behave under practical conditions
Praxisversuch practical test
präzipitiert precipitated
präzis precise, accurate
Präzision precision, accuracy
Präzisionsarbeit precision work
Präzisionsdosiereinheit precision metering unit
Präzisionsform precision mould
Präzisionsnormteil precision standard unit
Präzisionsschleifen precision grinding
Präzisionsschneidmühle precision pelletiser *(sr)*
Präzisionsspritzgießverarbeitung precision injection moulding
Präzisionsspritzguß precision injection moulding
Präzisionsspritzgußteil precision injection moulded part, precision injection moulding
Präzisionsteil precision moulding, precision moulded part
Präzisionsteilfertigung precision moulding
Präzisionswerkzeug 1. precision mould *(im)*. 2. precision die *(e)*
Präzisionswickeltechnik precision filament winding *(grp)*
Preisanhebung price increase
Preisdruck downward pressure on prices
Preis-Durchsatzverhältnis cost-performance ratio
Preiserhöhung price increase
Preisgefüge price structure
preisgünstig inexpensive, low-cost, reasonably priced, economical
Preis-Lebensdauerrelation cost-service life ratio/factor
Preis-Leistungsindex cost-performance factor
Preis-Leistungsverhältnis cost-performance ratio/factor
Preisnachlaß price reduction
Preisniveau price (level)
Preisreduzierung price reduction
Preissteigerung price increase
Preisvorteil cost advantage
preiswert inexpensive, low-cost
Preis-Wirkungsrelation cost-benefit ratio
preiswürdig inexpensive, low-cost, reasonably priced
Preiswürdigkeit reasonable/low cost
Prepolymer pre-polymer
Prepreggewebe prepreg
Prepregreste prepreg waste/scrap
Prepregverfahren prepreg moulding
Preßartikel press mouldings, press moulded articles
Preßautomat automatic compression moulding machine
Preßbedingungen press operating conditions
Preßblech press platen
Preßdauer pressing time

Preßdruck 1. moulding pressure. 2. laminating pressure. 3. bonding pressure
Presse press
Presse, ölhydraulische hydraulic press
Pressen compression moulding
Pressen, isostatisches isostatic moulding *(ptfe)*
Pressenöffnung daylight
Pressenplatte press platen
Pressentisch press table
Pressenvulkanisation press vulcanisation
Presserei moulding shop
Presseurwalze backing roll
Preßformen compression moulding
Preßgrat flash
Preßholz compressed/densified wood
Preßkraft moulding pressure
Preßkuchen filter cake
Preßlaminat press-moulded laminate
Preßling moulding *(made by compression or transfer moulding, as opposed to* **Spritzling** *(q.v.)*
Preßluft compressed air
preßluftbetätigt pneumatically operated
Preßluftnetz compressed air supply
Preßluftstrahl compressed air jet
Preßluftstrom compressed air jet
Preßmasse 1. moulding compound, thermoset moulding compound: **Formteile aus Preßmassen** thermoset mouldings. 2. compression moulding compound: *The second alternative of version 1. should only be used where the word occurs on its own, e.g. as a column heading in a table. If it occurs in association with another word, the shorter alternative should be used, e.g.* **Polyester-Preßmasse** *should be translated as* polyester moulding compound
Preßmethode pressing procedure
Preßpaket stack of prepregs
Preßplatte 1. press platen *(cm)*. 2. pressed sheet
Preßrecken compressive stretching
Preßsitz press-fit
Preßspan presspahn
Preßspritzen transfer moulding
Preßstoff cured resin, moulded thermoset resin
Preßteil moulding *(see explanatory note under* **Preßling**)
Preßtemperatur moulding temperature
Preßverfahren 1. compression moulding. 2. press moulding
Preßvulkanisation press vulcanisation
Preßwerkzeug 1. compression mould. 2. press tool
Preßzeit press time
Preßzyklus moulding cycle
primär primary
Primärabzug main/primary take-off (unit)
Primärachse primary axis
Primärelektronik main electronics system
Primärextruder main/principal extruder
Primärkorn primary particle

Primärkühlung main cooling system
Primärpartikel primary particle
Primärradikal primary radical
Primärstabilisator main/primary stabiliser
Primärstruktur primary structure
Primärteilchen primary particle
Primärweichmacher primary plasticiser
Primat priority, most important consideration/factor
primerlackiert primer coated, coated with primer
Primerlackierung application of primer, coating with primer
Primern priming, application of a primer
Primerschicht primer coat
Primerstrich primer coat
Primerung priming, application of a primer
Prinzipskizze schematic drawing
prismenförmig prismatic, prism-shaped
Probe 1. sample. 2. test piece/specimen
Probeabspritzungen moulding trials
Probehalterung test piece support/clamp
Probekörper test piece/specimen
Probekörpergeometrie test piece dimensions/shape
Probekörperhalterung test piece support/clamp
Probekörpermitte test piece centre
Probekörperoberfläche test piece surface
Probekörpertemperatur test piece temperature
Probekörpervolumen test piece volume
Probenabmessungen test piece dimensions
Probenaufbereitung test piece preparation
Probendimensionen test piece dimensions
Probeneinlegevorrichtung test piece insertion device
Probenmengen sample quantities/amounts
Probenträger test piece support
Proberaum test chamber
Probespritzungen moulding trials
Probestab test piece/specimen
Probestreifen test strip
Problematik problems
problembehaftet problematic(al)
problemlos without any problem
problemorientiert problem-oriented *(me)*
produktabhängig product-dependent, depending on the product
Produktalternative alternative product
Produktbereinigung streamlining of product range
Produkteigenschaften end product properties
Produktenreihe range of products, product range
Produktentwicklung 1. product development 2. product development department
Produktgestaltung product design
Produktgruppe product group
Produktionsabfälle in-plant waste, plant scrap
Produktionsablauf production cycle
Produktionsabschnitt production stage/phase: **...mit dem jeder Produktionsabschnitt überprüft werden kann** ...with which each stage of the manufacturing process can be checked
Produktionsanforderungen production requirements
Produktionsanlage production line
Produktionsaufgaben production tasks
Produktionsausfall production downtime
Produktionsaustrittseite *place where the product leaves the machine:* **auf der Produktionsaustrittseite des Extruders** at the die end
Produktionsbedingungen production/manufacturing conditions
Produktionsbeginn start of production
Produktionsbereich production area *(of a factory)*
Produktionscharge production batch
Produktionsdaten production data
Produktionsdatenerfassung 1. production data collection/capture/acquisition. 2. production data acquisition/collecting unit
Produktionsdatenerfassungsgerät production data collecting/acquisition unit
Produktionsdatenprotokoll production data record
Produktionsdatenverarbeitung production data processing
Produktionsdatum production/manufacturing date
Produktionsdrehzahl production (screw) speed
Produktionseinheit production unit
Produktionserfordernisse production requirements
Produktionserhöhung increase in production, production increase
Produktionsextruder production extruder
Produktionsfluß production flow
Produktionsgeschwindigkeit production speed
Produktionshalle production shed
Produktionskapazität production capacity
Produktionskneter production compounder
Produktionskonstanz constant production conditions
Produktionskontrolle production control
Produktionskosten production/manufacturing costs
Produktionskreislauf production cycle
Produktionsleistung production capacity
Produktionsleiter production manager
Produktionsleitrechner central production computer *(me)*
Produktionslinie production line
Produktionslos production batch
Produktionsmaschine 1. production-scale machine. 2. machine
Produktionsmaßstab production scale
Produktionsmengen production-scale quantities
Produktionsmittel machine *(literally:* means of production, *although this should not be used)*
Produktionspalette product range, range of products

Produktionsplanungs- und Steuerungs-System production planning and control system
Produktionsprogramm 1. product range. 2. production schedule/plan/programme
Produktionsprozeß production/manufacturing process
Produktionsrate production rate
Produktionsrückstände reject mouldings, scrap (material)
Produktionsschema flow diagram
Produktionsschreiber production logger
Produktionsschritt production stage
Produktionsserien, große long runs
Produktionsserien, kleine short runs
Produktionsstandort production site
Produktionsstätte 1. plant, factory. 2. moulding shop. 3. extrusion shop
Produktionssteuerer production controller
Produktionsstillstand break in production
Produktionsstörung break in production, production breakdown
Produktionsstraße production line
Produktionsstunden production hours
Produktionsteilzeichnung part drawing, moulded-part drawing
Produktionsübersicht production overview
Produktionsüberwachung production control
Produktionsunterbrechung break in production, production breakdown
Produktionsunterbruch break in production
Produktionsverhältnisse production/manufacturing conditions
Produktionsversuch production trial
Produktionsvolumen production volume
Produktionswerkzeug 1. production mould *(im)*. 2. production die *(e)*
Produktionswerte production data *(e.g. film gauge and width, frost line temperature etc.)*
Produktionszahlen production figures
Produktionszahlen, kleine short runs
Produktionsziffern production figures
Produktionszustand production status
Produktionszuwachs production increase, increase in production
Produktionszyklus production cycle
Produktivität productivity
Produktivitätssteigerung productivity increase
Produktivzeit productive time *(of a machine)*
Produktkategorie product group
Produktkennzahl product constant/property
Produktklasse product group
Produktkonstanz product uniformity: **Reaktionsbedingungen zur Erzielung der Produktkonstanz** reaction conditions to achieve a uniform product
Produktnorm product specification
Produktpalette product range, range of products
Produktprogramm product range, range of products
Produktrückgewinnung product recovery

Produktschwankungen product variations
Produktsortiment product group, group of products, product range
Produktspezifikation product specification
produktspezifisch product-oriented
Produktstrom 1. material flow. 2. melt stream
Produktvielfalt multitude of products: **Das Problem der großen Produktvielfalt...** The problem is that there are so many different products...
Produzent producer, manufacturer
Profil profile
Profilabschnitt profile section
Profilabzug profile take-off (unit) *(e)*
Profilanlage profile extrusion line
Profilaufweitung die swell *(e)*
Profildüse profile die *(e)*
Profilextrusionsstraße profile extrusion line
profilgebend this word expressses the idea of shaping, usually of extrudates but not necessarily of profiles, and cannot be translated literally. The word can, in fact, often be omitted as in the following example: **profilgebendes Werkzeug** die; **profilgebendes Werkzeug** should not, incidentally, be confused with **Profilwerkzeug** *(q.v.)*
Profilkalibrierung 1. profile calibration/sizing. 2. profile calibrator, profile calibrating/sizing unit *(e)*
Profilkopf profile die *(e)*
Profilmundstück profile die *(e)*
Profilnachfolge profile extrusion downstream equipment
Profilschneidmühle profile scrap granulator
Profilschweißautomat automatic profile welding unit
Profilspritzkopf profile die *(e)*
Profilstempel profile punch
Profilstrang profile
Profilstraße profile extrusion line
Profilwerkzeug profile die *(e)*
Profilziehverfahren pultrusion *(grp)*
progammierbar programmable *(me)*
Prognose prognosis, forecast
prognostiziert forecast, predicted
Programm 1. program *(me)*: Because of the exclusively American origin of computer language, the American spelling of this word has been adopted in the UK. 2. range *(of products)*
Programm, fremdbezogenes ready-made program *(me)*
Programm, käufliches ready-made program *(me)*
Programmablauf program operation
Programmabwandlungen program modifications *(me)*
Programmarchivierung program archiving *(me)*
Programmausdruck program print-out *(me)*
Programmbaustein program module *(me)*
Programmbefehl program instruction *(me)*

Programmbibliothek program library *(me)*
Programmdokumentation program documentation *(e)*
Programmeingabe program input *(me)*
Programmentwicklung program development *(me)*
Programmerstellung program preparation *(me)*
Programmgeber programmer *(me)*
programmgesteuert program-controlled *(me)*
Programmierarbeitsplatz programming workstation *(me)*
Programmierbarkeit programmability *(me)*
Programmierbaustein programming module *(me)*
Programmiereinheit programming unit *(me)*
programmieren to program *(me)*
Programmiergerät programming instrument, programmer *(me)*
Programmiersprache programming language *(me)*
Programmierung programming *(me)*
Programmierungsmodul programming module *(me)*
Programmimplementierung program implementation *(me)*
Programmodul program module *(me)*
Programmpaket program package *(me)*
Programmpflege program maintenance *(me)*
Programmregler program controller, program control unit *(me)*
Programmschritt program step *(me)*
Programmspeicher program store *(me)*
Programmspeicherkapazität program storage capacity *(me)*
Programmspeicherung 1. program store. 2. program storage *(me)*:
Programmspeicherung auf Kassette ist möglich programs may be stored on cassettes
Programmsprache program/programming language *(me)*
Programmsteuerung 1. program control. 2. program control unit *(me)*
Programmtaster program key *(me)*
Programmwahl program selection *(me)*
Programmwahlschalter program selection switch *(me)*
Programmwahltaste program selection key *(me)*
Programmzentraleinheit central program unit *(me)*
Progressivspindel increasing pitch screw
Projektierung configuring *(me)*
Projektierungsstadium planning stage
Projektionsfläche projected area
Projektphase planning stage
Pro-Kopf-Verbrauch per capita consumption
Propionaldehyd propionaldehyde
Propionsäure propionic acid
Proportionalhydraulik proportional hydraulics, proportional hydraulic system
Proportionalitätsgrenze proportionality limit

Proportionalitätskonstante proportionality constant
Proportionalmagnet proportional solenoid
Proportionalstellventil proportional control valve
Proportionalventil proportional valve
Propylen propylene
Propylenglykol propylene glycol
Propylenglykolether propylene glycol ether
Propylenglykolmaleat propylene glycol maleate
Propylenglykolmonoalkylether propylene glycol monoalkyl ether
Propylenglykolmonomethylether propylene glycol methyl ether
Propylenkette propylene chain
Propylenoxid propylene oxide
Propylenoxidkautschuk propylene oxide rubber
Protokoll 1. record, log. 2. print-out, hard copy *(me)*
Protokollausdruck hard copy print-out *(me)*
Protokolldrucker hard copy printer *(me)*
Protokollerstellung production of hard copy *(me)*
protokollierbar capable of being printed out *(me*
protokollieren 1. to document, to record, to log. 2. to print-out *(me)*
Protokollierung 1. documentation, recording, logging. 2. print-out *(me)*
Protokollschreibmaschine console typewriter *(me)*
Proton proton
protoniert protonised
Protonierung protonisation
Prototyp prototype
Prototypenfertigung prototype production
Prototypwerkzeug 1. prototype mould *(im)*. 2. prototype die *(e)*
Provenienz origin *(of materials)*
Prozentanteil percentage (amount)
Prozentpunkte percentage points
Prozentsatz percentage (amount)
prozentualer Umsatz 1. percentage conversion *(e.g. of monomer into polymer)*. 2. percentage turnover
Prozentzahl percentage
Prozeßablauf process, course of the process
Prozeßablauf, zyklischer moulding cycle
Prozeßabschnitt process segment *(me)*
Prozeßanalyse process analysis
Prozeßautomatisierung process automation
Prozeßbedingungen process conditions
Prozeßbeherrschung process management *(me)*
Prozeßdaten process data/variables *(me)*
Prozeßdatenausdruck process data printout *(me)*
Prozeßdatenausgang process data output *(me)*
Prozeßdatenerfassung 1. process data collecting/acquisition. 2. process data collecting/acquisition unit *(me)*

Prozeßdatenüberwachung 1. process data control/monitoring. 2. process data control unit, process data monitor *(me)*
Prozeßeinheit, zentrale central processing unit *(me)*
Prozeßfähigkeit process efficiency
prozeßführende Parameter processing conditions
Prozeßführung process control
Prozeßführungseinrichtung process control system
Prozeßführungsinstrument process control unit
Prozeßführungssystem process control system
prozeßgeführt process controlled
prozeßgeregelt 1. process controlled *(general term)*. 2. closed-loop process controlled: **prozeßgeregelte Maschine** machine equipped with closed-loop process controls; Version 2. should be used when the text distinguishes between machines which are **prozeßgesteuert** *(q.v.)* and **prozeßgeregelt**. See also explanatory note under **regeln**
prozeßgesteuert 1. process controlled *(general term)*. 2. open-loop process controlled: Version 2. should be used when the text distinguishes between machines which are **prozeßgesteuert** and **prozeßgeregelt**. See also explanatory note under **regeln**
Prozeßgröße process parameter/variable
Prozeßkenngröße process parameter/variable
Prozeßkontrolle 1. process control. 2. process control unit
Prozeßleitsystem process control system
Prozeßmodell process model
Prozeßoptimierung upgrading a process
Prozessor central processing unit, CPU, computer *(me)*
prozessorgesteuert computer controlled
Prozeßparameter process parameter/variable
Prozeßrechner process computer
Prozeßregelung 1. process control *(general term)*. 2. closed-loop process control *(see explanatory notes under* **prozeßgeregelt** *and* **regeln**)
prozeßrelevant process-related
Prozeßschema flow fiagram
Prozeßschritte process stages
Prozeßsimulation process simulation
Prozeßsteueranlage process control equipment
Prozeßsteuerung 1. process control *(general term)*. 2. open-loop process control *(see explanatory notes under* **prozeßgesteuert** *and* **regeln**)
Prozeßüberwachung 1. process monitoring. 2. process monitor, process monitoring unit/system
Prozeßvariablen process variables
Prozeßwarte control desk/panel/console
Prozeßzustand process status
Prüfablauf test sequence
Prüfanordnung test set-up

Prüfanstalt, amtliche official test establishment
Prüfapparatur test apparatus
Prüfbedingungen test conditions
Prüfbericht test report
Prüfbestimmung test specification
Prüfdaten test data
Prüfdauer test duration
Prüfdruck test pressure
Prüfeinrichtung test apparatus, test(ing) equipment
prüfen to test, to check
Prüfergebnis test result
Prüffehler test error
Prüffeld test bed
Prüffläche test surface
Prüfflüssigkeit test liquid
Prüfformulierung test formulation
Prüfgelände test bed/site
Prüfgerät test instrument/apparatus
Prüfgeschwindigkeit testing speed
Prüfklima test atmosphere
Prüfkörper test piece/specimen
Prüfkörperentlastung removal of stress from the test piece
Prüfkraft applied load
Prüflehre test jig
Prüfling test piece/specimen
Prüflösung test solution
Prüfmaschine test apparatus
Prüfmedium test medium
Prüfmethode test method/procedure
Prüfmethodik test method/procedure
Prüfmischung test compound
Prüfmittel 1. test medium *(if liquid)*. 2. test atmosphere *(if gaseous)*. 3. test substance *(if solid)*
Prüfparameter test parameter
Prüfprogramm test programme
Prüfresultat test result
Prüfrezeptur test formulation
Prüfrichtlinie test guideline
Prüfsieb test sieve
Prüfsiebgewebe test filter cloth
Prüfspannung 1. test voltage. 2. test stress
Prüfstab test piece/specimen
Prüfstand test bed/rig
Prüfstandversuch test bed trial
Prüfstreifen test strip
Prüfstück test piece/specimen
Prüfsubstanz test substance, substance under test
Prüftemperatur test temperature
Prüfumfang test scope
Prüfung, qualitätssichernde quality assurance test
Prüfungs- *see* **Prüf-**
Prüfverfahren test method/procedure
Prüfvorschrift test specification
Prüfvulkanisat test vulcanisate
Prüfwerkzeug 1. experimental mould *(im)*. 2. experimental die *(e)*
Prüfwert test figure

Prüfzeit test period/duration
Prüfzeitraum test period/duration
Prüfzeugnis test certificate
Prüfzyklus test cycle
PSA *abbr. of* **Phthalsäureanhydrid**, phthalic anhydride
pseudoplastisch pseudoplastic
Pseudoplastizität pseudoplasticity, non-Newtonian flow
PS-O *abbr. of* **orientiertes Polystyrol**, oriented polystyrene
PTC *abbr. of* permanent temperature control
Pufferwirkung buffer effect/action
Pufferzone buffer zone
Pulsation surging, pulsation *(e)*
pulsationsfrei surge-free *(e)*
Pulsationsfrequenz surge/pulsation frequency
Pulsieren surging, pulsation *(e)*
pulsierend 1. intermittent *(e.g. blasts of air)*. 2. surging, pulsating *(e)*
pulverbeschichtet powder coated
Pulverbeschichtung powder coating
Pulverchemikalien powdered chemicals
Pulvercompound dry blend
Pulverextruder dry blend extruder
Pulverextrusion dry blend extrusion
Pulverfarbe powder coating (material)
Pulverfliesenkleber powdered tile adhesive
Pulverform, in powdered
pulverförmig powdered coating (material)
Pulverfüllstoff powdered filler
Pulvergesamtmenge total amount of powder
Pulverharz powdered resin
pulverisieren to pulverise
Pulverkleber powdered adhesive
Pulverkorn particle
Pulverlack powder coating (material)
Pulverlackierung powder coating
Pulvermischung dry blend
Pulverruß powdered carbon black
Pulverschnecke dry blend screw *(e)*
Pulverteilchen particle
Pulververarbeitung dry blend processing
pumpbar pumpable
pumpen 1. to pump. 2. to surge, to pulsate *(e)*
Pumpenaggregat pump unit
Pumpenantriebsleistung pump drive power
Pumpenantriebsmotor pump motor
Pumpendrehzahl pump speed
Pumpenfördermenge amount conveyed by the pump
Pumpenkreis pump circuit
Pumpenleistung pump capacity
Pumpenmotor pump motor
Pumpenverdrängerraum pump displacement chamber
Pumperscheinungen surging, pulsation *(e)*
Pumpzone metering section/zone *(e)*
Punktanguß pin gate *(im)*
Punktanguß mit Vorkammer antechamber-type pin gate *(im)*
Punktangußdüse pin gate nozzle *(im)*

Punktangußkegel pin gate sprue *(im)*
Punktanschnitt pin gate *(im)*
Punktanschnitt, Tunnelanguß mit tunnel gate with pin point feed *(im)*
Punktanschnittdüse pin gate nozzle *(im)*
Punktanschnittkanal pin gate *(im)*
punktförmig point-like
punktförmiger Anschnittkanal pin gate *(im)*
punktiert dotted *(curve)*
Punktkleben spot bonding
Punktschweißen spot welding
Punktüberheizung local overheating
Punktverschweißen spot welding
Puppe 1. dolly *(c)*: *This is a rolled-up piece of hot compound cut from milled strip for passing to the calender feed.* 2. doll
PUR-Hartschaum rigid PU/polyurethane foam
PUR-Integralhartschaum rigid integral/structural PU foam
PUR-Integralschaum integral/structural PU foam
PUR-Kleber polyurethane adhesive
PUR-Lack polyurethane lacquer/varnish
PUR-Reaktionswerkstoff PU/polyurethane material
PUR-Weichschaum flexible PU/polyurethane foam
Putz plaster, rendering (coat), rendering mix
Putzfläche plaster surface
Putzmörtel rendering mix
PVB-Folie polyvinyl butyral film
PVB-Harz polyvinyl butyral resin
PVC hart rigid/unplasticised PVC, uPVC
PVC weich plasticised PVC
PVC-Abfall scrap PVC, PVC scrap
PVC-Bodenreste PVC floorcovering waste
PVC-fremd non-PVC **PVC-fremde Anwendungen** non-PVC applications
PVC-Granulat PVC pellets
PVC-Hartfolie rigid PVC film
PVC-Hartschaumstoff rigid PVC foam
PVC-Klarpaste clear PVC paste
PVC-Mischung PVC compound
PVC-Paste PVC paste
PVC-Pastenbeschichtung PVC paste coating
PVC-Pastentyp PVC paste resin, paste-making PVC
PVC-Plastisol PVC plastisol, PVC paste
PVC-Pulverkorn PVC particle
PVC-Regenerat PVC reclaim, recycled PVC
PVC-Schnecke PVC screw *(i.e. screw designed for extruding PVC)*
PVC-Teilchen PVC particle
PVC-Weichfolie flexible PVC film/sheeting
PVC-Weichschaumstoff flexible PVC foam
PVK *abbr. of* **Pigment-Volumen-Konzentration**, pigment volume concentration, p.v.c.
Pyknometer pyknometer
pyramidenförmig pyramid-shaped
Pyridin pyridine
Pyridinkautschuk pyridine rubber

pyroelektrisch pyroelectric
pyrogen pyrogenic
Pyrogramm pyrogram
Pyrolyse pyrolysis
Pyrolysefragment pyrolysis fraction
Pyrolyseprodukt pyrolysis product
pyrolysieren to pyrolyse
pyrolytisch pyrolytic
Pyromellitsäure pyromellitic acid
Pyromellitsäureanhydrid pyromellitic anhydride

Q

Q *symbol for* silicone rubber
QM *abbr. of* **Quarzmehl**, silica flour
QS *abbr. of* **Qualitätssicherung**, quality assurance
QS-System quality assurance system
quaderförmig square
quadratisch square
Quadratmetergewicht weight per square meter, weight/m²
Quadratwurzel square root
qualitativ qualitative
qualitativ hochwertig high quality
Qualitätsabweichungen quality deviations
Qualitätsanalyse quality analysis
Qualitätsanforderungen quality requirements
Qualitätsansprüche quality requirements
Qualitätsbeurteilung quality assessment: **Für die Qualitätsbeurteilung von Profilen...** to assess profile quality...
Qualitätsbewußtsein quality awareness
qualitätsbewußt quality-conscious
Qualitätsdokumentation quality record
Qualitätseinbuße loss of quality
Qualitätsgarantie guaranteed quality
qualitätsgerecht high-quality
qualitätskonstant of constant quality
Qualitätskonstanz constant quality **verbesserte Qualitätskonstanz** more constant quality
Qualitätskontinuität constant/consistent quality
Qualitätskontrolle quality control
Qualitätskriterien quality criteria
Qualitätslenkung quality control
Qualitätsmaßstab quality standard
Qualitätsmerkmal quality characteristic
Qualitätsminderung reduction/drop in quality
Qualitätsniveau (standard of) quality
qualitätsorientiert quality-orientated
Qualitätsplanung quality planning
Qualitätsprüfung quality testing/control
Qualitätsrichtlinien quality guidelines
Qualitätsschwankungen quality variations, variations in quality
Qualitätssicherheit quality assurance

qualitätssichernd *relating to quality assurance*: **qualitätssichernde Prüfung** quality assurance test
Qualitätssicherung quality assurance
Qualitätssicherungsprogramm quality assurance programme
Qualitätsspritzguß precision injection moulding
Qualitätsstandard standard of quality
qualitätssteigernd quality-enhancing
Qualitätssteigerung improvement in quality
Qualitätsstrategie quality policy
Qualitätsüberprüfung quality check, check on quality
Qualitätsüberwachung quality control
Qualitätsüberwachungspaket quality control package
Qualitätsverantwortung responsibility for quality
qualitätsverbessernd quality-enhancing
Qualitätsverbesserung improvement in quality
Quanteneffekt quantum effect
Quantifizierung quantification
quantitativ quantitative
quartär quaternary
Quarz quartz
Quarzfeinstmehl extremely fine particle silica flour
Quarzgut synthetic silica
Quarzgutmehl synthetic silica flour
Quarzkristallaufnehmer quartz transducer
Quarzkristalldruckaufnehmer quartz pressure transducer
Quarzkristallkraftaufnehmer quartz force transducer
Quarzkristallmeßwertaufnehmer quartz transducer
Quarzmehl silica flour
Quarzpulver powdered silica
Quarzsand silica sand
Quarzschiffchen silica combustion boat
Quarztiegel silica crucible
quasielastisch quasi-elastic
quasiisotrop quasi-isotropic
quasistatisch quasi-static
quaternär quaternary
Quecksilberdampfhochdrucklampe high pressure mercury vapour lamp
Quecksilberdampflampe mercury vapour lamp
Quecksilbersäule mercury column
Quellbeständigkeit swelling resistance
Quelldehnung swelling
Quellenangabe indication of source
Quellfaktor die swell *(e)*
Quellfließen frontal/laminar flow
Quellfluß frontal/laminar flow
Quellgrad amount of swelling
Quellmittel swelling agent
Quellschweißen solvent welding
Quellschweißmittel welding solvent
Quellschweißnaht solvent weld
Quellströmung frontal/laminar flow
Quellung swelling

Quellung an der Düse die swell *(e)*
Quellung beim Spritzen die swell *(e)*
Quellungsbeständigkeit resistance to swelling
Quellungsvorgang swelling process
Quellverhalten swelling behaviour
Quellvermögen swelling power
Quellversuch swelling test
Quellwirkung swelling effect
quer transverse(ly)
Querbelastbarkeit transverse strength
Querdehnung transverse expansion
Querfestigkeit transverse strength
Querkontraktion transverse shrinkage
Querkopf crosshead (die) *(e)*
Querkraft transverse force
Quermischung transverse mixing
Querorientierung transverse orientation
Querrichtung transverse direction
Querscherung transverse shear
Querschliff microtome section
Querschneidemaschine transverse cutter
Querschnitt 1. cross-section, diameter. 2. flow channel. 3. runner *(im)*. 4. transverse section
Querschnittsänderungen changes in diameter. *For translation example see under* **Querschnittsübergänge**
Querschnittsfläche cross-sectional area
Querschnittsübergänge changes in diameter: **Die Fließkanäle eines Profilwerkzeugs sind mit nur allmählichen Querschnittsübergängen zu gestalten** the flow channels of a profile die should be designed with only gradually changing diameters
Querschnittsveränderungen changes in diameter *(for translation example see under* **Querschnittsübergänge***)*
Querschnittsverengungen 1. narrowing cross-sections. 2. narrowing flow channels. 3. narrowing runners *(im)*
Querschrumpf transverse shrinkage
Querschwindung transverse shrinkage
Querspannung transverse stress
Querspritzkopf crosshead (die) *(e)*
Querstift transverse pin
Querströmung transverse flow
Quertrenneinrichtung transverse cutter
querverstreckt transversely stretched/oriented *(see explanatory note under* **Recken***)*
Querverstreckung transverse stretching/orientation *(see explanatory note under* **Recken***)*
Querverteiler hot runner unit, hot runner manifold block *(im)*
Querwellmaschine transverse corrugating machine
Querzugbeanspruchung transverse tensile stress
Querzugfestigkeit transverse tensile strength
Querzugspannung transverse tensile stress

∗**Quetschkanten** pinch-off/nip-off edges *(bm)*
Quetschnaht pinch-off weld *(bm)*
Quetschspalt nip *(bfe,c)*
Quetschspannung compressive yield stress
Quetschtasche flash chamber *(bm)*
Quetschwalzen nip/pinch rolls *(bfe)*
Quetschwalzenpaar nip/pinch rolls *(bfe)*
Quetschzone pinch-off area *(bm)*
Quietschgeräusch squeaking noise
quittieren to acknowledge *(me)*
Quotient quotient, ratio

R

Rad wheel
Radblende hub cap
Radfelge (wheel) rim
Radhaus wheel arch
radial radial
Radialkolbenmotor radial piston motor
Radialkolbenpumpe radial piston pump
Radialspiel radial clearance
Radialsteghalter spider, spider-type mandrel support *(e)*
Radikal radical
Radikalangebot radical availability
Radikalangriff radical attack
Radikalausbeute radical yield
radikalbildend radical-forming
Radikalbildner radical former
Radikalbildung radical formation
Radikaleinfang radical interception
Radikalfänger radical interceptor
radikalisch radical
Radikalkette free-radical chain
Radikalkettenmechanismus free-radical chain growth mechanism
Radikalkettenpolymerisation free-radical (chain growth) polymerisation
Radikalkonzentration radical concentration
Radikalpolymerisation free-radical polymerisation
radikalspendend radical-donating
Radikalspender radical donor
radioaktiv radioactive
radiographisch radiographic
Radioindustrie audio sector/industry
Radiometrie radiometry
radiometrisch radiometric
Radius radius
Radkappe hub cap
Radkasten wheel arch
Radlagerdichtung wheel bearing seal
Radlauf wheel race

∗ *Words starting with* **Quetsch-**, *not found here, may be found under* **Abquetsch-**

Radnabe hub
Radom radome
Radstand wheel base
Radzierblende hub cap
Raffination refining
Raffinerie refinery
Rahmen frame
Rahmenbauweise framework construction
Rahmengestell frame
Rahmenkonfektionierung window frame assembly
Rahmenrezeptur starting formulation
Rahmenrichtlinien general guidelines
Rakel doctor knife
Rakelauftrag knife application
Rakelblatt (doctor) knife blade
Rakelmesser doctor knife
Rakelmesserauftrag knife application
Rakeln knife application
Ramextruder ram extruder
Ramextrusion ram extrusion
RAM-Speicher random access memory *(me)*
Rand edge, rim
Randbedingungen boundary conditions
Randbeschneidung 1. edge trimmer, edge trimming unit. 2. trimming the edge *(of film or sheeting)*
Randbeschnitt 1. edge trim, edge trimmings. 2. edge trimmer, edge trimming unit. 3. trimming the edges *(of film or sheeting)*
Randfaserbereich outer fibre zone
Randfaserdehnung outer fibre strain
Randfaserspannung outer fibre stress
randnah near the edge
Randomcopolymerisat random copolymer
Randschicht outer layer
randschichtbehandelt surface treated
Randschichtbehandlung surface treatment
Randschnittabsaugung 1. edge trim removal 2. edge trim removal unit
Randstreifen edge trim, edge trimmings
Randstreifenabfall edge trim waste
Randstreifenabsauganlage edge trim removal unit
Randstreifenaufwicklung edge trim wind-up (unit)
Randstreifenbeschneidstation edge trimming unit
Randstreifenbeschnitt 1. edge trim, edge trimmings. 2. edge trimmer, edge trimming unit. 3. trimming the edges *(of film or sheeting)*: **Die Schwierigkeiten können umgangen werden, wenn der Randstreifenbeschnitt am noch plastischen Material erfolgt** these difficulties can be avoided by trimming the edges whilst the material is still plastic
Randstreifenrezirkulierung 1. edge trim recycling *(for translation example see under* **Randstreifenrückspeisung**). 2. edge trim recycling unit/system

Randstreifenrückführanlage edge trim recycling unit
Randstreifenrückführsystem edge trim recycling unit/system
Randstreifenrückführung 1. edge trim recycling *(for translation example see under* **Randstreifenrückspeisung**). 2. edge trim recycling unit/system
Randstreifenrückspeiseeinrichtung edge trim recycling unit
Randstreifenrückspeisung 1. edge trim recycling: **die unmittelbare Randstreifenrückspeisung in den Extruder** the immediate return of edge trims to the extruder
Randstreifenschneidvorrichtung edge trimmer, edge trimming unit
Randstreifenzerhacker edge trim shredder
Randwinkel angle of contact
Randwülste edge build-up *(resulting from uneven reeling of film)*
Randzone outer zone
rasch aushärtbar fast curing
Raschelgewirke raschel-knit fabric *(txt)*
Raschelmaschine raschel machine *(txt)*
raschhärtend fast curing
raschlaufend 1. high speed *(when describing a continuous process such as extrusion, calendering, etc.)*. 2. fast cycling *(when describing an intermittent process such as injection or blow moulding)*
raschplastifizierend fast gelling
Rasterelektronenmikroskop scanning electron microscope
Rasterelektronenmikroskopaufnahme scanning electron micrograph
Rasterelektronenmikroskopie scanning electron microscopy
rasterelektronenmikroskopische Aufnahme scanning electron micrograph
Rasterwalze halftone roller
rationalisieren to rationalise
Rationalisierung rationalisation, streamlining
Rationalisierungsgründe economic considerations
Rationalisierungsmaßnahmen economy measures
rationell efficient, economical
Rauch smoke, flue gas
Rauchdetektor smoke detector
Rauchdichte smoke density
Rauchgas flue gas
Rauchgasdichte smoke density
Rauchgasentschwefelungsanlage flue gas desulphurisation plant
Rauchgaswäscher flue gas scrubber
Rauchgaswert smoke density
Rauchmelder smoke detector
rauchunterdrückend smoke-suppressant
rauh 1. rough 2. severe, extreme *(conditions)*

Rauhigkeit 1. roughness: **bei hohen Rauhigkeiten** if the surface is very rough. 2. roughness height
Rauhtiefe roughness height
Raumbedarfsplan 1. space requirements. 2. machine dimensions *(e.g. in a diagram of a machine with all dimensions indicated)*
Raumdichte density
Räume, fließtote dead spots
Raumfahrt 1. space travel. 2. aerospace industry
Raumfahrtindustrie aerospace industry
Raumfahrttechnik aerospace engineering
Raumfahrzeug spacecraft
Raumgewicht density
Raumkapsel space capsule
Raumklima room temperature
räumlich spatial, three-dimensional
Raumsonde space probe
raumsparend space saving
Raumstruktur three-dimensional structure
Raumteile parts by volume
Raumtemperatur room temperature
Raumtemperaturbereich room temperature
raumtemperaturhärtend room temperature curing
Raumtemperaturhärtung room temperature curing
Raumtemperaturlagerung room temperature ageing
Raumtemperaturtrocknung room temperature drying **für Raumtemperaturtrocknung** for drying at room temperature
raumtemperaturvernetzend room temperature vulcanising/curing
Raumtemperaturvernetzung room temperature vulcanisation/curing
Raupenabzug caterpillar take-off (unit)
rautenförmig diamond-shaped, rhombic
Rautenmischteil faceted mixing section *(of screw) (e)*
Reagenzglas test tube
Reagglomeration re-agglomeration
Reaktand reactant
Reaktion erster Ordnung first order reaction
Reaktion zweiter Ordnung second order reaction
Reaktionsbedingungen reaction conditions
Reaktionsbereitschaft reactivity
reaktionsbeschleunigend reaction-accelerating: **Alkohole wirken reaktionsbeschleunigend, Ketone hingegen reaktionshemmend auf das Epoxidharzsystem** alcohols tend to accelerate curing of epoxy systems, ketones on the other hand to retard it
Reaktionsdauer reaction time
Reaktionsenthalpie reaction enthalpy
reaktionsfähig reactive
Reaktionsfähigkeit reactivity
reaktionsfreudig reactive
Reaktionsgefäß reaction vessel

Reaktionsgemisch reaction mixture
Reaktionsgeschwindigkeit reaction rate, speed of reaction
Reaktionsgießen reaction injection moulding, RIM
Reaktionsgießharz reactive casting resin
Reaktionsgießmaschine reaction injection moulding machine, RIM machine
Reaktionsgießverarbeitung reaction injection moulding, RIM
Reaktionsgleichung equation, chemical equation: **chemische Reaktionsgleichung** chemical equation
Reaktionsguß reaction injection moulding, RIM
Reaktionsharz 1. reactive resin *(a generally applicable version)*. 2. epoxy/unsaturated polyester/methacrylate/ isocyanate resin. *(to be used when the text is specifically devoted to one or more of these resins)*. 3. casting/laminating/impregnating resin: *(to be used when a text deals with any one or all three processes)*. Since epoxies and polyester resins are the most commonly encountered **Reaktionsharze,** it will usually suffice to use the names of either or both these resins in a translation, as in this example: **Glasfaserverstärkte Kunststoffe sind Verbundwerkstoffe deren Hauptkomponenten Textilglasfasern und Reaktionsharze bilden** glass fibre reinforced plastics are composites based mainly on synthetic resins such as epoxies and unsaturated polyesters, and glass fibre. *Where the word occurs in the title of an article, the translation will depend on the subject.* **Reationsharzgebundene Produkte in der Elektrotechnik**, *for example, should be rendered as* The use of expoxy resins in electrical engineering *since these are by far the most widely used resins in this field. In fact, the article in question dealt exclusively with epoxies.*
Reaktionsharzbeton polymer concrete
Reaktionsharzblend thermoset blend
Reaktionsharzformstoff cured casting resin
Reaktionsharzmasse catalysed resin, resin-catalyst mix
Reaktionsharzmischung catalysed resin, resin-catalyst mix
reaktionshemmend reaction-retarding *(for translation example see under* **reaktionsbeschleunigend***)*
Reaktionsinhibitor reaction inhibitor
Reaktionskinetik reaction kinetics
reaktionskinetisch reaction-kinetic
Reaktionskleber two-pack adhesive
Reaktionsklebstoff two-pack adhesive
Reaktionskomponente reaction component
Reaktionskunststoff 1. reactive resin *(a generally applicable version)*. 2. epoxy/polyester/methacrylate/isocyanate resin *(to be used when a text is specifically devoted to one*

or more of these resins). 3. casting/laminating/ impregnating resin *(to be used when a text deals with any one or all three of these processes). For translation example see under* **Reaktionsharz.**
Reaktionslack catalysed lacquer, two-pack lacquer
Reaktionslacksystem catalysed surface coating system
Reaktionsmaschine reactor
Reaktionsmasse catalysed resin, resin-catalyst mix
Reaktionsmechanismus reaction mechanism
Reaktionsmedium reaction medium
Reaktionsmilieu reaction environment
Reaktionsmittel 1. hardener, catalyst. 2. accelerator: *The English version of this word given in DIN 16945 part 1* (reactant) *has a totally different connotation and to use it here would be wrong as well as misleading*
Reaktionsparameter reaction conditions
Reaktionspartner reactant
Reaktionsprodukt reaction product
Reaktionsschaumgießverfahren reaction foam moulding (process)
Reaktionsschema general reaction
reaktionsschnell fast-reacting
Reaktionsschritte reaction stages
reaktionsspritzgegossen reaction injection moulded
Reaktionsspritzgießen reaction injection moulding, RIM
Reaktionstemperatur reaction temperature
reaktionsträg low-reactivity
reaktionsträger less reactive
Reaktionsverzögerung delaying the reaction
Reaktionswärme heat of reaction
Reaktionswerkstoff thermoset (material)
Reaktionszeit 1. reaction time. 2. vulcanising/ cure time. 3. response time *(me)*
reaktiv reactive
reaktiv härtend chemically curing
Reaktivität reactivity
Reaktivitätsunterschiede reactivity differences, differences in reactivity
Reaktivivitätsminderung reduction of reactivity
Reaktivlöser reactive solvent
Reaktivverdünner reactive thinner
Reaktivweichmacher reactive plasticiser
Reaktor reactor
Reaktorausstoß reactor output
Reaktorkopf top of the reactor
Realisierbarkeit feasibility
realisieren to realise, to put into practice
Realzeit real time *(me)*
Realzeiteigenschaften real-time properties *(me)*
Realzeitkommunikation real-time communication *(me)*
Realzeitrechner real-time computer *(me)*
Realzeitübertragung real-time transfer/ transmission *(me)*
Rechenanlage computer installation/system

Rechenansatz mathematical equation
Rechenaufwand number of calculations needed/required
Rechenbeispiel mathematical example
Rechenchip chip *(me)*
Recheneinheit computer *(me)*
Rechenformel mathematical formula
Rechenmethode mathematical method
Rechenmodell mathematical model
Rechenprogramm computer program *(me)*
Rechenschieber slide rule
Rechenwerk arithmetic unit *(me)*
Rechenwerk, zentrales central processing unit *(me)*
Rechenwert calculated value
Rechenzeitbedarf time needed for calculations
Rechenzentrum computer centre *(me)*
Rechner computer *(me)*
rechneransteuerbar computer controlled
Rechnerausdruck computer print-out
Rechnereinsatz use of computers
rechnergeführt computer controlled
rechnergesteuert computer controlled
rechnergestützt computer aided
rechnerisch 1. calculated: **rechnerisches Spritzvolumen** calculated shot volume. 2. mathematical
Rechnerkarte computer card
Rechnernetz computer network
Rechnerprogramm computer program *(me)*
Rechnersteuerung computer control
rechnerunterstützt computer aided *(me)*
rechnerunterstütztes Konstruieren computer- aided design, CAD
rechnerunterstütztes Planen computer aided planning, CAP
rechnerunterstütztes Zeichnen computer aided drafting, CAD
Rechnervernetzung computer networking
Rechteckdüse rectangular die *(e)*
Rechteckfeder rectangular spring
rechteckförmig rectangular
rechteckig rectangular, right-angled
Rechteckkanal 1. rectangular runner *(im)*. 2. rectangular flow channel *(general term)*
Rechteckkerbe rectangular notch
Rechteckprofildüse rectangular profile die *(e)*
Rechteckstab rectangular test piece/specimen
Rechteckstreifen rectangular strip
Rechtsgewinde right-hand thread
Rechtsvorschriften legal provisions, legislation
rechtwinklig rectangular, right-angled
Reckanlage stretching unit
Recken stretching, drawing *(of film; synonymous with* **Strecken** *and* **Verstrecken**): *Since the purpose of stretching a film is to orient it, the verb* **recken** *and its synonyms can also be rendered as* to orient. **Gestreckte Folie** *can thus be translated as* stretched *or as* oriented film
Recken, biaxiales biaxial stretching/orienting *(of film)*

Reckgeschwindigkeit stretching rate
Reckmaschine stretching unit
Recktemperatur stretching temperature
Reckverhältnis stretch/draw ratio
Reckwalze stretching roll
recycelfähig recyclable
Recyceln recycling
Recyclat recycled material(s)
Recyclatteile parts/components made from recycled material
Recycler recycling firm/contractor
recyclieren to recycle
Recycling, biologisches biological recycling
Recycling, chemisches chemical recycling
Recycling, energetisches energy recycling/recovery
Recycling, materielles material recycling
Recycling, stoffliches material recycling
Recycling, thermisches energy recovery/recycling
Recyclingbetrieb recycling firm/contractor
recyclingfähig recyclable
Recyclingfähigkeit recyclability
Recyclingfasern recycled fibres
recyclingfeindlich difficult to recycle
Recyclingfirma recycling firm/contractor
recyclingfreundlich easy to recycle, easily recycled
Recyclingfreundlichkeit ease of recycling
Recyclingfüllstoff recycled filler
Recyclinggerät recycling unit
recyclinggerecht easy to recycle
Recyclingglasfasern recycled glass fibres
Recyclinggranulat recycled moulding compound
Recyclingharzmatte recycled SMC
Recyclingkonzept recycling concept
Recyclingmaterial recycled material
Recyclingmöglichkeit recyclability, recycling option/possibility
Recyclingpotential recycling potential
Recyclingprodukte recycled products
Recyclingquote recycling quota
Recyclingstoffe recycled materials
Recyclingsystem recycling system
Recyclingunternehmen recycling firm/contractor
Recyclingverfahren recycling process
Recyclingvorgang recycling process
Recyclingware recycled products
Recyclisieren recycling
redispergierbar redispersible
Redispergierbarkeit redispersibility
redispergieren to redisperse
Redispersionspulver redispersible powder
Redoxinitiator redox initiator
Redoxpotential redox potential
Redoxsystem redox system
Reduktion reduction
Reduktionsmittel reducing agent
reduzierend reducing
Reduziergetriebe reducing gear

reduzierte Viskosität reduced viscosity, viscosity number
Reduzierung reduction, lowering
reemulgiert re-emulsified
Referenzdruckbereich reference pressure range
Referenzgröße reference quantity
Referenzmuster reference sample
Referenzprobe reference sample
Referenzsubstanz reference substance
Referenztemperatur reference temperature
reflektieren to reflect
Reflektor reflector
Reflexion reflection
Reflockulation re-flocculation
Refraktionsindex refractive index
Refraktometer refractometer
Regelalgorithmus control algorithm
regelbar adjustable
regelbar, stufenlos infinitely/steplessly variable
Regeleinrichtung control unit
Regelelektronik electronic control system, electronic controls
Regelelemente controls
Regelgenauigkeit control accuracy: **große Regelgenauigkeit** very accurate controls
Regelgerät 1. control instrument *(general term)*. 2. closed-loop control instrument *(if the text differentiates between* **Steuer-** *and* **Regelgerät***)*
Regelgröße controlled variable
Regelgüte control accuracy/quality
Regelhydraulik hydraulic controls
Regelinstrument control unit
Regelkreis 1. control circuit *(general term)*. 2. closed-loop control circuit *(if the text differentiates between* **Steuer-** *and* **Regelkreis,** *as in this example)*: **die Steuer- und Regelkreise sind häufig miteinander verknüpft** the open- and closed-loop circuits are often linked
Regelkreis, geschlossener closed-loop control circuit
Regelmagnet control solenoid
regelmäßig regular
Regelmöglichkeiten control options
regeln to control, to regulate: *In English, the word* to control *is generally applied both to* **steuern** *and* **regeln.** *In German however, there is a clear distinction which is explained in DIN 19226. In an article which differentiates between* **Steuern** *and* **Regeln,** *the former must be translated as* open-loop control, *the latter as* closed-loop control *(or* feedback control*). Where no such distinction is made, both words should be rendered as* to control. *It should be remembered that* to regulate *is synonymous with* to control *and that the word does not convey the difference that exists between* **steuern** *and* **regeln.** *To translate* **Steuern und Regeln** - *in cases where it is essential to make a distinction between the*

two words - as control *and* regulation *would be equivalent to saying* control and control. *See also explanatory note under* **Regelung**
Regeln 1. control *(general term)*. 2. closed-loop control, feedback control *(where the text differentiates between* **Steuern** *and* **Regeln**. *See also explanatory note under* **regeln**
Regelpumpe variable delivery pump
Regelschrank control cabinet
regeltechnisch *relating to control*:
 regeltechnisch schwieriger more difficult to control
Regelthermoelement control thermocouple
Regelung 1. closed-loop control, feedback control. 2. closed-loop control system, feedback control system: *Although* **Regelung** *also means* regulation, *this word should be avoided in a plastics machinery context since - especially in its plural form - it is apt to be ambiguous and result in some strange sounding sentences. To speak of* controls and regulations *for example, would raise serious doubts in the reader's mind as to what exactly was meant. See also explanatory note under* **regeln**
Regelung, gesetzliche law, regulation
Regelungen, immissionsschutzrechtliche pollution control regulations, anti-pollution regulations
Regelungsaufgabe control task
Regelungsaufwand *effort required to control a machine*: **ein geringer Regelungsaufwand** easy to control
Regelungsbaugruppe closed-loop control module
Regelungssystem closed-loop control system, feedback control system
Regelventil control valve
Regelverbund composite control system
Regelverstärker control amplifier
Regenabflußrohr downpipe
Regenbekleidung rainwear
Regenerat reclaim, regrind
Regeneratcellulose regenerated cellulose
Regeneratpulver powdered reclaim/regrind
Regeneratzugabe addition of reclaim/regrind
Regeneratzusatz addition of reclaim/regrind
Regenerieranlage reclaim plant
regenerierbar recyclable, recoverable
regenerieren to regenerate, to reclaim, to recycle
Regenerierluft recycled air
Regenfallrohr downpipe
Regenmantel raincoat, mackintosh
Regenperiode rainy season
Regenwald rain forest
Regenwasser rainwater
Regiepult control desk/panel/console
Register register *(me)*
Registriereinrichtung recording device/unit
Registriergerät recording instrument
Registrierung recording, logging *(me)*

Reglementierungen regulations
Regler regulator, controller
Reglerstruktur controller configuration *(me)*
Regranulat reclaim, regrind
Regranulieranlage scrap repelletising line *(sr)*
regranulieren to repelletise *(scrap) (sr)*
Regranulierextruder reclaim extruder
Regressionsanalyse regression analysis
regulierbar adjustable
Reguliereinrichtung control equipment *(plural: controls)*
Reguliersonde control sensor/probe
Regulierung control
Regulierventil control valve
Reibbelag friction lining
reibbeständig rub fast
Reibechtheit rub fastness
Reibeputz scraped finish plaster
reibfest rub fast
Reibgut millbase
Reibkraft frictional force
Reibrad abrasive wheel
Reibradgetriebe friction gear
Reibradverfahren Taber abrasion test
Reibschweißen friction welding
Reibschweißgerät friction welding instrument
Reibschweißmaschine friction welding machine
Reibstuhl roll mill, rolls
Reibtester abrasion tester
Reibung friction
Reibung, gleitende sliding friction
Reibung, innere internal friction, shear
Reibung, trockene dry friction
reibungsabhängig friction-dependent
reibungsarm low-friction
reibungsbedingt due to friction
Reibungsbeiwert coefficient of friction
Reibungsenergie frictional energy/heat
Reibungskoeffizient coefficient of friction
Reibungskraft frictional force
reibungslos 1. trouble-free, smooth, without a hitch. 2. frictionless, without (any) friction
Reibungspartner mating surfaces
Reibungsverhalten frictional behaviour/properties
Reibungsverlust frictional loss
reibungsvermindernd anti-friction
Reibungswärme frictional heat
Reibungswert coefficient of friction
Reibungswiderstand frictional resistance
Reibungszahl coefficient of friction
Reibverhalten frictional behaviour/properties
Reibwert coefficient of friction
Reibwiderstand frictional resistance
Reichweite range
Reifen tyre
reifen to mature
Reifenindustrie tyre industry
Reifenkarkasse tyre carcass
Reifenkord tyre cord
Reifenlauffläche tyre tread

Reifenlaufflächenmischung tyre tread compound
Reifenschaden tyre damage
Reifenseitenwand tyre sidewall
Reifenseitenwandmischung tyre sidewall compound
Reifensektor tyre industry
Reifentrennmittel inside tyre paint *(although this sounds strange, it is, in fact, correct usage)*
Reifevorgang maturing process *(e.g. of a PVC paste)*
Reifezeit maturing period: **Die Reifezeit entfällt** no maturing is necessary
Reifung maturing
Reihe, in in series
Reihenanordnung, in arranged in series
Reihenfolge sequence
Reihenpunktanschnitt multi-point pin gate, multiple pin gate *(im)*
Reihenschaltung series arrangement
Reihenverteiler runners arranged side by side *(im)*
Reingewinn net profit
Reinharz pure resin
Reinharzschicht gel coat *(grp)*
Reinharzstab pure-resin test piece/specimen
Reinheit 1. purity. 2. cleanliness
Reinheitsgrad (degree of) purity
reinigen 1. to clean *(e.g. a surface)* 2. to purify *(e.g. a chemical compound)*
Reinigungsaufwand amount of cleaning: **Der Reinigungsaufwand nach Produktionsende ist sehr gering** very little cleaning is necessary when production has been completed
Reinigungsbad cleaning bath
Reinigungscompound purging compound
Reinigungscreme cleansing cream
Reinigungsflüssigkeit cleaning fluid
reinigungsfreundlich easy to clean
Reinigungsgranulat purging compound
Reinigungsmittel 1. cleaner, cleaning agent/solvent. 2. purging compound *(e, im)*
Reinigungsturm scrubbing tower
Reinigungszeit time required for cleaning
Reinpolymerisat 1. homopolymer. 2. pure polymer
Reinraumbedingungen clean room conditions
reinviskos Newtonian
Reißarbeit fracture energy
Reißbrett drawing board
Reißdehnung elongation at break, ultimate elongation
reißfest tear resistant
Reißfestigkeit tear strength, ultimate tensile strength
Reißfestigkeit, relative relative/residual tear strength
Reißkraft breaking stress, ultimate tensile stress
Reißlack crackle finish
Reißlänge breaking length

Reißlast breaking stress
Reißprüfung tensile test
Reißspannung yield stress
reizend irritant
Reklamationen complaints
Rekombination re-combination
Rekristallisation recrystallisation
Rekristallisationstemperatur recrystallising temperature
Relais relay
Relaisausgang relay output
Relaiskontakt relay contact
Relaissteuerung relay control
Relativbewegung relative movement
Relativdruck relative pressure
relative Dichte relative density, specific gravity
relative Dielektrizitätskonstante relative permittivity, dielectric constant
relative Luftfeuchte relative humidity
relative Reißfestigkeit relative tear strength
relative Viskosität relative viscosity
Relativgeschwindigkeit relative velocity
Relativmessung relative determination
Relaxation relaxation
Relaxationsalgorithmus relaxation algorithm
Relaxationskurve relaxation curve
Relaxationsmodul relaxation modulus
Relaxationsprozeß relaxation process
Relaxationsverhalten relaxation behaviour
Relaxationsvermögen ability to relax
Relaxationsversuch relaxation test
Relaxationsvorgang relaxation process
relaxieren to relax
relevant relevant, appropriate. *Since the word is being increasingly used in place of the less grand sounding word* **wichtig**, *it may sometimes be better to render it as* important.
REM abbr. of **Rasterelektronenmikroskop**, scanning electron microscope
REM-Aufnahme scanning electron micrograph
Renner runner *(im)*
Renovierung renovation
Rentabilität profitability
Reparaturaufwand repair costs
reparaturfreundlich easy to repair
Reparaturkosten repair costs
Reparaturmörtel repair mortar
Repetierbarkeit repeatability
Repetitionsgenauigkeit precise repeatability/reproducibility
Repräsentativbauten public buildings
reproduzierbar reproducible, repeatable
Reproduzierbarkeit reproducibility, repeatability: **hohe Reproduzierbarkeit** excellent reproducibility
Reproduzierfähigkeit reproducibility, repeatability
reproduziergenau precisely reproducible/repeatable
Reproduziergenauigkeit precise repeatability/reproducibility

Reproduziermöglichkeit possibility of reproducing: **Dadurch ergibt sich eine schnelle, exakte Reproduziermöglichkeit der Bedingungen** this means that conditions can be quickly and acccurately reproduced
Reserveradmulde spare wheel recess
Resistenz resistance
Resolharz resol resin
Resonanz resonance
Resonanzfrequenz resonance frequency
Resonanzkurve resonance curve
Resonanzpeak resonance peak
Resonanzspektroskopie resonance spectroscopy
Resorcin resorcinol
Resorcinharz resorcinol resin, resorcinol-formaldehyde resin
Ressort 1. department, section 2. responsibility
ressortfrei without specific responsibilities
Ressortleiter head of department
Ressortverteilung delegation of responsibilities
Ressourcen resources
Ressourcen, stoffliche material resources
Ressourcenschnonung conservation of resources
ressourcenschonend resource-conserving
Rest 1. radical, group. 2. residue
Restaurierung restoration
Restaurierungsarbeiten restoration work
Restbiegefestigkeit retained/residual flexural strength
Restbruchdehnung retained/residual elongation at break, retained/residual ultimate elongation
Restdehnung retained/residual elongation
Restdoppelbindungen residual double bonds
Restdruck residual pressure
Restenergie residual energy
Restentgasung removal of residual gases/volatiles: **Die Restentgasung von flüchtigen Bestandteilen** the removal of residual volatiles
Rest-Epoxidgruppengehalt residual epoxide/epoxy group content
Restethylengehalt residual ethylene content
Restfestigkeit retained/residual strength
Restfeuchte 1. residual moisture. 2. residual moisture content
Restfeuchtegehalt residual moisture content
Restfeuchtigkeit 1. residual moisture. 2. residual moisture content
Restfeuchtigkeitsgehalt residual moisture content
Resthaftkraft residual adhesive force
Restklebrigkeit residual tack
Restmenge residual amount: **Restmengen an Feuchtigkeit** residual moisture
Restmonomer residual monomer
Restmonomerentfernung removal of residual monomer
Restmonomergehalt residual monomer content
Restmüll refuse, waste, rubbish, garbage
Restorientierung residual orientation

Restpendelhärte retained/residual pendulum hardness
Restreißdehnung residual/retained elongation at break, retained/residual ultimate elongation
Restreißfestigkeit retained/residual tear strength
Restreißkraft retained/residual breaking strength, retained/residual ultimate beaking strength
Restsäuregehalt residual acid content
Restspannungen residual stresses
Reststabilität residual stability
Reststauchung residual compression
Reststoff waste
Reststoffrecycling waste recyclng
Reststyrolanteil residual styrene content
Reststyrolgehalt residual styrene content
Restwärme residual heat
Restwasser residual water
Restzugfestigkeit retained/residual tensile strength
Retardation retardation
Retardationsverhalten retardation behaviour
Retardationsversuch 1. retardation test. 2. creep test
Retardationsvorgang retardation process
Retardationszeit retardation period
retardieren to retard
Retardierungs- *see* **Retardations-**
Retention retention
reversibel reversible
reversierend reversing
Reversiersystem reversing system *(bfe)*
Reversion reversion
Reversionsbeständigkeit reversion resistance
Reversionsfestigkeit reversion resistance
Reversionsschutzeffekt anti-reversion effect
Reversionsschutzmittel anti-reversion agent
reversionsstabilisiert reversion stabilised
Revolvereinheit carousel unit, rotary-table unit
Revolvermaschine carousel-type machine, rotary-table machine
Revolverspritzgießautomat automatic carousel-type injection moulding machine, automatic rotary-table injection moulding machine
Revolverspritzgießmaschine carousel-type injection moulding machine, rotary-table injection moulding machine
Revolverteller rotary table
Reynoldzahl Reynolds number
Rezeptbestandteil formulation component
Rezeptdatei formulation data file
rezeptiert formulated
Rezeptierung formulation **Compounds mit spezieller Rezeptierung** specially formulated compounds
Rezeptpalette range of formulations
Rezeptur formulation
rezepturabhängig formulation-dependent, depending on the formulation
Rezepturbestandteil formulation component

Rezepturentwicklung developing a formulation
Rezepturfehler formulation mistake
Rezepturgestaltung developing a formulation
Rezepturhinweis guide formulation
Rezepturkomponente formulation component
Rezepturoptimierung perfecting/upgrading a formulation
Rezepturvorschlag suggested formulation
Rezepturzusammenstellung formulation
Rezepturzusammenstellung formulation
Rezession recession
reziprok reciprocal
Rezirkulieren recirculation
Rezykl-, rezykl- *see* **Recycl-, recycl-**
r.F. *abbr. of* **relative Feuchte**, relative humidity
RFA *abbr. of* **Röntgenfluoreszenz-Analyse**, X-ray fluorescent analysis
RF-Ruß reinforcing furnace black
RG *abbr. of* **Raumgewicht**, density
Rheogoniometer rheogoniometer
Rheologie rheology
Rheologieberechnung rheological calculation
Rheologiebetrachtung rheological calculation/analysis
Rheologiedaten rheological data
rheologisch rheological
Rheometer rheometer
Rheometrie rheometry
rheometrisch rheometric
rheopex rheopectic
Rheopexie rheopexy
rhombusförmig rhomboid
RIC *abbr. of* runnerless injection compression
Richtdosierung approximate amount to be added
Richtgröße guide value; *for translation example see under* **Richtwert**
Richtlinie guideline
Richtpreis guide price
Richtrezeptur starting formulation
Richtungsabbhängigkeit anisotropy, direction dependence
richtungsabhängig direction-dependent
Richtungsänderung change of direction
Richtungsorientierung orientation
Richtwert guide value **Die angegebenen Werte sind als Richtwerte zu betrachten** the figures quoted should be taken as a guide
Ricinenalkydharz dehydrated castor oil alkyd resin
Ricinenöl dehydrated castor oil
Ricinensäure dehydrated castor acid
Ricinusöl castor oil
Riefe groove
Riegelbolzen locking bolt
Riemenscheibe belt pulley
rieselfähig free-flowing
Rieselfähigkeit free-flowing properties
Rieselverhalten free-flowing properties
Riesenmolekül giant molecule
RIM *abbr. of* reaction injection moulding
RIM-Anlage reaction injection moulding plant, RIM plant
RIM-Dosiergerät RIM dispensing unit
RIM-Maschine reaction injection moulding machine, RIM machine
RIM-Mischgerät RIM mixing unit
RIM-Prozeß reaction injection moulding (process), RIM (process)
RIM-Teil reaction injection moulded part, RIM part
RIM-Verarbeitung reaction injection moulding, RIM
RIM-Verfahren reaction injection moulding (process), RIM (process)
RIM-Werkstoff reaction injection moulding compound, RIM compound
RIM-Werkzeug RIM mould
Ring 1. ring 2. ball race *(of a ball bearing)*
Ringanschnitt ring gate *(im)*
Ringbilding ring formation
Ringdüse annular/ring-shaped die *(e)*
ringförmig 1. annular. 2. cyclic
Ringheizelement ring heater *(im)*
Ringkanal 1. circular runner *(im)*. 2. annular groove *(bm)*
Ringkolben tubular ram *(im)*
Ringkolbeninjektion tubular ram injection *(im)*
Ringkolbenspeicher tubular ram accumulator *(bm)*
Ringkolbenspeicherkopf tubular ram accumulator head *(bm)*
Ringlochplatte perforated disc
Ringnut ring-shaped groove
Ringraum annular space
Ringrillendorn mandrel with a ring-shaped groove *(bm)*
ringschließend ring-closing
Ringschlitzdüse annular/ring-shaped die *(e)*
Ringschlitzdüsenwerkzeug annular/ring-shaped die *(e)*
Ringschluß ring closure
Ringspaltdüse annular/ring-shaped die *(e)*
Ringspalt(e) annular slit
Ringspaltöffnung annular slit
Ringspaltwerkzeug annular/ring-shaped die
Ringstruktur ring/cyclic structure
Ringsystem ring/cyclic system
Ringverbindung cyclic compound
Ringverteiler ring-type distributor *(bfe)*
Ringwendelverteiler ring-type spiral distributor *(bfe)*
Rippentorpedo ribbed torpedo
rippenversteift rib-reinforced
Risiko risk
risikobereit ready/prepared to take risks
Risikobereitschaft readiness to take risks
 geringe Risikobereitschaft reluctance to take risks
risikofreudig ready/prepared to take risks
Risikofreudigkeit readiness to take risks *(for translation example see under* **Risikobereitschaft***)*
Riß crack

Rißanfälligkeit susceptibility to cracking
Rißausbreitung crack propagation
Rißausbreitungsgeschwindigkeit crack growth/propagation rate
Rißausbreitungskraft crack propagation force
Rißausbreitungswiderstand crack propagation resistance
rißauslösend crack-initiating
Rißauslösung crack initiation
Rißbildung cracking, checking, crazing
Rißbildungsneigung tendency to crack
Rißbildungsresistenz crack resistance
Rißerweiterungsgeschwindigkeit crack growth/propagation rate
Rißfläche crack surface
Rißfortpflanzung crack propagation
Rißfortpflanzungsgeschwindigkeit crack propagation rate
Rißfortpflanzungsversuch crack propagation test
Rißfortschritt crack growth/propagation
Rißfront crack front
rißfüllend crack/gap filling
Rißgeschwindigkeit crack growth propagation rate
Rißinitiierung crack initiation
Rißlänge crack length
Rißrichtung crack direction
Rißspitze crack tip
rißüberbrückend crack bridging
Rißüberbrückung crack bridging
Rißvergrößerung crack growth
Rißverlängerung crack growth
Rißwachstum crack growth/propagation
Rißwachstumsgeschwindigkeit crack growth/propagation rate
Rißwachstumsrate crack growth/propagation rate
Rißwachstumsrichtung crack growth direction
Rißwiderstand crack resistance
Rißzähigkeit fracture toughness/resistance
Ritzbeanspruchung scratching
Ritzel pinion
Ritzfestigkeit scratch resistance
Ritzhärte scratch hardness
Ritzwerkzeug scratching tool/device
Rizinusöl castor oil
Rizinusölfettsäure castor oil fatty acid
RLF abbr. of **relative Luftfeuchtigkeit**, relative humidity
Roboter robot
Robotgerät robot
Robotsystem robotic system
robust rugged, ruggedly constructed, sturdy
Robustheit ruggedness, sturdiness
Rockwellhärte Rockwell hardness
Rohbau shell *(of a vehicle)*
Rohdichte density
Rohfestigkeit green strength
Rohkarosserie shell *(of a vehicle)*
Rohkautschuk crude rubber
Rohlaufstreifen tread rubber, camelback

Rohling preform, blank
Rohmaterial 1. raw material 2. moulding compound
Rohmaterialersparnis raw material savings: **Aus Gründen der Rohmaterialersparnis** to save raw material
Rohmaterialfilter melt filtration unit
Rohmaterialkosten raw material costs
Rohmischung unvulcanised (rubber) mix
Rohöl crude oil
Roh-PVC PVC resin/polymer
Rohr pipe
Rohrabfälle pipe scrap
Rohrabschnitt pipe section
Rohrabzug pipe take-off (unit)
Rohrabzugswerk pipe take-off (unit)
Rohranlage 1. pipe extrusion line. 2. pipe production line *(general term) (see explanatory note under* **Rohrfertigungsstraße**; *the word should never be translated as pipe line which would cause confusion since this refers to something entirely different*
Rohraußendurchmesser outside pipe diameter
Rohrbeschichtungswerkzeug pipe sheathing die *(e)*: *This is a die used to extrude a polyethylene sleeve on to a steel pipe*
Rohrbogen pipe bend
Rohrdüsenkopf pipe die *(e)*
Röhrenverdampfer tubular vaporiser
Rohrextrusionsanlage pipe extrusion line
Rohrextrusionswerkzeug pipe die *(e)*
Rohrfabrikation 1. pipe production. 2. pipe production line
Rohrfertigungsstraße 1. pipe extrusion line. 2. pipe production line *(if the word is obviously intended to include other types of pipe making equipment, e.g. filament winding lines)*
Rohrfitting pipe fitting
rohrförmig tubular
Rohrgranulat pipe extrusion compound
Rohrgüte pipe quality
Rohrheizkörper tubular heater
Rohrherstellungsanlage 1. pipe extrusion line. 2. pipe production line *(general term; see explanatory note under* **Rohrfertigungsstraße***)*
Rohrinnendruckversuch long-term/sustained internal pressure test for pipes
Rohrinnendurchmesser inside pipe diameter
Rohrinnenfläche pipe bore
Rohrinnenwand inside pipe wall
Rohrkalibrierung 1. pipe calibration/sizing *(e)*. 2. pipe calibrating/sizing unit *(e)*
Rohrkopf pipe die head, pipe extrusion head, pipe die
Rohrkrümmer pipe bend
Rohrleitung pipeline
Rohrleitungsdämmung pipeline insulation (material)
Rohrleitungssystem pipeline system
Rohrleitungsteil pipe fitting

Rohrlinie 1. pipe extrusion line. 2. pipe production line *(see explanatory note under* **Rohrfertigungsstraße***)*
Rohrofen tubular furnace
Rohrprobe pipe specimen
Rohrprüfling pipe specimen
Rohrrezeptur pipe (compound) formulation
Rohrschneidmühle pipe scrap granulator *(sr)*
Rohrschweißgerät pipe welding instrument
Rohrschweißvorrichtung pipe welding unit/ equipment
Rohrspritzkopf pipe die head, pipe extrusion head, pipe die
Rohrstraße 1. pipe extrusion line. 2. pipe production line *(general term; see explanatory note under* **Rohrfertigungsstraße***)*
Rohrstück parison *(bm)*
Rohrtrennautomat automatic pipe cutter
Rohrtrennvorrichtung pipe cutter, pipe cutting device
Rohrumfang pipe circumference
Rohrverbindung pipe joint
Rohrverteiler runner *(im)*
Rohrwand pipe wall
Rohrwandbeanspruchung hoop stress *(of pipes)*
Rohrwanddicke pipe wall thickness
Rohrwanddickenmeßanlage device for measuring pipe wall thickness
Rohrware pipe extrusion compound
Rohrwerkstoff pipe extrusion compound, pipe material
Rohrwerkzeug pipe die *(e)*
Rohrwerkzeugkonstruktion pipe die design
Rohrwickelanlage pipe winding plant *(grp)*
Rohrzuschnitt pipe section
Rohstoff 1. raw material, feedstock. 2. polymer
Rohstoffbedarf raw material requirements: **Damit kann auch der Rohstoffbedarf größer werden** this means that more raw material may be required
Rohstoffdatei raw material data file
Rohstoffeigenschaften (raw) material properties
Rohstoffeingangskontrolle incoming raw materials control
Rohstoffhersteller raw material producer/ manufacturer
Rohstoffkonfektionierung polymer compounding
Rohstoffkosten raw material costs
Rohstofflieferant raw material supplier
Rohstoffmaterial raw material, feedstock
Rohstoffquelle raw material source
Rohstoffressourcenschonung conservation of raw material resources
Rohstoffrückgewinnungsanlage raw material recovery plant
Rohstoffschmelze polymer melt
Rohstoffschwankungen raw material variations
Rohstofftagesmenge amount of raw material required for a day's production

Rohstoffverfügbarkeit raw material availability
Rohstoffverknappung raw material shortage
Rohstoffzuführgerät raw material feed unit
Rolladen roller blinds
Rolladenführung shutter guide
Rolladenprofil roller blind profile
Rolladenstäbe roller blind slats
Rolle roll, roller
Rollen roller application
Rollen (des Schmelzeschlauches) parison curl *(bm)*
Rollenapplikation roller application *(e.g. of paint)*
Rollenführung guide rolls
Rollenkorb calibrating/sizing basket *(bfe)*
Rollenlager roller bearing
Rollenoffsetdruck rotary offset printing
Rollenoffsetfarbe rotary offset printing ink
Rollenreckstrecke stretching roll section
Rollenreckwerk stretching roll unit
Rollenschneidmaschine reel trimmer
Rollenwechsel 1. changing reels *(of film)*: **Damit ist der Folientransport bei Rollenwechsel gewährleistet** this means that film transport is not interrupted when the reel has to be changed. 2. reel changing mechanism/system/unit
Rollenwechselsystem reel changing mechanism/system/unit
Rollenwickelmaschine reel winder
Rollreibung rolling/sliding friction
Rollschälversuch climbing drum peel test
Rollwiderstand rolling reistance
ROM *abbr. of* read only memory *(me)*
ROM-Speicher read-only memory *(me)*
Röntgenanalyse X-ray analysis
Röntgenaufnahme X-ray photograph
Röntgenbeugung X-ray diffraction
Röntgenbeugungsbild X-ray diffraction photograph
Röntgenbeugungsmessung X-ray diffraction analysis
Röntgenfluoreszenzanalysator X-ray fluorescence analyser
Röntgenfluoreszenzanalyse X-ray fluorescence analysis
röntgenmikroskopische Aufnahme X-ray micrograph
röntgenographisch X-ray photographic
Röntgenspektroskopie X-ray spectroscopy
röntgenstrahldurchlässig permeable to X-rays
Röntgenstrahlen X-rays
Röntgenstrahlendurchlässigkeit X-ray permeability
röntgenstrahlundurchlässig impermeable to X-rays
Rost 1. rust. 2. screen, grid, grating
rostbeständig non-rusting
Rostbildung rusting, rust formation
rostfreier Stahl stainless steel
Rostschutzanstrich anti-corrosive coating/finish
Rostschutzgrundierung anti-corrosive primer

Rostschutzlack anti-corrosive paint
Rostschutzmittel rust inhibitor
Rostschutzpigment anti-corrosive pigment
Rostschutzwirkung anti-corrosive effect
Rotationsachse axis of rotation
Rotationsbewegung rotary movement
Rotationsdosiereinrichtung rotary feed unit
Rotationsformen rotational moulding, rotomoulding
rotationsgegossen rotationally moulded, rotomoulded
Rotationsgießen rotational moulding, rotomoulding
Rotationsgießmaschine rotational moulding machine, rotomoulder
Rotationsguß rotational moulding, rotomoulding
Rotationshohlkörper rotational moulding
Rotationsquerschneider rotary transverse cutter
Rotationsreibschweißen spin welding
Rotationsrheometer rotary rheometer
Rotationsschmelzanlage, kreisförmige carousel-type rotomoulder
Rotationsschmelzen rotational moulding, rotomoulding
Rotationsschweißen spin welding
Rotationssiebdruck rotary screen printing
Rotationssintern rotational sintering
Rotationsspritzgußautomat automatic rotary injection moulding machine
rotationssymmetrisch axially symmetrical
Rotationsverdrängerpumpe rotary positive displacement pump
Rotationsviskosimeter rotational viscometer
rotierend rotating
rotierend, entgegengesetzt counter-rotating
rötlich reddish
rötlichbraun reddish brown
Rotor rotor *(sr)*
Rotordrehzahl rotor speed *(sr)*
Rotorlagerung rotor bearing *(sr)*
Rotormesser rotor knife *(sr)*
Rotormesserbalken rotor knife block *(sr)*
Rotormesserkreis rotor knife cutting circle *(sr)*
Rotorscheibe rotor knife *(sr)*
Rotorwelle rotor shaft
Routine routine, program *(me)*
Routinekontrolle routine check
Routinemessung routine determination
Routineprüfung routine test
Routineuntersuchung routine test
Rovinggewebe woven roving
Rovinggewebeverstärkung woven roving reinforcement
Rovinglaminat roving laminate
Rovingstrang roving strand
RRIM *abbr. of* reinforced reaction injection moulding
RRIM-Anlage reinforced reaction injection moulding plant, RRIM plant
RRIM-Technologie reinforced reaction injection moulding, RRIM

RRIM-Teil reinforced reaction injection moulded part, RRIM part
RSG *abbr. of* **Reaktions-Schaum-Guß** reaction foam moulding
RSM *abbr. of* reaction spray moulding
RT *abbr. of* **Raumtemperatur**, room temperature
RTV *abbr. of* **Raum-Temperatur-Vernetzung/vernetzend,** room temperature vulcanisation/vulcanising
Rückbildung re-formation
Rückdeformation recovery
Rückdeformation, elastische elastic recovery
Rückdruck back pressure *(e)*
Rückdruckkraft back pressure force
Rückdrucklagerung thrust bearing assembly *(e)*
Rückdrückstift ejector plate return pin *(im)*
Rückenlehne back rest
Rückenstrich backing coat
Rückfederung recovery
Rückfederung, elastische elastic recovery
Rückfederungsvermögen resilience
Rückfluß reverse flow, backflow: **bei Rückfluß der Schmelze in den Zylinder** when the melt flows back into the cylinder
Rückfluß, unter under reflux
Rückflußkühler reflux condenser
Rückformvermögen resilience
Rückführmaterial reclaim, regrind
Rückführsystem feedback system
Rückführung 1. recycling. 2. feedback
 elektronische Rückführung electronic feedback
rückgewinnbar reclaimable, recoverable
Rückgewinnung reclamation, recovery
rückgewonnen reclaimed, recovered
Rückgrat backbone
Rückhub return stroke
Rückkopplung feedback
Rückkühlaggregat closed-circuit cooling unit
Rückkühlung closed-circuit cooling system/unit
Rückkühlwerk closed-circuit cooling unit
Rücklagen reserves
Rücklauf 1. retraction: **bei Rücklauf der Schnecke** when the screw retracts. 2. reflux
Rücklaufgeschwindigkeit retraction speed
Rücklaufquote return rate
Rücklaufsperre non-return valve *(im)*
Rücklauftemperatur return temperature
Rücklaufware reject articles/goods/mouldings
Rücklaufwasser returning water
Rückleuchte rear light
Rückleuchtengehäuse rear light housing
Rückluft returning air
Rückmeldung feedback
Rückmonomeranlage monomer recovery plant
Rücknahme taking back *(e.g. used plastics bottles and containers, film scrap etc. for reprocessing)*
Rücknahmekette take-back chain

Rücknahmepflicht take-back provision *(part of the* **Verpackungsverordnung** *(q.v.) which obliges manufactures and retailers to take back used pakaging)*
Rücknahmestelle take-back point
Rückplatte back plate *(im)*
Rückprall rebound
Rückprallelastizität rebound/impact resilience
Rückprallhöhe rebound height
Rückprallversuch rebound test
Rückschlagventil check/non-return valve *(im)*
Rückschrumpf recovery
Rückseite rear, back
Rückspiegel rear view mirror
Rücksprunghärte rebound/impact resilience
Rückspülung back flushing *(mf)*
Rückstand residue
rückstandslos without leaving a residue
Rückstau back pressure *(e)*
Rückstaudruck back pressure *(e)*
Rückstausperre non-return valve *(im)*
Rückstellelastizität rebound resilience
rückstellfähig resilient
Rückstellfähigkeit resilience
Rückstellkraft resilience
Rückstellung recovery *(after compression)*
Rückstellungen reserves *(e.g. allocated for retirement pensions etc.)*
Rückstellvermögen resilience
Rückstreuelektronen back-scattered electrons
Rückströmsperre non-return valve *(im)*
Rückströmung backflow, reverse flow *(e)*
Rücktitration back titration
Rückverformung recovery
Rückware reclaim, regrind
Rückwärtsbewegung backward movement
Rückwärtsentgasung 1. back venting *(e)*. 2. back venting system *(e)*
Rückzug retraction *(of screw; for translation example see under* **Rücklauf***)*
Rückzuggeschwindigkeit retraction speed
Rückzugkraft retraction force
ruhende Beanspruchung static stress
ruhende Belastung static stress
Ruhepenetration static penetration
Ruhereibung static friction
Ruhezeit downtime
Ruhezone stagnation zone
Ruhezustand state of rest
Ruhpenetration unworked penetration
rühren to stir
Rührer stirrer
Rührgefäß mixing tank/vessel
Rührgerät stirrer
Rührreaktor reactor equipped with stirrer
Rührstab 1. stirring rod 2. impeller shaft
Rührwelle impeller shaft
Rührwerk mixer, stirrer
Rührwerkeinrichtung mixer, stirrer
Rührwerkskugelmühle attrition mill
Rührwerkzeug impeller
rund round, circular, spherical

Runddüse round-section die *(e)*
Runderneuerung retreading
Runderneuerungsmaterial retreading compound/material
Rundfunkgerät radio
Rundgewebe circular-woven fabric
Rundkolben round-bottom flask
Rundläufer rotary-table/carousel-type machine
Rundläuferanlage rotary-table/carousel-type machine
Rundläuferautomat automatic rotary-table/carousel-type machine
Rundläufermaschine rotary-table/carousel-type machine
Rundlauffehler true-running error
Rundlaufgenauigkeit true-running characteristics
Rundlaufspaltmaschine rotary slitter
Rundlochkanal 1. round-section flow channel. 2. round-section runner *(im)*
Rundlochkapillare round-section capillary
Rundrohr round-section pipe
Rundschnurring O-ring
Rundstab round-section rod
Rundstange round-section rod
Rundstrangdüse round-section die, round-section profile die *(e)*
Rundstrangdüsenkopf round section die *(e)*
Rundtisch rotary table, roundtable
Rundtischanlage rotary-table/carousel-type machine
Rundtischanordnung rotary-table/carousel-type arrangement/system
Rundtischmaschine rotary-table/carousel-type machine
Rundvollstab round-section rod
Rundwebmaschine circular loom
Rundwebstuhl circular loom *(txt)*
Runzelbildung wrinkling
Runzellack wrinkle-finish paint
Ruß 1. carbon black. 2. soot
rußgefüllt carbon black-filled, filled with carbon black
Rußkautschuk carbon black masterbatch
Rußpaste carbon black concentrate
Rußpigment carbon black
Rüstvorgang set-up operation, setting-up
Rüstzeit set-up time
Rüstzeitaufnahme set-up time record
Rutil rutile
Rutilpigment rutile pigment
Rutsche chute
rutschfest non-skid, skidproof
Rutschfestbeschichtung non-skid coating/screed
Rutschfestigkeit 1. non-skid properties. 2. skid resistance *(of a tyre)*
Rutschkalibrierung friction calibration *(e)*
Rütteldichte packing density
Rüttelgewicht packing density
Rüttelmotor vibrator motor
Rütteltisch vibrating table

Rüttelverdichten compacting by vibration (concrete)
Rüttelverdichter (concrete) vibrator *(used to compact concrete by vibration)*

S

Sachanlage material/fixed asset
Sacheinlage subscription in kind
Sackloch blind hole
Sacklochbohrung blind hole
SAF-Ruß super abrasion furnace black
Sägeblatt saw blade
Salicylsäure salicylic acid
Salicylsäureester salicylate, salicylic acid ester
Salpetersäure nitric acid
Salzausblühungen efflorescence
Salzlösung salt solution
Salzsäure hydrochloric acid
Salzsprühbelastung exposure to salt spray
Salzsprühbeständigkeit salt spray resistance
Salzsprühnebelbeständigkeit salt spray resistance
Salzsprühtest salt spray test
Salzwasserlagerung immersion in salt water
Sammelbegriff collective name/term
Sammelbehälter 1. collecting tank 2. recycling bank *(for the roadside collection of different kinds of waste)*
Sammelbohrung manifold *(e)*
Sammelraum 1. (space) in front of the screw. 2. accumulator chamber *(e,bm)*
Sammelschiene busbar
Sammelstelle collecting point
Sammelsystem collecting system
Sammlung collection
samtartig velvety
sandgestrahlt sandblasted
Sandguß sand casting
Sandmühle sand mill/grinder
Sandpapier sandpaper
Sandstrahlen sandblasting
Sandwichaufbau sandwich construction
Sandwichbauweise sandwich construction
Sandwichplatte sandwich sheet/panel
Sandwichschäumverfahren sandwich moulding (process)
Sandwichspritzgießverfahren sandwich moulding (process)
Sandwichspritzgußteil sandwich moulding, sandwich moulded part
Sandwichteil sandwich component
Sandwichverbundweise sandwich construction
Sandwichverfahren sandwich moulding (process) *(im)*
sanieren to restore, to renovate, to refurbish

Sanierung restoration, renovation, refurbishment
Sanierungsarbeiten restoration work
Sanitärarmaturen sanitary fittings
Sanitärartikel sanitaryware, sanitary fittings
Sanitärbauteile sanitary fittings
Sanitäreinrichtungen sanitaryware, sanitary fittings
Sanitärindustrie sanitaryware industry
Sanitärtechnik sanitary engineering
Sanitärteile sanitaryware, sanitary fittings
Sanitärzelle bathroom
Satellitentechnik satellite technology/engineering
satt auftragen to apply liberally *(paint, adhesive etc.)*
Sattdampf 1. saturated steam. 2. saturated vapour
Sattdampfdruck saturated vapour pressure
Sättigung saturation
Sättigungsdampfdruck saturation vapour pressure
Sättigungsdruck saturation pressure
Sättigungsgrad degree of saturation
Sättigungskonzentration saturation concentration
Sättigungswassergehalt saturation moisture content
Sättigungszustand state of saturation
Säuberung cleaning
sauer acid
Sauerkrautmasse dough moulding compound, DMC
Sauerstoff oxygen
Sauerstoffabsorption oxygen absorption
Sauerstoffangebot amount of oxygen available
Sauerstoffatom oxygen atom
Sauerstoffausschluß, unter under the exclusion of oxygen
Sauerstoffbarriere oxygen barrier
Sauerstoffdurchlässigkeit oxygen permeability
Sauerstoffempfindlichkeit oxygen sensitivity
sauerstoffhaltig oxygen-containing
Sauerstoffindex oxygen index
Sauerstoffmangel lack of oxygen
Sauerstoffpermeabilität oxygen permeability
Sauerstoffüberschuß excess oxygen
saugend absorbent
Sauger suction device
saugfähig absorbent
Saugfähigkeit absorbency
Saugfilter suction filter
Saugfördergerät suction conveyor/feeder
Sauggebläse suction blower
Saugleitung suction line
Saugloch suction hole
Saugpneumatik pneumatic suction system/mechanism
Saugrohr suction tube/pipe
Saugschlauch suction hose
Saugschlitz suction slit
Saugventil suction valve

Saugvermögen absorbency
Säule tie bar/rod, column *(im)*
Säulen, lichte Weite zwischen den distance between tie bars *(im)*
Säulen, lichter Abstand zwischen den distance between tie bars *(im)*
Säulenabstand, lichter distance between tie bars *(im)*
Säulendiagramm bar graph
Säulenvorspannung tie bar pre-stressing *(im)*
Säure acid
Säureamid acid amide
Säureamidgruppe acid amide group
Säureanhydrid acid anhydride
Säureanhydridhärter acid anhydride catalyst
Säurebedarf amount of acid required
Säurebeizen pickling
Säurebelastung exposure to acids
säurebeständig acid resistant
Säurebeständigkeit acid resistance
Säurechlorid acid chloride
Säuredämpfe acid fumes
Säuregehalt acid content
Säuregruppe acid radical
säurehärtbar acid curing
säurehärtend acid curing
Säurehärter acid catalyst
Säurekatalysator acid catalyst
Säurekatalyse acid catalysis
säurekatalysiert acid-catalysed
säurelöslich acid-soluble, soluble in acids
Säureradikal acid radical
Säureresistenz acid resistance
Säurerest acid radical
säureunlöslich acid-insoluble, insoluble in acids
Säurezahl acid value
SBR abbr. of styrene-butadiene rubber
Scangeschwindigkeit scanning speed
Schablone template
Schadensanalyse fracture analysis
Schadenstoleranz resistance to damage
Schädigung damage, fracture
Schädigung, thermische thermal degradation
Schädigungsanfälligkeit susceptibility to damage
Schädigungsarbeit fracture energy
Schädigungsenergie fracture energy
Schädigungsgrad amount of damage
Schädigungskraft fracture force
schädigungslos without (suffering) damage
Schädigungsprozeß fracture process
Schadstelle damaged area: **örtliche Schadstellen** local damage
Schadstoff harmful substance
Schadstoffemission emission of harmful substances
Schadstoffverpackungsverordnung harmful substances packaging directive
Schaftschnecke variable-geometry screw, variable-design screw *(e) (see also*

explanatory note under **Einstückschnecke***)*
schälbeansprucht under peel stress
Schälbeanspruchung peel stress
Schälbelastung peel stress
Schalenbauweise shell-type construction
Schalenformguß investment casting, lost wax casting
Schalengießverfahren slush moulding
Schalenguß slush moulding
Schälfestigkeit peel strength
Schälgeschwindigkeit peel rate
Schälkraft peel force
***Schallabsorption** sound absorption
Schallabsorptionsgrad degree of sound absorption, amount of sound absorbed
Schallabstrahlung sound reflection
schallarm quiet in operation, quiet-running *(machine, pump etc.)*
Schallausbreitung sound propagation
Schalldämmaß degree of sound insulation
schalldämmend soundproof, sound insulating
Schalldämmplatte sound-insulating panel
Schalldämmschurre soundproof chute *(sr)*
Schalldämmstoff sound insulating material
Schalldämmung sound insulation, soundproofing
Schalldämmwert sound insulation factor
Schalldämpfung sound damping
Schalldämpfungsmaßnahmen noise reduction measures
Schalldruck sound pressure *(sr)*
Schalldruckpegel 1. sound pressure intensity/level. 2. noise level: *Version 1. should be used in scientific texts, version 2. in a general context, e.g. in connection with granulators and other noisy machines*
Schalldruckstärke 1. sound pressure intensity/level. 2. noise level *(general term; see explanatory note under* **Schalldruckpegel***)*
Schalldurchgang sound transmission
Schalleitfähigkeit sound conductivity
Schallemission acoustic emission
Schallemissionsanalyse acoustic emission analysis
Schallemissionskennwert acoustic emission constant
Schallemissionskurve acoustic emission curve
Schallemissionsmessung acoustic emission determination
Schallemissionsprüfung acoustic emission test
Schallemissionsrate acoustic emission rate
Schallemissionsverhalten acoustic emission behaviour
Schallenergie sonic energy
Schallfortpflanzung sound propagation
schallgedämpft soundproofed
schallgeschützt soundproofed
Schallgeschwindigkeit velocity of sound
Schallimpuls acoustic impulse
Schallisolationsvermögen sound insulation

* *Words starting with* **Schall-***, not found here, may be found under* **Geräusch-** *or* **Lärm-**

properties
schallisolierend sound insulating
schallisoliert soundproofed
Schallpegel noise level: *sie haben einen relativ hohen Schallpegel* they are relatively noisy
Schallpegelabsenkung noise level reduction *(for translation example see under* **Schallpegelreduzierung***)*
Schallpegelreduzierung noise level reduction: *Derartige Maßnahmen brachten Schallpegelreduzierungen bis zu 15 dBA* such measures reduced the noise level to 15 dBA
Schallplatte (gramophone) record, LP record, disc
Schallplattenindustrie record industry
Schallplattenmasse (LP) record moulding compound
Schallplattenpresse (LP) record moulding press
Schallplattenpreßmase (LP) record moulding compound
Schallplattenprodukt (LP) record moulding compound
Schallplattenrohstoff (LP) record moulding compound
Schallplattentyp (LP) record moulding compound
Schallschluckauskleidung sound absorbent lining
Schallschluckdeckenplatte sound absorbent ceiling tile
Schallschluckeigenschaften sound-absorbent properties
schallschluckend sound-absorbent
Schallschluckplatte sound abosrbent tile/sheet
Schallschutz sound insulation, soundproofing
Schallschutzelemente soundproofing elements
Schallschutzhaube soundproof hood *(sr)*
Schallsignal acoustic signal
Schallübertragung sound transmission
Schallverkleidung sound barrier, engine sound shield
Schallwandler sonic converter *(w)*
Schallweiterleitung sound propagation
Schallwelle sound wave
Schalöl mould oil
Schalt- und Regelelemente 1. controls. 2. open and closed loop controls *(see explanatory note under* **regeln***)*
Schalt- und Steuergeräte controls
Schaltbild circuit diagram
Schalteinrichtungen switchgear, controls
Schalter switch
Schaltergehäuse switch housing
Schältest peel test
Schaltfolge switch sequence
Schaltgeräte switchgear, controls
Schaltgetriebe control mechanism
Schaltgetriebedichtung manual transmission seal

Schaltkasten switch box
Schaltkreis switching/control circuit
Schaltkreis, integrierter integrated circuit
Schaltplan 1. circuit diagram. 2. circuitry
Schaltpult control desk/panel/console
Schaltpunkt switching point
Schaltschema 1. circuit diagram. 2. circuity
Schaltschrank control cabinet
Schaltschwelle switch limit
Schalttableau control panel
Schalttafel 1. control panel. 2. dashboard, instrument panel
Schaltuhr timer, time switch
Schaltung circuit
Schaltung, gedruckte printed circuit
Schaltungssystem control system
Schaltungstechnik circuitry
Schaltventil control valve
Schaltwarte control desk/panel/console
Schaltzeit switching time
Schaltzustand control status
Schälversuch peel test
Schälwerkzeug planing tool
Schälwert peel strength
Schälwiderstand peel strength
Schamotte fireclay
Schamottstein fireclay brick
scharfkantig sharp edged
Scharnier hinge
Scharte notch, crack
Schätzkosten estimated cost
Schätzwert estimated value/figure
Schaufel (mixing) blade
Schaufelrührer paddle mixer
Schauglas sight glass, window
Schaum foam
Schaum, chemischer chemically blown foam
Schaum, mechanischer mechanically blown foam
schaumanfällig liable to foam
Schäumanlage 1. foam moulding plant. 2. foaming plant
schäumbar foamable, expandable
Schäumbarkeit foamability
Schäumbedingungen foaming conditions
Schaumbekämpfung combating foam
Schaumbeton aerated/cellular concrete
Schaumbildner blowing/foaming agent
Schaumbildung foaming, foam formation
Schaumblasverfahren foam blow moulding (process)
Schaumdichte foam density
Schäumdruck foaming pressure
Schäumen foaming, expansion
schäumfähig expandable, foamable
Schaumfolienanlage expanded film extrusion line
Schaumfolienextruder expanded film extruder
Schäumform foaming mould
Schaumgerüst foam skeleton
Schaumgummi foam rubber
schauminhibierend anti-foam, foam-inhibiting

Schaumkunstleder foamed leathercloth
Schaumkunststoff plastics foam
Schäummaschine foam moulding machine
Schäummittel blowing/foaming agent
Schaumneigung tendency to foam
Schaumpaste expandable (PVC) paste
Schaumplastisol expandable (PVC) paste/plastisol
Schaumpolystyrol expanded polystyrene
Schaumprobleme foaming problems
Schäumprozeß foaming process/operation
Schaum-PVC PVC foam
Schaumqualität foam quality
Schaumregulator foam regulator
Schaumrücken foam backing *(of carpet)*
Schaumrückenbeschichtung foam backing *(of carpet)*
Schaumspritzgießen structural/integral foam moulding
Schaumstabilisator foam stabiliser
schaumstabilisierend foam stabilising
Schaumstabilisierung foam stabilisation
Schaumstoff foam
Schaumstoffbahn foam web
Schaumstoffblock slabstock foam
Schaumstoffblockmaterial slabstock foam
Schaumstoffgefüge foam structure
Schaumstoffgerüst foam skeleton
Schaumstoffkern foam core
Schaumstoffplatte foam sheet
Schaumstoffrezeptur foam formulation
Schaumstoffrohdichte foam density
Schaumstoffstruktur foam structure
Schaumstrich foam coating
Schaumstruktur foam structure
Schaumsystem foam system
schäumtechnisch *relating to foaming*:
  schäumtechnische Bedingungen foaming conditions
Schäumtemperatur foaming temperature
schaumverhindernd antifoam
Schäumzeit foaming time
Schaumzeitpunkt foaming temperature
schaumzerstörend antifoam
Scheckkarte cheque card
Scheibe disc
Scheibenanguß diaphragm gate *(im)*
Scheibenbremse disc brake
Scheibenextruder Diskpack plasticator/processor *(trade name of a machine made by the Farrel machinery group)*
Scheibenfilter disk filter
Scheibenpaket calibrating/sizing plate assembly *(e)*
Scheibenplastifizieraggregat Diskpack plasticator/processor *(see explanatory note under **Scheibenextruder**)*
Scheibenwaschbehälter windscreen wash bottle
scheinbare Viskosität apparent viscosity
Scheindreherbindung mock Leno weave *(txt)*
Scheinwerfer headlamp
Scheinwerfergehäuse headlamp housing
Scheinwerferreflektor headlamp reflector
Scheinwerferstreuscheibe headlamp diffuser
Scheitelpunkt vertex (*plural*: vertexes *or* vertices)
Schellak shellac
Schema 1. scheme, system. 2. schematic drawing/diagram
Schemarezeptur guide/suggested formulation
Schemaskizze schematic drawing/diagram
schematisch schematic
Schemazeichnung schematic diagram/drawing
Schenkelhärte Sward hardness
Schenkelhärteprüfung Sward hardness test
Schenkelweiterreißversuch trouser tear test
Scherbeanspruchung shear stress
Scherbelastung shear stress
Scherbereich shear range **im höheren Scherbereich...** at higher shear rates...
Schercraze shear craze
Scherdeformation shear deformation
Scherdehnung shear strain
Scherelement shear element *(e)*
scherempfindlich shear sensitive
Scherempfindlichkeit shear sensitivity
scheren to shear
Scherenergie shear energy
scherentzähend pseudoplastic, non-Newtonian
Scherermüdung shear fatigue
Schererwärmung heat produced/generated through shear
Scherfestigkeit shear strength
Scherfestigkeit, interlaminare interlaminar shear strength
Scherfließen shear flow
Schergefälle shear rate
Schergeschwindigkeit shear rate
schergeschwindigkeitsabhängig depending on the shear rate
Schergeschwindigkeitsbereich shear rate range
schergeschwindigkeitsunabhängig independent of the shear rate
scherinduziert shear-induced
scherintensiv shear-intensive
Scherkopf smear head, torpedo *(of screw)* *(e)*
Scherkraft shear force
Scherkrafteinleitung introduction of shear forces
scherkraftreich high-shear
Schermischschnecke shearing-mixing screw
Schermischteilschnecke shearing-mmixing screw
Scherniveau shear
Scherplastifizierung shear plasticisation/plastication
Scherprüfung shear test
Scherspalt (radial) screw clearance
Scherspannung shear stress
Scherstabilität shear resistance
Scherstandfestigkeit shear strength
Scherstandvermögen shear strength

Scherströmung shear flow
Scherteil shear section (of screw) (e)
Scherteilschnecke screw equipped with a shear section
Schertorpedo smear head torpedo (of screw) (e)
scherunempfindlich shear insensitive, unaffected by shear
Scherunempfindlichkeit shear insensitivity
Scherung shear
scherungsarm low-shear
Scherverbrennung charring due to shear
Scherverformung deformation due to shear
Scherversuch shear test
Schervorgeschichte previous shear history
Scherwalzen shearing rolls
Scherwalzwerk shearing rolls
Scherwirkung shear effect
Scherzone shear section (e)
Scherzonenschnecke screw equipped with a shear section
Scheuerbeständigkeit scrub resistance
Scheuerfestigkeit scrub resistance
Scheuerresistenz scrub resistance
Scheuertest scrubbing test
Schicht 1. layer. 2. coat. 3. shift
Schicht, dampfbremsende vapour barrier
Schichtablösung (paint) film detachment
Schichtabtrennung (paint) film detachment
Schichtbetrieb shift operation
schichtbildend film forming
Schichtdicke coating thickness
Schichtenströmung laminar flow
Schichtenströmung, laminare laminar flow
schichtförmig laminar
Schichtholz plywood
Schichtlage layer
Schichtlaminat laminate
Schichtmeister shift foreman
Schichtpreßstoff laminate
Schichtpreßstoffplatte laminate sheet
Schichtpreßstoffrohr laminate pipe
Schichtpreßstofftafel laminate sheet
Schichtprotokoll shift log/record (me)
Schichtsilikat layer-lattice silicate
Schichtstärke coating thickness
Schichtstoff laminate
Schichtstoffharz laminating resin
Schichtstoffplatte laminate sheet
Schichtstoffplatten, technische industrial laminates
Schichtverbund 1. laminar structure. 2. multi-layer film
Schiebebewegung sliding movement
Schiebedach sliding roof
Schiebedüse sliding shut-off nozzle (im)
Schiebeeinsatz sliding insert
Schiebegitter sliding safety guard, sliding guard door
Schieber 1. slide bar, guide bar. 2. split (im). 3. slide valve
Schieberform sliding split mould (im)

Schieberplatte slide plate (of a screen changer) (mf)
Schieberplattenwerkzeug sliding split mould (im)
Schieberwerkzeug sliding split mould (im)
Schiebeschutzgitter sliding safety guard, sliding guard door
Schiebetisch sliding table
Schiebetischmaschine sliding table machine
Schiebetür sliding guard door, sliding safety guard
Schiebeverschlußdüse sliding shut-off nozzle (im)
Schiebeverschlußspritzdüse sliding shut-off nozzle (im)
Schiefermehl slate powder/flour
Schienentankfahrzeug rail tanker
Schiffsanstrich marine finish/coating
Schiffsanstrichsystem marine paint
Schiffsbauindustrie shipbuilding industry
Schiffsbewuchs marine fouling
Schiffsfarbe marine paint
Schiffskörper hull
Schiffskörperaußenfarbe marine paint
Schiffsrumpf hull
Schikane baffle plate
Schimmelbefall mould attack
Schimmelpilz mould
Schirmanguß diaphragm gate (im)
Schirmanschnitt diaphragm gate (im)
Schlachthof slaughterhouse, abattoir
Schlag impact
Schlagarbeit impact energy
schlagartig sudden, abrupt
schlagartige Beanspruchung 1. impact stress. 2. sudden stress
schlagbeansprucht under impact stress
Schlagbeanspruchung impact stress
Schlagbelastbarkeit impact resistance
schlagbeständig impact resistant
Schlagbeständigkeit impact resistance
Schlagbiegebeanspruchung flexural impact stress
Schlagbiegefestigkeit flexural impact strength
Schlagbiegeprüfung flexural impact test
Schlagbiegestab flexural impact test piece, flexural impact bar
Schlagbiegeverhalten flexural impact behaviour
Schlagbiegeversuch flexural impact test
Schlagbiegezähigkeit flexural impact strength
Schlagenergie impact energy
schlagfest impact resistant, toughened (the second version is used mainly for polystyrene)
Schlagfestigkeit impact strength
Schlagfestmachen making impact resistant, toughening
Schlagfestmacher impact modifier, toughening agent
Schlagfestmodifikator impact modifier
Schlaggeschwindigkeit impact speed, speed of impact
Schlaghammer pendulum

Schlaghammerwerkzeug drop hammer tool
Schlagintensität impact force
Schlagkraft impact energy
Schlaglast impact stress
Schlagmesser impact cutter
Schlagpendel pendulum
Schlagpendelgeschwindigkeit pendulum impact speed
Schlagpressen solid phase forming, forging
Schlagprüfstab impact specimen/bar, impact test piece
Schlagprüfung impact test
Schlagregen driving rain
Schlagregendichte imperviousness to driving rain
schlagregensicher resistant to driving rain
Schlagregensicherheit resistance to driving rain
Schlagrichtung direction of impact
Schlagschaum mechanically blown foam
Schlagschaumpaste paste for making mechanically blown foam
Schlagschaumverfahren mechanical blowing using air
Schlagschere impact cutter
Schlagtest impact test
schlagverformbar resistant to deformation through impact
Schlagverformbarkeit resistance to deformation through impact
Schlagverhalten impact behaviour, behaviour on impact
Schlagversuch impact test
Schlagwerk impact tester
Schlagwinkel angle of impact
schlagzäh impact resistant, toughened *(the second version is used mainly for polystyrene)*
extrem schlagzäh extremely high impact
schlagzäh, erhöht with enhanced impact resistance/strength, high impact
Schlagzähbeanspruchbarkeit impact strength/resistance
Schlagzähigkeit impact strength/resistance
Schlagzähigkeitsleistung impact strength/resistance
Schlagzähigkeitsminderung impact strength reduction: **Jede Versprödung drückt sich als Schlagzähigkeitsverminderung aus** embrittlement always reduces impact strength
Schlagzähigkeitsoptimum maximum impact strength: *It is accepted that the higher the impact strength of a material, the better it is. It follows that the literal translation - optimum - is, in this case, synonymous with* maximum
Schlagzähigkeitsuntersuchung impact test
Schlagzähigkeitsverbesserer impact modifier, toughening agent
Schlagzähigkeitsverbesserung impact modification, toughening
Schlagzähkomponente impact modifier, toughening agent
Schlagzähkonzentrat impact modifying concentrate
Schlagzähmacher impact modifier, toughening agent
Schlagzähmischung high impact compound
Schlagzähmodifier impact modifier, toughening agent
Schlagzähmodifikator impact modifier, toughening agent
Schlagzähmodifizierharz impact modifier, toughening agent
Schlagzähmodifiziermittel impact modifier, toughening agent
schlagzähmodifiziert impact modified, toughened
Schlagzähmodifizierung impact modification
Schlagzugversuch tensile impact test
Schlagzugzähigkeit tensile impact strength
Schlämme slurry
Schlangenrohrverdampfer spiral coil vaporiser
Schlankheitsgrad slenderness ratio
Schlauch 1. parison *(bm)*. 2. film bubble *(bfe)*. 3. tube, hose
Schlauchabschneider parison cutter *(bm)*
Schlauchabschnitt parison *(bm)*
Schlauchabzug parison take-off *(bm)*
Schlauchaufweitung bubble expansion *(bfe)*
Schlauchausstoßgeschwindigkeit parison delivery rate *(bm)*
Schlauchaustrittsgeschwindigkeit parison delivery rate *(bm)*
Schlauchbildungszone bubble expansion zone *(bfe)*: *The part between the and frost line, where the extruded tube is inflated into a* **Folienschlauch** *(q.v.)*
Schlauchblase film bubble *(bfe)*
Schlauchblasenstabilität film bubble stability *(bfe)*
Schlauchboot dinghy
Schlauchdickenregelung 1. parison wall thickness control *(bm)*. 2. parison wall thickness controller *(bm)*
Schlauchdurchmesser 1. parison diameter *(bm)*. 2. film bubble diameter *(bfe)*
Schlauchdurchmesserregelung 1. parison diameter control *(bm)*. 2. parison diameter controller *(bm)*
Schlauchdüse 1. parison die *(bm)*. 2. blown film die, film blowing die *(bfe)*. 3. tube/pipe die *(e)*
Schlauchextrusionsdüse 1. parison die *(bm)*. 2. blown film die, film blowing die *(bfe)*. 3. tube/pipe die *(e)*
Schlauchfolie blown film, tubular film *(bfe)*
Schlauchfolienanlage blown film line, film blowing line
Schlauchfolienanlage, coextrudierende blown film coextrusion line
Schlauchfolienblasanlage blown film line, film blowing line
Schlauchfolienblasen blown film extrusion, film blowing
Schlauchfoliendüse blown film die, film blowing die

Schlauchfolienextruder blown film extruder
Schlauchfolienextrusion blown film extrusion, film blowing
Schlauchfolienextrusionsanlage blown film line, film blowing line
Schlauchfolienextrusionswerkzeug blown film die, film blowing die
Schlauchfolienfertigung blown film extrusion, film blowing
Schlauchfolienherstellung blown film extrusion, film blowing
Schlauchfolieninnenkühlung 1. internal bubble cooling system. 2. cooling the inside of the film bubble: *Ein Schlauchfolienwerkzeug mit der Möglichkeit zur Schlauchfolieninnenkühlung a blown film die with facilities for cooling the inside of the film bubble*
Schlauchfolienkonzept 1. blown film process. 2. blown film plant: *Context must be the guide to which version is appropriate. See also explanatory note under* **Konzept**
Schlauchfolienkopf blown film die, film blowing die
Schlauchfolienkühlung 1. blown film cooling. 2. blown film cooling system
Schlauchfolienverfahren blown film extrusion, film blowing
Schlauchfolienwerkzeug blown film die, film blowing die
Schlauchformeinheit 1. parison die *(bm)*. 2. blown film die, film blowing die *(bfe)*
Schlauchführung bubble guide *(bfe)*
Schlauchgreifer parison gripper *(bfe)*
Schlauchgreifvorrichtung parison gripping mechanism *(bm)*
Schlauchkontur bubble contour *(bfe)*
Schlauchkopf 1. parison die *(bm)*. 2. blown film die, film blowing die *(bfe)*. 3. tube die *(e)*
Schlauchkühlung 1. cooling of film bubble *(bfe)*: *Die Schlauchkühlung ist ungenügend the film bubble is cooled insufficiently.* 2. film bubble cooling system *(bfe)*
Schlauchkühlvorrichtung film bubble cooling system *(bfe)*
Schlauchlängenprogrammierung 1. parison length programming. 2. parison length programming device
Schlauchlängenregelung 1. parison length control *(bm)*. 2. parison length controller *(bm)*
Schlauchleitung pipeline
Schlauchliegebreite layflat width *(bfe)*
Schlauchreckverfahren blown film stretching/ orientation (process)
Schlauchrohling parison *(bm)*
Schlauchspreizvorrichtung parison stretching mandrel: *This is used in the production of flat canisters with side openings, where the blowing air would be insufficient to fully stretch the parison*

Schlauchspritzkopf 1. parison die *(bm)*. 2. blown film die, film blowing die *(bfe)*. 3. tube/ pipe die *(e)*
Schlauchstreckverfahren blown film stretching/ orientation (process)
Schlauchstück parison *(bm)*
Schlauchübernahmestation parison receiving station *(bm)*
Schlauchumfang 1. parison circumference *(bm)*. 2. film bubble circumference *(bfe)*
Schlauchvorformling parison *(bm)*
Schlauchwerkzeug 1. parison die *(bm)*. 2. blown film die, film blowing die *(bfe)*. 3. tube/ pipe die *(e)*
schlecht 1. bad. 2. poor *(when describing properties, performance, etc.)*
schlecht löslich having poor solubility
Schleichgang, im slow(ly)
Schleifband abrasive belt
Schleifen grinding
Schleifgewebe abrasive/emery cloth
Schleifkorn abrasive (material)
Schleifkörper abrasive/grinding wheel
Schleifleinen abrasive/emery cloth
Schleifmaterial abrasive (material)
Schleifmehl wood flour
Schleifmittel abrasive (material)
Schleifpapier sandpaper, emery paper
Schleifring slip ring
Schleifscheibe abrasive/grinding wheel
Schleifscheibenharz grinding wheel resin
Schleifstaub abrasive dust
Schleifwolle steel wool
Schleimhäute mucous membranes
schleimhautreizend irritating the mucous membranes
Schlempeschicht laitance
Schleppfluß drag flow *(e)*
Schleppstange drag link
Schleppstopfen floating plug/bung
Schleppströmung drag flow *(e)*
Schleuderanlage centrifugal casting plant
Schleudergießen centrifugal casting
Schleudergußteil centrifugally cast part/ component
Schleudergußverfahren centrifugal casting (process)
Schleuderlegierung centrifugal casting alloy
Schleudermaschine centrifugal casting machine
Schleuderrad impeller
Schleuderrohr centrifugally cast pipe
Schleuderverfahren centrifugal casting (process)
Schlichte size *(txt)*
Schlichteablagerungen size deposits *(txt)*
Schlichtemittel size *(txt)*
Schlieren streaks
Schlierenbildung streaking
schlierenfrei free from streaks
schlierig streaky

Schließ- und Öffnungsbewegungen (mould) closing and opening movements
Schließbewegung mould closing movement *(im)*
Schließdruck (mould) clamping pressure *(im)*
Schließe clamping unit
Schließeinheit clamp(ing) unit *(im)*
Schließeinrichtung clamping mechanism *(im)*
Schließenoberteil upper part of the clamping unit
Schließensteuerung clamping unit controls
Schließgeschwindigkeit 1. closing speed. 2. mould closing speed *(im)*
Schließgestell mould clamping frame *(im)*
Schließglied clamping element *(im)*
Schließhälfte clamp(ing) unit *(im)*
Schließhubsicherung mould safety mechanism
Schließkolben clamping unit ram, clamping cylinder ram, clamp ram *(im)*
Schließkraft 1. clamp(ing) force *(im)*. 2. locking force *(im)*: *The custom of writing* **Schließkraft** *instead of* **Zuhaltekraft** *(or the various alternative versions of these two words) and vice versa often creates translation problems which can only be solved by considering the word in context. For example, in a table quoting the* **Formschließkraft** *as 500 kN and* **Formzufahrkraft** *(q.v.) as 22 kN, it is obvious that the higher figure refers to the locking force and the lower one to the clamping force. Thus, although the two words appear to be the same, the figures indicate that* **schließ** *in this case is equivalent to* **zuhalte***, so that the first word must be translated as* mould locking force *and the second as* mould clamping force
Schließkraftaufbau clamping force build-up *(im)*
Schließkraftmeßplatte clamping force transducer
Schließkraftregelung 1. clamping force control *(im)*. 2. clamping force control device/system/mechanism
Schließkraftverlauf clamping force profile *(im)*
Schließkraftverteilung clamping force distribution
Schließmechanismus (mould) clamping mechanism *(im)*
Schließnadel needle shut-off mechanism *(im)*
Schließplatte moving platen *(im)*
Schließplatte, bewegliche moving platen *(im)*
Schließraum space pressurised during closing movement *(of toggle clamp unit)*
Schließseite 1. moving mould half *(im)*. 2. (mould) clamping unit *(im)*
schließseitig on the moving mould half *(im)*
schließseitige Werkzeughälfte moving mould half *(im)*
Schließstellung closed position
Schließsystem (mould) clamping mechanism *(im)*
Schließvorgang (mould) closing movement *(im)*

Schließzahl 1. clamp(ing) force *(im)*. 2. locking force *(im)*
Schließzeit mould closing time
Schließzylinder clamping cylinder *(im)*
Schliff 1. ground-glass joint. 2. ground-glass stopper
Schliffbild 1. roll contour diagram *(c)*. 2. photomicrograph of a polished section
Schliffstopfen ground-glass stopper
Schliffverbindung ground glass joint
Schlitten carriage, moving carriage
Schlitzdruckfestigkeit bursting strength *(particularly of roofing sheet)*
Schlitzdüse 1. slit die, slot die. 2. flat film (extrusion) die *(for gauges below 0.25 mm)*. 3. sheet (extrusion) die *(for thicker gauges)*
schlitzförmig slit-shaped/-like
Schlitzkapillare slit capillary
Schlitzvorrichtung slitting device, slitter
Schluckvolumen flow rating *(of pump)*
Schlupf slippage
Schlupflänge die land *(e)*
Schlüsseleigenschaften key properties
Schlüsselprodukt key product
Schlüsselrolle key role
Schlüsselschalter key-operated switch
Schlüsselstellung key position
Schlußfolgerung conclusion
Schlußlack finishing lacquer *(e.g. for PVC leathercloth)*
Schlußstrich top/finishing coat
schluüsselfertig turnkey *(meaning that a plant is handed over to the client ready for operation)*
Schmalbauweise narrow construction: **Einstationenmaschinen in Schmalbauweise** narrowly constructed single-station machines
Schmälze size *(txt)*
schmelzbar fusible
Schmelzbarkeit fusibility
Schmelzbereich melting range
Schmelzdehnbarkeit melt elasticity/extensibility
∗**Schmelze** melt
Schmelze, verarbeitbar aus der melt processable
Schmelzeaustrag melt delivery/discharge: **um einen Schmelzeaustrag innerhalb von 20-30 Minuten zu gewährleisten** to ensure that the melt is discharged within 20-30 minutes
Schmelzeaustragsextruder melt (fed) extruder, hot melt extruder
Schmelzeaustragsschnecke melt delivery screw
Schmelzeaustritt 1. melt delivery. 2. denoting the place where the melt leaves the machine. 3. escape of melt *(for translation examples see under* **Masseaustritt***)*
Schmelzebahn melt strand, extrudate

∗ *Words starting with* **Schmelze-***, not found here, may be found under* **Masse-**

Schmelzebehälter melt accumulator *(bm)*
Schmelzebohrung melt flow-way, runner *(im)*
Schmelzebruch melt fracture
Schmelzedehnung melt extensibility/elasticity
Schmelzedekompression 1. melt decompression, melt devolatilisation. 2. melt decompression system, melt devolatilising system
Schmelzedepot melt accumulation
Schmelzedosierpumpe melt metering/dispensing pump
Schmelzedosierung 1. melt metering/dispensing. 2. melt metering/dispensing unit
Schmelzedruck melt pressure
Schmelzeeinspeisung melt feed system
Schmelzeelastizität melt elasticity/extensibility
Schmelzeendtemperatur final melt temperature
Schmelzeentgasung 1. melt degassing/devolatilisation. 2. melt degassing/devolatilising system, melt degassing/devolatilising unit
Schmelzeextruder melt extruder, hot melt extruder, melt fed extruder
Schmelzefaden melt thread
Schmelzefestigkeit melt elasticity/extensibility: *This German word is misleading since what is actually meant is the extensibility or elasticity of a polymer melt under tensile stress (see Stoeckhert, Kunststoff-Lexikon, 7th ed. p.457)*
Schmelzefilter melt filter, melt filtration unit
Schmelzefiltrierung melt filtration
Schmelzefluß melt flow: **....während des gesamten Schmelzeflusses** ...whilst the melt is in motion
schmelzeflüssig melted, molten
Schmelzeförderung melt transport
Schmelzefront melt/flow front
Schmelzeführung passage of melt
  Schmelzeführung im Speicherkopf passage of the melt through the accumulator head
Schmelzeführungsbohrung melt flow-way, runner *(im)*
Schmelzeführungskanal melt flow-way, runner *(im)*
Schmelzehomogenisierextruder compounding extruder
Schmelzehomogenität melt homogeneity
Schmelzekanal 1. melt flow-way, runner *(im)*. 2. screw channel *(e)*
Schmelzeleitsystem melt flow-way system
Schmelzeleitung melt flow-way
Schmelzemischen melt mixing
Schmelzenfiltereinrichtung melt filtration unit
Schmelzenfilterung 1. melt filtration. 2. melt filtration unit
Schmelzenrückspülung melt back flushing *(mf)*
Schmelzenstromunterbrechung melt flow/stream interruption: **Der kontinuierliche Siebwechsler gestattet den Siebwechsel ohne Schmelzenstromunterbrechung** the continuous screen changer enables screens to be changed without interrupt ing the melt stream
Schmelzenthalpie melt enthalpy
Schmelzentropie entropy of fusion
Schmelzepfad melt stream
Schmelzepolster melt cushion
Schmelzequalität melt quality
Schmelzerückfluß melt back-flow, reverse melt flow
Schmelzescherung melt shear
Schmelzeschicht melt layer
Schmelzeschlauch 1. molten tube *(bfe)* (*as it comes out of the extruder, before being expanded into a* **Folienschlauch**, *q.v.*). 2. parison *(bm)*
Schmelzeschlauches, Rollen des parison curl *(bm)*
Schmelzespeicher melt accumulator *(bm)*
Schmelzespeicherkolben melt acumulator ram *(bm)*
Schmelzestabilität melt stability *(e.g. at high temperatures)*
Schmelzestillstand melt flow standstill **...bis zum Schmelzestillstand** ... until the melt comes to a halt OR: ...until the melt stops flowing
Schmelzestrang melt strand, extrudate
Schmelzestrom 1. melt stream, melt. 2. melt throughput, melt flow rate *(for translation examples see under* **Massestrom***)*
Schmelzestromführung passage of melt *(for translation example see under* **Schmelzeführung***)*
Schmelzestromteiler device which divides the melt stream
Schmelzeteilchen melt particle
schmelzeteilend melt dividing:
  schmelzeteilende Hindernisse obstacles which divide the melt stream
Schmelzeteilströme separate melt streams *(e)*: *(a melt stream divided by a spider)*
Schmelzetemperatur melt temperature
Schmelzetemperaturfühler melt thermocouple
Schmelzeumlagerung melt circulation
Schmelzeumlenkventil melt deflector
Schmelzeverteiler manifold *(e)*
Schmelzeverteilung melt distribution:
  Isolierkanal zur Schmelzeverteilung insulated runner for distributing the melt
Schmelzeviskosität melt viscosity
Schmelzewendelverteiler spiral mandrel (melt) distributor *(e,bm)*
Schmelzewendelverteilerkopf 1. spiral mandrel die *(e)*. 2. spiral mandrel blown film die *(bfe)*
Schmelzewendelverteilerwerkzeug 1. spiral mandrel die *(e)*. 2. spiral mandrel blown film die *(bfe)*
Schmelzewirbel melt vortex
Schmelzezufluß 1. melt feed. 2. melt feed system
Schmelzezustand condition of the melt
Schmelzfiltrieranlage melt filtration plant

schmelzflüssig melted: **schmelzflüssige Masse** polymer melt
schmelzförmig melted
schmelzgesponnen melt-spun
Schmelzhaftklebstoff hot melt pressure sensitive adhesive
Schmelzindex melt flow index, MFI
Schmelzindexwert melt flow index
Schmelzintervall melting range
Schmelzkerntechnik investment casting, lost wax casting
Schmelzklebefolie hot-melt film adhesive
Schmelzkleber hot melt adhesive
Schmelzklebstoff hot melt adhesive
Schmelzklebstoffauftragsgerät hot melt adhesive applicator
Schmelzmasse hot melt compound
Schmelzpolymerisat bulk polymer
Schmelzpunkt melting point
Schmelzpunktanhebung increase in melting point
Schmelzspinnen melt spinning
Schmelztellerextruder sinter plate extruder
Schmelztemperatur melting point
Schmelztemperaturbereich melting range
Schmelzviskosität melt viscosity
Schmelzwalzenbeschichtung hot melt coating
Schmelzwalzenbeschichtungsverfahren hot melt coating
Schmelzwalzenkalander laminating calender
Schmelzwalzenmaschine hot melt coater, hot melt coating machine
Schmelzwärme heat of fusion
Schmelzzone transition/plasticating section *(of screw)*
Schmiereffekt 1. lubricating effect. 2. smearing *(e.g. when polishing)*
Schmiereigenschaften lubricating properties
schmierend lubricating
Schmierfähigkeit lubricating properties, lubricity
Schmierfett lubricating grease
Schmierfilm lubricating film
Schmierkopf smear head, torpedo *(of screw) (e)*
Schmierkraft lubricating effect
Schmiermittel lubricant
Schmiermittelfilm lubricating film
Schmiermittelwirkung lubricating effect
Schmieröl lubricating oil
Schmieröltank lubrication oil reservoir
Schmierstellen lubricating points
Schmierstoff lubricant
Schmiersystem lubricating system
Schmierung lubrication
schmierungsfrei not requiring lubrication
Schmierwirkung 1. lubricating effect. 2. smearing *(e.g. when polishing)*
Schmirgelleinen emery cloth
schmirgeln to sand (down), to rub down
Schmirgelpapier emery paper
Schmirgeltuch emery cloth
Schmirgelwolle steel wool
schmutzabweisend dirt repellent

Schmutzpartikel dirt particle
Schmutzspeichervermögen dirt capacity *(of a filter) (mf)*
Schmutzstoff contaminant
Schmutzteilchen dirt particle
Schnappverbindung snap-in joint
Schnecke screw *(e)*
Schnecke, durchgeschnittene fully-flighted screw
Schnecke, flache shallow-flighted screw *(e)*
Schnecke, kernprogressive constant taper screw *(screw with constantly increasing root diameter)*
Schnecke, kompressionslose 1. screw *(general term)*. 2. transport/conveying screw *(for translation examples see under* **Förderschnecke**)
Schnecke, steigungsdegressive decreasing pitch screw
Schnecken- und Knetelemente screw flights and kneader discs
Schneckenabnutzung screw wear
Schneckenabschnitt screw section
Schneckenabstützung screw support
Schneckenachse screw axis
Schneckenaggregat screw unit
Schneckenantrieb screw drive
Schneckenantriebsart type of screw drive
Schneckenantriebsleistung screw drive power
Schneckenantriebsmotor screw drive motor
Schneckenantriebssystem screw drive mechanism
Schneckenantriebswelle screw drive shaft
Schneckenarbeitslänge effective screw length
Schneckenart type of screw
Schneckenausbau removal of screw: **Um einen Schneckenausbau zu ermöglichen** to enable the screw to be taken out
Schneckenausbildung screw design/configuration
Schneckenausführung screw design/configuration: **verschiedene Schneckenausführungen** different types of screw
Schneckenauslegung screw design/configuration *(for translation example see* **Schneckenausführung**)
Schneckenausrüstung screw assembly *(e)*
Schneckenaußendurchmesser external screw diameter
Schneckenaustausch changing the screw
Schneckenballen flattened screw tip
Schneckenbauart type of screw
Schneckenbauform screw configuration
Schneckenbaukasten 1. screw assembly. 2. modular screw, screw built up from segments
Schneckenbaukastensystem modular screw system
Schneckenbaulänge screw length *(for translation example see under* **Schneckenlänge**)

Schneckenbereich section/part of a screw: **Es sollte untersucht werden, wie dieser Schneckenbereich weiter verbessert werden kann** tests were carried out to find out how this part of the screw could be further improved
Schneckenbewegung screw movement
Schneckenbruch screw fracture: **Um beim Anfahren einen Schneckenbruch zu vermeiden...** to prevent the screw breaking/fracturing when starting-up
Schneckenbruchsicherung 1. preventing screw fracture. 2. device for preventing screw fracture
Schneckenbuchse screw bushing
Schneckendekompressionseinrichtung screw venting/decompression system *(e)*
Schneckendirekteinspritzung direct screw injection system *(im)*
Schneckendosiereinheit screw feeder
Schneckendosierer screw feeder
Schneckendosiergerät screw feeder
Schneckendosiervorrichtung screw feeder
Schneckendrehbewegung screw rotation
Schneckendrehmoment screw torque
Schneckendrehrichtung direction of screw rotation
Schneckendrehsinn direction of screw rotation
Schneckendrehung screw rotation
Schneckendrehzahl screw speed
Schneckendrehzahlbereich screw speed range
Schneckendrehzeit time during which the screw is rotating: **Nach Beendigung der Schneckendrehzeit** when the screw has stopped rotating
Schneckendurchmesser screw diameter
Schneckeneigenschaften 1. screw characteristics. 2. screw performance
Schneckeneingriff screw intermeshing: **Die infolge des Schneckeneingriffs bedingte Kammerbildung** the chambers formed by the intermeshing of two screws
Schneckeneinheit screw unit, screw assembly
Schneckeneinspritzaggregat screw injection unit *(im)*
Schneckeneinzugszone feed section/zone *(e,im)*
Schneckenelement screw flight
Schneckenende screw tip
Schneckenentwurf screw design
Schneckenextruder extruder
Schneckenflanke (screw) flank
Schneckenflanke, aktive thrust/front face of flight, leading edge of flight *(e)*
Schneckenflanke, hintere rear face of flight, trailing edge of flight *(e)*
Schneckenflanke, passive rear face of flight, trailing edge of flight *(e)*
Schneckenflanke, treibende thrust/front face of flight, leading edge of flight *(e)*
Schneckenflanke, vordere thrust/front face of flight, leading edge of flight *(e)*

Schneckenflügel (screw) flight/thread
Schneckenflügelflanke screw flight flank
Schneckenförderer screw conveyor
Schneckenfördergerät screw conveyor
Schneckenförderspitze flighted screw tip
Schneckenführung screw alignment: **Wie wichtig eine exakte Schneckenführung gerade im Einzugsbereich ist, läßt sich erkennen...** the importance of accurately aligning the screw, especially in the feed section, is evident...
Schneckenführungszylinder (extruder) barrel
Schneckengang 1. (screw) channel *(where the screw and barrel are treated as one unit) (e)*. 2. (screw) flight *(where the screw is referred to in isolation, i.e. without the barrel) (e)* (For translation examples see under **Gang**)
Schneckengangfüllung screw flight contents, amount of material in the screw flight(s)
Schneckengangkanal screw channel
Schneckengangprofil (screw) channel profile
Schneckengangtiefe flight depth *(e) (see explanatory note under* **Kanaltiefe**)
Schneckengangvolumen (screw) channel volume *(e)*
Schneckengangzahl number of flights *(e)*
Schneckengarnitur screw (assembly)
Schneckengegendruck screw back pressure *(e)*
Schneckengehäuse (extruder) barrel
Schneckengehäuseschüsse barrel sections
Schneckengeometrie screw configuration/geometry
Schneckengesamtdrehmoment total screw torque
Schneckengeschwindigkeit screw speed
Schneckengestalt screw configuration
Schneckengestaltung screw configuration/design
Schneckengetriebe 1. screw drive. 2. worm gear
Schneckengewinde screw flights/thread
Schneckengröße screw size
Schneckengrund (screw) root surface *(e)*
Schneckengrundspalte flight clearance *(see explanatory note under* **Kopfspiel** *with which the word is synonymous)*
Schneckenhub screw stroke
Schneckenhubeinstellung 1. screw stroke adjustment. 2. screw stroke adjusting mechanism
Schneckeninnendurchmesser internal screw diameter
Schneckeninnenkühlung internal screw cooling system
Schneckenkamm flight land *(e)*
Schneckenkanal screw channel
Schneckenkanalgrund (screw) root surface *(e)*
Schneckenkanaloberfläche (screw) root surface *(e)*
Schneckenkanalprofil (screw) channel profile

Schneckenkanaltiefe (screw) channel depth *(see explanatory note under* **Kanaltiefe***)*
Schneckenkanalvolumen screw channel volume
Schneckenkennlinie screw characteristic
Schneckenkennzahl screw characteristic
Schneckenkern screw root
Schneckenkerntemperierung 1. screw root temperature control. 2. screw root temperature control system
Schneckenkneter screw compounder/plasticator
Schneckenkneter, einwelliger single-screw compounder/plasticator
Schneckenkneter, zweiwelliger twin screw compounder/plasticator
Schneckenkolben screw-plunger/reciprocating screw *(im)*
Schneckenkolbeneinheit screw-plunger/reciprocating-screw unit *(im)*
Schneckenkolbeneinspritzaggregat screw-plunger/reciprocating-screw injection unit
Schneckenkolbeneinspritzsystem screw-plunger/reciprocating-screw injection system
Schneckenkolbeneinspritzung 1. screw-plunger/reciprocating-screw injection. 2. screw-plunger/reciprocating-screw injection unit
Schneckenkolbeninjektion 1. screw-plunger/reciprocating-screw injection. 2. screw-plunger/reciprocating-screw injection unit
Schneckenkolbenmaschine screw-plunger/reciprocating-screw machine
Schneckenkolbenspeicher reciprocating-screw accumulator *(bm)*
Schneckenkolbenspritzgießmaschine screw-plunger/reciprocating-screw injection moulding machine
Schneckenkolbenspritzsystem screw-plunger/reciprocating-screw injection system
Schneckenkompression compression ratio *(e)*
Schneckenkonfiguration screw configuration
Schneckenkonstruktion screw design
Schneckenkonzept screw design
Schneckenkonzeption screw design
Schneckenkopf screw tip
Schneckenkühlung 1. screw cooling. 2. screw cooling system
Schneckenlagerung screw support
Schneckenlänge screw length: **kurze Schneckenlänge** short screw; **Doppelschneckenextruder mit übergroßer Schneckenlänge** twin-screw extruder with excessively long screws
Schneckenlauffläche screw wearing surface
Schneckenleistung screw performance, screw output
Schneckenmaschine 1. screw injection moulding machine. 2. extruder: **Es resultiert ein trockenes Produkt, das problemlos auf Schneckenmaschinen verarbeitet werden kann** the result is a dry material which can be extruded or injection moulded without difficulty; **zweiwellige Schneckenmaschine** twin screw extruder. 3. screw compounder. 4. screw devolatiliser
Schneckenoberfläche 1. screw surface. 2. screw surface area
Schneckenpaar twin screws, twin screw assembly
Schneckenplastifizieraggregat screw plasticising unit
Schneckenplastifiziereinheit screw plasticising unit
Schneckenplastifizierung 1. screw plasticisation/plastication. 2. screw plasticising unit
Schneckenplastifizierungszylinder screw plasticising cylinder *(im)*
Schneckenpresse extruder
Schneckenprofil screw profile
Schneckenrohr (extruder) barrel
Schneckenrotation 1. screw rotation. 2. screw rotating mechanism
Schneckenrückdrehsicherung device to prevent screw retraction
Schneckenrückdrehung screw retraction
Schneckenrückdruck screw back pressure
Schneckenrückdruckkraft screw back pressure
Schneckenrückholung screw retraction
Schneckenrückholvorrichtung screw retraction mechanism
Schneckenrückhub screw return stroke
Schneckenrücklauf screw retraction
Schneckenrücklaufgeschwindigkeit screw return speed
Schneckenrücklaufzeit screw return time
Schneckenrückzug screw retraction
Schneckenrückzugkraft screw retraction force
Schneckensatz screw assembly
Schneckensatzelemente screw assembly components
Schneckenschaft screw shank
Schneckenscherspalt radial screw clearance
Schneckenschub reciprocating screw
Schneckenschubaggregat reciprocating-screw unit
Schneckenschubmaschine reciprocating-screw machine
Schneckenschubplastifizieraggregat reciprocating-screw plasticising unit
Schneckenschubspeicher reciprocating-screw accumulator *(bm)*
Schneckenschubzylinder reciprocating-screw cylinder
Schneckensortiment screw range, range of screws
Schneckenspalt flight land clearance, inter-screw clearance *(e) (clearance between the flight lands of twin screws)*
Schneckenspeiseeinrichtung screw feeder
Schneckenspiel (radial) screw clearance
Schneckenspitze screw tip
Schneckenspitze, glatte unflighted screw tip

Schneckenspitzenstege flights near the screw tip
Schneckenspritzeinheit screw injection unit *(im)*
Schneckenspritzgießautomat automatic screw-plunger/reciprocating-screw injection moulding machine
Schneckenspritzgießmaschine screw injection moulding machine
Schneckenspritzzylinder screw injection cylinder
Schneckenstaudruck screw back pressure
Schneckenstauraum 1. space in front of the screw. 2. in front of the screw
Schneckensteg (screw) flight/thread
Schneckenstegfläche flight land *(e)*
Schneckensteigung pitch *(e)*
Schneckensteigungswinkel helix angle *(e)*
Schneckenstellung screw position
Schneckenstrangpresse extruder
Schneckenstufe screw section
Schneckensystem screw system/arrangement
Schneckenteil screw section
Schneckentemperierung 1. screw temperature control. 2. screw temperature control system
Schneckentiefe screw depth
Schneckenumdrehung screw revolution
Schneckenumdrehungszahl screw speed
Schneckenumfang screw circumference
Schneckenumfangsgeschwindigkeit peripheral screw speed/velocity
Schneckenumgang screw turn
Schneckenumlaufgeschwindigkeit screw speed
Schneckenverdampfer screw devolatiliser
Schneckenverschleiß screw wear
Schneckenvorgelege worm gear
Schneckenvorlauf screw advance: **Wichtig ist ein gleichmäßiger Schneckenvorlauf beim Einspritzen** it is important that the screw advances evenly during injection
Schneckenvorlaufbewegung screw forward movement
Schneckenvorlaufgeschwindigkeit 1. screw advance speed. 2. injection speed/rate
Schneckenvorplastifiziergerät screw preplasticiser, screw preplasticising unit
Schneckenvorplastifizierung 1. screw preplasticisation. 2. screw preplasticising unit
Schneckenvorraum 1. space in front of the screw. 2. in front of the screw
Schneckenvorschub screw advance
Schneckenwechsel changing the screw: **...die den Schneckenwechsel erleichtern** ...which make it easier to change the screw
Schneckenweg screw stroke
schneckenwegabhängig depending on the screw stroke
Schneckenwegaufnehmer screw stroke transducer

Schneckenwegeinstellung 1. screw stroke adjustment. 2. screw stroke adjusting mechanism
Schneckenwelle 1. screw shank. 2. screw: **Die Schneckenwelle ist eine verchromte 3-Zonenschnecke von 200 mm Durchmesser** the screw is chromium plated, divided into three sections and has a diameter of 200 mm
Schneckenwellendrehzahl screw speed
Schneckenwellenschaft screw shank
Schneckenwendel screw flight
Schneckenwindungen screw threads
Schneckenzone screw section
Schneckenzylinder 1. (extruder) barrel *(e)*. 2. (preplasticising) cylinder *(im)*
Schneidaggregat 1. slitting device *(for making film tape from film)*. 2. cutting device *(general term)*
schneiden to cut
Schneiden, funkenerosives spark erosion, electric discharge machining, EDM
Schneidgeschwindigkeit cutting speed
Schneidgranulator 1. granulator *(for cutting up scrap)*. 2. pelletiser *(for cutting up extruded strands to make pellets)*
Schneidkammer cutting chamber *(sr)*
Schneidkante cutting edge *(of knife)*
Schneidkanten 1. pinch-off, pinch-off bars/edges/inserts *(bm)*. 2. cutting edges *(of knives)*
Schneidmühle 1. granulator *(sr)*. 2. pelletiser *(see explanatory notes under* **Schneidgranulator***)*
Schneidöl cutting oil
Schneidraum cutting chamber *(sr)*
Schneidrotor rotor *(sr)*: Since the word is normally used in descriptions of granulators where the context is clear, the English version given is adequate
Schneidrovings chopping rovings
Schneidspalt blade/knife clearance *(sr)*: This is the gap between the bed knife and rotor knife of a granulator
Schneidvorrichtung 1. slitting device *(for making film tape from film)*. 2. cutting device *(general term)*
Schneidwalze rotor *(sr)* *(see explanatory note under* **Schneidrotor***)*
schnellabbindend fast-setting, fast-curing
Schnellarbeitsstahl high speed steel
schnellaufend 1. high speed *(when describing a continuous process such as extrusion, calendering, etc.)*. 2. fast cycling *(when describing an intermittent process such as injection or blow moulding)*
Schnelläufer high speed machine, fast cycling machine *(see explanatory note under* **schnellaufend***)*
Schnelläuferextruder high speed extruder
Schnelläuferspritzgießmaschine fast cycling injection moulding machine

Schnellbewitterung accelerated weathering test
Schnellbewitterungsgerät accelerated weathering instrument
Schnellbewitterungsversuch accelerated weathering test
Schnelldrucker high speed printer *(me)*
Schnelleinspritzen high speed injection *(im)*
schnellerstarrend fast-solidifying/-setting
Schnellfahrzylinder high speed (injection) cylinder *(im)*
schnellfließend fast flowing
schnellgelierend 1. fast-gelling *(UP, EP resins)*. 2. fast-solvating/gelling *(PVC in plasticisers)*
schnellhärtend fast curing/setting
Schnellhärter fast-reacting hardener, high speed hardener
Schnellhärtung accelerated cure
Schnelligkeit speed
Schnellkleber fast-setting adhesive
Schnellkupplung quick-action coupling
Schnellmischer high speed mixer
Schnellpolymerisation fast/high-speed polymerisation
Schnellrührer high speed stirrer/mixer
Schnellschließzylinder high speed clamping cylinder *(im)*
Schnellschliffgrundierung quick sanding primer
Schnellschweißdüse high speed welding nozzle
Schnellschweißen high speed welding
Schnellsiebwechseleinrichtung quick-action screen changer *(mf)*
Schnellspannsystem quick-action coupling/ mounting system
Schnellspannvorrichtung quick-action coupling/mounting mechanism
Schnellspritzgießautomat automatic fast cycling injection moulding machine
Schnelltest accelerated test
schnelltrocknend fast drying
Schnelltrockner high speed dryer, high speed drying oven
Schnellverbindung quick-action coupling
Schnellverfahren quick method
Schnellverschluß quick-action clamp
Schnellverschlußkupplung quick-action coupling
Schnellwechselfilter quick-change filter unit
Schnitt 1. cut. 2. section *(e.g. through a test piece)*
Schnittbeständigkeit cutting resistance
Schnittbild sectional drawing/diagram
Schnittdarstellung sectional drawing/diagram
Schnittfläche cut surface
Schnittgeschwindigkeit cutting speed
Schnittglasfasern chopped strands
Schnittgüte cutting quality
Schnittkante cut edge
Schnittkraft cutting force *(sr)*
Schnittlänge length *(e.g. of cut fibres)*
   Glasfasern in verschiedenen Schnittlängen glass fibres cut to different lengths
Schnittpräzision cutting accuracy *(sr)*
Schnittpunkt point of intersection
Schnittspiel cutting clearance
Schnittstelle interface *(me)*
Schnittwerkzeug cutting tool
Schnittwinkel cutting angle
Schnittzeichnung sectional drawing
schnitzelförmig chopped (up) *(cotton cloth etc., used as filler)*
Schnitzelmühle shredder *(e.g for film offcuts) (sr)*
Schnüffelventil relief valve
Schockabsorber shock absorber
schockfest shock resistant, shockproof
Schockfestigkeit shock resistance
schockgekühlt quenched
Schockkühlung sudden cooling
schockresistent shock resistant, shockproof
Schockzähigkeit shock resistance
schonend gentle, gently, careful(ly): **schonende Aufbereitung** gentle compounding
Schonung 1. gentle treatment **aus Gründen thermischer Schonung des Extrudats** to prevent the extrudate becoming overheated; **größtmögliche Schonung des Werkzeugs** maximum mould protection. 2. conservation: **Schonung der Rohstoffressourcen** conserving raw material resources
Schöpfraum pumping space
Schornstein chimney
schraffiert shaded *(area of a graph)*
Schrägbolzen angled bolt
Schrägkopf 1. crosshead (die) *(e)*. 2. angled extrusion die head *(e)*
Schrägspritzkopf angled extrusion die head
Schrägverstellung cross-axis roll adjustment *(c)*, axis/roll crossing *(c)*
schrägverzahnt helical
Schrägverzahnung helical gearing
Schraube screw
Schraube, gewindeformende self-tapping screw
Schraubenanschluß screw coupling/connection
Schraubendreher screwdriver
Schraubenfeder spiral/helical spring
Schraubenkopf screw head
Schraubenzieher screwdriver
Schraubkappe screw cap
Schraubkern threaded core *(im)*
Schraubverschluß screw cap
Schraubwerkzeug unscrewing mould *(im)*
Schraubzwinge screw clamp
Schreddermüll shredded waste
Schreiber write/writing/recording head *(me)*
Schreib-Lesespeicher write-read memory *(me)*
Schrifttum published literature
Schriftzeile display line, character row *(me)*
Schrittfolgesteuerung sequence control *(me)*
Schrittmotor stepper motor
schrittweise gradual
Schrotstrahlen grit blasting

Schrott 1. scrap. 2. scrap metal
Schrottschnecke worn screw *(the word is derived from* **Schrott**, *meaning scrap metal)*
Schrottstoßfänger scrap bumpers
Schrottwirtschaft scrap metal industry
Schrumpf shrinkage
schrumpfarm 1. low-shrinkage *(general term)*. 2. low-profile, LP *(applied to unsaturated polyester resins)*
Schrumpffolie shrink wrapping film
schrumpffrei non-shrink(ing)
schrumpfkompensiert low-profile *(UP resins)*
schrumpfmindernd shrinkage-reducing
Schrumpfneigung tendency *(of something)* to shrink
Schrumpftunnel shrink tunnel
Schrumpfung shrinkage
Schrumpfungskompensation shrinkage compensation
Schrumpfungswert shrinkage (value)
Schrumpfverpackung 1. shrink wrapping. 2. shrink wrapped pack
Schrumpfwert shrinkage (value)
Schubbeanspruchung shear stress
Schubbewegung reciprocating movement
Schubbruchspannung shear stress at break
Schubdeformation shear deformation
Schubfestigkeit shear strength
Schubflanke thrust/front face of flight, leading edge of flight *(e)*
Schubkraft shear force
Schubmodul shear modulus
Schubscherfestigkeit shear strength
Schubschnecke reciprocating screw
Schubschneckeneinheit reciprocating-screw unit
Schubschneckenextruder reciprocating-screw extruder
Schubschneckenmaschine reciprocating-screw machine
Schubschneckenspritzgießmaschine reciprocating-screw injection moulding machine
Schubschwellfestigkeit shear fatigue strength
Schubspannung shear stress
schubsteif shear resistant, resistant to shear forces/stress
Schubsteifigkeit shear strength
schubweich not resistant to shear, affected by shear
Schuhabsatz heel
Schuhindustrie shoe industry
Schuhsohle (shoe) sole
Schuhsohlenmaterial soling material
Schulterbutzen shoulder flash *(bm)*
Schulterprobe dumbbell test piece/specimen
Schulterquetschkante shoulder pinch-off *(bm)*
Schulterstab dumbbell test piece/specimen
Schulterzugstab dumbbell tensile test piece/specimen
Schulung training, instruction
Schulungsprogramm training programme
Schulungszeit training period
Schulungszentrum training centre
Schuppen flakes
Schurre chute *(sr)*
Schuß 1. shot *(im)*. 2. section
Schußdraht weft wire
Schußdruck injection pressure *(im)*
schußfest bulletproof
Schußfolge shot sequence: *The translation of this word very much depends on the context, as the following example shows:* **Mit beiden Maschinen läßt sich eine stündliche Schußfolge von 240 erreichen** both machines are capable of achieving 240 shots/min.
Schußgewicht shot weight *(im)*
Schußgewichtsschwankungen shot weight variations/fluctuations
Schußgröße shot weight *(im)*
Schußleistung shot capacity *(im)*
Schußmasse shot weight *(im)*
Schußrichtung weft direction *(txt)*
Schußvolumen shot volume *(im)*
Schußvolumen, mögliches shot capacity *(im)*
Schußzahl number of shots *(im)*
Schußzeit injection time, mould filling time *(im)*
schüttbar pourable
Schüttdichte apparent density
schütten to pour
Schüttfähigkeit pourability
Schüttgewicht apparent density
Schüttgut loose material, bulk solids, powder. *The word is used to describe any solid material which can be poured out of a bag, for example* (**schütten** = to pour). *It sometimes has to be translated very freely, e.g.:* **Materialtrichter zur Aufnahme des Schüttguts** *feed hopper to hold the moulding compound;* **...verhindert das Anbacken von Schüttgütern** *...prevents powders sticking (i.e. to the walls)*
Schütz contactor, relay
Schutzabdeckung safety hood
Schutzanstrich protective coating
Schutzbekleidung protective clothing
Schutzbeschichtung protective coating
Schutzbrille (safety) goggles
Schutzcreme barrier cream
Schütze shuttle *(txt)*
Schutzeinrichtung safety device
schützengesteuert relay-controlled
Schützensteuerung relay control
Schutzfolie protective film
Schutzfunktion protective function
Schutzgas inert gas
Schutzgasatmosphäre inert gas atmosphere
Schutzgitter guard door, safety guard
Schutzgittersicherung guard door interlock system/mechanism
Schutzglasur protective glaze
Schutzhandschuhe protective gloves
Schutzhaube safety hood

Schutzhelm 1. safety helmet. 2. crash helmet
Schutzhülse protective sleeve
Schutzkleidung protective clothing
Schutzkolloid protective colloid
Schutzlack protective enamel
Schutzmaske face mask
Schutzmaßnahmen 1. safety precautions. 2. protective action
Schutzrechte patent rights
Schutzschicht 1. protective coating/layer. 2. gel coat *(grp)*
Schutztür guard door
Schutztürsicherung guard-door interlock system/mechanism
Schutzüberzug protective coating
Schutzverdeck safety cover/hood
Schutzverkleidung safety shield/cover/hood
Schutzvorrichtung safety device
Schutzvorschriften safety regulations
Schwabbelscheibe buffing wheel
schwach vernetzt loosely crosslinked
schwachpigmentiert slightly pigmented, with a low pigment content
schwachpolar slightly polar
Schwachpunkt weak spot/point
schwachstabilisiert slightly stabilised, with a low stabiliser content
Schwachstelle weak spot/point
Schwalbenschwanzausnehmung swallowtail recess
Schwammgummi foam rubber
Schwankungen fluctuations, variations
Schwankungsbreite variation range
Schwärzung blackening
Schwebstoffilter filter for suspended particles
Schwebstoffteilchen suspended particles
Schwefel sulphur
schwefelarm low-sulphur, with a low sulphur content
Schwefelbeschleuniger sulphur accelerator
Schwefeldosierung 1. addition of sulphur 2. amount of sulphur
schwefelfrei sulphur-free, non-sulphur
Schwefelgehalt sulphur content
Schwefelglied sulphur segment
schwefelhaltig sulphur-containing, containing sulphur
Schwefelkohlenstoff carbon disulphide
Schwefelmenge amount of sulphur
Schwefelsäure sulphuric acid
Schwefelspender sulphur donor
Schwefelverbindung sulphur compound
schwefelvernetzbar sulphur-vulcanisable
schwefelvernetzt sulphur-vulcanised/-cured
Schwefelvernetzung vulcanisation with sulphur
Schwefelvulkanisat sulphur-vulcanised/-cured rubber
Schwefelvulkanisation sulphur vulcanisation/cure
Schwefelvulkanisationssystem sulphur vulcanising/curing system
schwefelvulkanisiert sulphur-vulcanised/-cured

Schwefelwasserstoff hydrogen sulphide
Schweißamplitude welding amplitude
Schweißaufnahme welding fixture
Schweißbacken welding bars
schweißbar weldable
Schweißbarkeit weldability
Schweißbedingungen welding conditions
Schweißbereich weld zone
Schweißdraht welding rod
Schweißdruck welding pressure
Schweißdüse welding nozzle
Schweißelektrode welding electrode
schweißen to weld
Schweißenergie welding energy
Schweißfaktor welding factor
Schweißfuge weld, welded joint
Schweißgas welding gas
Schweißgasversorgung welding gas supply
Schweißgerät welding instrument/torch
Schweißgeschwindigkeit welding speed/rate
Schweißgut 1. material being welded. 2. material to be welded
Schweißgutstrang welding rod
Schweißkanten pinch-off, pinch-off bars/edges/inserts *(bm)*
Schweißkonstruktion welded: **Das Maschinenbett ist als Schweißkonstruktion ausgeführt** the machine frame is welded
Schweißkraft welding pressure
Schweißlehre welding jig
Schweißnaht 1. weld, seam. 2. pinch-off weld *(bm)*. 3. weld/flow line *(im)* (for explanatory note see under **Bindenaht**)
Schweißnahtfestigkeit weld strength
Schweißnahtgüte weld quality
Schweißnahtqualität weld quality
Schweißparameter welding conditions
Schweißpreßdruck welding pressure
Schweißpresse welding press
Schweißprozeß welding (operation/process)
Schweißraupe bead *(w)*
Schweißrestspannungen residual welding stresses
Schweißrestspannungsverteilung distribution of residual welding stresses
Schweißschnur welding rod
Schweißspiegel hot plate welding tool
Schweißstab welding rod
Schweißtechnik welding technology
Schweißverbindung welded joint
Schweißvorgang welding operation/process
Schweißwerkzeug welding tool
Schweißwulst weld bead
Schweißzeit welding time
Schweißzone weld zone
Schweißzusatz welding rod
Schweißzusatzdraht welding rod
Schweißzusatzstoff welding rod
Schweißzusatzwerkstoff welding rod
schwelen to smoulder
Schwellbeanspruchung fatigue stress
Schwellbelastbarkeit fatigue endurance

Schwellbelastung fatigue stress
schwellende Belastung fatigue stress
Schwellenländer developing countries
Schwellenwert limit
Schweller door sill
Schwellfestigkeit 1. fatigue strength. 2. resistance to swelling
Schwellverhalten 1. swelling characteristics. 2. die swell *(e)*. 3. fatigue properties
Schwellwert limit
Schwenkaggregat tilting unit
schwenkbar swing-back, swing-hinged, swivel-mounted, swivel-type, hinge-mounted, hinged *(machine unit)*
Schwenkbarkeit *denotes that a machine or part thereof can be tilted or swivelled. For translation example see under* **Drehbarkeit**
Schwenkbewegung turning/rotating/swivel movement: **Schwenkbewegung der Maschine um jeweils 180° ist möglich** the machine can be rotated by 180°
Schwenkeinrichtung tilting mechanism/device
Schwenktisch tilting table
Schwenkwinkel tilting angle
schwer löslich difficult to dissolve
schwer, spezifisch heavy
schwerbrennbar flame retardant
Schwerbrennbarkeit flame resistance
schwerentflammbar flame resistant/retardant
Schwerentflammbarkeit flame resistance, flame retardant properties
schwerfließend poor-flow, with poor flow
schwerflüchtig low-volatility
Schwerflüchtigkeit low volatility
Schwerfolie thick sheeting, heavy gauge film
Schwerfolienwickler heavy-gauge film winder
Schwerfraktion heavy fraction
Schwergutsack heavy-duty sack
Schwergutsackfolie heavy-duty sack film
Schwergutteil heavy article
schwerklebbar difficult to bond/stick
Schwerkraft force of gravity
schwerlöslich sparingly soluble
Schwermetall heavy metal
schwermetallfrei free from heavy metals
schwermetallhaltig containing heavy metals
Schwermetallkomplex heavy metal complex
Schwermetallsalz heavy metal salt
Schwermetallverbindung heavy metal compound
Schweröl heavy oil
schwerplastifizierbar difficult to soften/plasticise/plasticate
Schwerpunkt 1. centre of gravity. 2. emphasis: **Der Schwerpunkt bei diesen Entwicklungsarbeiten war...** emphasis in this development work was on... OR: this development centred on...
Schwerpunktbereich most important part/point: **Die Schwerpunktbereiche sind die Elektrotechnik, allgemeiner Maschinenbau und die Textilindustrie** the most important industries are electrical and mechanical engineering and the textile industry
Schwersackfolie sheeting for heavy-duty sacks
Schwerspat barytes
Schwesterfirma affiliate
Schwierigkeiten, fördertechnische transport problems
Schwierigkeiten, verfahrenstechnische processing problems
Schwierigkeitsgrad degree of complexity
Schwimmbeckenfolie swimming pool liner
schwimmend floating
schwimmender Estrich floating screed/floor
Schwimmer float
Schwimmhaut flash
Schwimmkörper float
Schwindmaß (degree of) shrinkage
Schwindriß shrinkage crack
Schwindspannung shrinkage stress
Schwindung shrinkage
Schwindungsanisotropie anisotropic shrinkage
schwindungsarm low-shrinkage, low-profile *(especially used for UP resins)*
Schwindungsdifferenzen shrinkage differences/variations
Schwindungseigenschaften shrinkage properties/characteristics
Schwindungskatalog shrinkage record
Schwindungsspannung shrinkage stress
Schwindungsstab shrinkage test piece
Schwindungsunterschied shrinkage difference
Schwindungsuntersuchung shrinkage test
Schwindverhalten shrinkage characteristics
Schwingbeanspruchung cyclic/oscillating stress
schwingbelastet subjected to vibrational stress
Schwinge rocker arm
schwingend vibrating
Schwingfestigkeit fatigue strength
Schwingfestigkeitsverhalten fatigue strength
Schwingflügel swing casement
Schwingfrequenz vibration frequency
Schwingschleifen honing, superfinishing
Schwingschleifer honing machine
Schwingsieb vibrating screen
Schwingspiel (one complete) vibration
Schwingspielfrequenz vibration frequency
Schwingspielzahl number of (complete) vibrations
Schwingung vibration
Schwingungsamplitude vibration amplitude
Schwingungsbeanspruchung vibration stress
schwingungsdämpfend vibration-damping
Schwingungsdämpfer vibration damper
Schwingungsenergie vibrational energy
schwingungsfähig capable of vibration/oscillation
schwingungsfrei vibration-free
Schwingungsgerät torsion pendulum apparatus
Schwingungsversuch fatigue/dynamic/cyclic test
Schwingweite amplitude

Schwitzwasser condensed moisture, condensation
Schwitzwasserklima condensed moisture atmosphere
Schwitzwasserprüfung condensed moisture test
Schwund shrinkage
Schwundriß shrinkage crack
Schwungmasse centrifugal mass
scorchanfällig susceptible to scorching
Scorchgefahr risk of scorching, risk of premature vulcanisation
scorchsicher scorch resistant
Scorchsicherheit scorch resistance
Scorchtemperatur scorch temperature
Scorchzeit scorch value/time
SE abbr. of **Schallemision**, acoustic emission
SEA abbr. of **Schallemissionsanalyse**, acoustic emission analysis
Sebacat sebacate
Sebacinsäure sebacic acid
Sebacinsäureanhydrid sebacic anhydride
Sebacinsäureester sebacate
Sechsfachverteilerkanal six-runner arrangement *(im)*
Sechsfachwerkzeug six-cavity/-impression mould *(im)*
Sechskantrohr hexagonal-section pipe
Sechskantschraube hexagonal screw
Sechskantvollstab hexagonal-section solid rod
Sechszehn-Stationen-Drehtischmaschine 16-station rotary table machine
Sediment sediment
Sedimentation sedimentation, settling out
Sedimentationsanalyse sedimentation analysis
Sedimentationsneigung tendency to settle (out)
Sedimentationstendenz tendency to settle (out)
Sedimentbildung sedimentation, settling (out)
sedimentieren to settle out
Seele centre *(usually of a runner. See also entry under **Seele, plastische**)*
Seele, flüssige fluid centre
Seele, plastische fluid centre: **zwecks Erhaltung der plastischen Seele im Verteilerkanal...** in order to keep the material at the centre of the runner fluid...
Seeluft sea air
Seewasser seawater
Segmentbauweise, in constructed from units/modules
segmentiert segmented
Segmentpolymerisat block copolymer
seichtgeschnitten shallow-flighted *(screw) (e)*
seidenglänzend silk-finish
Seidenglanzlack silk-finish paint
Seife soap
Seifenlösung soap solution
Seil rope
Seilscheibe pulley
Seitenansicht side view
Seitenauswerfer side ejector/ejection mechanism *(im)*
Seitenbeschneidung edge trimmer/cutter
Seiteneinsprung necking *(of film)*
Seitenextruder ancillary/subsidiary extruder
Seitenfalteinrichtung side gussetting device
Seitenfaltenschlauchfolie side-gussetted blown/tubular film
Seitenführung lateral (film) bubble guide *(bfe)*
Seitengruppe side group
Seitenkette side chain
Seitenkettenflüssigkristall branched liquid cyrstal
Seitenkettenpolymer branched polymer
Seitenscheibe side window
Seitenschild side panel *(of a machine)*
Seitenschnecke ancillary/subsidiary screw
Seitenschneider edge trimmer/cutter
Seitenschneidvorrichtung edge trimming/cutting device
Seitenspeisevorrichtung lateral feed device
Seitenspoiler side spoiler
seitenständig lateral
Seitenverkleidungen side panels *(of a car)*
Seitenwand side wall
Seitenwandmischung side wall compound
seitlich angebunden side-gated, edge-gated, side-fed *(im)*
seitlich angespeist side-fed *(e, bm, bfe)*
seitlich ausschwenkbar can be swivelled/swung out sideways
sekundär secondary
Sekundärabzug secondary/subsidiary take-off (unit)
Sekundärachse secondary axis
Sekundärbindung secondary linkage
Sekundärextruder ancillary/subsidiary extruder
Sekundärkorn secondary particle
Sekundärkühlung secondary cooling system/unit
Sekundärpartikel secondary particle
Sekundärradikal secondary radical
Sekundärreaktion secondary reaction
Sekundärregister secondary register *(me)*
Sekundärrohstoff secondary raw material
Sekundärstruktur secondary structure
Sekundärteilchen secondary particle
Sekundärweichmacher secondary plasticiser
Sekundärwerkstoff recycled material
Sekundenbruchteil fraction of a second
Sekundenschnelle, in within seconds
selbstbeschleunigend self-accelerating
Selbstbindemittel self-curing binder
Selbstdiagnose self-diagnosis *(me)*
selbstdichtend self-sealing
selbstdosierend self-feeding
Selbsteinfärben in-plant colouring *(i.e. the addition of pigments by the moulder rather than by the moulding compound manufacturer)*
selbsteinstellend self-adjusting
selbstemulgierend self-emulsifying
Selbstentzündung self-ignition

Selbstentzündungstemperatur self-ignition temperature
selbsterklärend self-explanatory
selbsthaftend self-adhesive
selbsthärtend self-curing
selbstheilend self-healing
selbstisolierend self-insulating
Selbstklebemasse contact adhesive
selbstklebend self-adhesive
Selbstkleber contact adhesive
Selbstklebeschicht self-adhesive coating
selbstkonfigurierend self-configuring *(me)*
Selbstkontrolle self-regulation
selbstlöschend self-extinguishing
selbstnachstellend self-adjusting
selbstoptimierend self-adjusting
selbstregulierend self-regulating, self-adjusting
Selbstregulierung automatic adjustment
Selbstreinigung self-cleaning (effect)
selbstschmierend self-lubricating
Selbstschmierfähigkeit self-lubricating properties
Selbstschmierungseigenschaften self-lubricating properties
selbstschneidend self-tapping *(screw)*
selbstsichernd self-locking
selbsttätig automatic(ally)
Selbsttest self-test *(me)*
selbsttragend self-supporting
Selbsttrennung in-mould release
selbstüberwacht self-regulated
Selbstverfestigung self-reinforcement
selbstverlaufend self-levelling
selbstverlöschend self-extinguishing
selbstverlöschend ausgerüstet self-extinguishing
selbstvernetzend self-crosslinking, self-curing
selbstverriegelnd self-locking
Selbstverriegelung self-locking mechanism
selbstverschweißend self-sealing
selbstverstärkend 1. self-reinforcing. 2. liquid crystalline
selbstverstärkende Kunststoffe self-reinforcing polymers, liquid crystal polymers, LCPs
Selbstverstärkungseffekt self-reinforcing effect
selbstvulkanisierend self-vulcanising
selbstzentrierend self-centring
selektiv selective
Selektivität selectivity
SEM 1. *abbr. of* scanning electron microscope. 2. *abbr. of* scanning electron microscopy
Semicarbazid semicarbazide
semikristallin semi-crystalline
Semipermeabilität semi-permeability
Sender transmitter
Senkerodieren spark erosion, electric discharge machining, EDM
Senkgeschwindigkeit lowering speed
Senkrechtextruder vertical extruder
Senkrechtspritzkopf vertical extruder head, vertical die head

sensibel sensitive; *this word should never be translated as* sensible, *which has a totally different connotation. To speak of* sensible controls, *for example, would imply controls that are blessed with common sense.*
Sensibilisator sensitiser
Sensibilisierung sensitisation
Sensibilisierungsmittel sensitising agent
Sensor sensor, transducer, probe
Separationsmittel (mould) release agent
Separierung 1. separation. 2. segregation *(e.g. of plastics waste)*
Sequenzpolymer sequential polymer
Sequenzsteuerung sequential control
Sequenztyp sequential polymer
seriell serial, sequential *(me)*
Serien, grosse long runs
Serien, kleine short runs: **in kleinen Serien** in small quantities
Serien, mittlere medium runs
Serienaufgabeschurre standard feed chute *(sr)*
Serienausführung standard design: **Aufgabeschurren in Serienausführung oder Sonderkonstruktion** standard or specially designed feed chutes
Serienbedingungen 1. mass production conditions. 2. standard conditions
Serienfahrzeug 1. standard vehicle. 2. mass produced vehicle
Serienfertigung mass production
Seriengerät standard unit
Serienmaschine standard machine
serienmäßig standard, as a standard feature
serienmäßige Ausrüstung standard equipment
Serienprodukt standard product
Serienproduktion mass/large scale production
Serienprogramm standard range
Serienschaltung series arrangement
Serienteil standard part
serienüblich standard
Serienversion standard version/model
Serienversuche test series
Serienwerkzeug 1. mass production mould. 2. standard mould
servicefreundlich easy to service/maintain
Servicefreundlichkeit ease of servicing/maintenace
servohydraulisch servohydraulic
Servomotor servo-motor
Servoventil servo-valve
SFK *abbr. of* **synthetikfaserverstärkte Kunststoffe** synthetic fibre reinforced plastics
SFK-Prepregmaterial synthetic fibre prepreg
S-Form four-roll inclined Z-type calender
SH 1. *abbr. of* **selbsthärtend**, self-curing. 2. *abbr. of* **säurehärtend**, acid-curing
SH-Lack 1. self-curing paint. 2. acid-curing paint
Shoreänderung change in Shore hardness
Shore-Härte Shore hardness
Shore-Härte A Shore-A hardness
Shore-Härte D Shore-D hardness
Shredderabfall shredded waste

shreddergerecht suitable for shredding
Shreddermühle shredder *(sr)*
Shreddermüll shredded waste
Shreddern *German version of* shredding
Sicherheit security, safety, reliability
Sicherheit, statistische statistic certainty
Sicherheitsabstand safety margin/factor
Sicherheitsanforderungen safety requirements
Sicherheitsauflagen safety instructions
Sicherheitsbarriere safety barrier
Sicherheitsbedürfnis safety requirements
Sicherheitsbeiwert safety factor
Sicherheitsbestimmungen safety regulations
Sicherheitseinrichtung safety device
Sicherheitsfaktor safety factor
Sicherheitsforderungen safety requirements
Sicherheitsgitter guard door, screen guard
Sicherheitsglas safety glass
Sicherheitsglasfolie safety glass film
Sicherheitsgrenze safety margin
Sicherheitsgründen, aus for reasons of safety
Sicherheitshinweise safety instructions
Sicherheitskennzahlen safety data
Sicherheitskoeffizient safety factor
Sicherheitsmaßnahmen safety precautions
Sicherheitsnormen safety standards
Sicherheitsratschläge safety instructions
Sicherheitsregeln safety regulations
Sicherheitsreserve safety margin
Sicherheitsrichtlinien safety guidelines
Sicherheitsrisiko safety risk
Sicherheitsschalter safety switch
Sicherheitsschiebetür guard door
Sicherheitsspielraum safety margin
sicherheitstechnisch *relating to safety:* **Eine sicherheitstechnisch einwandfreie Abführung der freiwerdenden Dämpfe ist nur schwer zu realisieren** it is difficult to safely remove the vapours produced
sicherheitstechnische Hinweise safety instructions
Sicherheitsventil safety valve
Sicherheitsverriegelung safety interlock system/mechanism
Sicherheitsverschluß safety lock
Sicherheitsvorkehrungen safety precautions
Sicherheitsvorschriften safety regulations
Sicherheitszahl safety factor
Sicherungsautomat automatic circuit breaker, automatic cut-out
Sicherungspatrone fuse
sichtbar visible
Sichtbarmachen making visible
Sichtbeton exposed concrete
sichten to sift
Sichtfenster sight glass, window
Sichtfläche visible surface
Sichtgerät video display unit *(me)*
Sichtglas sight glass, window
Sichtkontrolle visual check/inspection
Sichtprüfstelle visual inspection unit
Sichtprüfstrecke visual inspection unit

Sichtprüfung visual inspection/check/ examination
Sichtscheibe sight glass, window
Sichtseite visible/exposed side, side exposed to view
Sickerwasser leachate
Sieb 1. screen, filter screen *(mf)*. 2. sieve
Siebanalyse sieve analysis
Siebband filter ribbon *(mf)*
Siebblock screen pack *(e)*
Siebdornhalter breaker plate-type mandrel support *(e)*
Siebdruckfarbe screen printing ink
Siebdruckmasse screen printing ink
Siebdruckverfahren screen printing (process)
Siebeinsatz screen pack *(e)*
Siebfiltration surface filtration *(mf)*
Siebfläche, freie effective screen area *(mf)*
Siebgewebe filter screen/cloth, screen fabric *(mf)*
Siebgewebepackung screen pack *(e)*
Siebkassette screen cassette *(mf)*
Siebkontrolle sieve analysis
Siebkorb screen pack *(e)*
Siebkorbdornhalter breaker plate-type mandrel support *(e)*
Siebkorbhalterung breaker plate-type mandrel support *(e)*
Siebkorbkopf breaker plate-type die *(e)*
Siebkorbrohrkopf breaker plate-type pipe die *(e)*
Siebkörper filter element/screen *(mf)*
Sieblochplatte breaker plate *(e)*
Siebpaket screen pack *(e)*
Siebronde filter disc *(mf)*
Siebrückstand sieving residue
Siebsatz screen pack *(e)*
Siebscheibe filter disc *(mf)*
Siebscheibenwechsler screen changer *(mf)*
Siebschnellwechseleinrichtung quick action screen changer
Siebstützplatte breaker plate *(e)*
Siebträger 1. breaker plate *(e)*. 2. screen support/carrier *(i.e. when it is a support other than a breaker plate, e.g. a plate carrying cartridge filters)*
Siebträgerscheibe breaker plate
Siebwechsel replacing/changing the screen: **leichter Siebwechsel** the screen is easily replaced
Siebwechseleinrichtung screen changer
Siebwechselgerät screen changer *(mf)*
Siebwechselkassette screen changer
Siebwechsler screen changer
Siebwechslerkörper screen changer body
Siedebereich boiling range
Siedeintervall boiling range
siedend boiling
Siedepunkt boiling point
Siedetemperatur boiling point
Siedewasserlagerung immersion in boiling water

Siegelbacke (heat) sealing bar
siegelbar heat sealable
siegelfähig heat sealable
Siegelfolie heat sealing film
Siegelpunkt gate sealing point *(im)*: *This is the point at which the gate is sealed due to solidification of the melt;* **Da über den Siegelpunkt hinaus keine Druckbeeinflussung mehr möglich ist...** since the pressure can no longer be influenced once the gate has been sealed...
Siegelschicht heat sealable coating
Siegelstation (heat) sealing station
Siegeltemperatur (heat) sealing temperature
Siegelwerkzeug (heat) sealing instrument
Siegelzeit gate open time *(im)*: *This is the time during which the gate remains open, i.e. during which the melt can pass through into the mould. The end of this period marks the* **Siegelpunkt** *(q.v.)*; **nach Ablauf der Siegelzeit** after the gate open time has passed OR: once the gate has been sealed
Siegelzeitpunkt gate sealing point: *(see explanatory note and translation example under* **Siegelpunkt**)
Sigmarotor sigma-type rotor
Sigmawert sigma value, safe working stress *(of pipes)*
Signalausgang signal output *(me)*
Signaleingang signal input *(me)*
Signalflußplan signal flowchart *(me)*
Signalgeber signal generator *(me)*
Signalintensität signal intensity
Signalisierung signalling
Signalumwandler signal transducer/converter
Signalvorgabe signal input
Signalzustand signal status *(me)*
Signalzustandsanzeige signal status display *(me)*
Signiergerät 1. marking device *(used for impressing identifying marks into extruded pipe).* 2. colour coding device
Sikkativ drier
Silan silane
silanbehandelt silanised, silane-treated
Silanharz silane resin
silanisiert silanised
Silanisierung silanisation
Silankette silane chain
silanolfunktionell silanol-functional
Silanolgruppe silanol group
Silanverbindung silane compound
silanvernetzt silane-crosslinked/-cured
Silanvernetzung silane crosslinkage/cure
Silberschlieren silver streaks
Silicium silicon
Siliciumchip silicon chip *(me)*
Siliciumdioxid silica
Siliciumeinkristallplättchen silicon chip *(me)*
Siliciumkarbid silicon carbide
siliciumorganisch organo-silicon

Silicium-Sauerstoffbindung silicon-oxygen linkage
Siliciumtetrachlorid silicon tetrachloride
Siliconalkydharz silicone alkyd resin
Siliconanteil silicone content
siliconbeschichtet silicone coated
Siliconelastomer silicone elastomer
Siliconfestharz solid silicone resin
Siliconfett silicone grease
siliconfrei silicone-free
Silicongehalt silicone content
Silicongießharz silicone casting resin
Silicongummi silicone rubber *(see explanatory note and translation expample under* **Kautschuk**)
Siliconharz silicone resin
Siliconharzpreßmasse silicone resin moulding compound
Siliconharzpulver powdered silicone resin
siliconisiert siliconised, silicone-treated
Siliconisierung siliconisation
Siliconkautschuk silicone rubber *(see explanatory note and translation example under* **Kautschuk**)
Siliconkautschukform silicone rubber mould
Siliconkautschukmischung silicone rubber compound
Siliconlack silicone paint
siliconmodifiziert silicone-modified
Siliconöl silicone fluid
Siliconpaste silicone paste
Siliconpolyether silicone polyether
siliconrückseitenbeschichtet silicone backed
Siliconschaum silicone foam
Silicontensid silicone surfactant
Silicontrennharz silicone release resin
Silicontrennmittel silicone release agent
Silicontrennpapier silicone release/backing paper
Siliconüberzug silicone coating
Siliconumhüllung 1. silicone coating 2. silicone encapsulation
silierbar suitable for storing in silos, suitable for transferring to silos
Silieren transfer to silos
silierfähig suitable for storing in silos, suitable for transferring to silos
Silikagel silica gel
Silikat silicate
Silikatfarbe silicate paint
Silikatfüllstoff silicate filler
Silikatgerüst silicate skeleton
Silikatglas ordinary glass
silikatisch silicate **silikatischer Füllstoff** silicate filler
Silikon- *see* **Silicon-**
Silikose silicosis
Silikosegefahr risk of silicosis
Silizium- *see* **Silicium-**
Silo silo
Siloanlage silo installation
Silofahrzeug road tanker

Siloinhalt silo contents
Silowagen road tanker
Siloxan siloxane
siloxanbasierend siloxane-based
Siloxanbaustein siloxane unit
Siloxanbindung siloxane linkage
Siloxangruppe siloxane group
Siloxankette siloxane chain
Siloxanpolymer siloxane polymer
Siloxanverbindung siloxane compound
Silylether silyl ether
Silylgruppe silyl group
Silylierung silylation
Simmerring sealing ring, oil seal
Simulationsrechnung simulation calculation
simuliert simulated
Simultanbetrieb simultaneous operation
Simultanreckverfahren simultaneous stretching (process)
Simultanstrecken simultaneous stretching/ orientation *(a special process of simultaneously stretching film transversely and in machine direction)*
Sinkmarkierungen sink marks *(im)*
sinnvoll meaningful, sensible **es ist nicht sinnvoll...** there is no point in... OR: nothing is gained by...
sinnvoll, ökonomisch making economic sense
sintern to sinter
Sinterpulver powder coating
sinusförmig sinusoidal
Sinuskurve sine curve
Sitz seat
Sitzbezug seat cover
Sitzfläche seat *(of bearing)*
Sitzgestell seat frame
Sitzpolster seat upholstery/cushion
Sitzschale seat shell
Sitzventil seating valve
Skala scale
S-Kalander four-roll inclined Z-type calender
Skalenteil scale division
Skalenteilung scale division
skimmen to skim coat *(c)*
Skinformmaschine skin packaging machine
Skinpackfolie skin packaging film
Skinpackmaschine skin packaging machine
Skinpackung skin pack
Skinverpackung skin pack
Skischuh ski boot
Skleroskop scleroscope
Slipmittel slip agent
SMC *abbr. of* sheet moulding compound
SMC-Abfall SMC scrap/waste
SMC-Harzmatte prepreg, sheet moulding compound, SMC
SMC-Produktionsabfall SMC production scrap/ waste
smektisch smectic
SMT *abbr. of* surface mounting technology
Sockel plinth
Sockelleiste skirting board

Soft-Feeling *German version* of soft feel
Software software *(me)*
Software, fremde packaged/third-party software, software package *(me)*
Softwarearchitektur software architecture *(me)*
Softwarebaustein software module *(me)*
Softwarehaus software house *(me)*
Softwaremenü software menu, (list of) software options *(me)*
Softwaremodul software module *(me)*
Softwarepacket software package *(me)*
Softwareprogramm software range *(me)*
Softwareschnittstelle software interface *(me)*
Sojaalkyd soya bean oil alkyd
Sojabohnenöl soya bean oil
Sojafettsäure soya bean oil fatty acid
Sojaöl soya bean oil
Sojaölfettsäure soya bean oil fatty acid
Solargenerator solar generator
Solarkollektor solar collector
Solarstrahlung solar radiation
Solarzelle solar cell
Sol-Gelumwandlung sol-gel transformation/ transition
Sollbruchstellen (artificially) weakened points *(literally: points which are intended to break, e.g. where part of a moulding is to be broken off at some later stage)*
Solldeformation required deformation
Solldruck required/set pressure
Solldruckkurve reference pressure curve
Solldurchmesser required diameter
Sollgewicht required weight
Sollgröße required/set value, setpoint *(for translation example see under* **Sollwert***)*
Sollkonzentration required concentration
Sollkurve setpoint/reference curve
Sollmaß required dimension
Solltemperatur required/set temperature
Sollverarbeitungsbedingungen pre-set processing conditions
Sollwanddicke required wall thickness
Sollwert required/set value, setpoint: **Die Massetemperatur wird auf den Sollwert gebracht von z.B.190°C** the melt is brought to the required temperature, e.g. 190°C
Sollwertabweichungen setpoint deviations
Sollwertanzeige setpoint display *(me)*
Sollwertauswahltaste setpoint selector key *(me)*
Sollwerteingabe setpoint input/entry *(me)*
Sollwerteinstellung setpoint adjustment
Sollwertkurve setpoint curve
Sollwertspeicher setpoint memory *(me)*
Sollwertverarbeitung required value processing *(me)*
Sollwertvorgabe 1. setpoint input *(me)*. 2. setpoint setting *(me)*
Solvatation solvation, gelation
Solvationsfähigkeit solvating/dissolving power
solvatisierend solvating, gelling
Solventnaphtha solvent naphtha

Sonde sensor, transducer, probe
Sonderabfall toxic/hazardous waste
Sonderausrüstung 1. special equipment. 2. extra: **Eine elektronische Steuerung kann als Sonderausrüstung geliefert werden** electronic controls can be supplied as an extra
Sonderdruck reprint *(e.g. of an article)*
Sondereinstellung special formulation
Sonderextruder special-purpose extruder
Sonderkautschuk special-purpose rubber, speciality rubber
Sondermaschine special-purpose machine
Sondermüll toxic/hazardous waste
Sondermüllverbrennungsanlage toxic/hazardous waste incinerator
Sonderqualität special-purpose grade
Sonderschnecke special-purpose screw
Sondertyp special-purpose grade
Sonnenbestrahlung solar irradiation: **Schutz vor direkter Sonnenbestrahlung** protection against direct sunlight
Sonnenblende sun visor
Sonnendach sun roof
Sonneneinstrahlung solar radiation
Sonnenkollektor solar collector
Sonnensimulation sun simulation
Sonnenstrahlung solar radiation: **direkte Sonnenbestrahlung** direct sunlight
Sonnenstunden hours of sunshine
Sonotrode horn *(w)*
Sorption sorption
Sorptionsmittel adsorbent *(used in chromatography)*
sortenähnlich segregated *(plastics waste)*; for translation example see under **sortenrein**
sortenrein sorted *(plastics waste)*; **sortenreine Kunststoffabfälle** segrated plastics waste (i.e. plastics waste that has been sorted into its different constituents such as PVC, PE, PS etc.); **sortenreine Wiederverwertung** recycling of sorted/segregated/separated waste
Sortentrennung segregation/sorting *(of refuse into its component parts e.g. metal, glass, plastic, paper etc.)*
Sortierung sorting
Sortiment range
Sortimentsübersicht product range, range of products
Sozialaufwendung social (insurance) contributions
Spachtel 1. putty knife, trowel. 2. knifing filler, stopper, surfacer
Spachtelkonsistenz trowelling consistency
Spachtelmasse knifing filler, stopper, surfacer
Spachteln knife/trowel application
Spachtelputz surfacing plaster
***Spalt** 1. slit, gap. 2. nip *(c)*
Spaltbreite 1. nip width *(c)*. 2. die gap width/thickness *(e)*

Spaltdruck nip pressure *(c)*
Spalten splitting
spaltfüllend gap filling
Spalthöhe (radial) screw clearance *(e)*
Spaltkraft 1. nip pressure *(c)*. 2. delaminating force
Spaltlast 1. nip pressure *(c)*. 2. delaminating pressure
Spaltlastprüfung delamination test
Spaltmaschine slitting machine/unit
Spaltprodukt decomposition product
Spaltverstellung 1. nip adjustment *(c)*. 2. nip adjusting mechanism/device *(c)*
Spaltversuch delamination test
Spaltwalzenpaar nip/pinch rolls *(bfe)*
Spaltweite 1. nip width *(c)*. 2. die gap width/thickness *(e)*
spanabhebende Bearbeitung machining
spanende Bearbeitung machining: **Die Platten können sehr gut spanend bearbeitet werden** the sheets are very easy to machine
spangebende Bearbeitung machining
spanlose Bearbeitung forming, shaping, thermoforming
spanlose Verarbeitung forming, shaping, thermoforming
spanloses Umformen forming, shaping, thermoforming
Spannbacke clamping jaw
Spannbeton pre-stressed concrete
Spanndruck clamping pressure
Spannelement clamp, clamping device
Spannglied tensioning element
Spannkeil clamp, clamping device
Spannklemme clamp
Spannkraft clamping force
Spannrahmen 1. clamping frame *(t)*. 2. tenter/stenter frame *(used in stretching film)*
Spannring clamping ring
Spannrolle tension roll
Spannstahl tensioned steel rod
Spannung 1. stress. 2. voltage, tension
Spannungen, eingefrorene frozen-in stresses
Spannungsabbau stress relaxation
Spannungsabbaugeschwindigkeit stress relaxation speed/rate
Spannungsabfall stress decrease
Spannungsabnahme stress decrease
Spannungsanalyse stress analysis
spannungsarm low-stress
Spannungsausfall power failure
Spannungsausgleich release of frozen-in stresses
Spannungsausschlag stress amplitude
spannungsbezogen stress-related
Spannungs-Dehnungsdiagramm stress-strain diagram, stress- elongation diagram
Spannungs-Dehnungslinie stress-strain curve
Spannungs-Dehnungsverhalten stress-strain behaviour

---

* *Words starting with* **Spalt-**, *not found here, may be found under* **Walzenspalt-**

Spannungseinschlüsse moulded-in stresses
spannungsfrei stress-free
Spannungsgeber voltage regulator
Spannungsgleichgewicht stress equilibrium
Spannungsgrenze stress limit
Spannungsinhomogenitäten uneven stresses
Spannungsinkrement stress increment
Spannungsintensität stress intensity
Spannungsintensitätsamplitude stress intensity amplitude
Spannungsintensitätsfaktor stress intensity factor
Spannungskontrolle 1. tension control. 2. tension control mechanism *(for film as it comes off the calender)*
Spannungskonzentration stress concentration
Spannungskorrosion (environmental) stress cracking
Spannungskorrosionsbeständigkeit (environmental) stress cracking resistance
Spannungskorrosionsriß stress crack
Spannungskorrosionsverhalten stress cracking behaviour
Spannungskraft stress
Spannungsoptik photoelasticity
spannungsoptisch photoelastic
Spannungsquelle voltage source
Spannungsrelaxation stress relaxation
Spannungsrelaxationsversuch stress relaxation test
Spannungsriß stress crack
Spannungsrißanfälligkeit susceptibility to stress cracking
spannungsrißauslösend stress crack-inducing
spannungsrißbeständig resistant to stress cracking
Spannungsrißbeständigkeit stress cracking resistance
Spannungsrißbildung stress crack formation: **Versuche zeigten, daß durch Schwefelsäure Spannungsrißbildung hervorgerufen wird** tests have shown that sulphuric acid causes stress cracking; **Beständigkeit gegen Spannungsrißbildung** stress cracking resistance
Spannungsrißbildung, umgebungsbeeinflußte environmental stress cracking
spannungsrißempfindlich susceptible to stress cracking
Spannungsrißkorrosion (environmental) stress cracking
Spannungsrißkorrosionsbeständigkeit stress cracking resistance
Spannungsrißverhalten stress cracking behaviour
Spannungsschwankungen voltage fluctuations
Spannungsspitze stress peak
Spannungsstabilisator voltage stabiliser
Spannungsumbildung, umgebungsbedingte environmental stress cracking
spannungsunabhänbgig independent of stress
Spannungs-Verformungsverhalten stress-strain behaviour
Spannungsverlauf stress pattern
Spannungsverteilung stress distribution
Spannungswandler voltage transformer
Spannungswert modulus
Spannungszustand state of stress
Spannvorrichtung jig, fixture, clamping device
Spannwalze tension roll
Spannzange clamp
Spanplatte (wood) chipboard
Spantiefe depth of cut
Spanungsauslösezeit stress crack initiation period, time needed to produce stress cracking
Spanwinkel rake angle
Sparte division, branch
Spartenleiter divisional manager
SPC-Anlage statistical process control system
Speicher 1. accumulator *(im)*. 2. melt accumulator *(bm)*. 3. memory, storage device *(me)*
Speicher, interner internal memory *(me)*
Speicher, optischer optic(al) disk *(me)*
Speicheradresse memory address *(me)*
Speicherausbau, maximaler maximum memory capacity *(me)*
Speicherbauart type of accumulator
Speicherbauelement memory chip *(me)*
Speicherbaustein memory chip *(me)*
Speicherchip memory chip *(me)*
Speicherdichte storage density *(me)*
Speicherelement memory chip *(me)*
Speicherinhalt memory content *(me)*
Speicherkapazität memory/storage capacity *(me)*
Speicherkopf accumulator head *(bm)*
Speichermaschine accumulator-type machine *(im)*
Speichermedium input medium *(me)*
Speichermodul 1. storage modulus. 2. memory module *(me)*
Speichern storage *(of data) (me)*
Speicherplatte 1. storage/magnetic/optic(al) disk *(me)*. 2 compact disc, CD
Speicherplatte, optische optic(al) disk *(me)*
speicherprogrammierbar stored program *(me)*: **speicherprogrammierbare Steuerung** stored program control
speicherprogrammiert stored program *(me)*: **speicherprogrammierte Steuerung** stored program control
Speicherraum accumulator chamber *(bm)*
Speicherung storage *(of data) (me)*
Speicherzylinder accumulator cylinder *(bm)*
Speiseaggregat feed unit
Speiseeinrichtung feed unit/mechanism
Speiseeis ice cream
Speiseextruder feed extruder
Speisefett edible fat
Speisegerät feed unit
Speisemaschine feed mill
Speiseöffnung feed throat/opening/port

Speiseöl edible oil
Speisepumpe feed pump
Speiseschnecke feed screw
Speiseschneckeneinheit feed screw unit
Speisespannung supply voltage
Speisetrichter (feed/material) hopper *(e,im)*
Speisevorrichtung feed mechanism/unit
Speisewalze feed roll
Speisewalzwerk feed mill
Speisungsanlage feed unit
spektral spectral
Spektralbereich spectral range
Spektralkurve spectral distribution curve
Spektrallinie spectral line
Spektralphotometer spectrophotometer
spektrographisch spectrographic
spektrometrisch spectrometric
Spektrophotometer spectrophotometer
spektrophotometrisch spectrophotometric
Spektroskopie spectroscopy
spektroskopisch spectroscopic
Spektrum 1. spectrum. 2. range
Sperrbolzen locking bolt
Sperreigenschaften barrier properties
Sperrfolie barrier sheet(ing)
Sperrholz plywood
Sperrpigment barrier pigment
Sperrschicht 1. barrier film/coating. 2. barrier layer
Sperrschichteigenschaft barrier property
Sperrschichtkunststoff barrier polymer/plastic
Sperrschieberpumpe rotary piston pump
Sperrsteg barrier flight *(e)*
Sperrventil shut-off valve, check valve
Sperrwirkung barrier effect
Spezialeinstellung special formulation
Spezialfahrzeug special purpose vehicle
Spezialformmasse special-purpose moulding compound
Spezialharz special-purpose resin, speciality resin
Spezialkautschuk special-purpose rubber, speciality rubber
Spezialkenntnisse specialised knowledge/experience
Spezialkunststoff special-purpose plastic, speciality plastic
Spezialmarke speciality/special-purpose grade
Spezialschnecke special-purpose screw
Spezialtype special grade
Spezialweichmacher special-purpose plasticiser, speciality plasticiser
Spezifikation specification
spezifikationsgerecht according to specification
spezifisch schwer heavy
spezifische Oberfläche specific surface area
spezifische Viskosität specific viscosity
spezifische Wärme specific heat
spezifischer Durchgangswiderstand volume resistivity
spezifischer Oberflächenwiderstand surface resistivity
spezifischer Widerstand resistivity
spezifisches Gewicht specific gravity
spezifisches Volumen specific volume
sphärisch spherical
Sphärolit spherulite
Sphärolitgefüge spherulite structure
sphärolitisch spherulitic
Sphärolitwachstumsgeschwindigkeit rate of spherulite growth
spiegelhochglanzpoliert polished to a mirror finish
Spiegelschweißgerät hot plate welding tool
Spiegelschweißverfahren hot plate welding (process)
Spiegelverfahren hot plate welding
Spiel clearance
Spiel, radiales 1. radial clearance. 2. screw clearance
spielfrei tight fitting
Spielgeräte, elektronische electronic games
Spielraum clearance
Spielwaren toys
Spielzeugartikel toys and games
Spinndüse spinneret *(e)*
Spinne 1. feed system, runner system *(im)*. 2. radial system of runners *(im)*. 3. sprues and runners *(sr) (when referring to thermoplastic scrap)*
Spinnextruder monofilament extruder
Spinnfaden strand
Spinnkopf spinneret *(e)*
Spiralbohrer twist drill
Spiraldorn spiral mandrel *(e,bfe)*
Spiraldornblaskopf 1. spiral mandrel die *(e)*. 2. spiral mandrel blown film die *(bfe)*
Spiraldornkopf 1. spiral mandrel die *(e)*. 2. spiral mandrel blown film die *(bfe)*
Spiralfeder spiral spring
spiralförmig spiral
spiralgenutet spirally grooved
Spirallänge spiral flow length
Spiraltest spiral flow test
Spiralverteilerblaskopf 1. spiral mandrel die *(e)*. 2. spiral mandrel (blown film) die *(bfe)*
Spiralwendelverteilerblaskopf spiral mandrel blown film die *(bfe)*
Spiritus spirit
Spitze, glatte unflighted screw tip
Spitzenbedarf maximum requirements
Spitzendruck peak/maximum pressure
Spitzenleistung maximum output/efficiency
Spitzenradius tip radius
Spitzenspeicher peak memory *(me)*
Spitzentemperatur maximum/peak temperature
Spitzenwert maximum/peak value
Spitzenwinkel point angle
Spitzkerbe pointed notch
spleißen 1. to split *(general term)*. 2. to fibrillate *(see explanatory note under* **Fibrillieren**)
Spleißfasern fibrillated film, film fibres
Spleißgarn fibrillated film, film fibre
Spleißneigung tendency to split

Spleißzement splicing cement/adhesive
Splitterbruch brittle failure/fracture
splittern to splinter
Spoiler spoiler
S-Polymerisationsverfahren suspension polymerisation (process)
spontan spontaneous
Sportartikel sports goods
Sportgeräte sports equipment/goods
Sportschuhe sports shoes, trainers
spreiten to spread
Spreitungsvermögen spreading power
Spreitvermögen spreading power
Spreizdorn parison stretching mandrel *(bm) (see explanatory note under* **Schlauchspreizvorrichtung***)*
Spreizdornanlage parison stretching mandrel *(bm) (see explanatory note under* **Schlauchspreizvorrichtung***)*
Spreizdornvorrichtung parison stretching mandrel *(bm) (see explanatory note under* **Schlauchspreizvorrichtung***)*
spreizen to spread (out), to stretch
Spreizvorrichtung parison stretching mandrel *(bm) (see explanatory note under* **Schlauchspreizvorrichtung***)*
***Spritzabstand** spraying distance
Spritzaggregat 1. injection unit *(im)*. 2. spraying unit
Spritzaggregatseite fixed/stationary mould half *(im)*
Spritzartikel 1. mouldings 2. extrusions
Spritzauftrag spray application
Spritzautomat 1. automatic injection moulding machine *(im)*. 2. automatic transfer moulding machine *(tm)*. 3. automatic spraying unit
spritzbar 1. mouldable. 2. sprayable
Spritzbarkeit 1. mouldability *(im)*. 2. extrudability. 3. sprayability
Sprltzbedingungen moulding conditions
Spritzblasautomat automatic injection blow moulder/moulding machine
Spritzblasdorn blowing mandrel/spigot, inflating mandrel/spigot *(bm): The prefix* **Spritz-** *merely indicates that the* **Blasdorn** *is being used for injection blow moulding. It does not, however, have any bearing on the translation because the design is the same, no matter whether one is dealing with extrusion blow moulding or injection blow moulding.*
Spritzblasen injection blow moulding
Spritzblasformen injection blow moulding
Spritzblasmaschine injection blow moulding machine
Spritzblasverfahren injection blow moulding (process)
Spritzblaswerkzeug injection blow mould
Spritzdorn blowing mandrel/spigot, inflating mandrel/spigot *(see also explanatory note under* **Spritzblasdorn***)*

Spritzdruck 1. injection pressure *(im)*. 2. extrusion pressure *(e)*
Spritzdruckbegrenzung 1. maximum injection pressure: *The word is sometimes used in a general context in place of* **Spritzdruck***, where it should be translated as* injection pressure. 2. limiting the injection pressure **...wirkt wie eine Spritzdruckbegrenzung** ...has the same effect as limiting the injection pressure
Spritzdruckregler injection pressure regulator
Spritzdruckstufe injection pressure stage *(im)*
Spritzdruckzeit injection pressure time *(im)*
Spritzdruckzylinder, hydraulischer injection cylinder
Spritzdüse 1. (extrusion) die *(e)*. 2. injection nozzle *(im)*. 3. spraying nozzle *(grp)*
Spritzeinheit injection unit *(im)*
spritzen 1. to extrude. 2. to injection mould. 3. to spray
Spritzen, angußloses sprueless injection moulding
Spritzer moulder
Spritzerei 1. moulding shop *(im)*. 2. extrusion shop *(e)*
spritzfähig 1. mouldable: **ABC Duroplaste sind leicht spritzfähig** ABC thermosets are easy to mould. 2. sprayable
Spritzfähigkeit 1. mouldability. 2. sprayability
Spritzfehler moulding fault
spritzfertig ready for spraying
Spritzfläche projected moulding area *(im)*
Spritzform injection mould, injection moulding tool
spritzfrisch freshly moulded
spritzgeblasen injection blow moulded
spritzgegossen injection moulded
spritzgepreßt transfer moulded
Spritzgeschwindigkeit 1. injection speed/rate, rate of injection *(im)*. 2. extrusion speed/rate, rate of extrusion *(e)*
Spritzgewicht shot weight *(im)*
Spritzgießaggregat injection moulding machine, injection moulder *(see explanatory note under* **Spritzgießmaschine***)*
Spritzgießanlage injection moulding line
Spritzgießaufgaben (injection) moulding jobs
Spritzgießautomat automatic injection moulding machine
spritzgießbar injection mouldable: **spritzgießbare Formmasse** injection moulding compound
Spritzgießbetrieb injection moulding shop
Spritzgießblasformen injection blow moulding
Spritzgießeinheit injection moulding machine
Spritzgießen injection moulding
Spritzgießer injection moulder *(person, not the machine)*
Spritzgießfabrik injection moulding shop

---

* *Words starting with* **Spritz-***, not found here, may be found under* **Einspritz-***.*

spritzgießfähig injection mouldable
Spritzgießfertigung injection moulding
Spritzgießform injection mould, injection moulding tool
Spritzgießformteil injection moulding, injection moulded part
Spritzgießgesenk mould cavity *(im) (see explanatory note under* **Spritzgießkavität***)*
Spritzgießharz injection moulding resin
Spritzgießheißkanalsystem hot runner injection moulding system
Spritzgießkavität mould cavity *(im): Since the word will invariably occur in an injection moulding context, there is no need to translate the first part*
Spritzgießkern mould core *(im)*
Spritzgießmarke injection moulding grade *(of moulding compound)*
Spritzgießmaschine injection moulding machine, injection moulder: *The latter expression is used where brevity is important, e.g. in headlines, captions, advertising slogans etc.*
Spritzgießmaschinenbaureihe range of injection moulders/moulding machines
Spritzgießmaschinenprogramm range of injection moulders/moulding machines
Spritzgießmaschinenschnecke injection screw *(im)*
Spritzgießmasse 1. injection moulding compound *(if used in isolation, e.g. as a heading for a table of different products).* 2. moulding compound *(if used in a continuous text where it invariably occurs in an injection moulding context, making the use of the first word in version 1. superfluous)*
Spritzgießmaterial 1. injection moulding compound. 2. moulding compound *(see explanatory note under* **Spritzgießmasse***)*
Spritzgießrezeptur injection moulding formulation
Spritzgießrundtischanlage rotary table injection moulding machine, carousel-type injection moulding machine
Spritzgießschließeinheit clamp(ing) unit *(im)*
Spritzgießschubschnecke reciprocating screw
Spritzgießstreckblasen injection stretch blow moulding
Spritzgießstreckblasmaschine injection stretch blow moulding machine
Spritzgießtechnik injection moulding (technology)
spritzgießtechnisch *relating to injection moulding*; **auf spritzgießtechnischem Wege hergestellt** injection moulded, made by injection moulding
Spritzgießteil injection moulding, injection moulded part/article
Spritzgießteilvolumen part/moulded-part volume
Spritzgießverarbeitung 1. processing by injection moulding: **Für PVC bestehen mehr Chancen bei genügend sorgfältiger Spritzgießverarbeitung** PVC stands a better chance if it is injection moulded with care. 2. injection moulding (process)
Spritzgießverfahren injection moulding (process) **im Spritzgießverfahren hergestellt** injection moulded
Spritzgießvorgang injection moulding process/operation
Spritzgießwerkzeug injection mould, injection moulding tool
Spritzgießwerkzeug, dreifaches three-cavity/-impression injection mould
Spritzgrat flash
Spritzguß injection moulding
Spritzguß- *see* **Spritzgieß-**
Spritzguß-Fittings injection moulded fittings
Spritzguttemperatur melt temperature
Spritzhaut flash
Spritzhub injection stroke *(im)*
Spritzkabine paint spraying booth
Spritzkapazität shot capacity *(im)*
Spritzkolben 1. injection plunger *(im).* 2. transfer plunger *(tm)*
Spritzkopf 1. extrusion die. 2. extruder head, die head *(see also explanatory note under* **Kopf** *to decide which version to use in a given context)*
Spritzkopfwiderstand die resistance *(e)*
Spritzkraft injection force *(im)*
Spritzlack 1. spraying paint. 2. spraying varnish/lacquer
Spritzlackierung spray painting
Spritzlaminat spray lay-up laminate
Spritzleistung injection capacity *(im)*
Spritzling 1. injection moulding, injection moulded part. 2. moulding. 3. parison *(bm)*: *Version 2. applies if the word occurs in an injection moulding context. Version 3. applies only when the word is used in an injection blow moulding context to describe the injection moulded parison.*
Spritzlingsfläche, projizierte projected moulding area
Spritzmaschine paint spraying machine
Spritzparameter moulding conditions
Spritzpistole spraygun
Spritzprägen injection-compression moulding
Spritzpreßanlage transfer moulding plant
Spritzpreßautomat automatic transfer moulding press
Spritzpresse transfer moulding press
Spritzpressen transfer moulding
Spritzpreßform transfer mould
Spritzpreßhalbautomat semi-automatic transfer moulding press
Spritzquellung die swell *(e)*
Spritzrichtung direction of moulding
Spritzseite 1. fixed/stationary mould half *(im).* 2. injection unit *(im)*
spritzseitig on the fixed mould half, on the injection side

spritzseitige Werkzeughälfte fixed/stationary mould half *(im)*
Spritzstation injection station
Spritzstreckblasanlage injection stretch blow moulding plant
Spritzstreckblasen injection stretch blow moulding
Spritzstreckblasmaschine injection stretch blow moulding machine
Spritztakt moulding cycle
spritztechnisch *relating to moulding in general, or to injection moulding in particular.* **spritztechnische Einzelheiten** moulding details; **Aus spritztechnischen Gründen** for technical reasons relating to the moulding process OR: for technical reasons; **Ein kurzer Stangenanguß ist spritztechnisch am günstigsten** a short sprue gate is to be preferred from the moulding point of view; **Rein spritztechnisch gehört der Anguß an eine möglichst zentrale Stelle des Teiles** from the purely technical point of view the sprue should be as close to the centre of the moulding as possible
∗**Spritzteil** 1. injection moulding, injection moulded part. 2. moulding: *Version 2. applies if the word is used in an injection moulding text, where further qualification is unnecessary*
Spritzteilfüllvolumen cavity volume *(im)*
Spritzteilgewicht shot weight *(im)*
Spritzteilmasse injection moulding compound
Spritzversuche moulding trials
Spritzviskosität spraying consistency/viscosity
Spritzvolumen injection volume, shot volume *(im)*
Spritzwerkzeug 1. injection mould, injection moulding tool *(im)*. 2. (extrusion) die *(e)*
Spritzzeit injection time, mould filling time *(im)*
Spritzzyklus moulding cycle
Spritzzylinder injection cylinder
Sprödbruch brittle fracture/failure
spröde brittle
sprödes Versagen brittle failure
spröd-hart glassy *(condition of a polymer below the glass transition temperature)*, hard and brittle
Sprödigkeit brittleness
Spröd-Zähbruchübergang brittle-ductile failure transition
Spröd-Zähbruchübergangsgebiet brittle-ductile failure transition range
Spröd-Zähübergang glass-rubber transition
Sprühdose aerosol can
Sprühen spraying
sprühgetrocknet spray dried
Sprühkühlstrecke spray cooling section
Sprühkülung 1. spray cooling. 2. spray cooling unit
Sprühpistole spraygun
Sprühtrocknen spray drying

Sprühturm spray drying tower
Sprühverfahren, luftloses airless spraying
sprunghaft sudden
SPS *abbr. of* **speicherprogrammierbare Steuerung**, stored-program control *(me)*
SPS-gesteuert stored-program controlled *(me)*
Spülbecken sink
spülen 1. to purge *(i.e. to clean out an extruder or injection moulding machine between changes of material or colour).* 2. to rinse, to flush
Spulenkörper coil former
Spülluft purging air
Spülmaschine dishwasher
Spülmittel washing-up liquid
Spulwerk reeling unit
Spurenanalyse trace analysis
Sputterätzen sputter etching
S-PVC *abbr. of* **Suspensions-PVC,** suspension PVC
SQC *abbbr. of* statistical quality control
SRA-Produkt scorch retarding agent
SRF-Ruß semi-reinforcing furnace black
St. *abbr. of* **Stahl**, steel
Staatshandelsland country with state monopoly trading
Staatshandelsländer, östliche Eastern bloc countries
Stab 1. rod, bar. 2. test piece/specimen
stäbchenförmig bar/rod shaped
stabförmig bar/rod shaped
Stabheizkörper heating rod
Stabilisator stabiliser
Stabilisatordosierung 1. addition of stabiliser 2. amount of stabiliser (added)
Stabilisatoreinsparung stabiliser saving
stabilisatorfrei unstabilised, free from stabiliser
Stabilisator-Gleitmittelkombination stabiliser-lubricant blend
stabilisatorhaltig stabilised, containing stabiliser
Stabilisatorkombination stabiliser blend
Stabilisatormenge amount of stabiliser
Stabilisatormischung stabiliser blend
Stabilisatorwirksamkeit stabiliser efficiency
Stabilisierbarkeit stabiliser tolerance
Stabilisierung stabilisation
Stabilisierungssystem stabiliser system
Stabilisierwirkung stabilising effect
Stabilität, thermische thermal stability
Stabilitätsberechnung stability calculation
Stabilitätsreserven stability reserves
Stadtgas town gas
Stagnationsstelle stagnation zone
Stagnationszone stagnation zone
Stahl, nichtrostender stainless steel
Stahl, rostfreier stainless steel
Stahlarmierung reinforcing steel (rods)
Stahlbeton reinforced concrete
Stahlbetonbau reinforced concrete construction

∗ *Words starting with* **Spritzteil-***, not found here, may be found under* **Formteil-, Teil-** *or* **Teile-**

Stahlblech steel plate, sheet steel
Stahldrahtbürste (steel) wire brush
Stahldrahtkord steel (wire) cord
Stahleinsatz steel insert
stahlelastisch energy-elastic
Stahlelastizität energy elasticity
Stahlfaser steel fibre
Stahlfeder steel spring
Stahlfederpaket steel spring assembly
Stahlfolie steel foil
Stahlgestell steel frame
Stahlguß cast steel
Stahlkalibrierung steel calibrator *(e)*
Stahlklebung 1. bonding of steel. 2. bonded steel joint
Stahlkonstruktion steel structure
Stahlkord steel cord
Stahlkordbindung steel cord reinforcement
Stahlkugel steel ball
Stahlkugelmühle steel ball mill
Stahlmembrane steel diaphragm
Stahlrahmen steel frame/chassis
Stahlschablone steel template
Stahlschweißkonstruktion welded steel construction: **Das Gestell, eine Stahlschweißkonstruktion, ist auf Rollen verfahrbar** the welded steel frame can be moved on rollers
Stahlserienwerkzeug standard steel mould
Stahlumflechtung steel braid
Stahlverklebung 1. bonding of steel. 2. bonded steel joint
Stahlwatte steel wool
Stahlwolle steel wool
Stammdatei master file *(me)*
Stammform standard mould unit/base *(im)*
Stammkapital ordinary capital
Stammlösung stock solution
Stammwerkzeug standard mould unit/base *(im)*
Stampfdichte compacted bulk density
Stampfvolumen compacted bulk volume
Stand der Technik state of the art
Standardabweichung standard deviation
Standardanlage standard system/equipment
Standardausführung 1. standard design. 2. standard model *(for translation example see under* **Serienausführung***)*
Standardausrüstung standard equipment
Standardbedingungen standard conditions
Standarddosierung standard amount
Standardeinrichtung standard equipment
Standardeinstellung 1. standard setting. 2. standard formulation
Standardempfehlung standard recommendation
Standardfarbenpalette standard colour range
Standardgeräteserie range of standard units
standardisiert standardised
standardisierte Werkzeugelemente standard mould units
Standardkunststoff commodity plastic
Standardlänge standard length

Standardlieferprogramm standard range
standardmäßig as a standard feature: **Standardmäßig ist die Maschine mit einer Abschaltautomatik ausgestattet** an automatic cut-out device is a standard feature of the machine
Standardmethode standard method
Standardpolystyrol standard polystyrene
Standardqualität standard grade
Standardreglerstruktur standard controller configuration
Standardrezeptur standard formulation
Standardschnecke general purpose screw
Standardsorte standard grade
Standardsortiment standard range
Standardspritzgußmarke standard injection moulding compound
Standardtype standard grade
Standardverpackung standard packaging
Standardweichmacher standard plasticiser
standfest firm, non-sag
Standfestigkeit 1. stability, rigidity. 2. non-sag properties
Standort location
Standsicherheit stability
Standvermögen 1. stability, rigidity. 2. non-sag properties
Standversuch creep test
Standzeit 1. working/service life *(e.g. of a machine)*. 2. residence/dwell time. 3. freeze/setting/cooling time *(im)*. 4. fibre time *(of PU foam)*. 5. shelf life *(e.g. of a resin)*. 6. pot life *(of catalysed resin)*. 7. time-to-failure *(pipe testing)*
Standzeitkurve time-to-failure curve *(in pipe testing)*
Stangenanguß 1. sprue gate *(im)* *(if referring to the mould)*. 2. sprue *(im)* *(if referring to the moulded part)*
Stangenangußkegel sprue *(im)*
Stangenanschnitt sprue gate *(im)*
Stangenauswerfer ejector rod *(im)*
Stangenpunktanguß antechamber-type pin gate *(im)*
Stanzabfall web scrap *(sr)*
Stanzbarkeit punchability, punching characteristics
Stanzen punching, stamping
Stanzfähigkeit punchability, punching characteristics
Stanzgitter web scrap *(sr)*
Stanzgitterabfall web scrap *(sr)*
Stanzgittermühle web scrap granulator *(sr)*
Stanzlöcher punched holes
Stanzverhalten punching qualities/characteristics
Stanzwerkzeug punching/blanking tool, cutting die
Stapelanlage stacking unit
Stapelautomat automatic stacker
Stapelbarkeit stackability
Stapelbehälter stacking container

Stapeleinrichtung stacking unit/device
Stapelfähigkeit stackability
Stapelfaser staple fibre
Stapelfasergewebe staple fibre fabric
Stapelmaschine stacker
stapeln to stack
Stapeltisch stacking table
Stapelung stacking
Stapelverarbeitung batch processing *(me)*
Stapelvorrichtung stacking device
stark vernetzt densely crosslinked
starkbasisch strongly basic/alkaline
Stärke 1. thickness. 2. strength. 3. starch
Stärkederivat starch derivative
starkpolar strongly polar
Starkstromkabel power cable
starkwandig thick-walled
starr rigid, stiff
starr-elastisch hard-elastic
Starrheit stiffness, rigidity
Startdrehzahl initial speed
Starterhülse starter housing
Startknopf start button
Startpunkt starting point
Startradikal initiator radical
Startreaktion initial reaction
Startreibung static friction
Startrezeptur starting formulation
Starttaste start button
Startzeit cream time *(in PU foaming)*
Statikmischelement static mixing element
stationär stationary
statisch static
statische Berechnungen stress analysis
statistisch statistic
statistische Sicherheit statistic certainty
Statormesser stationary/fixed/bed knife *(sr)*
Statormesserbalken bed knife block *(sr)*
Statormesserblock bed knife block *(sr)*
Statormesserkreis bed knife cutting circle *(sr)*
Statorscheibe fixed/bed knife *(sr)*
Status- *see* **Zustands-**
Staubabscheider dust separator
Staubalken restrictor/choke bar *(e)*
Staubalkenverstellung 1. restrictor bar adjustment. 2. restrictor bar adjusting mechanism
Staubbildung dust formation
staubfrei 1. dust-free *(e.g. atmosphere)*. 2. non-dusting *(e.g. moulding compound, in the sense that no dust is produced during handling)*
Staubmaske face mask
Staubschutz-Vorsichtsmaßnahmen precautions against inhalation of dust
Staubteilchen dust particles
staubtrocken dust dry
Staubtrockenheit dust dryness
Staubtrockenzeit dust dry time
Staubüchse flow restriction bush
Stauchbelastung compressive stress
Stauchdruck compressive force

Stauchdruckprüfung compression test
Stauchfestigkeit compressive strength
Stauchhärte compressive strength
Stauchhärteermüdung compressive failure
Stauchkraft compressive force
Stauchspannung compressive stress
Stauchung 1. compression. 2. compressive strain
Stauchung bei Bruch compressive strain at rupture
Stauchung bei Quetschspannung compressive strain at compressive yield stress
Stauchungsgeschwindigkeit rate of compression
Staudinger-Index limiting viscosity number, intrinsic viscosity
Staudruck 1. back pressure *(im)*. 2. dynamic pressure
Staudruckabbau back pressure reduction
Staudruckeinstellung 1. back pressure adjustment. 2. back pressure adjusting mechanism
Staudruckentlastung 1. back pressure relief. 2. back pressure relief mechanism
Staudruckprogrammverlauf back pressure programme
Staudruckregler back pressure controller
Stauelement baffle
Staukopf accumulator head *(bm)*
Staukopfplastifiziergerät accumulator head plasticising unit
Staukopfverfahren accumulator head blow moulding (process), blow moulding using an accumulator head
Stauleiste restrictor/choke bar *(e)*
Stauraum (space) in front of the screw
Stauring restrictor ring
Stauscheibe melt flow restrictor: **Man muß zusätzlich noch eine Lochscheibe als Stauscheibe hinter den Dornhalter einsetzen** an extra breaker plate must be fitted behind the mandrel support to restrict melt flow
Stauspalt (radial) screw clearance
Strauströmung pressure flow
Stauzone restricted flow zone *(e)*
Stauzonen dead spots
Stauzylinder melt accumulator *(bm)*
Stearat stearate
Stearinsäure stearic acid
Stearylstearat stearyl stearate
Steckanschluß plug-in connection
steckbar plug-in
Steckdose socket
Stecker plug
Steckerleiste edge connector
Steckkarte plug-in card *(me)*
Steckkupplung plug-in coupling
Steckmontage snap-on assembly
Steckverbindung 1. bayonet coupling. 2. socket joint

Steg 1. spider leg *(bm)*. 2. (screw) flight/thread *(e,im)*. 3. crosspiece, fillet, flange, bar
Stegbreite (flight) land width *(e,im)*
Stegdoppelplatte twin-wall sheet
Stegdornblaskopf spider-type blown film die
Stegdornhalter spider-type mandrel support, spider *(e,bm)*
Stegdornhalterblaskopf 1. spider-type blown film die *(bfe)*. 2. spider-type parison die *(bm)*
Stegdornhalterkopf 1. spider-type die *(general term)*. 2. spider-type blown film die *(bfe)*. 3. spider-type parison die *(bm)*
Stegdornhalterwerkzeug 1. spider-type die *(general term)*. 2. spider-type blown film die *(bfe)*. 3. spider-type parison die *(bm)*
Stegdornmarkierungen spider lines *(e,bfe)*
Stegdreifachplatte triple-wall sheet
Stegflanke flank
Stegflanke, aktive thrust/front face of flight, leading edge of flight *(e)*
Stegflanke, hintere rear face of flight, trailing edge of flight *(e)*
Stegflanke, passive rear face of flight, trailing edge of flight *(e)*
Stegflanke, treibende thrust/front face of flight, leading edge of flight *(e)*
Stegflanke, vordere thrust/front face of flight, leading edge of flight *(e)*
steggepanzert with hardened flights *(screw)*
Steghalterkopf 1. spider-type die *(general term)*. 2. spider-type blown film die *(bfe)*. 3. spider-type parison die *(bm)*
Steghaltermarkierungen spider lines *(e,bm)*
Stegkontur (screw) flight contours, (screw) thread contours *(e)*
Stegkopf flight land *(e)*
Stegmarkierungen spider lines *(e,bm)*
Stegoberfläche flight land *(e)*
Stegpaneel multi-wall sheet
Stegplatte multi-wall sheet
Stegsteigung pitch
Stegtragring spider-type mandrel support, spider
Stegverwischungseinrichtung smear/blending device *(e,bfe,bm) (used to obliterate spider lines in the melt flow)*
Stehkolben flat-bottom flask
Stehzeit downtime, shut-down period, non-productive time *(for translation example see under* **Stillstandzeit***)*
Steifheit stiffness, rigidity
Steifheitsplateau stiffness, rigidity
Steifigkeit stiffness, rigidity
Steifigkeitsanisotropie anisotropic rigidity
Steifigkeitsverlust loss of stiffness
Steiger riser
Steigerungsrate growth rate, rate of increase
Steigung 1. pitch *(e) (of screw)*. 2. slope *(of a curve)*
steigungsdegressiv with decreasing pitch *(screw)*

steigungsprogressiv with constantly increasing pitch *(screw)*
Steigungswinkel helix angle *(e)*
Steigzeit full rise time *(PU foaming)*
Steildach pitched roof
Steilheit slope *(of a curve)*
Steingut earthenware
Steinkohlenteer coal tar
Steinkonservierung stone conservation
Steinrestaurierung stone restoration
Steinsalz rock salt
Steinschlag flying stones
steinschlagfest resistant to flying stones
Steinschlagfestigkeit resistance to flying stones
Steinschlagresistenz resistance to flying stones
Steinschlagschutzwirkung protection against flying stones
Steinwolle mineral/rock wool
Steinzeug stoneware
Stelle, an anderer elsewhere
Stelleinrichtungen controls
Stellen, eingefallene sink marks *(im)*
Stellen, tote dead spots
Stellenwert (relative) importance
Stellflächenbedarf floor space requirement(s)
Stellglied control element
Stellgröße adjustable/variable
Stellmittel anti-flow additive
Stellmotor servo-motor
Stellschraube adjusting screw
Stellung position
stellvertretendes Vorstandsmitglied acting member of the Board of Directors
Stellwalze adjustable roller
Stellwert adjustable, variable
Stempel 1. male mould, punch *(t)*. 2. (mould) core *(im)*
Stempelkneter internal mixer
Stempelplatte core plate *(im)*
stereochemisch stereochemical
Stereoisomer stereo-isomer
Stereokautschuk stereo-rubber
stereospezifisch stereospecific
Sterilisation sterilisation
sterilisationsbeständig resistant to sterilising temperatures
Sterilisationsfestigkeit resistance to sterilising temperatures
Sterilisationszyklus sterilising cycle
sterilisierbar sterilisable
Sterilisierbarkeit sterilisability
Sterilisierfestigkeit resistance to sterilising temperatures
sterisch steric
sterisch gehindert sterically hindered
sterische Hinderung steric hindrance
Sternanordnung star-shaped (runner) arrangement *(im)*
Sterndreiecklanlauf star-delta start
sternförmig star-shaped

Sternverteiler 1. star-type distributor *(bfe)*. 2. feed system, runner system *(im)*. 3. radial system of runners *(im)*. 4. sprues and runners *(sr) (when referring to thermoplastic scrap)*
Stern-Wendelverteiler star-type spiral distributor *(bfe)*
Stetigregler continuous control unit
Steuer- und Regeleinrichtung 1. control system *(general term)*. 2. open and closed-loop controls *(see also explanatory note under* **Steuern** *and* **regeln***)*
Steueraggregat control unit
Steueranlage control system
Steueraufgabe control function
Steuerautomatik automatic controls
Steuerbefehl control signal/instruction *(me)*
Steuerblock control unit
Steuerbus control bus *(me)*
Steuerdaten control data
Steuereingang control input
Steuereinheit control unit
Steuereinrichtung control equipment *(plural: controls)*
Steuerelektronik electronic controls
Steuerelemente controls
Steuerfunktion control function
Steuergerät 1. control instrument *(general term)*. 2. open-loop control instrument *(if the text differentiates between* **Steuer-** *and* **Regelgerät***)*
Steuergröße control variable
Steuerkasten control cabinet
Steuerkommando control instruction *(me)*
Steuerkreis 1. control circuit *(general term)*. 2. open-loop control circuit *(if the text differentiates between* **Steuer-** *and* **Regelkreis***. For translation example see under* **Regelkreis***)*
Steuerkreis, offener open-loop control circuit
Steuern 1. control. 2. open-loop control: *Version 2. should be used only where the text makes a distinction between* **Steuern** *and* **Regeln***. See explanatory note under* **regeln***. In all other cases use version 1.*
Steuernocken cam
Steuerpult control desk/panel/console
Steuerschrank control cabinet
Steuersignal control signal
Steuerspannung control voltage
Steuersystem control system
Steuerung 1. control. 2. control unit
Steuerung, kontaktlose solid-state controls
Steuerung, numerische numeric control, NC
Steuerungen und Regelungen 1. controls, control systems *(general term)*. 2. open and closed-loop controls *(see also explanatory note under* **Steuern** *and* **regeln***)*
Steuerungsanweisung control instruction *(me)*
Steuerungsaufgabe control task
Steuerungsausgang control output
Steuerungseingang control input
Steuerungselektronik electronic controls

Steuerungsfunktion control function
Steuerungsrechner control computer *(me)*
Steuerungsreihenfolge control sequence
steuerungstechnisch *relating to control*:
 steuerungstechnische Komponente controls
Steuervarianten control options
Steuerventil control valve
Steuerwarte control desk/panel/console
Steuerwerk controller *(me)*
Stichausreißfestigkeit stitch tear strength
Stichmaße inside dimensions
Stichprobe random sample
Stichprobenprüfung 1. random test. 2. spot check
Stichsäge compass saw
Stickoxid nitrogen oxide
Stickstoff nitrogen
Stickstoffatmosphäre atmosphere of nitrogen
Stickstoffdurchlässigkeit nitrogen permeability
Stickstoffinnenkühlung internal cooling with liquid nitrogen
Stickstofflasche (liquid) nitrogen cylinder
Stickstoffstrom stream of nitrogen
Stickstoffzuleitung nitrogen feed line
Stifte, eingelegte pin inserts
Stiftzylinder pin-lined barrel *(e)*
Stillstand 1. stoppage. 2. downtime
Stillstandminimierung downtime minimisation: **Damit können keine Maßnahmen zur Stillstandminimierung getroffen werden** no measures can therefore be taken to minimise downtime
Stillstandzeit downtime, shut-down period, non-productive time: **bei Stillstandzeit der Maschine** when the machine is out of action
Stippe 1. speck. 2. fish eye
stippenarm 1. with a low speck count. 2. with a low fish eye content
Stippenarmut 1. low speck count. 2. low fish eye content
stippenfrei 1. free from specks. 2. free from fish eyes
Stirnfläche face
Stirnrad spur gear
Stirnraduntersetzungsgetriebe spur gear speed reduction mechanism
stirnseitig frontal
Stirnzahnrad spur gear
Stöchiometrie stoichiometry
stöchiometrisch stoichiometric
Stockpunkt setting/pour point
Stockwerkspritzgießwerkzeug multi-daylight injection mould
Stoffdaten material constants/properties
Stoffeigenschaften material properties
Stoffgefüge structure *(of a material, polymer etc.)*
Stoffgesetz materials law
Stoffgröße material constant/property
Stoffkenndaten material constants/properties
Stoffkennwert material constant/property

Stoffklasse material group, group of materials
Stoffkonstante material constant/property
Stoffkosten (raw) material costs
Stoffkreislauf material cycle
stoffliche Ressourcen material resources
stoffliche Verwertung material recycling
stoffliche Wiederverwertung material recycling
stoffliches Recycling material recycling
Stoffrecycling material recycling
Stoffverhalten material properties/performance: **rheologisches Stoffverhalten** rheological properties
Stopfaggregat stuffing unit, stuffer *(im)*
Stopfdichte compacted apparent density
Stopfdruck stuffing pressure
Stopfen, fliegender floating plug
Stopfkolben stuffing ram
Stopfschnecke stuffing screw
Stopftrichter feed hopper equipped with a stuffing device
Stopfvorrichtung stuffing device, stuffer
Stopfwerk stuffing unit, stuffer
Stöpseln, Widerstand zwischen insulation resistance
Stoptaste "stop" button
störanfällig liable to go wrong, liable to give trouble
Störanfälligkeit tendency to go wrong, tendency to give trouble: **Auswerfer sind wegen der Störanfälligkeit nicht zu empfehlen** ejectors are not recommended because of their tendency to go wrong
Störanzeige malfunction/warning signal
Störanzeige, akustische audible warning signal
Störanzeige, optische visual warning signal
Störeinflüsse disruptive/disturbing influences
störempfindlich liable to go wrong, liable to give trouble
störend troublesome
Störfaktor disruptive/disturbing influence
Störfall, im in case of malfunction
Störgeräusch (unwanted) noise
Störgröße disruptive/disturbing influence
störgrößenunabhöngig unaffected by disruptive/disturbing influences
Störmeldeeinrichtung malfunction alarm/indicator, alarm/warning signal
Störmeldung warning signal
Störmeldungsanzeige malfunction alarm/indicator, alarm/warning signal
Störsignal alarm/warning signal
Störsignalisierung malfunction indicator
Störstoffe contaminants, impurities
störunempfindlich unlikely to go wrong, unlikely to give trouble
Störung malfunction, disruption, disturbance, fault, interference
Störungsanzeige malfunction indicator
störungsarm almost trouble-free
Störungsbehebung elimination of faults *(in a moulding process)*: **Durch den übersichtlichen, logischen Aufbau ist eine schnelle Störungsbehebung möglich** thanks to the clear, logical design *(of the machine)* faults are quickly eliminated
Störungsblinkanzeige flashing light warning signal
Störungsdiagnose malfunction diagnostics *(me)*
Störungsfall, im if there is a fault/breakdown
störungsfrei trouble-free
Störungslokalisierung fault location: **Eine Störungslokalisierung kann rasch durchgeführt werden** faults can be located quickly
Störungsmeldegerät malfunction indicator
Störungsmeldung malfunction/alarm/warning signal: **Jede Störungsmeldung erscheint auf dem Bildschirm** every malfunction is displayed on the screen
Störungsortsignalisierung fault location indicator
störungssicher trouble-free
Störungssuche troubleshooting
Störungsüberwachung malfunction monitor, malfunction monitoring device
Störungsunanfälligkeit not subject to breakdowns/malfunction: **Wartungsfreiheit und Störungsunanfälligkeit sind die besonderen Vorzüge dieser Maschine** the special advantages of this machine are that it requires no maintenance and is not likely to break down
Störungsursache cause of malfunctioning
Störwertmeldung malfunction alarm/indicator, alarm/warning signal
Stoß 1. impact. 2. joint
stoßabsorbierend shock-absorbent
Stoßabsorption shock absorption
Stoßarbeit impact energy
stoßartige Beanspruchung impact stress
stoßbeansprucht subjected to impact (stress)
Stoßbeanspruchung impact stress
Stoßbelastbarkeit impact resistance
Stoßbelastung impact stress
stoßdämmend shock-absorbent
stoßdämpfend shock absorbent
Stoßdämpfer shock absorber
Stößel ram *(cm)*
Stoßelastizität (impact/rebound) resilience
Stößeldruck ram pressure *(cm)*
Stößelhub ram stroke *(cm)*
Stoßempfindlichkeit sensitivity to impact
Stoßenergie impact energy
Stoßenergieaufnahmevermögen impact energy absorbtive capacity
Stoßfänger bumper
Stoßfängerfunktion bumper performance
Stoßfängerhülle bumper cover
Stoßfängerrecyclat recycled bumpers
Stoßfängerschale bumper shell
Stoßfängerschürze bumper apron
Stoßfängersystem bumper system/assembly
Stoßfängerträger bumper bracket

Stoßfängerverkleidung bumper cover
Stoßfängerwerkzeug bumper mould
stoßfest impact resistant
Stoßfestigkeit impact strength/resistance
Stoßfläche impact surface
Stoßgeschwindigkeit impact speed, speed of impact
Stoßkraft impact force
Stoßprüfung impact test
stoßsicher impact resistant
Stoßstange bumper
Stoßstangenabdeckung bumper cover
Stoßverhalten impact behaviour, behaviour on impact
Stoßverklebung butt joint
Stoßversuch impact test
stoßweise discontinuous(ly), intermittent(ly)
Stoßzähigkeit impact strength/resistance
Strahl, freier jet of material *(im) (for translation example and explanatory note see under* **Freistrahl**)
Strahlanlage grit blasting unit
Strahleinfall beam incidence
strahlenabsorbierend radiation absorbent
strahlenbeständig radiation resistant
Strahlenbeständigkeit radiation resistance
strahlende Wärme radiant heat
Strahlendosis radiation dose
Strahlendurchlässigkeit radiation permeability
strahlenempfindlich radiation-sensitive
Strahlenenergie radiation energy
strahlenhärtend radiation curing
Strahlenhärtung radiation curing
strahleninduziert radiation-induced
Strahlenoxidation radiation oxidation
Strahlenquelle radiation source
strahlenresistent radiation resistant
Strahlenresistenz radiation resistance
Strahlenschutzmittel radiation stabiliser
Strahlenstreuung radiation scatter
strahlenvernetzbar radiation curable
strahlenvernetzt radiation crosslinked/cured
Strahlenvernetzung radiation crosslinkage/curing
Strahlerheizung radiant heaters/heating
Strahlung radiation
strahlungsbeständig radiation resistant
Strahlungsbeständigkeit radiation resistance
Strahlungsdose radiation dose
Strahlungsdosis radiation dose
Strahlungsenergie radiation energy
strahlungshärtbar radiation curable
Strahlungshärtung radiation curing/cure
Strahlungsheizung radiant heaters/heating
strahlungsinduziert radiation-induced
Strahlungsintensität radiation intensity
Strahlungsoxidation radiation-induced oxidation
Strahlungspyrometer radiation pyrometer
Strahlungsquelle radiation source
Strahlungsverlust heat loss due to radiation, radiant heat loss

Strahlungsvernetzung radiation crosslinkage/cure
Strahlungsvulkanisation radiation cure/vulcanisation
Strahlungswärme radiant heat
Strahlungswärmeverlust radiant heat loss
strainern to sieve, to strain
stramm stiff
Strang stand, extrudate *(e)*
Strangabschlagsystem strand pelletising system *(sr)*
Strangaufweitung die swell *(e)*
Strangaufweitungsverhältnis die swell *(e)*
Strangdüse strand die *(e)*
Strangdüsenkopf strand die head *(e)*
Strangextrusion extrusion
Strangextrusionsmaschine extruder
stranggezogen pultruded
Stranggranulat cylindrical pellets
Stranggranulator strand pelletiser *(sr)*
Stranggranulieranlage strand pelletising line *(sr)*
Stranggranulierung strand pelletisation *(sr)*
Strangpresse extruder
Strangpressen extrusion
Strangpreßmasse extrusion compound
Strangpreßverfahren extrusion (process)
Strangpreßwerkzeug (extrusion) die
Strangschneider strand pelletiser *(sr)*
Strangwerkzeug strand die *(e)*
Strangziehen pultrusion
Strangziehtechnik pultrusion
strapazierbar hard wearing
Straßenbau road building/making
Straßenbeleuchtung street lighting
Straßenmarkierungsfarbe road marking paint
Straßenmarkierungsmasse road marking paint
Straßensilofahrzeug road tanker
Straßentankfahrzeug road tanker
Straßentankwagen road tanker
Straßentankzug road tanker
Straßentransport road transport
Straßentransportwesen road transport (sector)
Straßenversuch road test
streckbar extensible
Streckbarkeit extensibility
Streckblasanlage stretch blow moulding plant
Streckblasen stretch blow moulding
Streckblasformen stretch blow moulding
Streckblasformmaschine stretch blow moulder/moulding machine
Streckblasmaschine stretch blow moulder/moulding machine
Streckblasverfahren stretch blow moulding (process)
Streckdehnung elongation
Streckeinrichtung stretching unit
Strecken stretching, drawing *(e.g. of film) (see explanatory note under* **Recken**)
Streckfolie stretch wrapping film
Streckfolienverpackung 1. stretch wrapping. 2. stretch wrapped pack

Streckgrenze yield point
Streckgrenze, Zugdehnung bei elongation at yield
Streckgrenze, Zugspannung bei tensile stress at yield
Streckmetall expanded metal
Streckmittel extender
Streckspannung yield stress, tensile stress at yield
Streckungsmittel extender
Streckverhältnis stretch/draw ratio
Streckverpackung 1. stretch wrapping. 2. stretch wrapped pack
Streckwalze stretching roll
Streckwerk stretching unit
Streckziehen stretch forming
Streckziehwerkzeug stretch forming tool
Streichanlage spreading plant, spread coating plant
Streichartikel coated products
streichbar spreadable
Streichbarkeit 1. spreadability *(e.g. PVC paste)* 2. brushability *(e.g. paint)*
Streichbeschichten spread coating
Streichen spreading, spread coating, brushing
streichfähig 1. spreadable *(e.g. PVC paste)* 2. brushable *(e.g. paint)*
Streichkunstleder leathercloth
Streichmaschine spread coating machine
Streichmasse spread coating compound
Streichmesser spreading knife
Streichpaste spread coating paste
Streichverfahren spread coating (process)
Streichversuch spread coating trial
streifenförmig streaky: **streifenförmige Markierungen** streaks
Streifigkeit striation
Streubeiwert scatter coefficient
Streubereich scatter range
Streubreite scatter range
Streueffekt scatter effect
Streulicht scattered light
Streulichtverhalten light scattering behaviour
Streusalz road salt, (road) de-icing chemicals
Streuscheibe (headlamp) diffuser
Streuung scattering *(of test results)*
Strichcode bar code *(me)*
strichpoliert with a brushed finish
Strichpolitur brushed finish
strichpunktiert dot-dash *(curve)*
stringerversteift stringer-stiffened/-braced
Strom 1. current 2. stream, flow
-strom *whilst this suffix can sometimes be translated as* -flow *or* -stream, *it often denotes quantity. Examples are given under* **Volumenstrom, Massestrom, Mengenstrom** *and* **Ölstrom** *q.v.*
stromabwärts downstream
Stromabwärtsentgasung 1. downstream venting/degassing/ devolatilisation. 2. downstream venting/degassing/devolatilising system
Stromanzeige ammeter
Stromaufnahme current input/consumption
stromaufwärts upstream
Stromausfall power failure: **bei Stromausfall** if the power supply breaks down
Strombegrenzungswiderstand current-limiting resistance
Stromdichte current density
Stromerzeugung production of electricity
Stromfluß current flow
Stromistwert actual current
Stromkosten electricity costs
Stromkreis (electric) circuit
Stromlaufplan circuit diagram
Stromquelle power supply/source
Stromregelventil flow control valve
Stromregler 1. current controller 2. flow control unit/device
Stromschiene busbar
Stromsollwert required current
stromsparend energy-saving
Stromstärke current intensity
Strömung flow
Strömung, laminare laminar flow
Strömung, turbulente turbulent flow
strömungsabhängig flow-dependent
Strömungsanalyse flow analysis
strömungsbedingt due to flow, caused by flow
∗**Strömungsbedingungen** flow conditions
Strömungsbild flow pattern
Strömungsgeschwindigkeit flow rate
Strömungsgeschwindigkeitsaufnehmer flow rate transducer
strömungsgünstig *favourable for flow*: **Die Bohrung muß strömungsgünstig gestaltet sein** the channel must be designed so as to ensure smooth and even flow
strömungsinduziert flow-induced
Strömungskanal 1. flow channel, melt flow-way. 2. runner *(im)*
Strömungslinie weld line, flow line *(see explanatory note under* **Bindenaht**)
Strömungsmesser flowmeter
Strömungspotential streaming potential
Strömungsprofil 1. flow diagram. 2. flow profile
Strömungsquerschnitt 1. flow channel. 2. runner *(im)*
Strömungsrichtung flow direction, direction of flow
Strömungsschatten spider lines *(e,bm)*
strömungstechnisch *relating to flow*: **strömungstechnische Verhältnisse** flow conditions; **Verbesserung der Rohrköpfe in strömungstechnischer Hinsicht** improvements in pipe die design to achieve better flow
strömungsteilend melt stream-dividing
Strömungsumverteilung flow redistribution

∗ *Words starting with* **Strömungs-**, *not found here, may be found under* **Fließ-**

Strömungsverhältnisse flow conditions
Strömungsverlagerung flow redistribution
Strömungsvorgänge flow processes
Strömungswächter flow monitor
Strömungsweg 1. (melt) flow path, (melt) flowway. 2. runner *(im)*
Strömungswiderstand flow resistance
Stromventil flow control valve
Stromverbrauch electricity consumption: **der tatsächliche Stromverbrauch** the actual amount of electricity used
Stromversorgung power supply
Stromwandler current transformer
Stromzufuhr power supply
Struktur 1. structure 2. texture 2. configuration *(me)*
Strukturanalyse structural analysis
Struktureinheit structural unit
Strukturelement structural element
strukturell structural
Strukturfehler structural defect
Strukturfestigkeit tear propagation resistance
Strukturformel structural formula
strukturgeschäumt structural foam: **strukturgeschäumte Profile und Platten** structural foam profiles and sheets
strukturgleich having the same structure
strukturiert 1. textured *(e.g. surface)*. 2. structured
Strukturkleber structural adhesive
Strukturmerkmale structural characteristics/features
Strukturparameter structural parameter
Strukturschaum structural/integral foam
Strukturschaummaschine structural foam moulding machine
Strukturschaumspritzgießmaschine structural foam moulding machine
Strukturschaumstoff structural/integral foam
Strukturschaumverfahren structural foam moulding
Strukturunterschiede structural differences
Strukturveränderungen structural changes
strukturviskos pseudoplastic, non-Newtonian
Strukturviskosität pseudoplasticity, pseudoplastic/non-Newtonian behaviour: *Literal translation of the word should be avoided since the result would be meaningless, and incomprehensible to a native English or American speaker*
Strukturviskositätseffekt pseudoplasticity, pseudoplastic/non-Newtonian behaviour *(see explanatory note under* **Strukturviskosität***)*
Strukturwerkstoff structural material
Stückprozeß discontinuous process
Stückzahlen, große large numbers
Stückzahlen, kleine small numbers, few
Stückzähler parts counter, parts counting unit/device
stückzahlunabhängig irrespective of the number of items/articles

stufenlos infinite(ly), stepless(ly)
stufenlos einstellbar steplessly/infinitely variable
stufenlos verstellbar steplessly/infinitely adjustable
Stufenschnecke multi-stage screw *(e)*
Stufenschnitt, Bandgranulator mit stair-step dicer *(sr)*
Stufenschnittgranulator stair-step dicer *(sr)*
Stufenskala graduated scale
Stufenwerkzeug multi-stage die *(e)*
stumpfgeschweißt butt welded
Stumpfnaht butt-welded joint
Stumpfschweißen butt welding
Stumpfschweißverbindung butt welded joint
Stumpfstoß butt joint
Stundendurchsatz hourly throughput, throughput per hour
Stundenleistung hourly output, output per hour
Sturzhelm crash helmet
Stützisolator post/pin insulator
Stützlochplatte breaker plate *(e)*
Stützluft 1. calibrating air *(e)*: *This is the air introduced into the extruded pipe to keep it in intimate contact with the calibrating die. The word is derived from* **stützen***, to support.* 2. bubble inflation air *(bfe)*: *Air used to inflate the tube to form the film bubble and subsequently keep it in shape, i.e. to support it.* 3. air: **Die Quetschwalzen verhindern gleichzeitig das Entweichen der Stützluft aus dem Folienschlauch** the pinch rolls at the same time prevent the air inside the film bubble from escaping. 4. blowing/inflation air *(bm)*
Stützluftbohrungen calibrating air holes *(e)*
Stützluftkalibrierung 1. air pressure calibration. 2. air pressure calibration unit *(e)*
Stützplatte 1. backing/supporting plate *(im)*. 2. breaker plate *(e)*
Stützring thrust/supporting ring
Stützrippen supporting ribs
Stützweite distance between supports *(of test piece)*
Stylinganforderungen styling requirements
Stylingvorstellung styling concept
Styrol styrene (monomer)
Styrolacrylatcopolymerisat styrene-acrylate copolymer
Styrol-Acrylnitril-Butadienkautschuk ABS rubber, acrylonitrile-butadiene- styrene rubber
Styrol-Acrylnitrilcopolymerisat styrene-acrylonitrile copolymer, SAN
Styrolalkyd styrene alkyd
Styrolantell styrene content
Styrolblock styrene block
Styrol-Butadienkaltpolymerisat cold polymerised styrene-butadiene rubber
Styrol-Butadienkautschuk styrene-butadiene rubber, SBR
Styrol-Butadienlatex styrene-butadiene latex/dispersion

Styrol-Butadienölkautschuk styrene-butadiene-oil masterbatch
Styrol-Butadienpfropfpolymerisat styrene-butadiene copolymer, SBR
Styrolcopolymerisat styrene copolymer
Styroldämpfe styrene vapours
Styroleinheit styrene unit
Styrolemission styrene emission
styrolfrei styrene-free
Styrolhomopolymerisat styrene homopolymer
styrolisiert styrenated
styrolmodifiziert styrene modified
Styrolpolymerisat styrene polymer
Styrolreste residual styrene (monomer)
Styrolrestgehalt residual styrene content
Styroltriblockcopolymer styrene triblock copolymer
styrolunempfindlich unaffected by styrene
Styrolzugabe addition of styrene: **eine Styrolzugabe von 5 Gew.-Tln.** the addition of 5 p.b.w. of styrene
Styrolzusatz addition of styrene *(for translation example see under* **Styrolzugabe***)*
subjektiv subjective
Sublimation sublimation
submikroskopisch submicroscopic
Subroutine subroutine *(me)*
Substanzpolymerisation bulk polymerisation
Substituent substituent
substituiert substituted
Substitutionsmöglichkeiten substitution possibilities
Substitutionsprodukt alternative product, substitute
Substitutionsreaktion substitution reaction
Substitutionswerkstoff alternative material
Substrat substrate
Substratbenetzer surface wetting agent
Suchprogramm search routine/program *(me)*
Sulfanilsäure sulphanilic acid
Sulfatasche sulphated ash
Sulfataschegehalt sulphated ash (content)
Sulfation sulphate ion
Sulfenamidbeschleuniger sulphenamide accelerator
Sulfolignin sulpholignin
Sulfonat sulphonate
Sulfonatgruppe sulphonate group
Sulfoneinheit sulphone group
Sulfongruppe sulphone group
Sulfonierung sulphonation
Sulfoniumzwitterion sulphonium zwitterion
Sulfoniumzwitterverbindung sulphonium zwitter compound
Sulfonsäure sulphonic acid
Sulfonsäureamid sulphonic acid amide
Sulfonsäureester sulphonate, sulphonic acid ester
sulzig gelatinous
Summe der kleinsten Fehlerquadrate sum of least squares
Summenformel empirical formula

Superbenzin super-grade petrol
Superkraftstoff super-grade petrol
Supervisorrechner supervisory computer *(me)*
superweich supersoft
supramolekular supramolecular
Surfbrett surfboard
suspendieren to suspend
Suspension suspension
Suspensionscopolymerisat suspension copolymer
Suspensionshilfsmittel suspending agent
Suspensionshomopolymer suspension homopolymer/polymer
Suspensionspfropfcopolymerisat suspension graft copolymer
Suspensionspolymerisat suspension polymer
Suspensionspolymerisation suspension polymerisation
Suspensionspolymerisationsverfahren suspension polymerisation (process)
Suspensions-PVC suspension PVC
Suspensionstype suspension PVC, suspension polymer
Suspensionsverfahren suspension polymerisation
S-vernetzt sulphur-vulcanised/-cured
SVK *abbr. of* **selbstverstärkende Kunststoffe**, self-reinforcing polymers, liquid crystal polymers, LCP
S/W-Videogerät monochrome visual display unit *(me)*
symbolisch symbolic *(me)*
Symboltaste symbolic key *(me)*
Synärese syneresis
Synchronisation synchronisation
Synchronomotor synchronous motor
Synchronschaltuhr synchronous timer
syndiotaktisch syndiotactic
synergetisch synergetic, synergistic
Synergismus synergism
Synergist synergist
synergistisch synergistic, synergetic
syntaktisch syntactic
Synthese synthesis
Synthesefaser synthetic/man-made fibre
Synthesefasergewebe synthetic fabric
Synthesegummi synthetic rubber
Synthesekautschuk synthetic rubber
Synthetikkautschuk synthetic rubber
synthetisch synthetic
synthetisieren to synthesise
Systemanalytiker system analyst
Systembus system bus *(me)*
Systemdruck hydraulic pressure
Systemdruck, hydraulischer hydraulic pressure
Systemerweiterungsfähigkeit ability of the system to expand: **Es gibt sich dadurch eine leichte Systemerweiterungsfähigkeit** as a result, the system can be easily expanded
Systemleitung hydraulic line
Systemsoftware system software *(me)*

Systemspeicher hydraulic accumulator
SZ *abbr. of* **Säurezahl**, acid value

T

Tabakwaren tobacco products
tabellarisch tabular
Tabellenform, in tabulated, in tabular form
Tabellenwert figure given in tables)
Taber Abrieb Taber abrasion
Taberprüfung Taber abrasion test
Tablett tray
Tablettenmaschine pelleting machine
Tablettenröhrchen tablet vial/container
Tablettierfähigkeit pelleting properties
Tablettierpresse pelleting press
Tablettierung pelleting
Tacho tachometer
Tafel- *see* **Platten-**
Tagesbehälter silo containing a day's supply of moulding compound
Tagesleistung daily output
Tagesleuchtfluoreszenz daylight fluorescence
Tageslicht daylight
Tagesordnung agenda *(for a meeting)*
Tagesschicht day shift
Tagessilo silo containing a day's supply of moulding compound
Tageszeit time of day
TA-Harz triazine A resin
Takt (moulding) cycle
Taktfolge, rasche fast cycling
Taktgenerator clock signal generator *(me)*
Taktizität tacticity
Taktsteuerung cycle control
Taktüberwachung cycle monitor
Taktverfahren, im intermittent(ly)
Taktzahl(en) number of (moulding) cycles
Taktzeit cycle time
Taktzeit im Trockenlauf dry cycle time *(see explanatory note under* **Trockenlaufzeit***)*
Täler valleys *(of an irregular surface)*
Talkum talc
talkumgefüllt talc filled
Talkumieren powdering/treatment with talc
talkumverstärkt talc-filled
Talkumvorkommen talc deposit
Tallöl tall oil
Tallölalkydharz tall oil alkyd (resin)
Tallölfettsäure tall oil fatty acid
TA-Luft Technical Directive for the Prevention of Air Pollution. *It is best to put the German term in inverted commas, followed by the English equiavalent in brackets. When the term occurs several times, it can be treated as above the first time it occurs, and can then be left in the original German, possibly in inverted commas.*
TAM triazine A monomer
Tandemanordnung tandem arrangement
Tandemextruder cascade extruder
Tangentenmodul tangential modulus
Tangentenrichtung tangential direction
Tangentialkraft tangential/shear force
Tangentialschurre tangential chute
Tangentialspannung tangential/shear stress
Tangentialströmung tangential flow
Tangentialwickler tangential-drive winder
tangierend non-intermeshing *(screws)*
Tankauskleidung tank lining
Tankdeckel petrol/fuel tank cap
Tankdeckeldichtung fuel tank cap seal
Tankeinfüllstutzen petrol/fuel tank neck
Tankgeber petrol/fuel gauge
Tanklastwagen road tanker
Tankschwimmer petrol/fuel tank float
Tankverschluß petrol/fuel tank cap
Tankwagen road tanker
Tankzug rail tanker
Tänzerarmfühler compensating arm sensor
Tänzerwalze compensating/dancing/idling roller
Tänzerwalzensteuerung compensating/dancing/idling roller control unit
TAO triazine A oligomer
Tapete wallpaper
Tapetenkleister wallpaper adhesive
Taragewicht tare weight
Tarierung taring
Taschenformat, im pocket-size
Taschenrechner pocket calculator/computer
Täschnerindustrie handbag and luggage industry
Täschnerwaren handbags and luggage
Tastatur keyboard
Tastaturfeld keyboard
Taste key, push-button
Tastendruck, auf at the touch of a button
Tastendruck, per by pressing a key, at the touch of a button
Tastenfeld keyboard
Tastenplatte keyboard
Tastkopf scanning head
Tauchapparatur dipping equipment
Tauchbad dip coating bath, dipping bath
Tauchbehälter 1. dipping tank 2. immersion tank
Tauchbeschichtung dip coating
Tauchblasen dip blow moulding
Tauchblasformen dip blow moulding
Tauchblasmaschine dip blow moulder/moulding machine
Tauchblasverfahren dip blow moulding (process)
Tauchdorn dipping mandrel
Tauchdüse extended/long-reach nozzle *(im)*
Tauchen 1. dipping. 2. dip coating *(e.g. of wirework articles with PVC paste)*. 3. dip moulding *(e.g. of PVC gloves; see explanatory note under* **Tauchform***)*

Tauchform former *(used in dip moulding of PVC paste, e.g. a hand-shaped metal former is dipped into the paste to make gloves)*
Tauchheizung immersed/immersion heater
Tauchimprägnierung dip impregnation
Tauchkammer accumulator cylinder *(bm)*
Tauchkammerkolben piston, plunger: *Since the expression is used only in connection with* **Tauchblasen** *(q.v.) and the* **Tauchkammer** *(q.v.), the piston's function and location will be obvious from the context, so that the word can be treated simply as* **Kolben**
Tauchkammerzylinder accumulator cylinder
Tauchkante vertical flash face
Tauchkantenspiel vertical flash face clearance
Tauchkantenwerkzeug positive mould *(cm)*
Tauchkolben piston, plunger *(see explanatory note under* **Tauchkammerkolben***)*
Tauchkolbenpumpe piston pump
Tauchlack 1. dipping paint. 2. dipping varnish/lacquer
tauchlackiert dip coated
Tauchmasse dip coating compound
Tauchpaste dipping paste
Tauchpumpe piston pump
Tauchverfahren 1. dipping. 2. dip coating. 3. dip moulding *(see explanatory notes under* **Tauchen***)*
Tauchversuch immersion test
Tauchwägeverfahren hydrometer method *(of determining density)*
Tauchwanne dipping tank
Taumelbewegung tumbling movement
Taumelmischer tumble mixer
Taumeltrockner tumble drier
Taumittel de-icing chemical, road salt
Taupunkt dew point
Taupunkttemperatur dew point
TCF abbr. of **Trikresylphosphat**, tricresyl phosphate, TCP
TDI abbr. of **Toluoldiisocyanat**, also called **Toluylendiisocyanat**, toluene/tolylene di-isocyanate, TDI
Technik 1. technique 2. technology 3. engineering
Technik, Stand der state of the art
Technikum pilot plant
Technikummaschine pilot plant machine
Technikummaßstab pilot plant scale
technikumsadäquat *adequate or suitable for pilot plant* **unter technikumsadäquaten Bedingungen** under pilot plant conditions
Technikumsanlage pilot plant
Technikumstreichanlage pilot spread coating plant
Technikumversuch pilot plant trial
technisch 1. technical. 2. industrial: **technische Gewebe** industrial fabrics; **technische Hohlkörper** industrial containers; industrial blow mouldings. 3. engineering. **technische Kunststoffe** engineering plastics; **technische Phenolharze** phenolic engineering resins; **technische Thermoplaste** engineering thermoplastics; *The term "technical plastics/ thermoplastics" etc., often encountered in translations, is not English and is neither used nor understood by native English/American speakers. It should therefore be avoided. The ONLY correct and universally accepted name for these materials is the one given above.*
-technisch this suffix is used to convert nouns into adjectives and is only rarely translated as technical. *For translation examples see under* **fertigungstechnisch, fließtechnisch, gestaltungstechnisch, maschinentechnisch, regeltechnisch, spritztechnisch, steuerungstechnisch, verarbeitungstechnisch** *and* **verfahrenstechnisch**
technisch machbar technically feasible
technische Chemiewerkstoffe engineering plastics/polymers
technische TPE engineering thermoplastic elastomers
Technologie technology
Technologietransfer technology transfer
Teer tar
Teerfraktion tar fraction
Teeröl tar oil
Teerpappe roofing felt
Teerpech tar pitch
Teerreste tar residues
Teichabdichtungsfolie pond liner sheeting
∗**Teilansicht** partial view
teilaromatisch partly aromatic
teilaufbereitet partly compounded
Teilausdruck partial print-out *(me)*
teilautomatisch semi-automatic(ally)
teilautomatisiert partly automated
teilbeheizt partly heated
Teilbereich specific area/part *(e.g. of an industry)*
∗∗**Teilchen** particle
Teilchendurchmesser particle size
Teilchendurchmesser, mittlerer mean/average particle size
Teilchenfluß particle flow
Teilchenform particle shape
Teilchengröße particle size
Teilchengröße, mittlere mean/average particle size
Teilchengröße, vorherrschende predominant particle size
Teilchengrößendurchmesser particle size
Teilchengrößenverteilung particle size distribution
Teilchenoberfläche particle surface
Teilchenphysik particle physics

∗ *Words starting with* **Teil-** *or* **Teile-**, *not found here, may be found under* **Formteil-** *or* **Spritzteil-**
∗∗ *Words starting with* **Teilchen-**, *not found here, may be found under* **Korn-** *or* **Partikel-**

Teilchenradius particle radius
Teilchenstruktur particle structure
Teilchenverteilung particle size distribution
Teilchenzwischenräume space between the particles
Teile- see **Formteil-**
Teile, blasgeformte blow mouldings
Teile, unvollständige short mouldings
Teilentladung partial discharge
Teilester partial ester
teilevakuiert partly evacuated
teilfluoriert partly fluorinated
teilgefüllt partly full/filled
Teilgewicht, maximales maximum shot weight
teilhalogeniert partly halogenated
Teilheißläufer partial hot runner *(im)*
teilkristallin partially crystalline
Teilkristallinität partial crystallinity
teilmethyliert partly methylated
teiloxidiert partially/partly oxidised
Teilqualität part/moulded-part quality
Teilströme separate melt streams *(e) (e.g. a melt stream divided by a spider)*
Teilungsebene (mould) parting surface/line *(im)*
Teilungsfläche (mould) parting surface/line *(im)*
Teilungsfuge (mould) parting surface/line *(im)*
Teilungslinie parting/joint line *(mark on moulded article where the two mould halves meet)*
Teilvakuum partial vacuum
teilvernetzt partly crosslinked
teilverseift partly saponified
teilverzweigt partly branched
teilweise vernetzt partly crosslinked
telechel telechelate
Telefonapparat telephone
Telefonkarte phonecard
Telleranguß diaphragm gate *(im)*
Tellerfeder disc spring
Tellerfederpaket disc spring assembly
TEM abbr. of **Transmissionselektronenmikroskop**, transmission electron microscope
TEM-Aufnahme transmission electron micrograph
Temperatur temperature
Temperatur, erhöhte elevated temperature
Temperaturabfall temperature decrease, drop/decrease in temperature
temperaturabhängig temperature-dependent, depending on the temperature
Temperaturabhängigkeit temperature dependence: **Die Temperaturabhängigkeit der Viskosität** the dependence of viscosity on temperature, the effect of temperature on viscosity
Temperaturabnahme temperature decrease
Temperaturabsenkung temperature decrease, drop/decrease in temperature
Temperaturalterung heat ageing
Temperaturalterungsverhalten heat ageing behaviour
Temperaturänderung change in temperature

Temperaturanhebung raising/increasing the temperature: **bei einer Temperaturanhebung von 10°C...** if the temperature is increased by 10°C...
Temperaturanstieg temperature increase, rise/increase in temperature
Temperaturanwendungsbereich operating temperature range
Temperaturanzeige temperature display
Temperaturanzeigegerät temperature indicator
Temperaturaufgabe application of heat
Temperaturaufnehmer thermocouple
Temperaturaufzeichnung recording of temperature
Temperaturausgleich temperature adjustment
Temperaturausgleichsystem temperature control/adjusting mechanism
Temperaturbandbreite temperature range
temperaturbeansprucht subjected to high temperatures
Temperaturbeanspruchung exposure to high temperatures
temperaturbedingt due to temperature
Temperaturbehandlung 1. heat treatment. 2. conditioning, annealing
Temperaturbelastbarkeit thermal endurance, the ability to withstand high temperatures: **Hohe Temperaturbelastbarkeit ist einer der Hauptvorteile von Fluorkunststoffen** one of the main advantages of fluorocarbon plastics is their ability to withstand high temperatures
Temperaturbelastung heat exposure, exposure to high temperatures: **extreme Temperaturbelastung** exposure to extremely high temperatures; **Mit Polyethersulfon sind Temperaturbelastungen bis 160°C möglich** Polyethersulphone will withstand temperatures of up to 160°C; **bei hoher Temperaturbelastung** at high temperatures
Temperaturbereich temperature range
temperaturbeständig 1. heat resistant. 2. resistant to high and low temperatures
Temperaturbeständigkeit 1. heat resistance. 2. resistance to high and low temperatures: **Die Temperaturbeständigkeit reicht von ca. -60 bis +230°C** it withstands temperatures from -60 to +230°C
Temperaturdifferenz temperature difference
Temperatureinfluß effect of temperature
Temperatureinsatzbereich operating temperature range
Temperatureinsatzgrenze maximum operating temperature
Temperatureinschwingen temperature build-up
Temperatureinstellung 1. temperature setting. 2. temperature setting mechanism
Temperatureinstufung temperature rating
Temperatureinwirkung action/effect/application of heat
temperaturempfindlich heat sensitive, affected by heat
Temperaturempfindlichkeit heat sensitivity

Temperaturerfassung temperature measurement
Temperaturerhöhung 1. temperature increase. 2. increasing/raising the temperature
Temperaturfeld temperature range
temperaturfest heat resistant
Temperaturfestigkeit heat resistance
Temperaturfühler thermocouple, temperature sensor/probe
Temperaturführung 1. temperature control: **In der Austragszone sorgt eine spezielle Heiz-Kühl-Einrichtung für eine konstante Temperaturführung** a special heating-cooling system in the metering section ensures that the temperature is kept constant; **Abb.19 zeigt, daß durch geeignete Temperaturführung...** Fig.19 shows that, if the temperature is controlled properly.... 2. temperature profile: **Das segmentweise Beheizen erlaubt eine differenzierte Temperaturführung über die Breite der Düse** sectional heating enables a gradual temperature profile to be achieved across the width of the die
Temperaturgeber thermocouple, temperature sensor/probe
Temperaturgefälle temperature gradient/difference
Temperaturgeschichte thermal history
Temperaturgleichgewicht temperature equilibrium
Temperaturgleichmäßigkeit even/uniform temperature
Temperaturgradient 1. temperature gradient, change in temperature. 2. temperature difference
Temperaturgrenze temperature limit
Temperaturgrenzwert limiting temperature, temperature limit
Temperaturhomogenität even/uniform temperature: **damit eine ausreichende Temperaturhomogenität gewährleistet ist** so that the temperature is sufficiently uniform throughout
Temperaturindikator temperature indicator
temperaturinduziert temperature-induced
Temperaturintervall temperature range
Temperaturkompensation temperature compensation
Temperaturkonstanz constancy of temperature
Temperaturkontrolle temperature control
Temperaturleitfähigkeit thermal diffusivity
Temperaturleitzahl thermal diffusivity
Temperaturmeßfühler thermocouple, temperature sensor/probe
Temperaturmeßgeber thermocouple, temperature sensor/probe
Temperaturmeßgerät temperature measuring device, thermocouple, thermometer
Temperaturmeßmethode method of measuring temperature

Temperaturmeßstelle temperature measuring point
Temperaturmessung 1. temperature measurement. 2. temperature measuring device
Temperaturmeßvorrichtung temperature measuring device
Temperaturmittelwert mean/average temperature
Temperaturniveau temperature level
Temperaturobergrenze upper temperature limit
Temperaturprofil temperature profile/pattern
Temperaturprogramm temperature programme
Temperaturregeleinschub slide-in temperature control module
Temperaturregelgerät temperature controller, temperature control instrument/unit
Temperaturregelkreis temperature control circuit
Temperaturregelung temperature control
Temperaturregler temperature controller, temperature control instrument
Temperaturreglerbaugruppe temperature control module
Temperaturregulierung 1. temperature control. 2. temperature control system
Temperaturrichtwert approximate temperature
Temperaturschockbeständigkeit heat shock resistance
Temperaturschreiber temperature recorder
Temperaturschwankungen temperature fluctuations
Temperaturschwelle temperature limit
Temperaturschwingungen temperature fluctuations
Temperatursenkung temperature decrease
Temperatursensor thermocouple, temperature sensor/probe
Temperatursollwert required temperature
Temperatursonde thermocouple, temperature sensor/probe
Temperaturspanne temperature range
Temperaturspielraum temperature range
Temperaturspitze temperature peak, peak/maximum temperature
Temperaturspitzenbelastbarkeit thermal endurance
temperaturstabil thermally stable
Temperaturstabilität thermal stability, heat resistance: **... besitzt eine Temperaturstabilität bis 300°C** ...will withstand temperatures of up to 300°C
Temperaturstandfestigkeit heat resistance
Temperatursteuereinheit temperature control unit
Temperatursteuerung 1. temperature control. 2. temperature control system
Temperaturtoleranzfeld temperature tolerance range
Temperaturüberwachung 1. temperature control. 2. temperature control unit, temperature monitor

temperaturunabhängig temperature-independent, independent of the temperature
Temperaturuntergrenze lower temperature limit
Temperaturunterschied temperature difference
Temperaturverlauf temperature profile, changes in temperature: *Die zeitabhängig gemessenen Temperaturverläufe* changes of temperature measured over a period of time
Temperaturverlauf, zeitlicher time-temperature relationship
Temperaturverteilung temperature distribution
Temperaturwechsel change in temperature
temperaturwechselbeansprucht subjected to changing temperatures
Temperaturwechselbeanspruchung exposure to alternating/changing temperatures
Temperaturwechselbeständigkeit resistance to changing/fluctuating temperatures
Temperaturwechselprüfung alternating temperature test
Temperaturzuführung heating: *Da der Schäumprozeß exotherm verläuft, ist eine Temperaturzuführung nicht erforderlich* Since foaming is exothermic, the mix need not be heated
Temperaturzunahme temperature increase
Temperaturzustände temperature conditions
Temperdauer conditioning/annealing period
Temperierbad constant temperature bath
Temperierband band heater, heater band
temperierbar *capable of having its temperature kept under control*: **Die Schnecken sollten temperierbar sein** it should be possible to control the screw temperature
Temperierbohrung heating-cooling channel
temperieren 1. to control the temperature, to keep the temperature constant *(by heating or cooling)*. 2. to heat: **Temperier- und Kühlgeräte** heating and cooling equipment; **Temperierstrecke mit IR Strahlern** IR heating tunnel; *Eine Lagerung des Materials in einem temperierten Raum ist zu empfehlen* it is advisable to keep the material in a warm room; *Sometimes, the word is used as padding, as in this sentence*: **Je nach Rohstoff muß das Werkzeug gekühlt, temperiert oder beheizt werden** Obviously, the word **temperiert** must be omitted, although the idea of mould temperature control could be incorporated. The English version would then be as follows: The mould temperature must be controlled by heating or cooling, depending on the type of moulding compound used
Temperierflüssigkeit 1. heating-cooling medium, constant temperature medium, temperature control medium. 2. heating medium
Temperiergerät 1. temperature control unit. 2. heating unit
Temperierkammer constant temperature chamber

Temperierkanal 1. heating-cooling channel *(im)*. 2. heating tunnel
Temperierkanalauslegung 1. heating-cooling channel layout. 2. heating tunnel layout
Temperierkanalsystem heating-cooling channel system
Temperierkreis(lauf) temperature control circuit, heating-cooling circuit
Temperiermedium 1. heating-cooling medium, constant temperature medium, temperature control medium. 2. heating medium
Temperiermittel 1. heating-cooling medium, constant temperature medium, temperature control medium. 2. heating medium
Temperierschutzplatte insulating plate
Temperierstrecke heating section
Temperiersystem 1. temperature control system. 2. heating system
Temperierung 1. temperature control: **Das Heiz-Kühl-Aggregat für die Temperierung der drei Glättwalzen** the heating-cooling unit which controls the temperature of the three polishing rolls. 2. temperature control system. 3. heating: **Während bei Ausführung B die Temperierung mit Heizpatronen erfolgt...** whereas, in design B, heat is supplied by cartridge heaters.... 4. cooling: **Nur für Schneckendurchmesser von über 100 mm ist eine zusätzliche Schneckentemperierung erforderlich, die Materialüberhitzungen im Schneckenspitzenbereich verhindert** Only if the screw diameter is greater than 100 mm does the screw have to be additionally cooled to prevent melt near the screw tip being overheated; **Dadurch wird eine Temperierung der Form mit einer Genauigkeit von ± 1°C moglich** This enables the mould temperature to be controlled with an accuracy of ±1°C.
Temperierwalzen 1. heating rolls. 2. cooling rolls. 3. constant temperature rolls *(which of the first two versions is used will depend on the context. If in doubt use version 3.)*
Temperierzone 1. heating zone. 2. cooling zone. 3. constant temperature zone *(see explanatory note under **Temperierwalzen**)*
tempern to condition, to anneal *(test specimens by keeping them at a certain temperature to relieve internal stresses)*
Temperofen 1. conditioning oven. 2. annealing furnace
Temperstrecke conditioning/annealing section
Tempertemperatur conditioning/annealing temperature
Temperung conditioning, post-curing, post-cure, annealing
Temperzeit conditioning/annealing period
Tennisschläger tennis racquet/racket
Tensid surfactant
Teppichauslegeware carpet tiles
Teppichfliese carpet tile

Teppichrückenbeschichtung carpet backing
Teppichrückseitenbeschichtung carpet backing
Teppichunterlage carpet underlay
Teppichunterschicht carpet backing
Terephthalat terephthalate
Terephthalsäure terephthalic acid
Terephthalsäuredimethylester dimethyl terephthalate
Termin target date, deadline, time limit
Terminal, intelligentes intelligent terminal *(me)*
Terminanforderung required/requested deadline
Terminbüro scheduling department *(me)*
termingemäß according to schedule
termingerecht according to schedule
Termintreue adherence/keeping to deadline
Terminüberwachung deadline control
ternär ternary
Terpen terpene
Terpenharz terpene resin
Terpenkohlenwasserstoff terpene hydrocarbon
Terpentin turpentine
Terpolymer(isat) terpolymer
Terpolymerisation terpolymerisation
tert.-Butylhydroperoxid tertiary-butyl hydroperoxide
tert.-Butylperbenzoat tertiary-butyl perbenzoate
tert.-Butylperoxid tertiary-butyl peroxide
tertiär tertiary
Tertiärteilchen tertiary particle
Testbenzin white spirit
Testversuch test
Tetraalkylorthosilikat tetra-alkyl orthosilicate
Tetraalkyzinverbindung tetra-alkyl tin compound
Tetraarylzinnverbindung tetra-aryl tin compound
tetrabasisch tetrabasic
Tetrabenzylthiuramdisulfid tetrabenzylthiuram disulphide
Tetrabromphthalsäureanhydrid tetrabromophthalic anhydride
Tetrabromphthalsäureimid tetrabromophthalimide
Tetrabutyltitanat tetrabutyl titanate
Tetracarbonsäure tetracarboxylic acid
Tetrachlorethan tetrachloroethane
Tetrachlorkohlenstoff carbon tetrachloride
Tetrachlorphthalsäure tetrachlorophthalic acid
Tetrachlorphthalsäureanhydrid tetrachlorophthalic anhydride
Tetrachlorstyrol tetrachlorostyrene
Tetrafluorethylen tetrafluoroethylene
Tetrafluorethylenharz tetrafluoroethylene resin
Tetrafluorkohlenstoff carbon tetrafluoride
tetrafunktionell tetrafunctional
Tetrahydrofuran tetrahydrofuran
Tetralin tetralin
Tetramer tetramer
tetramer tetrameric
Tetramethylsilan tetramethyl silane

Tetramethylthiuramdisulfid tetramethylthiuram disulphide
Tetraphenylethan tetraphenyl ethane
tetrasubstituiert tetra-substituted
Tetrasulfid tetrasulphide
Tetrayhdrophthalsäure tetrahydrophthalic acid
Tetrol tetrol
Teufelskreis vicious circle
textiles Bauen production of air-supported structures
Textilfaser (textile) fibre
Textilgewebe (woven) fabric
Textilglas glass fibre material
Textilglas, geschnittenes chopped strands
Textilglas, vorimprägniertes prepreg
Textilglaserzeugnisse glass fibre materials/products
Textilglasfaser glass fibre
Textilglasgewebe glass (fibre) fabric, glass cloth
Textilglasmatte glass (fibre) mat
Textilglasprodukte glass fibre materials/products
Textilglasroving roving
Textilglasspinnfaden glass filament
textilglasverstärkt glass fibre reinforced
Textilglasverstärkung glass fibre reinforcement
Textilglasvlies glass (fibre) mat
Textilimprägnierung fabric impregnation
Textilschlichte textile size
textilverstärkt fabric reinforced
Textilweichmacher fabric softener
Textzeile display line, character row *(me)*
tex-Zahl tex *(a unit for measuring the thickness of fibres, threads etc.)*: **Rovings in verschiedenen tex-Zahlen** rovings in different thicknesses
TG *abbr. of* **Thermogravimetrie**, thermogravimetry
TGA *abbr. of* **thermogravimetrische Analyse**, thermogravimetric analysis
Tg-Wert glass transition temperature
TH *abbr. of* **Technische Hochschule**, technical university
theoretisch begründet theoretically established
thermisch thermal
thermisch geschädigt charred *(polymer, during processing)*
thermisch hochbelastet subjected to high temperatures
thermisch verarbeitbar melt processable
thermische Alterung heat ageing
thermische Analyse thermoanalysis
thermische Beanspruchung thermal stress
thermische Belastbarkeit thermal stability, heat resistance
thermische Beständigkeit thermal stability, heat resistance
thermische Eigenschaften thermal properties
thermische Leistung thermal efficiency
thermische Nutzung energy recovery/recycling

thermische Oberflächenschäden surface charring
thermische Stabilität thermal stability
thermische Überlastung overheating
thermische Verwertung energy recovery/recycling
thermische Vorgeschichte thermal history
thermische Wechselbelastungen alternating exposure to heat and cold, changes in temperature, fluctuating temperatures
thermische Werte thermal properties
thermischer Ausdehnungskoeffizient coefficient of (thermal) expansion
thermischer Längenausdehnungskoeffizient coefficient of linear expansion
thermischer Volumenausdehnungskoeffizient coefficient of cubical/volume expansion
thermisches Lebensdauerprofil thermal endurance profile
thermisches Recycling energy recovery/recycling
thermoaktivierbar capable of being heat activated **Die Beschichtungen sind thermoaktivierbar** the coatings can be heat activated
Thermoanalyse thermoanalysis
thermoanalytisch thermoanalytical
Thermodiffusion thermodiffusion
Thermodraht thermocouple, temperature sensor/probe
thermoduktil thermoductile
Thermoduktilität thermoductility
Thermodur thermoset
Thermodynamik thermodynamics
thermodynamisch thermodynamic
thermoelastisch thermoelastic: *This must sometimes be translated freely, e.g.* **Die Rohrstücke werden bis in den thermoelastischen Bereich erwärmt** the parisons are heated until they have become soft and flexible
Thermoelastizität thermoelasticity
thermoelektrisch thermoelectric
Thermoelement thermocouple, temperature sensor/probe
Thermoelementbohrung thermocouple well
Thermoelementbruch, bei if the thermocouple breaks
Thermoelementfühler thermocouple, temperature sensor/probe
Thermofixierung heat setting *(of film, fibres etc. after stretching, to prevent shrinkage)*
Thermoformautomat automatic thermoforming machine
thermoformbar thermoformable, suitable for thermoforming
Thermoformen thermoforming
Thermofühler thermocouple, temperature sensor/probe: *One sometimes encounters two different German versions of thermocouple in one sentence, e.g.;* **Als Thermofühler werden Fe-Ko Thermoelemente verwendet** the temperature is measured with iron-constantan thermocouples; *Although this translation departs from the original, it avoids clumsiness without losing the meaning of the sentence*
Thermofühlerbohrung thermocouple well
thermogeformt thermoformed
Thermogramm thermogram
Thermographie thermography
Thermogravimetrie thermogravimetry
thermogravimetrisch thermogravimetric
thermogravimetrische Analyse thermogravimetric analysis, TGA
thermohärtend 1. thermosetting. 2. heat/hot curing
thermoinstabil thermally unstable
thermoisoliert thermally insulated
Thermokaschieranlage heat laminating plant
Thermokompression thermocompression
thermolabil thermally unstable
Thermolyse thermolysis, pyrolysis
Thermolysekonstante thermolysis/pyrolysis constant
thermomechanisch thermomechanical
thermomechanische Analyse thermomechanical analysis, TMA
Thermomikroskopie thermomicroscopy
Thermooxidation thermo-oxidation
thermooxidativ thermo-oxidative
thermooxidative Stabilität resistance to thermo-oxidation
Thermopaar thermocouple wire
Thermopapier thermosensitive paper
Thermoplast thermoplastic (material)
Thermoplastblend thermoplastic polymer blend
Thermoplaste, hochbeanspruchbare engineering thermoplastics
Thermoplaste, technische engineering thermoplastics
Thermoplast-Extrusions-Schaumblasen extrusion blow moulding of expandable thermoplastics
thermoplastisch 1. thermoplastic *(but only if the word refers to properties) e.g.:* **Das Material hat thermoplastische Eigenschaften** the material is thermoplastic; **Ein thermoplastischer Kunststoff** a thermoplastic polymer, a thermoplastic. 2. plastic, soft: **Die Einfriergrenze markiert den Übergang vom thermoplastischen in den festen Zustand** The frost line marks the point where the material changes from the plastic to the solid state; **Unter "Plastifizierung" ist die Umwandlung eines thermoplastischen Stoffes unter Einwirkung von Druck und Temperatur in den thermoplastischen Zustand zu verstehen.** *Here, obviously, the first* **thermoplastisch** *must be translated as 1. above, but not the second, so that the English version should read as follows*: Plasticisation is understood to be the softening of a solid thermoplastic material under the influence of

heat and pressure; **Beide Enden des Rohrstücks werden bis in den thermoplastischen Bereich erwärmt** both ends of the parison are heated until they are soft; *The use of version 2. in these three examples is dictated by the fact that thermoplastic is the property of a hard material to become soft on heating. In other words, it expresses the idea of "becoming soft" rather than "being soft". To say, in the last example above, that the parison is heated "until it is thermoplastic" would obviously be wrong - since it is, by its very nature, thermoplastic to start with. Sometimes, the word must be omitted altogether, e.g. in* **thermoplastisches Verformen**, *which simply means* thermoforming *and* **thermoplastisch verformbar**, *which means* thermoformable. **Tiefziehbarkeit im thermoplastischen Bereich** *can only be translated as* thermoformability *since, to qualify this by adding "in the plastic state" would be stating the obvious*; **...bis in den Bereich der thermoplastischen Erweichung der Oberfläche des Kunststoffpulvers...** ...until the surface of the plastics powder has softened...

thermoplastisch verarbeitbar melt processable
thermoplastische Verarbeitbarkeit melt processability **Elastomere mit thermoplastischer Verarbeitbarkeit** melt processable elastomers, thermoplastic elastomers
thermoplastische Verarbeitung melt processing: *whilst this expression has now become generally accepted, its use when dealing, for example, with PVC, is not always correct. The translation will depend on the context and the translator's judgement, as the following examples show:* **Stabilisatoren wie sie bei der thermoplastischen Verarbeitung von PVC üblich sind** *Here, the word* **thermoplastisch** *obviously contributes nothing whatever to the meaning, and should be omitted. The correct translation is therefore* Stabilisers such as those used in PVC processing; **Emulsions-PVC für die thermoplastische Hartverarbeitung**: *The same comments apply as in the previous example and the translation should therefore read as follows:* Emulsion PVC for processing in unplasticised form (OR: for processing without plasticiser); **Masse-PVC für die thermoplastische Verarbeitung** Bulk PVC for melt processing, **Während der thermoplastischen Verarbeitung von Polymeren**: *Here the adjective* **thermoplastisch** *can be converted into a noun:* when thermoplastics are being processed. *Finally, two further examples to show how important it is to take the context into account:* **Gleitmittel erleichtern die thermoplastische Verarbeitung von Kunststoffen** Lubricants make thermoplastics easier to process; **Durch Vorwärmen wird Kettenwachstum und Vernetzung bewirkt, das Material bleibt jedoch thermoplastisch verarbeitbar** Pre-heating causes chain growth and crosslinkage, but the material can still be processed like a thermoplastic. *The point here is that crosslinkage normally causes a material to lose its thermoplastic properties - something that has obviously not happened in this instance, so that this must be conveyed in the translation*; **...umfaßt homo- und copolymere PVC-Typen für alle Einsatzzwecke in der thermoplastischen Hart- und Weichverarbeitung, sowie in der Pastenverarbeitung** ...includes PVC homopolymers and copolymers which can be melt processed with and without plasticiser, as well as being suitable for PVC pastes
thermoplastischer Kunststoff thermoplastic
thermoplastisches Umformen thermoforming
Thermoplastizität thermoplasticity
Thermoplastschaumextrusion extrusion of expandable thermoplastics
Thermoplastschaumgießmaschine structural foam moulding machine
Thermoplastschaumguß structural foam moulding *(see explanatory note under* **TSG***)*
Thermoplastschaumteil thermoplastic foam moulding/moulded part
Thermoplastschmelze polymer melt
Thermoplastschnecke thermoplastic screw *(screw suitable for extruding thermoplastics)*
Thermoplastspritzguß injection moulding of thermoplastics
Thermoplastverarbeitung processing of thermoplastics
thermoreversibel thermoreversible
Thermoschockbeständigkeit thermal shock resistance
thermostabil thermally stable
Thermostabilisator heat stabiliser
thermostabilisierend heat stabilising
Thermostabilisierung heat stabilisation
Thermostabilität thermal stability
Thermostat thermostat
thermostatisierbar thermostatically controllable
thermostatisiert thermostatically controlled, thermostated
thermotrop thermotropic
thermoverformen to thermoform
Thermowaage thermobalance
Thermowächter temperature controller/monitor
THF *abbr. of* **Tetrahydrofuran,** tetrahydrofuran
Thiocarbamat thiocarbamate
Thiocarbamylsulfenamid thiocarbamyl sulphenamide
Thiodicarbonsäureester thiodicarboxylate
Thiodipropionsäureester thiodipropionate
Thioester thioester
Thioether thioether

Thioglykolat thioglycolate
Thioglykolsäureester thioglycolate
Thioglykolsäurerest thioglycolate group/radical
Thioharnstoff thiourea
Thioharnstoffharz thiourea-formaldehyde resin
Thiokol thiokol, polysulphide rubber
Thiokolkautschuk polysulphide rubber
Thioplast thioplast (*plural:* thioplasts), polysulphide rubber
Thioverbindung thio-compound
Thioxotropierung making thixotropic, imparting thixotropy
Thiuramverbindung thiuram compound
thixotrop thixotropic
Thixotropie thixotropy
Thixotropiegeber thixotropic agent
thixotropierendes Additiv thixotropic agent
Thixotropiermittel thixotropic agent
Thixotropierungsmittel thixotropic agent
Thixotropierungsverhalten thixotropic behaviour
Thixotropiewirkung thixotropic effect
thyristorgespeist thyristor controlled
thyristorgesteuert thyristor controlled
Thyristorregler thyristor control unit
TI *abbr. of* **technische Intelligenz**, technical intelligence
Tiefbau civil engineering
Tiefdruck intaglio/gravure/rotogravure printing
Tiefdruckfarbe gravure printing ink
Tiefenfilter depth filter *(mf)*
Tiefengrundierung deep-penetration primer
tiefgängig deep-flighted *(screw)*
Tiefgarage underground car park
tiefgeschnitten deep-flighted *(screw)*
tiefgezogen thermoformed *(see explanatory note under* **Tiefziehen***)*
Tiefpaßfilter low-pass filter
Tiefprägen deep embossing
Tiefpressen pressure forming
tiefschwarz jet black
tiefsiedend low-boiling
Tieftemperaturanwendungen low temperature applications
Tieftemperaturbereich low temperature range
...auch im Tieftemperaturbereich ...also at low temperatures
Tieftemperaturbeständigkeit low-temperature resistance
Tieftemperatureigenschaften low temperature properties
Tieftemperaturen low temperatures
Tieftemperaturfestigkeit low temperature resistance
Tieftemperaturflexibilität low temperature flexibility
Tieftemperaturhärtung low-temperature curing
Tieftemperaturinitiator low temperature initiator
Tieftemperaturperoxid low temperature peroxide
Tieftemperaturschlagzähigkeit low temperature impact strength
Tieftemperaturschmierfett low temperature lubricating grease
Tieftemperaturverhalten low temperature performance/behaviour
tieftemperaturzäh low-temperature impact resistant
Tieftemperaturzähigkeit low-temperature impact strength
Tiefungsversuch indentation test
Tiefziehautomat automatic thermoforming machine
Tiefzieheigenschaften thermoforming properties
Tiefziehen thermoforming: *The expression* deep drawing *should be avoided, since this is applied mainly to metals. Although* **Tiefziehen** *is commonly used instead of* **Thermoformen** *in German, this use is, strictly speaking, incorrect (see Stoeckhert, Kunststoff-Lexikon, 7th edition, p.505)*
tiefziehfähig thermoformable, suitable for thermoforming
Tiefziehfähigkeit thermoformability, thermoforming properties
Tiefziehfolie thermoforming sheet
Tiefziehfolienanlage thermoforming sheet plant
Tiefziehfolienextrusionsanlage thermoforming sheet extrusion line
tiefziehgünstig thermoformable
Tiefziehmaschine thermoforming machine
Tiefziehteil thermoformed part
Tiefziehverhalten thermoforming properties
Tiefziehverhältnis thermoforming ratio
Tiefziehverpackung vacuum formed pack
Tiefziehvorgang thermoforming (process)
Tiefziehwerkzeug thermoforming mould
Tiegel crucible
tierisch animal *(e.g. oils or fats)*
Tierversuche test on animals
Tintenstrahldrucker ink jet printer
TiO$_2$-PE-Konzentrat TiO$_2$-PE masterbatch
Tisch press table
Tischcomputer desk-top computer *(me)*
Tischgerät table-top model/unit/instrument
Tischleuchte table lamp
Tischmodell table-top model
Tischrechner desk-top computer *(me)*
Tischversion table-top version
Titanacetylacetonat titanium acetyl acetonate
Titandioxid titanium dioxide
Titankatalysator titanium catalyst
titanorganisch organo-titanium
Titanweiß titanium white
Titer titre
Titration titration
titrieren to titrate
TLV *abbr. of* threshold limit value
TMA *abbr. of* **thermomechanische Analyse**, thermomechanical analysis
Tochtergesellschaft subsidiary
TOF *abbr. of* **Trioctylphosphat**, trioctyl phosphate, TOP

Toleranz tolerance
Toleranzanforderungen tolerance requirements
Toleranzangaben tolerance details
Toleranzbänder tolerance limits
Toleranzbereich tolerance range
Toleranzbreite tolerance range
Toleranzfeld tolerance range
Toleranzgrenze tolerance limit
Toleranzüberwachungseinrichtung tolerance monitor
Toluol toluene
Toluoldiisocyanat toluene di-isocyanate
Toluolsulfonsäure toluene sulphonic acid
Tonabnehmergerät gramophone, record player
Tonband (recording) tape
Tonerde alumina
Tonerdehydrat aluminium hydroxide
Tonstudio recording studio
Topfzeit pot life
topologisch topological
Tor gateway *(me)*
TOR *abbr. of* trans-octenylene rubber
TOR-Gehalt trans-octenylene rubber content
TOR-haltig containing trans-octenylene rubber
Torpedo torpedo
Torpedobug torpedo tip
Torpedohalterung torpedo support
Torpedokopf centre-fed die *(e,bm)*
Torpedokörper torpedo
Torpedoschaft torpedo shank
Torpedoscherteil smear head, torpedo *(of screw) (e)*
Torpedospitze torpedo tip
Torquemotor torque motor
Torsion torsion, twisting
Torsionsausschwingungsversuch torsion pendulum test
Torsionsautomat automatic torsion pendulum apparatus/instrument
Torsionsbeanspruchung torsional stress
Torsionsbelastbarkeit resistance to torsional stress
Torsionsbelastung torsional stress
Torsionsmodul torsion modulus
Torsionsmoment torsional moment
Torsionspendel torsion pendulum
Torsionsscherbelastung torsional shear stress
Torsionsschwingung torsional vibration/oscillation
Torsionsschwingungsanalyse torsional vibration/oscillation analysis
Torsionsschwingungsgerät torsion pendulum (apparatus/instrument)
Torsionsschwingungsmessung torsion pendulum test
Torsionsschwingversuch torsion pendulum test
Torsionsspannung torsional stress
Torsionssteifheit torsional rigidity, stiffness in torsion, apparent modulus of rigidity
Torsionsversuch torsion pendulum test
Torsionswinkel angle of torsion

tote Ecken dead spots
tote Stellen dead spots
tote Winkel dead spots
tote Zonen dead spots
Toträume dead spots
Totzeit downtime, shut-down period, non-productive time
Tourenzahl 1. number of revolutions. 2. speed *(of a rotating element)*. 3. screw speed *(e)*
Toxikant toxic substance/compound
Toxikologie toxicology
toxikologisch toxicological
toxikologisch einwandfrei non-toxic
toxikologisch harmlos non-toxic
toxikologische Daten toxicological data
toxikologische Eigenschaften toxicological properties
toxisch toxic
Toxizität toxicity
Toxizitätsrisiko toxicity risk
TP *abbr. of* **Thermoplast**, thermoplastic
TPE *abbr. of* **thermoplastische Elastomere**, thermoplastic elastomers
TPE, technische engineering thermoplastic elastomers
TPE-E *symbol for* thermoplastic copolyesters
TPU *abbr. of* **thermoplastische Polyurethane**, thermoplastic polyurethanes
Trafobau transformer construction
Trafogehäuse transformer housing
Trafoöl transformer oil
tragbar 1. portable. 2. acceptable, reasonable *(e.g. costs)*
Tragegestell supporting frame
Trageigenschaften load-bearing properties
Träger carrier, support, backing
Trägerfolie supporting/backing film
Trägergas carrier gas
Trägergewebe backing fabric
Trägerharz binder resin
trägerlos unsupported
Trägermaterial supporting/backing material, substrate
Trägermedium (paint) medium
Trägerpapier backing paper
Trägerplatte platen *(im)*
Trägerschicht backing/supporting layer
Trägersubstanz 1. carrier (substance). 2. (paint) medium
Tragetasche carrier bag
tragfähig strong *(i.e. able to support heavy loads)*
Tragfähigkeit load-carrying capacity
Tragfähigkeitsminderung reduction in load-carrying capacity
trägheitsarm low-inertia
Trägheitshalbmesser radius of gyration
Trägheitsmoment moment of inertia
Traglufthalle air-supported structure
Tragring mandrel support *(bfe, e)*
Tragtaschenfolie film/sheeting for carrier bags
Tragverhalten load-carrying capacity

Tragwerk self-supporting structure, space frame
Tränkanlage impregnating plant
Tränkbad impregnating bath
Tränkharz impregnating resin
Tränklack impregnating varnish
Tränkmittel impregnating agent, impregnant
Tränkpaste (PVC) impregnating paste
Tränkung impregnation
Tränkungszeit impregnating time/period
Tränkverhalten impregnating behaviour
Tränkwanne impregnating bath
Transferform transfer mould
Transferpresse transfer moulding press
Transferpreßverfahren transfer moulding
Transferspritzpresse transfer moulding press
Transformator transformer
Transientenspeicher working/intermediate storage, work area *(me)*
Transistor transistor
transistorisiert transistorised
transkristallin trans-crystalline
transluzent translucent
Transluzenz translucency
Transmissionselektronenmikroskop transmission electron microscope
Transmissionsgrad transmission coefficient
Transmissionsriemen transmission belt
Transmissionswärmeverlust transmission heat loss
transparent transparent
transparent gemacht explained, made clear, clarified: **Spritzguß transparent gemacht** injection moulding explained
Transparentlack clear varnish/lacquer
Transparenz 1. transparency. 2. clarity *(e.g. of information)*
transportabel transportable
Transportanlage conveying unit
Transportband conveyor belt
Transportbehälter transit container
Transporter van
Transportkette chain conveyor
Transportkosten transport/shipping costs
Transportmittel means of transport
Transportteil conveying section *(of screw)*
Transportverpackung transit packaging
Transportwesen transport industry
Transversalströmung transverse flow
trapezförmig trapezoidal
Trapezspindel trapezoidal spindle
Träufelharz trickle impregnating resin
Treeing, elektrisches water treeing *(an electrical phenomenon)*
Treibgas blowing gas
Treibhauseffekt greenhouse effect
Treibmittel 1. blowing agent. 2. fuel. 3. propellant *(for aerosols)*
Treibmittelanteil amount of blowing agent, blowing agent content
Treibmittelart type of blowing agent
treibmittelfrei free from blowing agent
Treibmittelgehalt blowing agent content

treibmittelhaltig expandable *(containing a blowing agent)*
Treibmittelkonzentrat blowing agent concentrate
Treibmittelmenge amount of blowing agent
Treibmittelpumpe fuel pump
Treibmittelschaumstoff chemically blown foam
Treibreaktion expansion/foaming reaction
Treibstoff fuel
Treibstoffbedarf fuel requirements
Treibstoffbehälter fuel/petrol tank
treibstoffbeständig fuel resistant
Treibstofftank fuel tank
Treibstoffverbrauch fuel/petrol consumption
Treibsystem blowing agent
Trendrichtung trend
Trennaht (mould) parting line *(im)*
Trennahtschweißen hot wire welding
Trennbeschichtung release coating
Trennbruch brittle failure/fracture
Trennebene (mould) parting surface/line *(im)*
Trenneffekt non-stick/release effect
Trenneigenschaften non-stick/release properties
trennen 1. to separate. 2. to cut
Trennfähigkeit non-stick properties
Trennfestigkeit bond strength, (degree of) adhesion
Trennfilm release film/coating
Trennfläche 1. (mould) parting surface/line *(im)*. 2. glueline
Trennfolie release film
trennfreundlich easy-release
Trennfuge (mould) parting surface/line *(im)*
Trennkante (mould) parting line/surface *(im)*
Trennkraft peel force
Trennlageneinspritzung gating at the (mould) parting line *(im)*
Trennmethode method of separation
Trennmittel (mould) release agent
Trennmittelbehandlung application of release agent
Trennmittelfilm release film/coating
trennmittelhaltig containing release agent
Trennmittelreste release agent residues
Trennmittelrückstände release agent residues
Trennmittelwirkung non-stick/release effect
Trennpapier release paper
Trennsäule separating column *(in chromatography)*
Trennsubstanz (mould) release agent
Trennung separation
Trennung, thermische thermal isolation *(e.g. of the hot runner manifold block from the cavity plate in hot runner systems)*
Trennungsenergie separation energy
Trennungslinie (dividing) line: **Eine Trennungslinie zwischen Weichmachern und inneren Gleitmitteln ist schwer zu ziehen** it is difficult to draw a line between plasticisers and internal lubricants
Trennungsprozeß separation process

Trennverhalten release properties
Trennwalze separator rotor *(sr)*: This is a specially designed roller for separating the cut-off sprues and runners from the moulded part after demoulding
Trennwand partition wall
Trennwirkung non-stick/release effect
Treppenkantenprofil stair edging profile
Treufelharz trickle impregnation resin
Tri trichloroethylene
Trialkoxysilan trialkoxysilane
Trialkylphosphat trialkyl phosphate
Trialkylphosphin trialkyl phosphine
Trialkylzinnverbindung trialkyl tin compound
Triallylcyanurat triallyl cyanurate
Triarylphosphat triaryl phosphate
Triarylphosphin triaryl phosphine
Triazin triazine
Triazinharz triazine resin
Triazol triazole
tribasisch tribasic
Triblockcopolymer triblock copolymer
tribochemisch tribo-chemical
Tribologie tribology
tribologisch tribological
triborheologisch tribo-rheological
Tribromphenylmaleimid tribromophenyl maleimide
Tributylacrylat tributyl acrylate
Tributylmethacrylat tributyl methacrylate
Tributylthioharnstoff tributyl thiourea
Tributylzinnfluorid tributyl tin fluoride
Tricarbonsäure tricarboxylic acid
Tricarbonsäureanhydrid tricarboxylic acid anhydride
Trichlorbenzol trichlorobenzene
Trichlorethan trichloroethane
Trichlorethylen trichloroethylene
Trichlorethylphosphat trichloroethyl phosphate
Trichlorfluormethan trichlorofluoromethane
Trichlorstyrol trichlorostyrene
Trichter 1. funnel. 2. (feed/material) hopper *(e,im)*
Trichterentgasung 1. vented hopper system. 2. hopper devolatilisation *(of a moulding compound)*
trichterförmig funnel shaped
Trichterinhalt 1. hopper capacity. 2. hopper contents
Trichterstückbuchse feed bush
Trichtertrockner hopper drier
Trichterzone feed section/zone
Triebwerkskapselung engine sound shield
Triethanolamim triethanolamine
Triethylamin triethylamine
Triethylenglykol triethylene glycol
Triethylentetramin triethylene tetramine
Trifluorchlorethylen trifluorochloroethylene
Trifluornitromethan trifluoronitromethane
trifunktionell trifunctional
Triglycerid triglyceride
Triglycidylisocyanurat triglycidyl isocyanurate

Triisopropanolamin tri-isopropanolamine
Trikresylphosphat tricresyl phosphate, TCP
Trimellitat trimellitate
Trimellitsäure trimellitic acid
Trimellitsäureester trimellitate, trimellitic acid ester
Trimer trimer
trimer trimeric
Trimerisat trimer
Trimerisation trimerisation
trimerisiert trimerised
Trimerisierungsvorgang trimerisation reaction
Trimethylhexamethylendiamin trimethyl hexamethylenediamine
Trimethylolpropan trimethylol propane
Trimethylolpropandiallylether trimethylolpropane diallyl ether
Trimethylolpropantriacrylat trimethylol propane triacrylate
Trimethylolpropantrimethacrylat trimethylolpropane trimethacrylate
Trimethylsilylester trimethyl silyl ester
Trimm- und Entnahmestation trimming and take-off station
Trimmstation trimming unit
Trinkwasserleitung drinking water pipeline, drinking water supply line
Trinkwasserrohr drinking water pipe
Trinkwasserrohrleitung drinking water pipeline, drinking water supply line
Trinkwasserversorgung 1. drinking water supply. 2. drinking water supply system
Trioctylphosphat trioctyl phosphate
Triol triol
Triorganozinnverbindung triorgano-tin compound
Trioxyethylen trioxyethylene
Triphenylmethylperoxid triphenylmethyl peroxide
Triphenylmethylradikal triphenylmethyl radical
Triphenylphosphat triphenyl phosphate, TPP
Triphenylphosphit triphenyl phosphite
Tripropylenglykoldiacrylat tripropylene glycol diacrylate
Tripropylenglykoltriacrylat tripropylene glycol triacrylate
Trisulfid trisulphide
Trittschall solid-borne sound
Trittschallisoliermaterial solid-borne sound insulating material
Trivialnamen common name *(of a chemical compound)*
Trixylenylphosphat trixylenyl phosphate, TXP
Trockenbedingungen drying conditions
Trockendeckfähigkeit dry hiding power
trockene Reibung dry friction
Trockeneis dry ice
trockener Griff dry handle
Trockenfestigkeit dry strength
Trockenfilm dry film
Trockenfilmdicke dry film thickness
Trockenfilmstärke dry film thickness

Trockengehalt 1. dryness. 2. dry solids content
Trockengerät drying unit
Trockengewicht dry weight
Trockengleitverschleiß wear due to dry sliding friction
trockengleitverschleißarm resistant to dry sliding friction
Trockenharzgehalt dry resin content
Trockenkanal drying tunnel
Trockenklebstoff dry adhesive
Trockenlauf, bei when operating without lubrication
Trockenlauf, Taktzeit im dry cycle time *(im)* *(see explanatory note under* **Trockenlaufzeit**)
Trockenlaufzahl number of dry cycles *(usually per minute) (im)*
Trockenlaufzeit dry cycle time *(im)*: This is the time of one complete cycle, in seconds
Trockenlaufzykluszeit dry cycle time *(im) (see explanatory note under* **Trockenlaufzeit**)
Trockenluft 1. dry air. 2. drying air
Trockenmahlung dry grinding
Trockenmischung dry blend
Trockenmittel drying agent
Trockenofen drying oven
Trockenpigment dry pigment
Trockenputz dry plaster
Trockenreibung solid friction
Trockenrückstand dry residue
Trockenruß dry carbon black
Trockenschichtdicke dry film thickness
Trockenschichtstärke dry film thickness
Trockenschliff dry grinding
Trockenschrank drying cabinet/oven
Trockenspinnverfahren dry spinning (process)
Trockenstoff drier *(in a paint)*
Trockenstrecke drying section/tunnel
Trockentaktzeit dry cycle time *(im)*
Trockentrennen dry separation
Trockenverhalten drying characteristics
Trockenzeit drying time
Trockenzustand dry state
trocknend drying
trocknend, luftforciert force drying
trocknend, physikalisch physically drying
trocknend, wärmeforciert force drying
Trockner 1. drier. 2. drying unit
Trocknung, physikalische physical drying
Trocknungsablauf drying process
Trocknungsanlage drying unit
Trocknungsbedingungen drying conditions
Trocknungsbeschleuniger drier
Trocknungseinrichtung drying equipment/unit
Trocknungsgerät drying unit
Trocknungsluft drying air
Trocknungsmittel drier
Trocknungsschrumpf shrinkage on drying
Trocknungstemperatur drying temperature
Trocknungstrichter drying hopper
Trocknungsverhalten drying properties/characteristics
Trocknungsverzögerung delay in drying

Trocknungszeit drying time
Trommel drum
Trommelbremse drum brake
Trommelmischer drum mixer
Trommelplotter drum plotter *(me)*
Trommelschältest climbing drum peel test
Trommeltrockner drum dryer
Trommelwickler drum winder, drum wind-up unit
Tropenbedingungen tropical conditions
tropenbeständig resistant to tropical conditions
Tropenbeständigkeit resistance to tropical conditions
tropenfest resistant to tropical conditions
Tropenklima tropical conditions
Tropenklimabedingungen tropical conditions
tropenverwendungsfähig suitable for use in the tropics OR: under tropical conditions
Tröpfchen droplet
Tröpfchengrößenverteilung droplet size distribution
Tropfenbildung drooling *(im)*
Tropfkörper trickling filter
Tropfpunkt, ohne non-melting, infusible
trübe cloudy, turbid
Trübung 1. cloudiness, turbidity. 2. haze *(in film testing)*
Trübungspunkt cloud point
Trübungszone craze zone
Trübungszonenbildung crazing, craze formation
TSB *abbr. of* **Thermoplast-Schaum-Blasformen** extrusion blow moulding of expandable thermoplastics
TSE *abbr. of* **Thermoplast-Schaum-Extrusion** extrusion of expandable thermoplastics
TSG *abbr. of* **Thermoplast-Schaum-Guß** structural foam moulding: *The context will normally make the use of the word thermoplastic superfluous, e.g. in an article on the structural foam moulding of polystyrene*
TSG-Maschine structural foam moulding machine
TSG-Mehrkomponenten-Rundläufermaschine rotary table, multi-component structural foam moulding machine
TSG-Mehrkomponentenverfahren multi-component structural foam moulding (process)
TSG-Spritzgießmaschine structural foam moulding machine
TSG-Technik structural/integral foam moulding technology
TSG-Teile structural foam mouldings
TSG-Verarbeitung structural foam moulding
TSG-Verfahren structural foam moulding (process)
TSG-Werkzeug structural foam mould
T-Stück T-piece
Tuften tufting
Tuftingteppich tufted carpet
Tunnelanguß tunnel/submarine gate *(im)*

Tunnelanguß mit Punktanschnitt tunnel gate with pin-point feed *(im)*
unnelanschnitt tunnel/submarine gate *(im)*
Tunnelofen 1. drying tunnel. 2. fusion tunnel *(for PVC coated fabrics)*
Tunnelpunktanguß tunnel gate with pin-point feed *(im)*
Türabdichtung door seal
Türaußenhaut outer door panel
Turbinenrührer turbomixer
Turbolader turbocharger
Turbotrockner turbo-drier
Turbovakuumpumpe turbo-vacuum pump
Turboverdichter turbo-compressor
turbulent turbulent
turbulente Strömung turbulent flow
Turbulenz turbulence
Türinnenverkleidung door trim
Turmbauweise turret construction
Türschweller door sill
Türverkleidung door trim (panel)
Türzarge door case
Tuschieren bedding-down
Tuschierfarbe toolmaker's/engineer's/ micrometer blue
Tuschierlehre spotting aid
Tuschierpresse bedding-down press
Twist-off twist-off lid
Typ grade *(of material) A word that can often be omitted in translation, e.g.* **Für Formen aus RTV-2 Siliconkautschuk-Typen gibt es zahlreiche Einsatzmöglichkeiten** Moulds made from RTV-2 silicone rubbers have many uses
Typenbeschreibung description of products/ materials/grades
Typenbezeichnung product code
Typenmerkblatt product data sheet
Typenpalette range of grades/products
Typenprogramm range of grades/products
Typensortiment range of grades/products
Typenübersicht range of grades/products
typisiert typified, characterised, categorised
Typisierung typification, characterisation, categorisation

U

Überarbeitung re-designing, revision, working-over: **technische Überarbeitung** technical improvement
überbasisch superbasic
Überbeanspruchung overloading
Überbrennen overstoving, overbaking *(paint film)*
Überbrennstabilität overstoving/overbaking resistance
überbrücken to bridge
überdimensioniert too big, oversize, extra large
Überdosierung 1. overfeeding. 2. adding too much: **Eine Überdosierung des Vernetzers macht sich sehr bei der Bruchdehnung bemerkbar** if too much catalyst is added, this has a very marked effect on the elongation at break
Überdruck excess pressure
Überdruckkalibrator air pressure calibrating/ sizing unit *(e)*
Überdruckkalibrierung 1. air pressure calibration. 2. air pressure calibration unit *(e)*
Überdrucklack overprinting ink
Überdrucksicherung excess-pressure safety device
Überdruckventil excess pressure safety valve
überdurchschnittlich above average
Übereinstimmung agreement
Überfahreinrichtung override mechanism
überfahren to override
überfüllen to overfeed, to overpack *(a mould)*
überfüttern to overfeed, to overpack *(a mould)*
Übergang transition
Übergangsbereich transition section/zone *(e)*
Übergangslösung provisional/temporary solution
Übergangsmetallion transition metal ion
Übergangsstück adaptor
Übergangszeit transitional period
Übergangszone 1. transition section *(of screw)*. 2. transition zone *(of a curve)*.
übergeordnet higher-level, superordinate *(me)*
übergroß too big, oversize, extra large
überhärtet overcured
Überhärtung overcuring
Überheizung overheating
Überhitzung overheating
Überhitzung, örtliche local overheating
Überhitzungsgefahr risk/danger of overheating
Überhitzungsschutz protection against overheating
Überhitzungs-Sicherheitsthermostat thermostat to prevent overheating
überirdisch above ground
Überkapazität overcapacity
Überlackierbarkeit overpaintability
Überlackierechtheit overpaintability
Überlackierung overpainting
überladen 1. to overload. 2. to overfeed, to overpack *(a mould)*
überlagert 1. superimposed. 2. stored for too long
Überlagerung superimposition
überlappend overlapping
Überlappnaht lap welded joint
Überlappschweißen lap welding
Überlappstoß lap joint
Überlappung overlap
Überlappung, abgesetzte recessed lap joint

Überlappungsanschnitt overlap gate *(im)*
Überlappungslänge length of overlap
Überlappungsverklebung lap joint
Überlast overload
Überlastanzeige overload signal
überlastet overloaded
Überlastschutz overload prevention device
überlastsicher overload-proof, safe against overloading
Überlastsicherung 1. overload prevention. 2. overload prevention device
Überlastung overloading
Überlastung, thermische overheating
Überlastungsgefahr risk of overloading
Überlastungsschutz overload protection
Überlastungssicherung 1. overload prevention. 2. overload prevention device
Überlauf overflow
Überlaufwerkzeug flash/semi-positive mould *(cm)*
Überlebenschancen chances of survival
Überlegungen, verfahrenstechnische technical considerations
überprüfen to check
Überprüfung check
übersättigt supersaturated
Übersättigung supersaturation
Überscherung excessive shear
Überschlag flash-over, spark-over
Überschlagfestigkeit flash-over resistance
überschreiben to overwrite *(me)*
Überschreiten exceeding, rising above: **Abschalten der Steuerung bei Überschreiten der Maximum-Öltemperatur** the control unit is switched off if the oil temperature rises above the maximum
Überschuß excess
überschüssig excess
Übersetzung 1. transmission, gearing. 2. translation *(me)*
Übersicht general survey, overview *(e.g. of machine data)*
übersichtlich clearly distinguishable
Übersichtstabelle general table
Überspritzbarkeit overspray tolerance
überspritzen 1. to overfeed, to overpack *(a mould)*. 2. to overspray
Überstreichbarkeit overpainting tolerance
überstreichen to overpaint
Überstromventil overflow valve
Übertemperatur too high a temperature, excessive temperature: **örtliche Übertemperaturen** local overheating
Übertemperaturschutz protection against overheating
Übertragung transfer
Übertragungsgeschwindigkeit transmission speed *(me)*
Übertragungsmodul transmission module, modem *(me)*
Übertragungsrate transmission speed *(me)*
Übervulkanisation over-vulcanisation, overcuring
überwachen to monitor, to control, to supervise
Überwachung monitoring, control
Überwachungsarmaturen controls, monitoring devices
Überwachungsaufgabe monitoring task
Überwachungseinheit monitoring/control unit
Überwachungseinrichtung monitoring equipment
Überwachungselemente controls, monitoring devices
Überwachungsfunktion monitoring function
Überwachungsgerät monitoring instrument
Überwachungskontrolleuchte pilot light
Überwachungsmöglichkeiten monitoring/control options
Überwachungsorgan monitoring device
Überwachungsprüfung control test
Überwachungssystem monitoring system
Überwachungstafel control panel
Überwurfmutter cap nut
überzeugend impressive *(e.g. properties, advantages etc. The literal translation, convincing, can sometimes introduce an element of awkwardness - the version given is a perfectly acceptable, general-purpose alternative).*
überziehen to coat
Überzug coat(ing), film
Überzugmasse coating compound
Überzugsbildung film formation
Überzugsmittel 1. coating compound/composition. 2. paint
UBS *abbr. of* **Unterbodenschutz,** underbody sealant
UBS-Masse underbody sealant paste
UBS-Plastisol underbody sealant paste
UD *abbr. of* unidirectional
UD-Prepreg unidirectional prepreg
UHF *abbr. of* **ultrahochfrequent,** ultra-high frequency
Uhrzeit time, time of day
ultradünn ultrathin
Ultradünnschnitt ultra-thin (microtome) section
Ultradünnschnittmikrotomie ultra-thin section microtomy
Ultrafiltration ultrafiltration
ultrahochfrequent ultra-high frequency, UHF
ultrahochmolekular ultra-high molecular weight
Ultrahochvakuum ultra-high vacuum
ultrarein ultra-pure
ultrarot infra-red
Ultraschall ultrasound, ultrasonic waves
Ultraschallbereich ultrasonic range
Ultraschalldurchflußmeßgerät ultrasonic flowmeter
Ultraschallenergie ultrasonic energy
ultraschallerregt ultrasonically excited
Ultraschallfrequenz ultrasonic frequency
Ultraschallfügen ultrasonic welding
Ultraschallgenerator ultrasonic generator *(w)*
ultraschallgeschnitten ultrasonically cut

ultraschallgeschweißt ultrasonically welded
Ultraschallgeschwindigkeit ultrasonic speed
Ultraschallimpuls ultrasonic impulse
Ultraschallmikroskop ultrasonic/ultrasound microscope
Ultraschallnieten ultrasonic riveting
Ultraschallpunktschweißen ultrasonic spot welding
Ultraschallpunktschweißgerät ultrasonic spot welding instrument
Ultraschallschneiden ultrasonic cutting
Ultraschallschweißanlage ultrasonic welding plant/unit
Ultraschallschweißen ultrasonic welding
Ultraschallschweißmaschine ultrasonic welding machine
Ultraschallschwingungen ultrasonic vibrations
Ultraschallsensor ultrasonic probe/sensor/transducer
Ultraschallsignal ultrasonic signal
Ultraschallverbindung ultrasonic weld
Ultraschallwandler ultrasonic converter
Ultraschallwelle ultrasonic wave
Ultraschallwellenimpuls ultrasonic impulse
ultraviolett ultra-violet, UV
Ultraviolettspektroskopie UV spectroscopy
Ultrazentrifuge ultracentrifuge
umbauen to rebuild *(a machine)*
Umbäumstuhl re-winding unit *(txt)*
Umbausatz conversion unit
Umbruchfestigkeit cantilever strength, bending strength under two-point loading
Umdrehungsgeschwindigkeit rotational speed, speed of rotation
Umdrehungsgeschwindigkeit der Schnecke screw speed
Umdrehungskonstanz uniformity of rotation: **Umdrehungskonstanz der Extruderschnecke prüfen** check whether the screw is rotating evenly
Umdrehungszahl 1. number of revolutions. 2. speed *(of a rotating element)*. 3. screw speed *(e)*
Umdrehungszahlbereich der Schnecke screw speed range
Umesterung ester interchange
Umfang circumference
Umfangsgeschwindigkeit peripheral velocity
Umfangsrichtung circumferential direction
Umfeld environment
Umfeldbedingungen ambient conditions
Umformeinheit mould assembly: **Umformeinheiten mit mehr als einem Blaswerkzeug benutzt man für Hohlkörper von mehr als 5 l sehr selten** mould assemblies comprising more than one blow mould are used but very rarely for containers bigger than five litres
Umformen forming: *This covers all those operations which are performed on semi-finished materials such as sheets, pipes etc, e.g. vacuum forming, pipe bending and the like. The word is often used in conjunction with* **Urformen** *(q.v.)*
Umformen, spanloses forming, thermoforming
Umformen, thermoplastisches thermoforming
Umformungstemperatur 1. thermoforming temperature. 2. moulding temperature
Umformungstiefe depth of draw *(t)*
Umformvorgang 1. forming/thermoforming process, forming/thermoforming operation. 2. moulding process/operation
Umformwerkzeug forming tool
Umgebung environment
Umgebungsatmosphäre surrounding atmosphere
umgebungsbedingt due to OR caused by the environment: **umweltbedingte Einflüsse** environmental influences, due to environmental conditions/influences
umgebungsbedingte Spannungsumbildung environmental stress cracking
Umgebungsbedingungen ambient/environmental conditions
umgebungsbeeinflußte Spannungsrißbildung environmental stress cracking
Umgebungseinflüsse environmental influences
Umgebungsklima ambient/environmental conditions, surrounding atmosphere
Umgebungsluft surrounding atmosphere
Umgebungsmedium surrounding medium
Umgebungstemperatur ambient temperature
umgeformten Zustand, im in the unmoulded state
Umgießen potting/embedding, encapsulation
Umgranulieren re-pelletising *(sr)*
Umgriff throwing power *(in electrophoretic deposition)*
umhüllen to envelop
Umhüllmasse potting/embedding/encapsulating compound
Umhüllmaterial potting/embedding/encapsulating compound
Umhüllsystem potting/embedding/encapsulating compound
Umhüllung potting/embedding, encapsulation
umkehrbar reversible
Umkehrbeschichtung reverse roll coating
Umkehrosmose reverse osmosis
Umkehrwalzenbeschichter reverse roll coater
Umkonstruktion 1. re-designing. 2. rebuilding
Umkristallisation recrystallisation
umkrystallisieren to recrystallise
Umlage outer layer *(e.g. of a laminate)*
Umlaufgeschwindigkeit rotational speed
Umlaufkühlung 1. circulation cooling. 2. circulation cooling system
Umlaufmedium circulating medium/fluid
Umlaufölkühlung 1. circulating oil cooling. 2. circulating oil cooling system
Umlaufölung 1. circulating oil lubrication. 2. circulating oil lubricating system
Umlaufschmierung 1. circulating oil lubrication. 2. circulating oil lubricating system

Umlaufthermostat circulation thermostat
Umlaufverdrängerpumpe positive displacement circulating pump
Umlaufvermögen current assets
Umlaufwasser circulating water
Umlenkblaskopf side-fed blown film die *(bfe)*
Umlenkblech baffle plate *(bfe)*
Umlenkkopf crosshead (die) *(e)*
Umlenkung by-pass, diversion
Umlenkwalze deflecting roller
Umlenkwerkzeug crosshead (die) *(e)*
Umlenkwinkel deflection angle
Umluftautoklav circulating air autoclave
Umluftofen circulating air (drying) oven
Umluftschrank circulating air (drying) oven
Umlufttrockenschrank circulating air drying cabinet/oven
Umluftwärmeschrank circulating air drying oven/cabinet
ummantelt sheathed
Ummantelung sheathing
Ummantelungsanlage sheathing plant
Ummantelungsextruder cable sheathing extruder
Ummantelungswerkzeug cable sheathing die
Umorganisation reorganisation
Umorientierung re-orientation
Umprogrammierung re-programming *(me)*
Umrechnungsfaktor conversion factor
Umrechnungsformel conversion formula
umrüsten to change, to modify *(part of a machine, e.g. a mould)*
Umrüstzeit change-over time *(time taken to change machine components)*: **Alle obigen Einheiten können bei kurzen Umrüstzeiten untereinander gewechselt werden** all the above units can be quickly interchanged
Umsatz 1. conversion. 2. turnover
Umsatz, chemischer chemical reaction
Umsatz, prozentualer 1. percentage conversion *(e.g. of monomer into polymer)*. 2. percentage turnover
Umsatzausweitung turnover increase
Umsatzeinbruch drop in turnover
Umsatzeinbuße drop in turnover
Umsatzerlös sales proceeds
Umsatzgrad degree of conversion
Umsatzplus turnover increase
Umsatzrate conversion rate, rate of conversion
Umsatzrückgang drop in turnover
Umsatzvolumen volume turnover
Umsatzzuwachs turnover increase
Umschaltdruck change-over pressure
umschalten to change over
Umschaltniveau change-over point *(im)*
Umschaltpunkt change-over point
Umschaltschwelle change-over point/threshold
Umschaltung change-over
Umschaltzeit change-over time *(im)*
Umschaltzeitpunkt change-over point *(im)*
Umsetzung 1. reaction. 2. conversion. 3. implementation *(of regulations etc.)*

Umsetzung, chemische chemical reaction
Umsetzungsgrad degree of conversion
Umsetzungskennzahl degree of conversion
Umsetzungsprodukt reaction product
Umspannwerk transformer station
umspritzen 1. to sheathe *(cables, wires etc.)*. 2. to encapsulate by injection moulding *(inserts)*
umspritzt 1. sheathed. 2. moulded-in, encapsulated by injection moulding *(insert)*
umstrukturiert restructured
Umstrukturierung restructuring, reorganisation
Umverpackung 1. re-packaging. 2. secondary packaging
umwälzen to agitate
Umwälzkühlung 1. circulation cooling. 2. circulation cooling unit/system
Umwälzpumpe circulating pump
umwandeln to convert
Umwandlung transition
Umwandlungsbereich transition range
Umwandlungspunkt transition point
Umwandlungswärme heat of transformation
Umwandlungszone transition section *(of screw)*
Umwelt environment
Umweltanforderungen environmental requirements
Umweltaspekte ecological considerations
Umweltauflagen environmental regulations
Umweltauswirkung environmental effect, effect on the environment
umweltbedenklich a potential hazard/danger to the environment
Umweltbedingungen environmental conditions
Umweltbeeinträchtigung (environmental) pollution, pollution of the environment
umweltbelastend pollutant: **Dieser arbeitsintensive und umweltbelastende Prozeß...** This process, which is labour intensive and pollutes the environment. ...
Umweltbelastung (environmental) pollution, pollution of the environment
Umweltbewußtsein environmental awareness
umweltbewußt environment-conscious, environmentally aware
Umweltbilanz environmental audit
Umweltbundesamt Federal Environment Office
Umweltdiskussion discussion of environmental problems/questions
Umwelteinflüsse environmental influences
Umwelteinwirkungen environmental influences
Umwelteinwirkungen, schädliche (environmental) pollution
Umweltentlastung cleaning up the environment, keeping the environment clean
Umweltfaktoren environmental factors/considerations
Umweltfragen environmental/ecological considerations
umweltfreundlich environment-/eco-friendly, environmentally friendly
umweltgefährdend a danger/hazard to the environment

umweltgefährlich a danger/hazard to the environment
umweltgerecht environmentally acceptable
Umweltgesichtspunkte environmental considerations
Umwelthaftung environmental resonsibility/liability
Umwelthaftungsgesetz environmental liability act
Umweltkatastrophe environmental disaster
Umweltleitlinien environmental guidelines
umweltneutral inert, of no danger to the environment
Umweltökonom environmental economist
umweltorientiert environment-orientated
umweltpolitisch relating to environmental policy: **aus umweltpolitischen Gründen** for environmental reasons
Umweltprobleme environmental/ecological problems
Umweltqualität environmental quality
umweltrelevant environmentally important: *Like so many new words, this one is sometimes misused, as in example 44 in the appendix.*
Umweltrelevanz environmental importance/significance
Umweltrichtlinie environmental guideline
Umweltrisiko environmental risk
Umweltschäden environmental damage
Umweltschädlichkeit *The literal translation, using the word* harmfulness, *sounds awkward. The word should therefore be circumscribed, e.g.* **Es drängen sich immer mehr Stimmen in den Vordergrund, die Kunststoffen Umweltschädlichkeit nachsagen.** *There is an increasing body of opinion which claims that plastics harm the environment.*
umweltschonend environment-friendly, environmentally friendly, not harming/polluting: **...können sie umweltschnonend durch Recycling aufbereitet werden.** ...they can be recycled without polluting the environment
Umweltschutz environmental protection, protection of the environment, pollution control
Umweltschutzanlage pollution control equipment
Umweltschutzbedingungen pollution control regulations
Umweltschützer environmentalist
Umweltschutzgesetzgebung pollution control regulations
Umweltschutzmaßnahmen pollution control measures
Umweltschutzorganisation environmentalist organisation, environment protection agency
Umweltschutzprobleme pollution control problems
Umweltschutztechnik pollution control technology, environmental technology/engineering
umweltsicher environmentally safe
Umweltsicherheit environmental safety

Umweltsicht, aus from an environmental point of view
Umweltstatistik environmental statistics
Umweltstrategie environmental policy
umwelttechnisch environmental
Umweltvergiftung environmental pollution, pollution of the environment
Umweltverschmutzung environmental pollution, pollution of the environment
umweltverträglich environmentally compatible
Umweltverträglichkeit environmental compatibility
Umweltvorsorgepolitik environmental protection policy
Umweltzeichen environmental symbol, symbol for environmental friendliness *(usually associated with* **'Blauer Engel''**, *q.v.)*
unabgesättigt unsaturated
unaufgeschmolzen unmelted
unausgelastet non-utilised
unbedenklich, gesundheitlich non-toxic
unbedenklich, lebensmittelrechtlich food quality, suitable for food contact applications
unbedenklich, ökologisch ecologically safe
unbedenklich, physiologisch non-toxic
Unbedenklichkeit, gesundheitliche non-toxicity
Unbedenklichkeit, lebensmittelrechtliche suitability for food contact applications *(for translation example see under* **lebensmittelrechtliche Unbedenklichkeit***)*
Unbedenklichkeit, physiologische non-toxicity
Unbedenklichkeitsbescheinigung clearance certificate
unbedruckt unprinted
unbegrenzt unlimited
unbehandelt 1. untreated. 2. uncoated *(e.g. filler particles)*
unbelastet non-stressed
unbemannt unmanned, without human intervention
unbeschädigt undamaged
unbeschichtet uncoated
unbeständig not resistant *(material in presence of chemicals)*
unbestrahlt non-irradiated
unbewittert unweathered
unbrennbar non-flammable, non-combustible
Unbrennbarkeit non-flammability
undicht leaky
Undichtheit leak
undurchlässig impermeable
undurchsichtig opaque
uneben uneven
unempfindlich indifferent to, unaffected by
Unempfindlichkeit resistance/indifference (to) **thermische Unempfindlichkeit** resistance to high temperatures; **mechanische Unempfindlichkeit** indifference to mechanical influences
unendlich infinite

unerwünscht unwanted, unwelcome: *This word is often used unnecessarily and can then be omitted, as in this example:* **zur Vermeidung von unerwünschtem Polymerabbau** to prevent polymer degradation; *polymer degradation is never anything but unwanted - and one would hardly wish to prevent something "wanted" occurring*
Unfallschutz accident prevention
unfallsicher safe *(in operation)*
Unfallverhütung accident prevention
Unfallverhütungsmaßnahmen safety precautions, accident prevention measures
Unfallverhütungsvorschriften accident prevention regulations, safety regulations
ungealtert unaged, not aged
ungecoatet *German version of* uncoated
ungedämmt uninsulated
ungefärbt unpigmented
ungefüllt unfilled
ungehärtet uncured
ungekapselt not embedded/encapsulated
ungekerbt unnotched
ungeladen uncharged
ungeliert 1. ungelled *(UP, EP resins)*. 2. unfused *(PVC paste)*
Ungenauigkeit inaccuracy
ungeordnet 1. randomly arranged *(fibres)*. 2. disordered *(molecular structure)*
ungepaart unpaired
ungesättigt unsaturated
Ungesättigtheit degree of unsaturation
Ungesättigtheitsgrad degree of unsaturation
ungeschäumt unfoamed
ungetempert unannealed, unconditioned
ungetrocknet undried
ungiftig non-toxic
Ungiftigheit non-toxicity
ungleichmäßig uneven
unidirektional unidirectional
Unidirektionalverstärkung unidirectional reinforcement
Universalextruder general purpose extruder
Universalkopf general purpose extruder head
Universalmaschine general purpose machine
Universalmischer general purpose mixer
Universalplastifiziereinheit general purpose plasticising unit
Universalprobekörper general purpose test piece
Universalprodukt general purpose product
Universalprüfmaschine universal test apparatus, universal tester
Universalprüfmethode universal test (method)
Universalschnecke general purpose screw
Universaltyp 1. general purpose grade *(of moulding compound)*. 2. general purpose model *(of a machine)*
Universalzerkleinerer general purpose granulator *(sr)*
universell stabilisierbar with all-round stabiliser tolerance

universelle Stabilisierbarkeit all-round stabiliser tolerance
unkaschiert unlaminated
unkatalysiert uncatalysed
unkondensiert uncondensed
unkontrollierbar uncontrollable
unkontrolliert uncontrolled
unlackiert unpainted
unlösbare Verbindung permanent joint
unlöslich insoluble
Unlöslichkeit insolubility
unmodifiziert unmodified
Unordnungszustand state of disorder
unpigmentiert unpigmented
unplastifiziert unplasticised
unpolar non-polar
unporös non-porous
unregelmäßig irregular
unrund laufend eccentric
unschmelzbar infusible
Unschmelzbarkeit infusibility
unsortiert unsorted
unstabilisiert unstabilised
unsubstituiert unsubstituted
unsymmetrisch asymmetric(al)
untemperiert unheated
Unterbau 1. infrastructure. 2. sub-structure, supporting frame
Unterboden sub-floor
Unterbodenschutz underbody sealant
Unterbodenschutzmasse underbody sealant
Unterbodenschutzpaste underbody sealant paste
unterbrechungslos continuous, without a break, without interruption(s)
Unterbrennen understoving, underbaking *(paint film)*
unterdimensioniert too small, undersize
Unterdosierung underfeeding, starve feeding *(im)*
Unterdruck 1. reduced pressure 2. vacuum
Unterdruckpumpe vacuum pump
Unterdrückung suppression
unterfüttern to underfeed, to starve feed
untergeordnet secondary
Untergrenze lower/bottom limit
Untergrund substrate
Untergrundfestigkeit strength of the substrate
Untergrundmaterial substrate material
Untergrundoberfläche substrate surface
Untergrundvorbehandlung preparation/ pretreatment of the substrate
Untergrundvorbereitung preparation/ pretreatment of the substrate
Untergruppe sub-group *(me)*
Unterhalt upkeep, maintenance
unterhaltsaufwendig expensive/difficult to maintain
Unterhaltskosten 1. maintenance costs. 2. running/operating costs
Unterhaltung 1. servicing, maintenance. 2. entertainment

Unterhaltungselektronikindustrie home entertainment industry
unterhärtet undercured
Unterhärtung undercuring
Unterheizkörper bottom heater
unterirdisch below ground, underground
Unterkolbenpresse up-stroke press
Unterkühlung undercooling
unterlagert secondary
Unterlagsboden sub-floor
Unterlieferant sub-contractor
unternehmensintern in-house
Unternehmenspolitik company/corporate policy
Unternehmensstrategie corporate strategy, company/corporate policy
Unternehmensstruktur corporate/company structure
Unternehmungsberatung management consultancy
Unterprogramm sub-program *(me)*
Unterscheidungskriterien distinguishing features
Unterscheidungsmerkmale distinguishing features
Unterschneidungen undercuts
Unterschreiten dropping below:
Machinenabschaltung bei Unterschreiten des minimalen Ölstandes zum Vermeiden von Pumpenschäden machine cut-out if the oil level drops below the minimum, to prevent damage to the pump
Unterschriftstreifen signature strip
Unterschuß less *(of something)*: **Inwieweit ein Unterschuß an DBP Veränderungen im mechanischen Eigenschaftsbild mit sich bringt...** How far the mechanical properties are affected by using less DBP...
Untersetzung speed reduction
Untersetzungsgetriebe speed reduction gear
Untersetzungsverhältnis reduction (gear) ratio
Untersicht view/seen from below
Unterstation sub-station *(me)*
Unterstruktur sub-structure
Untersuchungsbericht test report
Untersuchungsergebnisse test results
Untersuchungsprogramm test/research programme
Untersuchungstemperatur test temperature
Untersuchungszeitraum test period/duration
Untertagedeponie underground landfill site
Untertauchversuch immersion test
Unterteil bottom part
Untervernetzung undercuring
Untervulkanisation undercure, under-vulcanisation
Unterwalze bottom roll
Unterwasseranstrich underwater coating/finish
Unterwasserfarbe underwater paint
Unterwassergranulator underwater pelletiser *(sr)*
Unterwassergranulieranlage underwater pelletiser *(sr)*
Unterwassergranuliersystem underwater pelletising system *(sr)*
Unterwassergranuliervorrichtung underwater strand pelletiser *(sr)*
Unterwasserstranggranulator underwater pelletiser *(sr)*
Unterwerkzeug bottom (mould) force *(cm)*
Unterziehen draw-down *(e)*
Unterzug draw-down *(e)*
Unterzugsverhältnis draw-down ratio *(e)*
untoxisch non-toxic
unumgesetzt unreacted
unverdünnt undiluted
unverestert unesterified
unverjüngt parallel-sided
unvernetzt 1. uncured, un-crosslinked. 2. not linked/(inter)connected
unverrostbar rust-proof, non-rusting
unverrottbar rot-proof
unverschnitten on its own, unblended
unversehrt undamaged
unverseifbar unsaponifiable
unverstärkt unreinforced
unverstreckt unstretched
unverträglich incompatible
Unverträglichkeit incompatibility
unvertretbar unjustifiable, unacceptable
unverzweigt unbranched
unvorbehandelt not pre-treated
unvulkanisiert unvulcanised, uncured
unwirksam ineffective
unwirtschaftlich uneconomic
Unwucht out-of-balance
Unwuchtkräfte out-of-balance forces
unzellig non-cellular, solid
unzerbrechlich unbreakable, indestructible
Unzerbrechlichkeit indestructibility
unzersetzt undecomposed *(e.g. peroxide)*
unzulässig inadmissible
UP-GF-SMC polyester sheet moulding compound, polyester SMC
UP-GF-Teil glass fibre reinforced polyester part/component
UP-Harz (unsaturated) polyester resin, UP resin
UP-Harzformstoff cured polyester/UP resin
UP-Harzkleber polyester adhesive
UPM, Upm abbr. of **Umdrehungen pro Minute**, revolutions per minute, rpm
UP-Reaktionsharz (unsaturated) polyester resin, UP resin
UP-Reaktionsharzmasse catalysed polyester resin, polyester resin mix
Urazolgruppe urazole group
Urethan urethane
Urethanacrylat urethane acrylate
Urethanalkyd urethane alkyd (resin)
Urethanalkydharz urethane alkyd (resin)
Urethanbindung urethane linkage
Urethanbindungsanteil urethane linkage content
Urethanethoxylat urethane ethoxylate
urethanfunktionell urethane-functional

Urethangruppe urethane group
urethanmodifiziert urethane modified
Urformen moulding: *This covers all those operations which start out from a moulding compound, such as extrusion, injection and blow moulding etc. The word is often used in conjunction with* **Umformen** *(q.v.)*
Urformtemperatur moulding temperature
Urmodell master model/pattern
Urmodellpaste modelling paste
Urmodellplatte master pattern plate
Ursache-Wirkungsanalyse cause-effect analysis
Ursache-Wirkungsforschung cause-effect investigation
Ursache-Wirkungszusammenhänge cause-effect relation
Ursprungsdruck original pressure
Ursprungsform original shape
Ursprungsmaß original dimension
Ursprungstemperatur original temperature
US *abbr.of* **Ultraschallschweißen**, ultrasonic welding
UV-Absorber UV absorber
UV-Absorption UV absorption
UV-Anteil UV content
UV-Ausrüstung UV stabiliser
UV-härtend UV-curing
UV-Schutzmittel UV stabiliser
UV-Spektroskopie UV spectroscopy
UV-Stabilisator UV stabiliser
UV-Stabilität UV stability/resistance
UV-Strahlung UV radiation
UV-Strecke UV heating tunnel
U.V.V. *abbr. of* **Unfallsverhütungsvorschrift(en)**, safety regulations

V

v.a. *abbr. of* **vor allem**, above all, especially
VA stainless steel
V2A stainless steel
VAE *abbr. of* **Vinylacetat-Ethylen,** *the English equivalent of which is EVA. See explanatory note under* **Vinylacetat-Ethylen-Copolymerisat**
Vakuum, im in vacuo
Vakuumanschluß vacuum connection
Vakuumapparatur vacuum apparatus/equipment
Vakuumaufdampftechnik vacuum deposition (process)
Vakuumdestillation vacuum distillation
vakuumdicht vacuum tight
Vakuumdoppeltrichter twin vacuum hopper

Vakuumdoppeltrichteranlage twin vacuum hopper assembly
Vakuumeinspritzverfahren vacuum injection moulding (process)
Vakuumentgasung 1. vacuum venting. 2. vacuum venting system. 3. vacuum deaeration: *Use versions 1. and 2. when the text refers to a machine. Version 3. applies when the text refers to the removal of gas or air from a PVC paste or moulding compound*
Vakuumextruder vented extruder
Vakuumfolienverfahren vacuum bag moulding *(grp)*
Vakuumfördergerät vacuum conveyor
Vakuumformautomat automatic vacuum forming machine
Vakuumformmaschine vacuum forming machine
Vakuumformung vacuum forming
Vakuumgießen vacuum casting
Vakuuminjektionsverfahren vacuum injection (process)
Vakuumkalibrierbecken vacuum calibrating/sizing tank *(e)*
Vakuumkalibrieren vacuum calibration/sizing *(e)*
Vakuumkalibrierstrecke vacuum calibration/sizing section *(e)*
Vakuumkalibriersystem vacuum calibration/sizing system *(e)*
Vakuumkalibrierung 1. vacuum calibrating/sizing. 2. vacuum calibrating/sizing unit *(e)*
Vakuumkammer vacuum tank/chamber
Vakuumkessel vacuum tank/chamber
Vakuum-Kühltankkalibrierung waterbath vacuum calibrating unit *(e)*
Vakuumofen vacuum oven
Vakuumpreßverfahren vacuum press moulding
Vakuumpumpe vacuum pump
Vakuumsack vacuum bag
Vakuumsauglöcher vacuum suction holes
Vakuumsaugteller vacuum suction plate
Vakuumschnecke vented screw
Vakuumschrank vacuum cabinet
Vakuumspeisetrichter vacuum/vented (feed) hopper *(e, im)*
Vakuumtankkalibrierung 1. vacuum calibration/sizing. 2. vacuum calibrating/sizing unit *(e)*
Vakuumteil vented section *(of screw)*
Vakuumtiefziehen vacuum forming
Vakuumtiefziehform vacuum forming tool
Vakuumtiefziehmaschine vacuum forming machine
Vakuumtränkverfahren vacuum impregnation (process)
Vakuumtrichter vacuum/vented (feed) hopper *(e,im)*
Vakuumtrichterentgasungsanlage vented hopper degassing unit
Vakuumtrockenschrank vacuum drying cabinet
Vakuumtrommelfilter vacuum drum filter
Vakuumversorgung vacuum supply

Vakuumzelle vacuum cell
Vakuumziehverfahren vacuum forming (process)
Vakuumzone vent zone/section
Valenz valency
Valeriansäure valeric acid
Vanadinbeschleuniger vanadium accelerator
van-der-Waals-Bindung van der Waals linkage
van-der-Waalssche Kräfte van der Waals' forces
variabel 1. variable. 2. variable displacement/ delivery *(pumps)*
Variable variable
VbF *abbr.* of **Verordnung brennbarer Flüssigkeiten**, German regulations covering flammable liquids
VC-Copolymerisat vinyl chloride copolymer
VCM *abbr. of* vinyl chloride monomer
VC-Restmonomergehalt 1. residual vinyl chloride monomer. 2. residual vinyl chloride (monomer) content
VE-Harz vinyl ester resin
Vektorisierung vectorisation
vektororientiert vector orientated
Ventil valve
Ventilator fan
Ventilauswerfer valve ejector *(im)*
Ventilsitz valve seat
Venturisystem venturi system
veränderbar, stufenlos steplessly/infinitely variable
veränderlich variable, changeable
Verankerung anchorage
Verankerungsaufstrich primer coat
Verantwortung responsibility
verarbeitbar processable: **verarbeitbar aus der Schmelze** melt processable; **thermisch verarbeitbar** melt processable
verarbeitbar, thermoplastisch melt processable
Verarbeitbarkeit 1. processability. plasticity, workability *(e.g. of a polymer-concrete mix)*
Verarbeitbarkeit, thermoplastische melt processability
Verarbeiter processor, fabricator
Verarbeitung 1. processing: *This must sometimes be translated rather freely, e.g.* **Bei der Verarbeitung technischer Thermoplaste zu Formteilen...** when moulding engineering thermoplastics.... 2. fabrication *(when referring to semi-finished materials)*
Verarbeitung, spanlose thermoforming, shaping, forming
Verarbeitung, thermoplastische melt processing; *for explanatory notes and translation examples see under* **thermoplastische Verarbeitung**
Verarbeitungaggregat 1. processing unit. 2. machine. 3. extruder, injection moulding machine, etc. *(depending on context)*

Verarbeitungsautomat automatic processing equipment
Verarbeitungsbedarf required pot life
Verarbeitungsbedingungen 1. processing conditions. 2. moulding conditions. 3. extrusion conditions
Verarbeitungsbereich processing range
verarbeitungsbereit ready for processing, ready to use
Verarbeitungsbreite 1. processing latitude. 2.effective width *(e.g. of a sheet extrusion die)*
Verarbeitungsdruck moulding pressure
Verarbeitungseigenschaften processing characteristics
Verarbeitungseinfluß effect of processing
Verarbeitungseinheit processing unit
Verarbeitungsempfehlungen processing recommendations/guidelines
verarbeitungsfähig 1. usable, capable of being processed. 2. having a pot life of...
Verarbeitungsfehler processing fault
Verarbeitungsfenster processing latitude
verarbeitungsfertig ready for processing, ready to use
verarbeitungsfreundlich 1. easy to process. 2. easy to use
Verarbeitungsgeräte processing equipment
verarbeitungsgerecht *suitable or correct for given processing conditions*: **verarbeitungsgerechte Auslegung der Profile** profiles designed for easy fabrication; **Die Kurzkompressionsschnecke wird weitgehend durch die verarbeitungsgerechtere Langkompressionsschnecke abgelöst** the short-compression zone screw is largely being replaced by the long-compression zone screw, which is the more suitable from the processing point of view
Verarbeitungsgröße processing parameter
verarbeitungsgünstig 1. easy to process. 2. easy to use
Verarbeitungsgut *This is a blanket expression covering anything that is being processed. It can be translated by any of the following words - and others, depending on the context:* material, compound, paste, adhesive, paint etc.
Verarbeitungshilfsmittel processing aid
Verarbeitungshilfsstoff processing aid
Verarbeitungshinweise processing guidelines
Verarbeitungskonsistenz working consistency
Verarbeitungsmangel processing fault
Verarbeitungsmaschine 1. machine (plural: machines, processing equipment). 2. moulding machine. 3. extruder
Verarbeitungsmerkmale processing characteristics
Verarbeitungsmöglichkeiten processing possibilities
Verarbeitungsparameter processing parameter
Verarbeitungsperiode pot life

Verarbeitungsphasen processing stages
Verarbeitungsprogramm processing program *(me)*
Verarbeitungsprotokoll processing record
Verarbeitungsprozeß 1. process, moulding process *(general terms)*. 2. extrusion, injection moulding etc. process *(depending on the type of process under discussion)*: **Während des Verarbeitungsprozesses** during processing
Verarbeitungsrichtlinien processing guidelines
Verarbeitungsrichtung, in in machine direction
Verarbeitungsrichtwerte approximate processing conditions
Verarbeitungsschritte processing stages
Verarbeitungsschrumpf moulding shrinkage *(see explanatory note under* **Verarbeitungsschwindung***)*
Verarbeitungsschwierigkeiten processing difficulties
Verarbeitungsschwindung moulding shrinkage: *Defined in BS 2782 and DIN 16901 as the difference in dimensions, at room temperature, between a moulded part and the mould in which it was made*
Verarbeitungsspannungen moulding stresses
Verarbeitungsspielraum 1. processing latitude. 2. pot life
Verarbeitungsstabilisator process stabiliser
Verarbeitungsstabilität stability during processing
verarbeitungstechnisch *relating to processing*: **verarbeitungstechnische Eigenschaften** processing characteristics; *See also under* **verfahrenstechnisch**
Verarbeitungstechnologie processing technique
Verarbeitungstemperatur processing temperature
Verarbeitungstemperaturbereich processing temperature range
Verarbeitungsunterbrechung break in production
Verarbeitungsunterschiede processing differences
Verarbeitungsverfahren method of processing, processing technique
Verarbeitungsverhalten processing characteristics/performance
Verarbeitungsviskosität working consistency *(e.g. of a paint or adhesive)*
Verarbeitungsvorteile processing advantages
Verarbeitungszeit pot life
Verarbeitungszyklus processing cycle
Veraschung ashing
verbacken to cake
verbessert improved
Verbesserung improvement
Verbesserungen, konstruktive design improvements
Verbilligung price reduction
Verbindlichkeit liability, obligation, commitment
Verbindung 1. connection, link, linkage. 2. (chemical) compound. 3. runner *(im)*
Verbindung, feste permanent joint
Verbindung, lösbare temporary joint
Verbindung, unlösbare permanent joint
Verbindungsfläche adherend surface, surface to be bonded
Verbindungshülse connecting bush
Verbindungskabel connecting cable
Verbindungsklasse compound group
verbindungsprogrammierbar hard-wired
Verbindungsrohr 1. connecting pipe/tube. 2. runner *(im)*
Verbindungsverfahren method of joining
verblockfrei non-blocking
Verbrauch consumption
Verbraucher 1. consumer. 2. consumer unit
Verbrauchergewohnheiten consumer habits
verbrauchernah consumer-oriented
Verbrauchsaufgliederung consumption breakdown
Verbrauchschätzung estimated consumption
Verbrauchseinbuße drop in consumption
verbrauchsfertig ready-to-use
Verbrauchsgüter consumer goods
Verbrauchsprognose consumption forecast
Verbrauchssteigerung increase in consumption
Verbrauchszahlen consumption figures
Verbrauchszunahme consumption increase, increase in consumption
Verbrauchszuwachs increase in consumption
Verbrennung 1. combustion. 2. charring. 3. incineration
Verbrennungsanlage incinerator, incineration plant
Verbrennungserscheinungen signs of charring
Verbrennungsgas flue gas
Verbrennungskosten incineration costs
Verbrennungskraftmaschine combustion engine
Verbrennungsprodukt combustion product
Verbrennungsprozeß combustion process
Verbrennungsschiffchen combustion boat
Verbrennungsschlieren charred streaks
Verbrennungswärme heat of combustion
Verbund composite
Verbundbetrieb coupled operation *(of several machines)*: **im Verbund betrieben** operated in conjunction with each other
Verbundfenster sealed unit
Verbundfestigkeit bond strength
Verbundfolie composite/multi-layer film
Verbundglasscheibe safety glass
Verbundglaswindschutzscheibe laminated windscreen
Verbundgüte bond quality
Verbundplatte composite sheet/panel
Verbundsicherheitsglas laminated safety glass
Verbundspritztechnik sandwich moulding *(im)*
Verbundstoff composite (material)
Verbundsystem composite system
Verbundwerkstoff composite (material)
verchromt chrome plated, chromed

Vercracken 1. *German version of "cracking" as applied to petroleum distillates.* 2. degradation
verdampfbar volatile
Verdampfbarkeit volatility
verdampfen to evaporate
Verdampfer 1. evaporator. 2. carburettor
Verdampfergehäuse carburettor housing
Verdampfung evaporation
Verdampfungsgeschwindigkeit evaporation rate
Verdampfungskühlung cooling due to evaporation
Verdampfungsrate evaporation rate
Verdampfungsverlust evaporation loss, loss through evaporation
Verdampfungswärme heat of evaporation
verdeutlichen to clarify, to explain, to illustrate
verdichten to compress, to compact
Verdichtung 1. compression, consolidation, compaction. 2. packing *(im)*
Verdichtungsdruck packing pressure *(im)*
Verdichtungsgrad degree of packing *(im)*
Verdichtungsphase packing phase *(im)*
Verdichtungsprofil packing profile *(im)*
Verdichtungsschnecke compression screw *(e)*
Verdichtungsverhältnis compression ratio *(e)*
Verdichtungszone compression/transition section *(of a screw) (e)*
verdickt thickened
Verdickung thickening
Verdickungsmittel thickener, thickening agent
Verdienstspanne profit margin
verdrahtet wired
Verdrahtung wiring
verdrahtungsprogrammiert hard-wired
Verdrallung twisting
Verdränger torpedo
Verdrängerform compression mould
Verdrängerkörper torpedo
Verdrängerprinzip displacement principle
Verdrängerpumpe positive displacement pump
Verdrängertorpedo torpedo
Verdrängervakuumpumpe positive displacement vacuum pump
Verdrängung displacement
Verdrängungsprinzip displacement principle
Verdrehung twisting
Verdrillung twisting
verdrillungsfrei non-twisting, without twisting
verdünnbar dilutable
Verdünnbarkeit dilutability
verdünnen to dilute
Verdünner 1. diluent. 2. thinner *(specifically for paints)*
Verdünneranteil amount of thinner: **Verdünneranteile werden lösemittelfreien Systemen vorwiegend zur Viskositätsminderung zugesetzt** thinners are added to solvent-free systems mainly to reduce their viscosity
Verdünnerharz extender resin

verdünnt 1. diluted. 2. dilute *(if referring to acids and alkalis)*
Verdünnung dilution
Verdünnungsmittel 1. diluent. 2. thinner *(specifically for paints)*
Verdunstung evaporation
Verdunstungsgeschwindigkeit evaporation rate
Verdunstungsverlust evaporation loss, loss through evaporation
Verdunstungszahl evaporation index
Veredeln 1. surface finishing *(i.e. the application of special colour or texture effects to coated products to improve their appearance).* 2. upgrading *(general term)*
Vereinfachung simplification, making something easier
verestert esterified
Veresterung esterification
Veresterungsgrad degree of esterification
verethert etherified
Veretherung etherification
-verfahren -process: *This suffix, attached to a word for a process such as extrusion, injection moulding, calendering etc. can often - indeed should - be omitted, as in the following example*: **PVC Folien werden zu 90% nach dem Kalanderverfahren hergestellt** 90% of PVC film and sheeting is made by calendering
Verfahrensablauf (course of the) process
Verfahrensanalyse process analysis
Verfahrensbedingungen processing conditions
Verfahrensdaten process data
Verfahrenseinflußgrößen factors influencing the process
Verfahrenseinheit processing unit: *This word cannot always be taken at face value, as the following example shows*: **Eine längere Verfahrenseinheit bis zu 30D** a longer screw, up to 30D long
verfahrensgerecht right for a particular process: **verfahrensgerechte Gestaltung von Werkzeugen** correct mould design; *See also explanatory note under* **-gerecht**
Verfahrenshilfsmittel processing aid
Verfahrenskurzbeschreibung brief/outline description of method
Verfahrensmaschine machine
Verfahrensparameter processing parameter
Verfahrensschritte process stages
verfahrensspezifisch specific to a particular process: **verfahrensspezifische Vorteile** processing advantages
Verfahrenstechnik process engineering/technology
Verfahrenstechnik, chemische chemical process engineering
Verfahrenstechniker process engineer
verfahrenstechnisch *relating to a process*: **Die auf dem Markt befindlichen Anlagen genügen in den meisten Fällen sowohl verfahrenstechnisch wie auch**

leistungsmäßig den derzeitigen Ansprüchen Most of the equipment on the market meets present-day requirements as regards processing technology and performance; **Je besser es gelingt, den Spritzgießprozeß verfahrenstechnisch zu beherrschen...** the more one can master the technicalities of the injection moulding process...; *Where neither* processing *nor* technical *contribute anything to the meaning, it is best to omit the word altogether, as in these two examples*: **Die verfahrenstechnische Aufgabe des Einspeisens, Plastifizierens, Homogenisierens und Einspritzens der Formmasse ist auf verschiedene Wege gelöst worden** The problem of feeding, plasticising, homogenising and injecting the moulding compound has been solved in various ways; **Optimierung der verfahrenstechnischen Auslegung von Einschneckenextrudern** optimising the design of single-screw extruders;
verfahrenstechnisch einfach technically simple/easy;
verfahrenstechnisch interessant technically interesting, interesting from the processing point of view;
verfahrenstechnisch nicht möglich technically impossible;
verfahrenstechnisch relevante Parameter technically important parameters/factors;
verfahrenstechnisch vorstellbar technically feasible;
verfahrenstechnisch wesentlich technically important, important from the processing point of view;
verfahrenstechnische Anforderungen processing requirements;
verfahrenstechnische Auslegung technical layout/design;
verfahrenstechnische Bedingungen processing conditions;
verfahrenstechnische Bedürfnisse processing requirements;
verfahrenstechnische Einheit processing unit;
verfahrenstechnische Entwicklung technical development(s);
verfahrenstechnische Forderungen technical demands/requirements;
verfahrenstechnische Hilfestellung technical assistance;
verfahrenstechnische Maßnahmen technical measures;
verfahrenstechnische Merkmale technical features;
verfahrenstechnische Möglichkeiten processing possibilities;
verfahrenstechnische Nachteile technical limitations/disadvantages;
verfahrenstechnische Schwierigkeiten technical problems;
verfahrenstechnische Überlegungen technical considerations;
verfahrenstechnische Vorteile technical advantages;
verfahrenstechnischen Gründen, aus owing to the nature of the process;
verfahrenstechnischer Ablauf processing sequence
Verfahrensteil 1. processing unit. 2. barrel. 3. screw: *The word is used to describe whatever part of the machine is under discussion. In addition to the examples given above, it could be a mould, calender rolls, granulator blades etc. depending on the context.* **Ein Verfahrensteil von 30D Länge** *obviously refers to a 30D screw*
Verfahrensvariablen process variables
Verfahrensweise procedure
Verfalldatum 1. use-by date. 2. sell-by date.
Verfälschung falsification *(eg. of test results)*
verfärbend staining
Verfärbung discolouration
Verfärbungsneigung tendency to discolour
verfeinert improved
Verfestigungsphase solidifying phase
Verfilmung film formation
Verfilmungshilfsmittel film-forming aid
verflüssigen to liquify
Verflüssigungsleistung plasticising capacity
Verflüssigungsmittel liquifying agent
verformbar, thermoplastisch thermoformable, suitable for thermoforming
Verformbarkeit 1. thermoformability. 2. plasticity *(e.g. of a polymer concrete mix)*. 3. deformability
Verformung 1. deformation. 2. forming, shaping
Verformung, bleibende residual/irreversible/permanent deformation
Verformung, thermoplastische thermoforming
Verformung unter Last deformation under load
Verformungsamplitude deforming amplitude
Verformungsarbeit deformation energy
Verformungsbereich crumple zone *(in a car)*
Verformungsbruch ductile failure/fracture
Verformungsenergie deformation energy
Verformungsfähigkeit 1. deformability 2. thermoformability 3. plasticity, workability *(e.g. of polymer concrete)*
Verformungsgeschwindigkeit deformation rate, rate of deformation
Verformungsgrad 1. depth of draw *(t)*. 2. amount/degree of deformation
Verformungsgrenze deformation limit
Verformungskraft (thermo)forming pressure
Verformungskurve deformation curve
Verformungsrest compression set
Verformungsstabilität deformation resistance
Verformungstemperatur 1. moulding temperature. 2. thermoforming temperature
Verformungsverhalten deformation behaviour
Verformungsvermögen 1. mouldability, moulding characteristics. 2. thermoformability.

3. plasticity, workability *(e.g. of a polymer concrete mix)*
Verformungsvorgang deformation process
Verformungswärme heat of deformation
Verformungswerkzeug 1. (moulding) tool. 2. die *(e)*. 3. mould *(im)* *(see explanatory note under* **Werkzeug**)
Verformungswert plasticity
Verformungswiderstand warp/deformation resistance
Verformungszone crumple zone *(in a car)*
Verformungszustand state of deformation
Verfügbarkeit 1. availability. 2. utilisation (of available capacity), efficiency: **Die Maschine hat eine Verfügbarkeit von 98%** the machine runs to 98% of its capacity; **Die Maschinen haben eine hohe Verfügbarkeit** the machines are highly efficient
Verfugen jointing
Verfugungsmasse 1. jointing filler/compound. 2. (tile) grout
Verfüllmasse jointing/filler compound
Vergaserschwimmer carburettor float
Vergelung gelation
vergilben to become yellow
Vergilbung 1. yellowing. 2. yellowness index
vergilbungsbeständig resistant to yellowing
vergilbungsfrei non-yellowing
Vergilbungsgrad yellowness index
Vergilbungszahl yellowness index
Verglasung glazing
Verglasungsmaterial glazing material
Verglasungsprofil glazing strip
Vergleichbarkeit comparability
Vergleichmäßigung smoothing out: **Eine Vergleichmäßigung der Teilströme im Massefluß** a smoothing out of the separate melt streams
Vergleichsberechnung comparative calculation
Vergleichslösung reference solution
Vergleichsmedium reference medium
Vergleichsmessung comparative measurement
Vergleichsmuster reference sample
Vergleichsprobe reference sample/specimen
Vergleichsprüfung comparative test
Vergleichsspannung hoop stress *(of pipes)*
Vergleichstabelle comparative table
Vergleichstest comparative test
Vergleichsversuch comparative test
Vergleichswert comparative figure
Vergleichszahl comparative figure
Vergrauung greying
Vergrößerung 1. magnification 2. enlargement
Verguß potting, embedding
Vergußharz potting/embedding resin
Vergußmasse potting/embedding compound
vergüten to improve, to enhance
Vergütungsstahl tempered steel
Verhalten behaviour, performance, properties
Verhalten, physiologisches toxicological properties
verharzt resinified

Verhütungsmaßnahmen preventive measures
verjüngt tapered, tapering
Verjüngung narrowing
Verjüngungswinkel taper angle
Verkabelung wiring
Verkapselung encapsulation
Verkaufsbedingungen conditions of sale
Verkaufspreis selling price
Verkaufsprogramm range of products, product range
Verkaufssortiment range of products
Verkaufsverpacking sales packaging
Verkehrsgewerbe transportation
Verkehrsmittel form(s) of transport
Verkehrswesen transport sector, transport and communication(s)
Verkehrszeichen road/traffic sign
verketten to link
Verkettung 1. chain formation 2. linkage
verklebbar bondable
Verklebeeigenschaften bonding properties/characteristics
verkleben to bond
Verklebung 1. bonding. 2. bonded joint
Verklebungsfläche bonded area
Verklebungspartner parts to be bonded, adherends
Verkleidung 1. cladding. 2. trim *(of a car)*
Verkleidungsplatte cladding panel
verklemmen to jam
Verklumpungsneigung tendency to form lumps
verknäuelt entangled *(molecule)*
Verknäuelung entangling *(of molecules)*
verknüpft linked
Verknüpfung linkage
Verknüpfungsmöglichkeiten possible linkages: **die Vielfalt der inter- und intramolekularen Verknüpfungsmöglichkeiten** the many different possible intermolecular and intramolecular linkages
Verknüpfungsreaktion crosslinking reaction
Verknüpfungsstelle linkage point
Verkochen cooking
verkohlt charred
Verkohlung charring
Verkohlungsneigung tendency to char
Verkoken carbonisation
verkräckt 1. *German version of* "cracked" *as applied to petroleum distillates*. 2. degraded
Verlängerung extension, increase in length
Verlängerungskabel extension cable
Verlauf flow
Verlauf, zeitlicher variation with time *(e.g. of a parameter)*: **Zeitlicher Verlauf des Werkzeuginnendrucks** variation of cavity pressure with time
Verlaufmittel flow control agent
Verlaufprobleme flow problems
Verlaufsbelag self-levelling screed
Verlaufschwierigkeiten flow problems
Verlaufseigenschaften flow properties/characteristics

verlaufsfördernd flow promoting
Verlaufshilfsmittel flow control agent
Verlaufsmörtel self-levelling mortar
Verlaufstörungen flow problems
Verlegeeinheit film gauge equalising unit *(see explanatory note under* **Folienverlegegerät**)
Verletzungsgefahr risk of injury
Verletzungsrisiko risk of injury
Verlustfaktor 1. dissipation factor *(electrical)*. 2. loss factor *(mechanical)*
Verlustfaktor, dielektrischer dissipation factor
Verlustfaktor, mechanischer loss factor
Verlustleistung power/energy loss
Verlustmodul loss modulus
Verlustströmung leakage flow
Verlustwärme lost heat: **schon bei normaler Betriebsbelastung entsteht Verlustwärme** heat is lost even under normal operating conditions
Verlustwinkel loss angle
Verlustzahl, dielektrische loss index
Verlustziffer loss index
Vermahlung grinding
vermarkten to market
Vermarktung marketing
vermessingt brass plated
vermindert reduced
vermischt mixed
Vermischung mixing
Vermischungsdruck mixing pressure
vernachlässigbar negligible
vernetzbar crosslinkable
Vernetzbarkeit crosslinkability: **Vernetzbarkeit mit reaktiven Lackrohstoffen ist gegeben** can be crosslinked with reactive surface coating resins
Vernetzer curing/crosslinking/vulcanising agent
Vernetzerart type of curing/crosslinking/vulcanising agent
Vernetzerkonzentration curing/crosslinking agent concentration
Vernetzermenge amount of curing/vulcanising agent
Vernetzerpaste curing/vulcanising agent paste
Vernetzerspaltprodukte curing/vulcanising agent decomposition products
vernetzt 1. crosslinked, cured, vulcanised. 2. linked, (inter)connected. 3. networked *(me)*
vernetzt, engmaschig closely crosslinked
vernetzt, schwach loosely crosslinked
vernetzt, stark densely crosslinked
vernetzt, teilweise partly crosslinked
vernetzt, weitmaschig loosely crosslinked
Vernetzung 1. crosslinkage, curing, vulcanisation. 2. networking *(me)*
Vernetzung, peroxidische peroxide crosslinkage/cure
Vernetzungsbedingungen curing/vulcanising conditions
Vernetzungsdichte crosslink density
Vernetzungseffizienz crosslinking efficiency
vernetzungsfähig crosslinkable

Vernetzungsgeschwindigkeit crosslinking/vulcanising/curing rate/speed
Vernetzungsgrad degree of crosslinking
Vernetzungshilfe curing/crosslinking/vulcanising agent
Vernetzungskatalysator crosslinking/curing agent
Vernetzungskomponente curing/crosslinking/vulcanising agent
Vernetzungsmechanismus crosslinking/vulcanising mechanism
Vernetzungsmittel curing/crosslinking/vulcanising agent
Vernetzungsmitteldosierung amount of curing/vulcanising agent
Vernetzungsprozeß crosslinking/curing/vulcanising process
Vernetzungsrate curing/vulcanising rate
Vernetzungsreaktion crosslinking/curing/vulcanising reaction
Vernetzungsstelle crosslink point
Vernetzungsstruktur crosslink structure
Vernetzungssystem curing/crosslinking/vulcanising system
Vernetzungstemperatur curing/vulcanising temperature
Vernetzungsumsatz percentage cure
Vernetzungsverhalten curing/vulcanising behaviour/characteristics
Vernetzungsweise crosslinking/curing mechanism
Vernetzungswirksamkeit crosslinking efficiency
vernickelt nickel-plated
Verordnung directive, ordinance, regulation
Verordnung brennbarer Flüssigkeiten German regulations covering flammable liquids
Verpackung 1. packaging. 2. pack
Verpackungsabfall packaging waste
Verpackungsanlage packaging plant
Verpackungsband packaging tape
Verpackungsbecher tub, cup
Verpackungsbehälter packaging container
Verpackungsfolie packaging film
Verpackungsgebiet packaging sector
Verpackungsgebinde packaging container
Verpackungsgüter products *(to be packaged)*: This translation will be perfectly adequate in context
Verpackungshohlkörper packaging container
Verpackungskleber packaging adhesive
Verpackungsklebstoff packaging adhesive
Verpackungsmaterial packaging material
Verpackungsmittel packaging container
Verpackungsrohstoff packaging material
Verpackungsspritzgießen injection moulding of packaging containers
Verpackungsstraße packaging line
verpackungstechnisch *relating to packaging*: **Die vom verpackungstechnischen Standpunkt aus so wichtigen**

Durchlässigkeitswerte... permeability, which is so important in packaging materials...
Verpackungsteil packaging container
Verpackungsverordnung packaging directive/ordinance
Verpackungswerkstoff packaging material
Verpackungswesen packaging industry
verpastbar paste-making *(PVC resin)*
Verpastung paste mixing
Verpreßbarkeit mouldability
Verpressen 1. press moulding *(grp)*. 2. laminating
verpreßte Dicke pressed thickness *(e.g. of a multi-layer laminate)*
Verpuffung explosion
Verriegelkraft locking force, lock *(im) (see explanatory note under* **Zuhaltekraft**)
verriegelt locked
Verriegelung locking mechanism
Verriegelungseinrichtung locking mechanism
Verriegelungssystem locking mechanism
Verringerung reduction
Verrippung application of ribs
verrottungsbeständig rot-proof
Verrottungsbeständigkeit rot resistance
verrottungsfest rot-proof
Verrottungsfestigkeit rot resistance
Versagen failure, breakdown
Versagen, duktiles ductile failure
Versagen, katastrophales catastrophic failure
Versagen, sprödes brittle failure
versagensauslösend failure-initiating
Versagenskriterien failure criteria
Versagensmechanismus failure mechanism
Versagensprozeß breakdown process
Versammlungsstätte place of assembly
Versand dispatch
Versandlager dispatch store
verschachtelt interlocked, interlocking
verschäumbar expandable, foamable
Verschäumbarkeit foaming characteristics, foamability
Verschäumen foaming, expansion
Verschäumung foaming, expansion
Verschäumung, mechanische mechanical foaming
Verschäumungseigenschaften foaming properties/characteristics
Verscheibung glazing
verschiebbar, axial axially movable
Verschiebeweg stroke
Verschiebung, axiale axial displacement
verschlauft entangled
Verschlaufung entanglement *(of molecule chains)*
verschlechtert deteriorated, made worse
Verschlechterung deterioration
Verschleiß wear (and tear)
verschleißanfällig susceptible to wear
verschleißarm hard wearing, wear resistant
Verschleißarmut hard wearing properties
verschleißbeansprucht subject to wear

verschleißbedingt due to wear
Verschleißbelag wear resistant coating
verschleißbeständig wear resistant
Verschleißbeständigkeit wear resistance
Verschleißbetrag amount of wear
Verschleißbüchse wear resistant bushing
Verschleißeigenschaften wear characteristics
verschleißen to wear out
Verschleißerscheinungen signs of wear
verschleißfest hard wearing, wear resistant
Verschleißfolie wear resistant film
Verschleißfrei wear resistant
verschleißgefährdet likely to be damaged by wear
verschleißgeschützt protected against wear
Verschleißgrad amount of wear
verschleißintensiv wear-intensive, producing a lot of wear
Verschleißmechanismus wear mechanism
verschleißmindernd wear resistant *(this is more acceptable than a literal translation)*
Verschleißminderung reduction of wear: **ein neuer Weg zur Verschleißminderung wurde von F. Braun vorgeschlagen** a new way of achieving wear resistance OR reducing wear has been suggested by F. Braun
Verschleißneigung tendency to become worn: **Die Größe der Verschleißneigung ist abhängig von...** the amount of wear to be expected depends on...
Verschleißplatte wear plate
Verschleißprüfung abrasion test
Verschleißrate rate of wear
verschleißreduzierend wear-reducing
verschleißreduzierende Maßnahmen measures to reduce wear
Verschleißschäden damage due to wear
Verschleißschicht 1. wear resistant coating. 2. wear resistant layer *(if nitrided - see explanatory note under* **Nitrierschicht**)
Verschleißschutz protection against wear
Verschleißstellen areas subject to wear
Verschleißteile parts/components subject to wear
Verschleißtiefe depth of wear
Verschleißverhalten wear characteristics
verschleißverursachend causing wear
Verschleißwiderstand wear resistance
verschleißwiderstandsfähig hard wearing, wear resistant
Verschleißwiderstandsfähigkeit wear resistance
Verschließanlage sealing machine
verschlissen worn
verschlungen entangled
Verschluß closure, cap
Verschlußdüse shut-off nozzle *(im)*
Verschlüsselung encoding *(me)*
Verschlußmechanismus shut-off mechanism *(lm)*
Verschlußnadel shut-off needle *(im)*
Verschlußventil shut-off valve *(im)*

verschmolzen fused
verschmutzt soiled, contaminated, polluted
Verschmutzung contamination, pollution
Verschmutzungsgefahr risk of contamination
Verschmutzungsgrad degree of contamination, amount of dirt *(e.g. in an oil filter or a moulding compound)*; degree of pollution *(e.g of the atmosphere)*
Verschmutzungsüberwachung contamination/pollution monitor
Verschneiden blending
Verschnitt blend: **im Verschnitt mit...** blended with
verschnitten blended
Verschnittharz extender resin/polymer, blending resin
Verschnittkomponente 1. extender *(general term)*. 2. extender plasticiser. 3. extender resin/polymer, blending resin
Verschnittlösemittel diluent
Verschnittmittel 1. extender *(general term)*. 2. extender plasticiser. 3. extender resin/polymer, blending resin. 4. thinner, diluent
Verschraubung screw coupling
verschrotten to scrap
verschweißbar weldable
Verschweißbarkeit weldability
Verschweißen 1. welding. 2. fusing *(e.g. of polymer particles)*
verseifbar saponifiable
Verseifbarkeit saponifiability
verseift saponified
Verseifung saponification
verseifungsbeständig saponification resistant
Verseifungsbeständigkeit saponification resistance
Verseifungsgeschwindigkeit saponification rate
Verseifungsrate saponification rate
Verseifungsresistenz saponification resistance
Verseifungszahl saponification number
versetzt staggered
Versetztstegdornhalter spider with staggered legs *(e, bm)*
Versiegeln sealing
Versiegelungsemulsion sealant emulsion
Versiegelungszeit gate opening time *(im) (see explanatory note and translation example under* **Siegelzeit***)*
Versintern sintering, caking
Versorgung supply (system)
Versorgungseinheit supply unit
Versorgungsleitung feed/supply line
Verspachteln trowelling
Verspritzen extrusion
verspröden to become brittle: **versprödende Wirkung** embrittling effect
Verspröden embrittlement
versprödet embrittled
Versprödung embrittlement
Versprödungserscheinungen signs of embrittlement
versprödungsstabil resistant to embrittlement
Versprödungstemperatur brittleness temperature, brittle point
Versprödungstendenz tendency to become brittle
Versprühen spraying
verstärkend reinforcing
Verstärker amplifier
Verstärkereffekt reinforcing effect
Verstärkerfüllstoff reinforcing filler
Verstärkerruß reinforcing carbon black
Verstärkerstoff reinforcing filler
Verstärkerwirkung reinforcing effect
Verstärkung reinforcement
Verstärkungsadditiv reinforcing filler
Verstärkungseffekt reinforcing effect
Verstärkungsfasern reinforcing fibres
Verstärkungsfüllstoff reinforcing filler
Verstärkungsgewebe reinforcing fabric
Verstärkungsmaterial reinforcing material *(grp)*
Verstärkungsmittel reinforcing agent
Verstärkungsring reinforcing ring
Verstärkungsrippen reinforcing ribs
Verstärkungsstoff reinforcing material
Verstärkungswirkung reinforcing effect
Versteifung stiffening, reinforcement, bracing
Versteifungsrippen reinforcing ribs
Versteifungswirkung stiffening effect
verstellbar adjustable
verstellbar, stufenlos steplessly/infinitely adjustable
Verstellbereich setting range
Verstelleinrichtung adjusting mechanism
Verstellimpuls servo-impulse
Verstellmöglichkeit 1. adjustment mechanism. 2. possibility of adjusting
Verstellmotor servomotor
Verstellpumpe variable displacement pump
Verstellschraube adjusting screw
Verstellung adjustment
Verstopfung clogging
verstrammt stiffened
Verstrammung stiffening *(in consistency)*
Verstrammungseffekt stiffening effect
Verstrecken stretching, drawing *(e.g. of film) (see explanatory note under* **Recken***)*
verstreckt stretched
verstreckt, biaxial biaxially oriented
verstreckt, monoaxial uniaxially oriented
Verstreckungsbruch ductile fracture/failure
Verstreckungsgrad stretch/draw ratio
Verstreckungsverhältnis stretch/draw ratio
Versuch test, trial
Versuchsanlage experimental plant
Versuchsanordnung test set-up/arrangement
Versuchsaufbau experimental set-up
Versuchsausrüstung experimental equipment/machine
Versuchsauswertung interpretation of test results
Versuchsbedingungen test conditions
Versuchsbericht test report

Versuchsdaten experimental/test data
Versuchsdauer test period/duration
Versuchsdurchführung experimental procedure
Versuchseinrichtung test set-up
Versuchsende, nach after/upon completion of the test
Versuchsergebnis test result
Versuchsextruder experimental extruder
Versuchsfehler experimental error
Versuchsfeld test bed
Versuchsformnest experimental/test cavity *(im)*
Versuchsformulierung test formulation
Versuchskabel experimental cable
Versuchskessel pilot reactor
Versuchskosten experimental costs
Versuchsmaßstab pilot plant scale
Versuchsmaterial test substance/material
Versuchsmenge experimental quantity
Versuchsmischer experimental mixer
Versuchsmuster 1. test sample. 2. prototype
Versuchsmustermengen experimental sample quantities
Versuchsparameter test parameter
Versuchsplatte test panel
Versuchsprobe test specimen
Versuchsprodukt experimental product
Versuchsprogramm test programme
Versuchsprotokoll test record
Versuchsreihe test series
Versuchsrezeptur test formulation
Versuchsschnecke experimental screw
Versuchsserie test series
Versuchsspritzungen moulding trials
Versuchsstadium experimental stage
Versuchstemperatur test temperature
Versuchswalzwerk experimental roll mill
Versuchswerkstoff test material
Versuchswerkzeug 1. experimental die *(e)*. 2. experimental mould *(im)*
Versuchswert test result
Versuchszeit test period/duration
Verteilbarkeit dispersibility
Verteileffekt dispersing effect
verteilen 1. to disperse. 2. to distribute
Verteiler 1. runner *(im)*. 2. manifold *(e)*
Verteilerbalken manifold block *(e,im)*
Verteilerblock manifold block *(e,im)*
Verteilerbohrung 1. runner *(im)*. 2. manifold *(e)*
Verteilerfahrzeug delivery van
Verteilerkanal 1. runner *(im)*. 2. manifold *(e)*
Verteilerkanaldüse manifold-type die *(e)*
Verteilerkanalfläche, projizierte projected runner surface area *(im)*
Verteilerkanalquerschnitt runner profile/shape/cross-section *(im)*
Verteilerkreuz 1. sprues and runners *(when referring to thermoplastic scrap)*. 2. feed/runner system *(im)*. 3. radial system of runners *(im)*
Verteilerplatte 1. hot runner plate *(im)*. 2. feed plate *(im)*
Verteilerquerschnitt runner profile/shape/cross-section *(im)*
Verteilerrohr 1. runner *(im)*. 2. manifold *(e)*
Verteilerröhrensystem 1. runner system *(im)*. 2. manifold system *(e, bfe)*. 3. feed system *(im)*
Verteilerspinne 1. sprues and runners *(sr)* *(when referring to thermoplastic scrap)*. 2. feed system, runner system *(im)*. 3. radial system of runners *(im)*
Verteilerstern 1. sprues and runners *(when referring to thermoplastic scrap)*. 2. feed/runner system *(im)*. 3. radial system of runners *(im)*
Verteilerstück manifold block *(e,im)*
Verteilersystem 1. runner system *(im)*. 2. manifold system *(e,bfe)*. 3. feed system *(im)*. 4. distribution system *(e.g. for cooling air)*
Verteilerwerkzeug manifold-type die *(e)*
Verteilgetriebe distributor drive
verteilt, feindispers finely dispersed
Verteilung 1. distribution. 2. dispersion
Verteilungschromatographie partition chromatography
Verteilungsgrad degree of dispersion
Verteilungsgüte efficiency of dispersion
Verteilungskanal 1. runner *(im)*. 2. manifold *(e)*
Verteilungsprobleme dispersion problems
Verteilvorgang dispersing process
Vertiefung depression, hollow, recess
vertikal vertical
Vertikalbauweise vertical (construction): **Ein Schlauchfolienextruder in Vertikalbauweise** a vertical blown film extruder
Vertikaldoppelschneckenextruder vertical twin screw extruder
vertikales Ausschwimmen flooding
vertikales Pigmentausschwimmen flooding
Vertikalextruder vertical extruder
Vertikalhub vertical stroke
Vertikalkraft vertical force
Vertikalschneidmühle vertical granulator *(sr)*
Vertikalspeiseapparat vertical feeder
Vertikalverstellung 1. vertical adjustment. 2. vertical adjusting mechanism
verträglich compatible
Verträglichkeit 1. compatibility. 2. tolerability *(in a medical context, involving the living organism, as in* **Hautverträglichkeit***, q.v.)*
Verträglichkeitseigenschaften compatibility (characteristics)
Verträglichkeitsgrad degree of compatibility
Verträglichkeitsgrenze compatibility limit
Verträglichkeitsmacher compatibiliser
Verträglichkeitsprüfung compatibility test
Verträglichkeitsschwierigkeiten compatibility problems
Verträglichkeitsuntersuchung compatibility test
Verträglichkeitsverhalten compatibility (characteristics/behaviour)
verträglich, begrenzt limited compatibility

Vertragshändlernetz authorised dealer network
vertretbar justifiable, acceptable
Vertrieb sale
Vertriebsförderung sales promotion
Vertriebsorganisation sales organisation
verunreinigt contaminated
Verunreinigung contamination, impurity
Verunreinigungsgefahr risk of contamination
Verunreinigungsgrad degree of contamination/pollution
Verunreinigungsrisiko risk of contamination
Verwaltung management *(e.g. of data) (me)*
Verwaltungskosten administrative costs
Verwaltungsrat board of directors *(composed of* **Vorstand** *(q.v.) and* **Aufsichtsrat** *(q.v.)*
Verwaltungssoftware management software *(me)*
Verweilzeit residence/dwell time
Verweilzeitdauer residence dwell time
Verweilzeitschwanz residual residence time
Verweilzeitspektrum residence time profile
Verweilzeitverteilung residence time distribution
Verwendungszweck (intended) application
Verwerfung warping, deformation
Verwertung recycling, recovery, reclamation
Verwertung, chemische chemical recycling
Verwertung, stoffliche material recycling
Verwertung, thermische energy recovery/recycling
Verwiegeanlage weighing equipment
Verwiegung weighing
Verwindung twisting
verwindungssteif torsion resistant
Verwischgewinde smear/blending device *(e,bm,bfe) (device for obliterating spider lines in the melt flow)*
verwittert weathered
Verwitterung weathering
Verzahnung 1. gear tooth system. 2. gear cutting
verzerrt distorted
verziehen to warp
verzinkt galvanised
Verzinsung interest
Verzögerer retarder, retarding agent
verzögernd retarding
verzögert 1. delayed. 2. retarded
Verzögerungseinrichtung delaying mechanism
verzögerungsfrei without delay, immediately
Verzögerungszeit delay time
Verzug warpage
verzugsarm low-warpage
Verzugsarmut low-warpage properties
Verzugserscheinungen warping, warpage
verzugsfrei non-warping, warp-free
Verzugsfreiheit non-warping properties, freedom from distortion
Verzugsneigung tendency to warp
Verzugsspannung warpage stress
verzugsstabil warp-free
verzweigt branched

Verzweigung branching
Verzweigungsgrad degree of branching
Verzweigungsreaktion branching reaction
Verzweigungsstelle branch point
Verzweigungsstruktur branched structure
VFF-Ruß very fine furnace black
Vibrationseinfülltrichter vibratory feed hopper
vibrationsempfindlich vibration sensitive
Vibrationsrinne vibrating chute
Vibrationsschweißen vibration welding
Vibratordosierung 1. vibratory feeding. 2. vibratory feed unit
vibrierend vibrating
Vicat-Erweichungspunkt Vicat softening point
Vicat-Erweichungstemperatur Vicat softening point
Vicatgrad Vicat softening point
Vicatnadel Vicat indentor
Vicatstift Vicat indentor
Vicat-Wärmeformbeständigkeit Vicat softening point
Vicatwert Vicat softening point
Vicatzahl Vicat softening point
Vickershärte Vickers hardness
Videoaufzeichnungsanlage video recording equipment
Videoband video tape
Videokassette video cassette
Videoplatte video disc
Videoterminal video terminal *(me)*
vieladrig multi-core (cable)
Vielfachstrangdüse multi-strand die *(e)*
Vielfachverbundsicherheitsglas multi-layer safety glass
Vielfachwerkzeug 1. multi-cavity/-impression mould *(im)*. 2. multi-orifice die *(e)*
Vielschichtextrusion coextrusion
Vielschneckenmaschine multi-screw extruder
Vielseitigkeit versatility
vierbindig quadrivalent
Vieretagenspritzen four-daylight moulding *(im)*
Vierfachform four-cavity/-impression mould
Vierfachkopf four-die (extruder) head
Vierfachpunktanschnitt four-point pin gate *(im)*
Vierfachschlauchfolienextrusionsanlage four-die blown film extrusion line
Vierfachschlauchkopf four-parison die *(bm)*
Vierfachschnecke quadruple screw
Vierfachwerkzeug four-cavity/-impression mould *(im)*
Vierholmbauweise four-column construction
vierholmig four-column *(im)*
Vierholmschließeinheit four-column clamp unit
Vierkanallinienschreiber four-channel line recorder
Vierkantmesser square-section knife *(sr)*
Vierkantstab 1. square-section test piece. 2. square-section rod
Vierkantstift square-section pin
Vierkantvollstab square-section solid rod
Vierpunktbelastung four-point loading
Vierpunktbiegeversuch four-point bending test

Vierpunktbiegung four-point bending
Vierpunktkniehebel four point toggle
Viersäulenkonstruktion four-column design
Viersäulenschließe four-column clamping unit *(im)*
Viersäulenschließeinheit four-column clamp unit *(im)*
Vierwalzen-F-Kalander four-roll inverted L-type calender
Vierwalzenglättwerk four-point polishing stack
Vierwalzen-I-Kalender four-roll vertical/ superimposed calender
Vierwalzen-I-Kalender mit schräggestellter Oberwalze four-roll offset calender
Vierwalzenkalander four-roll calender
Vierwalzen-L-Kalander four-roll L-type calender
Vierwalzen-S-Kalander four-type inclined Z-type calender
Vierwalzen-Z-Kalender four-roll Z-type calender
vierwellig four-screw
vierwertig 1. quadrivalent. 2. quadrihydric *(if an alcohol)*
Vierzonenschnecke four-section screw *(e)*
Vinylacetal vinyl acetal
Vinylacetat vinyl acetate
Vinylacetatanteil vinyl acetate content
Vinylacetat-Ethylencopolymerisat ethylene-vinyl acetate copolymer, EVA copolymer: *In German, one distinguishes between EVA copolymers with high and those with low vinyl acetate contents, the former being referred to as* **Vinylacetat-Ethylencopolymerisat**, *abbreviated VAE. No such distinction exists in English, where the above version applies, whatever the vinyl acetate content - and the abbreviation is* **always** EVA
Vinylalkohol vinyl alcohol
Vinyl-Asbestbelag vinyl-asbestos flooring
Vinyl-Asbestfliese vinyl-asbestos tile
Vinyl-Asbestplatte vinyl-asbestos tile
Vinylbutyral vinyl butyral
Vinylcarbazol vinyl carbazole
Vinylchlorid vinyl chloride
Vinylchlorid, monomeres vinyl chloride monomer
Vinylchloridhomopolymer vinyl chloride homopolymer
Vinylchloridpolymerisat vinyl chloride polymer
Vinylchlorid-Vinylacetatcopolymerisat vinyl chloride-vinyl acetate copolymer
Vinylester vinyl ester
Vinylesterharz vinyl ester resin
Vinylether vinyl ether
vinylfunktionell vinyl-functional
Vinylgruppe vinyl group
Vinylharz vinyl resin
Vinylharzlack vinyl paint
Vinylidenchlorid vinylidene chloride
Vinylidenfluorid vinylidene fluoride
Vinylidenfluorid-Tetrafluorethylencopolymerisat vinylidene fluoride-tetrafluoroethylene copolymer

Vinylisobutylether vinyl isobutyl ether
Vinyllactam vinyl lactam
Vinyllaurat vinyl laurate
Vinylmonomer vinyl monomer
Vinylnaphthalin vinyl naphthalene
Vinylpolymerisat vinyl polymer
Vinylpropionat vinyl propionate
Vinylpyridin vinyl pyridine
Vinylpyrrolidon vinyl pyrrolidone
Vinylrest vinyl radical
Vinylseitengruppe vinyl side group
Vinylsilan vinyl silane
Vinyltoluol vinyl toluene
Vinyltrialkoxysilan vinyl trialkoxysilane
Vinyltrimethoxysilan vinyl trimethoxysilane
Visier visor
viskoelastisch viscoelastic
Viskoelastizität viscoelasticity
viskos viscous
Viskose viscose
viskoses Fließen viscous flow
Viskosimeter viscometer
Viskosimetrie viscometry
viskosimetrisch viscometric
Viskosität viscosity
Viskosität, dynamische dynamic/absolute viscosity
Viskosität, kinematische kinematic viscosity
Viskosität, reduzierte reduced viscosity, viscosity number
Viskosität, relative relative viscosity
Viskosität, scheinbare apparent viscosity
Viskosität, spezifische specific viscosity
Viskositätsabfall viscosity decrease/reduction, drop in viscosity
viskositätsabhängig viscosity-dependent
Viskositätsabnahme viscosity decrease/ reduction, drop in viscosity
Viskositätsabsenkung viscosity decrease/ reduction, drop in viscosity
Viskositätsänderungen viscosity changes, changes in viscosity
Viskositätsansatz viscosity equation
Viskositätsanstieg 1. viscosity increase, rise in viscosity. 2. increasing the viscosity
Viskositätsaufbau viscosity increase
Viskositätsbereich viscosity range
Viskositätsberg viscosity peak
Viskositätseinstellung 1. viscosity. 2. viscosity adjustment
viskositätsenkend viscosity-reducing/-lowering
Viskositätserhöhung 1. viscosity increase, rise in viscosity. 2. increasing the viscosity
viskositätserniedrigend viscosity-reducing/-lowering
Viskositätserniedriger viscosity depressant
Viskositätserniedrigung 1. viscosity decrease/ reduction. 2. lowering/reducing the viscosity. 3. drop in viscosity
Viskositätsfunktion viscosity function
Viskositätsgrad viscosity
Viskositätsgradient viscosity gradient

Viskositätsgrenze limiting viscosity, viscosity limit
Viskositätsinhomogenitäten viscosity fluctuations/variations
Viskositätskoeffizient viscosity coefficient
Viskositätskurve viscosity curve
Viskositätsmessung viscosity determination
Viskositätsminderung viscosity reduction/decrease, drop in viscosity
Viskositätsmittel average viscosity
Viskositätsniveau viscosity
Viskositätsregelung viscosity control
Viskositätsregler viscosity regulator
viskositätsstabil having a constant/stable viscosity
Viskositätsstreuung viscosity variation
Viskositätsunterschiede viscosity differences
Viskositätsverhalten viscosity behaviour
Viskositätsverhältnis viscosity ratio, relative viscosity
Viskositätszahl viscosity number
Viskositätszunahme viscosity increase
visuell visual
visuelle Beurteilung visual inspection
VI-Verfahren vacuum injection (process)
Vlies mat, surfacing mat *(grp)*
Vliesstoff bonded/non-woven fabric *(plural:* nonwovens*)*
VMQ *symbol for* vinyl methyl polysiloxane
VMQ-Mischung silicone rubber mix
V-Naht single butt weld
VOC *abbr. of* volatile organic compound
Vol.-% percent by volume, % v/v
vollaromatisch completely aromatic
vollausgelastet run to capacity *(machine)*
Vollautomat fully automatic machine
Vollautomatik fully automatic system
vollautomatisch fully automatic(ally)
vollautomatisiert fully automated
vollelektronisch completely/fully electronic
vollfluoriert completely fluorinated
vollgeschlossen totally enclosed *(machine)*
Vollgummi solid rubber
Vollguß solid casting
vollhalogeniert completely halogenated
vollhydraulisch fully hydraulic(ally)
vollhydriert completely hydrogenated
Vollkunststoffauto all-plastics car
Vollkunststofflehnengestell all-plastics back rest frame
Vollkunststoffrad all-plastics wheel
Vollkunststoffsitz all-plastics seat
Vollkunststoffsitzgestell all-plastics seat frame
Vollkunststoffstoßfänger all-plastics bumper
vollölhydraulisch fully hydraulic(ally)
Vollprofil solid profile
Vollschaumstoffpolster all-foam upholstery
Vollstab solid rod
Vollstrangdüse strand die *(e)*
vollsynthetisch fully/completely synthetic
vollverethert completely etherified
vollverseift fully/completely saponified

Vollwärmeschutz exterior wall insulation
Vollzylinder solid cylinder
Voltmeter voltmeter
Voltzahl voltage
Volumen volume
Volumen, nutzbares effective volume
Volumen, spezifisches specific volume
Volumenabnahme volume decrease
Volumenänderung volume change, change in volume
Volumenanteil volume content
Volumenausdehnung volume/volumetric expansion
Volumenausdehnungskoeffizient coefficient of cubical/volume expansion
Volumenausdehnungskoeffizient, thermischer coefficient of cubical/volume expansion
Volumenbruch volume fraction
Volumendehnung volume expansion
Volumendilatation volume expansion
Volumendosieraggregat volumetric feeder
Volumendosierung 1. volumetric feeding. 2. volumetric feeder
Volumendurchsatz 1. volume throughput, volume flow rate, volumetric flow rate. 2. volumetric extrusion rate *(e)*
Volumenerhöhung volume increase, increase in volume
Volumenfließindex melt volume index
Volumengehalt volume content
volumengleich of equal volume, having the same volume
volumenintensiv bulky
Volumenkompressionsverhältnis compression ratio *(e)*
Volumenkonstanz constant volume
Volumenkontraktion volume shrinkage
Volumenkonzentration volume concentration
Volumenleistung volume throughput
volumenorientiert volume-orientated
Volumenpreis price per unit volume
Volumenpreisvorteil volume price advantage
Volumenprozent percent by volume, % v/v
Volumenschrumpfung volume shrinkage
Volumenschwindung volume shrinkage
Volumenschwund volume shrinkage
Volumensteuerung 1. volume control. 2. volume control unit/device
Volumenstrom volume throughput, volume/volumetric flow rate: *There are occasions when the word must be translated like* **Schmelzestrom**, *e.g.:* **Als Zielsetzung gilt die Abzweigung des Volumenstroms** the aim is to deflect the melt stream
Volumenstromregler volume control unit
Volumenveränderung change in volume
Volumenvergrößerung volume increase
Volumenverkleinerung volume reduction
Volumenverminderung volume decrease
Volumenwiderstand volume resistance
Volumenzunahme volume increase
volumetrisch volumetric

Vorabinformation advance information
Vorabmischen pre-mixing
Voranstrich primer
Vorarbeiten preparatory work
Vorausberechnung advance calculation/ computation
Vorbau front end
vorbehandelt pre-treated
Vorbehandlung pre-treatment
Vorbehandlungsgerät pre-treating unit *(e.g. for film)*
Vorbereitung preparation
vorbeschleunigt pre-accelerated
Vorbeugemaßnahmen precautionary measures
Vorblähen pre-expansion
Vorblaseinrichtung pre-blowing/-expansion device
Vorblasen pre-blowing, pre-expanding
Vorderseite front
vordispergiert pre-dispersed
Vordispergierung pre-dispersion
vordosiert, genau accurately measured out: **eine genau vordosierte Schmelzemenge** an accurately measured amount of melt
Voreilung, mit faster, more quickly: **Walze B läuft gegenüber Walze A mit Voreilung** roll B rotates more quickly than roll A
voreingestellt pre-set
vorexpandiert pre-expanded, pre-foamed
vorfabriziert prefabricated
Vorformeinheit pre-forming unit
Vorformling 1. parison *(bm)*. 2. preform *(grp)*
Vorformling, schlauchförmiger parison *(bm)*
Vorformlingslänge parison length *(bm)*
Vorformlingslängenregelung 1. parison length control *(bm)*. 2. parison length control mechanism *(bm)*
Vorformlingsträger parison support *(bm)*
Vorformlingswand parison wall
Vorformlingswerkzeug parison die *(bm)*
Vorgabe setting
vorgebbar *that which can be pre-set, input, etc*: **Temperaturabsenkung auf einen vorgebbaren Wert** lowering the temperature to a pre-set/pre-determined value
vorgeben to set, to pre-set, to specify, to prescribe, to program
vorgefertigt prefabricated
vorgegeben given, pre-selected, pre-set, specified, prescribed, programmed: **vorgegebene Sollwerte** given setpoints; **vorgegebene Grenzen** prescribed limits; **vorgegebene Steuerungsreihenfolge** programmed control sequence
vorgeheizt pre-heated
Vorgehensweise procedure
Vorgelierkanal pre-gelling tunnel
vorgeliert 1. pre-gelled, partly gelled *(UP, EP resins)*. 2. gelled *(PVC paste)*
Vorgeliertemperatur 1. pre-gelation temperature *(UP, EP resins)*. 2. gelation temperature *(PVC paste)*

Vorgelierung 1. partial gelation *(UP, EP resins)*. 2. gelation *(PVC pastes)*
vorgemischt pre-mixed
vorgereckt pre-stretched
vorgeschaltet upstream
vorgeschäumt pre-foamed, pre-expanded
Vorgeschichte previous history *(i.e. the thermal or mechanical history of a test piece or moulded part)*
Vorgeschichte, thermische thermal history
vorgeschrieben specified, prescribed
vorgesteuert servo-controlled
vorgewählt pre-selected
Vorhärtungstemperatur pre-curing temperature
vorheizen to preheat
Vorheizofen preheating oven
Vorheiztrommel preheating drum
vorherrschende Teilchengröße predominant particle size
Vorhersagen prediction
vorhomogenisieren to pre-homogenise
Vorhub initial stroke
vorimprägniertes Halbzeug prepreg
vorimprägniertes Textilglas prepreg
Vorinformation advance information
Vorjahr previous year
Vorkammer ante-chamber, hot well *(im)*
Vorkammer, Punktanguß mit ante-chamber type pin gate *(im)*
Vorkammerangießtechnik ante-chamber gating
Vorkammeranguß ante-chamber feed system
Vorkammerangußbuchse ante-chamber sprue bush *(im)*
Vorkammerbohrung ante-chamber, hot well *(im)*
Vorkammerbuchse ante-chamber bush *(im)*
Vorkammerdurchspritzverfahren sprueless injection moulding, ante-chamber direct feed injection
Vorkammerdüse ante-chamber nozzle *(im)*
Vorkammerkegel ante-chamber contents, hot well contents *(im)*: **...damit der Vorkammerkegel nicht einfriert...** so that the contents of the hot well do not solidify
Vorkammerpunktanguß ante-chamber pin gate *(im)*
Vorkammerraum ante-chamber, hot well *(im)*
Vorkenntnisse previous experience/knowledge
Vorkommen deposit *(e.g. of mineral oil, natural gas etc.)*
Vorkondensat pre-condensate
Vorkondensation pre-condensation
vorkondensiert pre-condensed
Vorlaufbehälter feed tank
Vorlaufgeschwindigkeit speed/rate of advance
Vorlauftemperatur starting/initial temperature
Vormastizieren pre-mastication
Vormauerziegel facing tile
Vormauerziegelstein facing brick
vormischen to pre-mix
Vormischer pre-mixing unit
Vormischstation pre-mixing station

Vornorm provisional specification
Vorplastifizierschnecke pre-plasticising screw
vorplastifiziert pre-plasticised
Vorplastifizierung 1. pre-plasticisation. 2. pre-plasticising unit *(im)*
Vorplastifizierungsaggregat pre-plasticising unit *(im)*
Vorplastifizierungssystem pre-plasticising system *(im)*
Vorpolymer pre-polymer
Vorpreßling preform *(grp)*
Vorprimern priming, applying a primer
Vorprodukt intermediate
vorprogrammiert pre-programmed
Vorprozeß preceding process
Vorprüfung preliminary test
Vorrat 1. stock. 2. inventory *(me)*
Vorratsbehälter 1. silo *(usually for solids)*. 2. storage tank *(usually for liquids)*. 3. accumulator *(of an hydraulic unit)*
Vorratstank storage tank
Vorratstrichter (feed/material) hopper *(e,im)*
Vorrichtung 1. device, mechanism, arrangement. 2. jig, fixture
Vorsatzscheibe secondary glazing
Vorschäumer pre-foaming unit
Vorschäumkügelchen pre-expanded beads
Vorschaumperlen pre-expanded beads
Vorschäumprozeß pre-foaming (process): **Die Zeit zwischen den einzelnen Vorschäumprozessen** the time between the various pre-foaming stages
Vorschäumvorgang pre-foaming
Vorschrift regulation, directive, ordinance, order, provision
Vorschubautomat automatic feed unit
Vorschubeinheit feed unit
Vorschubgerät feed unit
Vorschubgeschwindigkeit speed/rate of advance
Vorserienfertigung pilot plant production
Vorserienwerkzeug experiemntal mould
Vorsichtsmaßnahmen safety measures/precautions
Vorspanneinrichtung 1. pre-stressing. 2. hydraulic pull-back system *(c)*
Vorspannkabel pre-tensioning cable
Vorspannung 1. pre-stressing 2. hydraulic pull-back system *(c)*
Vorspannventil pressurising valve
vorstabilisiert pre-stabilised
Vorstabilisierung pre-stabilisation
Vorstand managing directors
Vorstandsmitglied, ordentliches full member of the Board of Directors
Vorstandsmitglied, stellvertretendes acting member of the Board of Directors
Vorstandsvorsitzender Chairman (of the Board of Directors)
vorstellbar, verfahrenstechnisch technically feasible
Vorsteuereinheit servo-control unit

Vorsteuerung servo-control
Vorsteuerventil servo-valve
Vorstrich primer
Vorstufe preliminary stage
Vorteile, verfahrenstechnische technical/processing advantages
vortemperieren to pre-heat
Vortest preliminary test/trial
Vortesten preliminary testing
Vortriebskraft forward thrust
Vortrockengerät pre-drying unit
Vortrocknung pre-drying
Voruntersuchung preliminary test
Vorverdichtung initial compression/consolidation
Vorvernetzung scorching, premature vulcanisation
vorverstreckt prestretched
Vorversuch preliminary test
vorwählbar pre-selectable
Vorwahlschalter pre-selector switch
vorwärmen to pre-heat
Vorwärmgerät pre-heating unit
Vorwärmofen pre-heating oven
Vorwärmwalzwerk pre-heating rolls *(c)*
Vorwärmzeit pre-heating period
Vorwärtsbewegung forward movement
Vorwärtsentgasung 1. forward venting. 2. forward venting system
vorzeitig premature
Vorzerkleinerer preliminary granulator, preliminary size reduction unit
Vorzerkleinern preliminary granulation, preliminary size reduction
vorzerkleinert rough-cut *(hard material)*, roughly shredded *(e.g. film scrap)*
Vorzerkleinerungsmühle preliminary granulator, preliminary size reduction unit
Vorzugsfarbe preferred colour
Vorzugsorientierung preferential orientation
Vorzugsprüfkörper preferred test piece/specimen
Votator heat exchanger *(**Votator** is a trade name, but the English version given is preferable)*
VPE abbr. of **vernetztes Polyethylen,** crosslinked polyethylene
VS 1. abbr. of **Vibrationsschweißen**, vibration welding. 2. abbr. of **Verarbeitungsschwindung,** moulding shrinkage
VSG abbr. of **Verbundsicherheitsglas,** laminated safety glass
VTMOS abbr. of vinyl trimethoxysilane
Vulkameter cure-meter
Vulkametrie vulcametry
Vulkanfiber vulcanised fibre
Vulkanisat vulcanisate
Vulkanisateigenschaften vulcanisate properties
Vulkanisation vulcanisation, cure, curing
Vulkanisationsbedingungen vulcanising conditions

Vulkanisationsbeschleuniger vulcanisation accelerator
Vulkanisationschemikalien vulcanising agents
Vulkanisationsdämpfe vulcanisation vapours
Vulkanisationsgeschwindigkeit vulcanising rate/speed
Vulkanisationsgrad degree of vulcanisation
Vulkanisationskatalysator vulcanising agent/catalyst
Vulkanisationsmittel vulcanising/curing agent
Vulkanisationsreaktion vulcanising/curing reaction
Vulkanisationssystem vulcanising system
Vulkanisationstemperatur vulcanising/curing temperature
Vulkanisationstunnel vulcanising tunnel
Vulkanisationsverhalten vulcanising behaviour
vulkanisationsverzögernd vulcanisation-retarding
Vulkanisationszeit vulcanising/cure time
Vulkanisationszone vulcanising section
vulkanisierbar vulcanisable
Vulkanisiermittel vulcanising/curing agent
vulkanisiert vulcanised, cured
Vulkanisierungssystem vulcanising system
Vulkanisierverhalten vulcanising characteristics/behaviour
VZ abbr. of **Viskositätszahl**, viscosity number

W

Waage 1. balance. 2. scales
Waagedosierung 1. weigh feeding. 2. weigh feeder
waagerecht horizontal
Waagerechtextruder horizontal extruder
Wabenkern honeycomb core
Wabenstruktur honeycomb structure
Wachs wax
wachsartig waxy
Wachsausschmelzguß investment casting, lost wax casting
Wachsemulsion wax emulsion
wachsmodifiziert wax-modified
Wachsreste wax residues
Wachstum growth
Wachstumseinflüsse factors influencing growth, factors which influence growth
Wachstumserwartungen growth expectations
Wachstumsgeschwindigkeit growth rate, rate of growth
Wachstumsmarkt growing market
Wachstumsperspektiven growth prospects
Wachstumsrate growth rate
Wägebehälter weighing container
Wägebereich weighing range
Wägedaten weighing data
Wägeeinrichtung weighing equipment
Wägefehler weighing error
Wägemechanismus weighing mechanism
wägen to weigh
Wägeplattform weighing platform
Wägeraum weighing compartment
Wägesystem weighing system
Wägung weighing
Wahlausrüstung optional equipment
wählbar selectable, as required: **Der Programmablauf ist frei wählbar** the program sequence can be selected as required
Wählmöglichkeit *possibility of choosing*: **Durch die Wählmöglichkeit zwischen zwei unterschiedlichen Maschinen** since one can choose one of two different machines
Wahlschalter selector switch
Wahltaster selector switch/key
wahlweise optional, if required, as required: **...zur wahlweisen Betätigung eines pneumatischen Ausstoßers oder einer pneumatischen Kernzugvorrichtung** ...for the operation of a pneumatic ejector or a pneumatic core puller, as required
Wahrscheinlichkeit probability
Walken pressing and squeezing *(e.g. tube of resin or adhesive)*
Walkpenetration worked penetration
Walzauftrag roller application *(e.g. of a paint or adhesive)*
Walze roll
Walzen, unrund laufende eccentric rolls
Walzenabstand nip width
Walzenabzug take-off rolls
Walzenachse roll axis
Walzenanordnung roll configuration
Walzenanstellung 1. roll adjustment *(c)*. 2. nip adjusting gear *(c)*
Walzenauftrag roller application *(e.g. of an adhesive or surface coating)*
Walzenauftragswerk roll applicator
Walzenballen roll face *(c)*
Walzenballenbreite roll face width *(c)*
Walzenballenlänge roll face width *(c)*
Walzenballenmitte roll face centre *(c)*
Walzenballenrand roll periphery *(c)*
Walzenbezug roll covering
Walzenbiegeeinrichtung roll bending mechanism *(c)*
Walzenbombage convex grinding *(c)*
Walzenbreite roll width
Walzendurchbiegevorrichtung roll bending mechanism *(c)*
Walzendurchbiegung roll deflection *(c)*
Walzendurchmesser roll diameter
Walzenextruder planetary gear extruder
Walzenfräser milling cutter
Walzengegenbiegeeinrichtung roll bending mechanism *(c)*
Walzengegenbiegung roll bending *(c)*

walzengetrocknet drum dried
Walzenklebrigkeit tendency to stick to the rolls (*e.g. a rubber compound*)
Walzenlager roll bearing *(c)*
Walzenlagerung roll bearing *(c)*
Walzenlast nip pressure/load *(c)*
Walzenmantel 1. roll surface. 2. roll covering
Walzenoberflächentemperatur roll surface temperature
Walzenpaar pair of rolls
Walzenrakel knife-roll coater
Walzenreibstuhl roll mill, rolls
Walzenrotor roller-type rotor
Walzenschliff roll grinding/contouring *(c)*
Walzenschmelzverfahren hot-melt roller application (*of an adhesive*)
Walzenschrägeinstellung cross-axis roll adjustment *(c)*, axis/roll crossing *(c)*
Walzenschrägstellung cross-axis roll adjustment *(c)*, axis/roll crossing *(c)*
Walzenschrägverstellung cross-axis roll adjustment *(c)*, axis/roll crossing *(c)*
∗**Walzenspalt** 1. nip *(c)*. 2. flight land clearance, inter-screw clearance (*i.e. the clearance between the flight lands of twin screws*)
Walzenspaltdruck nip pressure *(c)*
Walzenspalteinstellung 1. nip setting/adjustment. 2. nip setting mechanism/device *(c)*
Walzenspaltkraft nip pressure *(c)*
Walzenspaltweite nip width
Walzenstuhl roll mill, rolls
Walzentrennkraft nip pressure *(c)*
Walzentrockner drum dryer
Walzenverbiegung roll deflection *(c)*
Walzenverstellung roll adjusting mechanism
Walzenvorspanneinrichtung hydraulic pull-back system *(c)*
Walzenvorspannung hydraulic pull-back system *(c)*
Walzenzapfen roll journal *(c)*
Walzenzylinder barrel (*of a* **Planetwalzenextruder** *q.v.*)
Walzfell sheeted-out compound, strips, milled sheet (*for translation example see under* **Fell**)
Walzfellbildung banding on the mill
Wälzkolbenpumpe generated-rotor pump, gerotor pump
Walzlack coil coating paint/enamel/lacquer
Wälzlager roller/anti-friction bearing
Wälzlagerfett anti-friction bearing grease
Wälzlagerkäfig roller bearing cage
Wälzlagerkranz roller bearing cage
Walzprozeß milling (process/operation)
Wälzreibung rolling friction
Walztemperatur milling temperature
Walztest milling test
Walzwerk roll mill, rolls
Walzzeit milling time

Wandbeanspruchung hoop stress (*of pipes*)
Wandbeanspruchung, zulässige safe working stress (*of pipes*)
Wandbekleidung wall covering/cladding
Wandbelag wall covering/cladding
Wanddicke wall thickness
wanddickenabhängig depending on (the) wall thickness
Wanddickenabweichungen wall thickness variations/deviations
Wanddickenkontrolle 1. wall thickness control. 2. wall thickness control system
Wanddickenmeßgerät wall thickness gauge
Wanddickenmessung 1. wall thickness measurement. 2. wall thickness gauge
Wanddickenprogrammierer wall thickness programming device
Wanddickenprogrammiergerät wall thickness programming device
Wanddickenprogrammierung 1. wall thickness programming. 2. wall thickness programming device
Wanddickenregelgerät wall thickness control unit
Wanddickenregulierung 1. wall thickness control. 2. wall thickness control unit
Wanddickenregulierungssystem wall thickness control mechanism
Wanddickenschwankungen wall thickness variations
Wanddickensollwert required wall thickness
Wanddickensprünge sudden/abrupt changes in wall thickness
Wanddickensteuerung 1. wall thickness control. 2. wall thickness control unit
Wanddickenstreuungen wall thickness variations/deviations
Wanddickenunterschiede wall thickness differences, differences in wall thickness
Wanddickenverteilung wall thickness distribution
Wandern migration
wanderungsbeständig non-migrating
Wanderungsbeständigkeit migration resistance
Wanderungsfestigkeit migration resistance
Wanderungstendenz migration tendency
Wanderungsverhalten migration behaviour
Wandgleiten wall slippage
wandgleitend wall-slipping (*e.g. polymer melt*)
Wandhaften wall adherence/adhesion
wandhaftend wall-adhering (*e.g. polymer melt*)
Wandlampe wall light
Wandler 1. transformer, transducer. 2. converter *(me)*
Wandspachtelmasse wall surfacer
Wandstärken- *see* **Wanddicken-**
Wandverkleidung wall covering/panelling/cladding
Ware, braune brown goods (*consumer electronics such as video recorders, cam-*

∗ *Words starting with* **Walzenspalt-**, *not found here, may be found under* **Spalt-**

corders, fax machines, satellite dishes etc.)
Ware, weiße white goods, kitchen equipment
Warenbahn web
Warenbahnführung web guide
Warenbahnspannung 1. web tension. 2. web tensioning device/mechanism
Warenbahnsteuerung web guide
Wareneingangsbereich incoming goods section
Wareneingangskontrolle 1. incoming goods control. 2. incoming goods control department
Wareneingangsparameter incoming goods quality/properties
Wareneingangsprüfvorschrift incoming goods test specification
Warenhaus department store
Warenlaufrichtung machine direction
Warenname trade name
Warenspannung 1. web tension. 2. web tensioning device/mechanism
Warenzeichen trade mark
warmabbindend hot setting
Warmarbeitsstahl hot worked steel
Warmbehandlung 1. heat treatment. 2. heat ageing
Warmbiegen hot bending
Warmdach single-shell roof
Wärme heat
Wärme, spezifische specific heat
Wärme, strahlende radiant heat
Wärmeabfluß heat loss
Wärmeabfuhr 1. removal/dissipation of heat. 2. cooling
Wärmeabgabe heat dissipation, dissipation of heat
Wärmeableitfett heat sink grease
Wärmeableitpaste heat sink paste
Wärmeableitung heat dissipation
Wärmeableitungsverlust heat dissipation loss
Wärmeabschirmschild heat shield
wärmeabsorbierend heat absorbent
Wärmeabsorption heat absorption
Wärmeabstrahlfläche heat reflecting surface
Wärmeabstrahlung heat reflection
Wärmeaktivierung heat activation
Wärmealterung heat ageing
Wärmealterungsbeständigkeit heat ageing resistance
Wärmealterungsverhalten heat ageing behaviour
Wärmealterungsversuch heat ageing test
Wärmealterungswerte heat ageing properties
Wärmeaufbau heat build-up
Wärmeausdehnung thermal expansion
Wärmeausdehnungsausgleich compensation for thermal expansion
Wärmeausdehnungskoeffizient coefficient of thermal expansion *(usually, but not necessarily, linear expansion)*
Wärmeausdehnungskoeffizient, linearer coefficient of linear expansion
Wärmeausdehnungsverhalten thermal expansion behaviour

Wärmeausdehnungszahl coefficient of thermal expansion *(usually, but not necessarily, linear expansion)*
Wärmeaushärtung heat cure/curing
Wärmeaustausch heat exchange
Wärmeaustauscher heat exchanger
Wärmebeanspruchung exposure to high temperatures, thermal load
Wärmebedarf heat requirement
wärmebehandelt heat treated
Wärmebehandlung 1. heat treatment. 2. heat ageing
Wärmebelastbarkeit thermal endurance
wärmebeständig heat resistant
Wärmebeständigkeit heat resistance
Wärmebilanz heat balance
Wärmedämmeigenschaften thermal insulating properties
wärmedämmemd heat/thermally insulating
Wärmedämmplatte thermal insulation board
Wärmedämmputz heat insulating plaster
Wärmedämmschicht heat insulating layer
Wärmedämmstoff heat/thermal insulating material
Wärmedämmung thermal insulation
Wärmedämmverbundsystem composite thermal insulation system
Wärmedämmvermögen thermal insulation properties
Wärmedauerfestigkeit long-term heat resistance
Wärmedehnbolzen thermal expansion piece
Wärmedehnung thermal expansion
Wärmedehnungszahl coefficient of thermal expansion
Wärmedehnzahl coefficient of thermal expansion
Wärmedissoziierung thermal dissociation
Wärmedurchgang heat transmission/transfer
Wärmedurchgangskoeffizient heat transfer coefficient, coefficient of heat transmission
Wärmedurchgangswiderstand heat transfer resistance
Wärmedurchgangszahl heat transfer coefficient
Wärmedurchlaßkoeffizient heat transfer coefficient
Wärmedurchlaßwiderstand heat transmission resistance
Wärmedurchschlag thermal breakdown
Wärmeeinbringung application of heat
Wärmeeinwirkung heat exposure, action/effect of heat: **Sie sind gegen Wärmeeinwirkung stabilisiert** they are heat stabilised
wärmeempfindlich heat sensitive, affected by heat
Wärmeenergie thermal energy
Wärmeentwicklung heat evolution, evolution of heat
Wärmeentwicklungsgeschwindigkeit heat evolution rate, rate of heat evolution
Wärmeentzug removal of heat
wärmeerzeugend heat-producing

Wärmeerzeugung production of heat
wärmefest heat resistant
Wärmefestigkeit heat resistance
Wärmefluß heat flow
wärmeforciert force dried
wärmeforciert trocknend force drying
wärmeformbeständig heat resistant
Wärmeformbeständigkeit 1. deflection temperature, heat distortion temperature. 2. heat resistance: *Use version 1. in tables and test descriptions, otherwise use version 2.*
Wärmeformbeständigkeit nach ISO R 175 deflection temperature under load according to ISO R 175
Wärmeformbeständigkeit nach Martens Martens heat distortion temperature
Wärmeformbeständigkeit nach Vicat Vicat softening point
wärmeformstabil heat resistant
Wärmefühler thermocouple
wärmegedämmt thermally insulated
wärmehärtend heat/hot curing
Wärmehaushalt heat balance
Wärmeimpulsschweißautomat automatic heat impulse welding instrument
Wärmeimpulsschweißen heat impulse welding
Wärmeimpulsschweißmaschine heat impulse welding machine
Wärmeimpulssiegelung heat impulse welding
Wärmeinhalt heat content, enthalpy
wärmeintensiv very/extremely hot
Wärmeisolation thermal insulation
Wärmeisolator heat insulator
Wärmeisoliereigenschaften heat insulating properties
wärmeisolierend heat insulating
Wärmeisolierung thermal insulation
Wärmekapazität heat capacity
Wärmekontaktschweißen thermal sealing
Wärmelagerung heat ageing
Wärmeleitdüse thermally conductive nozzle *(im)*
wärmeleitend heat-conducting
Wärmeleiter conductor of heat
wärmeleitfähig thermally conductive
Wärmeleitfähigkeit thermal conductivity
Wärmeleitfähigkeitsdetektor thermal conductivity detector
Wärmeleitpaste heat sink paste
Wärmeleittorpedo thermally conductive torpedo *(im)*
Wärmeleitung thermal conductivity, conduction of heat
Wärmeleitungsgleichung thermal conductivity equation
Wärmeleitungsvermögen thermal conductivity
Wärmeleitvermögen thermal conductivity
Wärmeleitzahl thermal conductivity
Wärmemenge amount of heat
Wärmemenge, freigesetzte amount of heat released
Wärmenachbehandlung annealing
Wärmenest heat accumulation: **Die Folge kann sein, daß sich in schlecht gekühlten Werkzeugzonen Wärmenester bilden** this may result in local overheating of insufficiently cooled parts of the mould
Wärmeoxidation thermal oxidation
Wärmepumpe heat pump
Wärmequelle heat source
wärmereaktiv heat reactive
wärmereflektierend heat-reflecting
Wärmereserve heat reserve
Wärmerohr heat pipe
Wärmerückgewinnung heat recovery
Wärmeschild heat shield
Wärmeschockbeständigkeit thermal shock resistance
Wärmeschockverhalten thermal shock resistance
Wärmeschrank drying oven/cabinet, laboratory oven
Wärmeschutz thermal/heat insulation
Wärmeschutzmantel heat insulating jacket
wärmeschutztechnisch *relating to thermal insulation:* **Aus wärmeschutztechnischer Hinsicht...** from the thermal insulation point of view; **wärmeschutztechnische Eigenschaften** thermal insulation properties
Wärmesicherheit heat resistance
Wärmespannung thermal stress
Wärmespeicher heat storage unit
Wärmespeicherkapazität heat storage capacity, heat retention
Wärmespeicherofen storage heater
Wärmespeichervermögen heat storage capacity, heat retention
Wärmesperre heat barrier
wärmestabil heat resistant
Wärmestabilisator heat stabiliser
wärmestabilisiert heat stabilised
Wärmestabilisierung 1. heat stabilisation. 2. heat setting *(of film, fibres etc. after stretching, to prevent shrinkage)*
Wärmestabilisierungsvermögen heat stabilising properties
Wärmestabilität thermal stability
wärmestandfest heat resistant
Wärmestandfestigkeit heat resistance
Wärmestau heat build-up, accumulation of heat
Wärmestrahler 1. radiant heater. 2. health lamp *(e.g. UV or IR lamp)*
Wärmestrahlung heat radiation
Wärmestrom 1. amount of heat: **Der zwischen Einfriergrenze und Abquetschwalzen abzuführende Wärmestrom läßt sich ansetzen zu...** the amount of heat to be dissipated between the frost line and pinch rolls is given by the following equation... . 2. heat flow
Wärmestrombilanz heat balance
Wärmetauscher heat exchanger
Wärmetauschrohr heat exchanger pipe
Wärmetönung heat effect, heat produced, change in temperature

Wärmeträger 1. heat carrier. 2. heat transfer medium
Wärmeträgermedium heat transfer medium
Wärmeträgermittel heat transfer medium
Wärmeträgeröl heat transfer oil: ...**mit Wärmeträgeröl als Temperiermedium** ...with oil as heating medium
Wärmetrennung thermal isolation *(see explanatory note under* **Trennung, thermische***)*
Wärmeübergang heat transfer
Wärmeübergangskoeffizient heat transfer coefficient
Wärmeübergangszahl heat transfer coefficient
Wärmeübertragung heat transfer
Wärmeübertragungseigenschaften heat transfer properties/characteristics
Wärmeübertragungsmechanismus heat transfer mechanism
Wärmeübertragungsmedium heat transfer medium
Wärmeübertragungsmittel heat transfer medium
wärmeunempfindlich unaffected by heat
wärmeverbrauchend heat-consuming
Wärmeverbraucher heat consumer
Wärmeverlust heat/thermal loss
Wärmeverteilung heat distribution
Wärmewert calorific value
Wärmewiderstand thermal resistor, thermistor
Wärmezufuhr 1. heating, application of heat. 2. heat input: **Diese Werte machen deutlich, wie schwierig es wird, das Material über äußere Wärmezufuhr aufzuheizen** these figures show how difficult it is to heat the material from outside; **Die Wärmezufuhr an das System** the heat put into the system
Warmfestigkeit heat resistance
Warm-Feuchtbedingungen conditions of high temperature and humidity
Warmformanlage thermoforming machine/equipment
Warmformautomat automatic thermoforming machine
warmformbar thermoformable
Warmformbarkeit thermoformability
warmformbeständig dimensionally stable at elevated temperatures
Warmformeigenschaften thermoforming properties
Warmformen thermoforming
Warmformmaschine thermoforming machine
Warmformpressen hot press moulding
Warmformtemperatur thermoforming temperature
Warmformverfahren thermoforming (process)
Warmformwerkzeug thermoforming mould
Warmfütterextruder hot feed extruder
Warmgasschnellschweißen high speed hot air welding
Warmgasschweißautomat automatic hot gas welding unit
Warmgasschweißen hot air/gas welding
Warmgasschweißgerät hot gas welding instrument
Warmgasziehschweißung high speed hot air/gas welding
warmgeformt thermoformed
warmgefüttert hot-fed *(extruder)*
warmhärtbar hot/heat curing: **warmhärtbare Preßmasse** thermoset moulding compound
warmhärtend heat/hot curing
Warmhärtung heat/hot curing
Warmkleben hot bonding
Warmlagerung heat ageing
Warmlagerungstemperatur heat ageing temperature
Warmlagerungsversuch heat ageing test
Warmlagerungszeit heat ageing period
Warmluft hot air
Warmluftgebläse hot air blower
Warmlufttrockner hot air drier
Warmpressen hot press moulding, matched metal moulding *(grp)*
Warmpreßlaminat hot press moulded laminate
Warmpreßverfahren hot press moulding, matched metal moulding *(grp)*
Warmtauchverfahren hot dipping (process)
Warmverformung thermoforming, vacuum forming
Warmzugversuch high temperature tensile test
Warneinrichtung warning device
Warnlampe warning light
Warnleuchte warning light
Warnsignal warning/alarm signal
Warnzeichen warning sign
Warte- und Einlegezeit closed assembly time
Wartepositionen points requiring servicing *(on a machine)*: **Der Aufbau gewährleistet eine leichte Zugänglichkeit aller wichtigen Wartepositionan** the construction of the machine ensures that all the important points requiring servicing are easily accessible
Wartezeit, geschlossene closed assembly time
Wartung servicing, maintenance
Wartungsansprüche maintenance requirements: ...**und stellt deshalb minimale Wartungsansprüche** ...and therefore requires only minimum maintenance
Wartungsanweisungen servicing instructions
Wartungsarbeit servicing, maintenance
wartungsarm low-maintenance, requiring little maintenance
Wartungsaufwand *effort required to keep a machine serviced*: **Die Vorteile dieser Antriebsart liegen in der besseren Drehzahlkonstanz und im geringeren Wartungsaufwand** the advantages of this type of drive are that it makes it easier to keep the screw speed constant, as well as making maintenance/servicing easier
wartungsfrei maintenance-free, requiring no maintenance

Wartungsfreiheit freedom from maintenance, requiring no maintenance *(for translation example see under* **Störungsunanfälligkeit***)*
wartungsfreundlich easy to service/maintain
Wartungskosten maintenance/servicing costs
Wartungsmaßnahmen servicing, maintenance
Wartungspersonal maintenance personnel
Wartungstechniker service engineer
Waschbecken wash basin
Waschbeständigkeit washing resistance
Waschbeton exposed aggregate concrete
Wäscher scrubber
Waschküche laundry room
Waschmaschine washing machine
Waschmittellauge detergent solution
Waschprimer wash primer, etch primer
Waschturm scrubbing tower
Wasser, chemisch gebundenes water of crystallisation
Wasserabfluß water outlet
Wasserabsorption water absorption
Wasserabspaltung separation of water
wasserabstoßend water repellent
wasserabweisend water repellent
Wasserabweisung water repellency
Wasseranschluß water inlet
Wasseraufnahme water/moisture absorption
Wasseraufnahmefähigkeit ability to abosrb water
Wasseraufnahmegeschwindigkeit water absorption rate
Wasseraufnahmekapazität water absorptive capacity
Wasseraufnahmevermögen water absorptive capacity
Wasserauslaß water outlet
Wasserauslauf water outlet
Wasseraustritt water outlet
Wasseraustrittsöffnung water outlet
Wasserbad waterbath
Wasserbadkühlung 1. cooling in a water-bath. 2. water-bath cooling unit
Wasserbasis, auf water-based, waterborne *(e.g. paint or adhesive)*
Wasserbau hydraulic engineering
Wasserbäumchen water trees *(an electrical phenomenon)*
Wasserbäumchenbildung water treeing *(an electrical phenomenon)*
Wasserbelastung 1. exposure to water, immersion in water. 2. waterway pollution
wasserbeständig water resistant
Wasserbeständigkeit water resistance
Wasserdampf water vapour, steam
Wasserdampfdiffusion water vapour diffusion
Wasserdampfdiffusionswiderstand water vapour diffusion resistance
Wasserdampfdruck water vapour pressure
wasserdampfdurchlässig water vapour-permeable
Wasserdampfdurchlässigkeit water vapour permeability

wasserdampfgesättigt saturated with water vapour
Wasserdampflagerung exposure to steam, exposure to water vapour
Wasserdampfpartialdruck partial water vapour pressure
Wasserdampfpermeabilität water vapour permeability
Wasserdampfsättigungsdruck saturation vapour pressure
Wasserdampfstabilität water vapour resistance
Wasserdampfteildruck partial (water) vapour pressure
wasserdicht waterproof, watertight
Wasserdichtheit water impermeability, impermeability to water
wasserdispergierbar water dispersible
Wasserdispersionslack emulsion paint
wasserdurchlässig water permeable
Wasserdurchlässigkeit water permeability
Wassereindringverhalten water penetration (behaviour)
Wassereinlagerung immersion in water
Wassereintritt water inlet
Wassereintrittsöffnung water inlet
Wassereliminierung elimination of water *(e.g. during a chemical reaction)*
wasserempfindlich affected by water
Wasserempfindlichkeit sensitivity to water
wasseremulgierbar water emulsifiable
Wasserenthärtungsmittel water softener
Wasserentmischung separation of water *(e.g. in an emulsion paint)*
wasserfest water resistant
Wasserfestigkeit water resistance
Wasserfließweg water flow-way
wasserfrei anhydrous, free from water
wasserfreundlich hydrophilic
wasserführend in contact with water
Wasserführungsrohr water flow-way
Wassergehalt water/moisture content
wassergekühlt water cooled
Wasserglas waterglass, sodium silicate
wasserhaltig containing water, aqueous
wasserhell water-white
Wasserhydraulik water-hydraulic system
Wasserinhalt water/moisture content
Wasser-in-Öl-Emulsion water-in-oil emulsion, W/O emulsion
Wasserkasten radiator tank
Wasserkocher kettle
Wasserkontaktkühlung water cooling ring *(bfe)*
Wasserkonzentration water/moisture content
Wasserkühlung 1. water cooling. 2. water cooling unit/system
Wasserlack waterborne/water-based paint
Wasserlagerung immersion in water
Wasserlagerungsversuch water immersion test
wasserlöslich 1. water soluble. 2. water miscible *(e.g. solvents)*
Wasserlöslichkeit water solubility
Wassermantel water-filled cooling jacket

wassermischbar water miscible, miscible with water
Wassernetz water mains/supply
Wassernetzdruck mains water pressure
Wasserphase aqueous phase
Wasserringgranulierung 1. water-cooled die face pelletisation. 2. water-cooled die face pelletiser *(sr)*
Wasserringpumpe liquid-ring pump
Wasserringvakuumpumpe liquid-ring vacuum pump
Wasserrohr water supply pipe
Wasserrohrleitung water supply line
Wasserrückhaltevermögen water retentivity
Wasserückstand residual moisture
wassersaugend water absorbent
Wassersaugfähigkeit water absorbency
Wassersauggeschwindigkeit water absorption rate
wassersperrend water impermeable
Wasserstand water level
Wasserstoff hydrogen
Wasserstoffabstraktion removal of hydrogen
wasserstoffaktiv hydrogen-active
Wasserstoffakzeptor hydrogen acceptor
Wasserstoffatom hydrogen atom
Wasserstoffbindung hydrogen bond
Wasserstoffbrücke hydrogen bridge
Wasserstoffbrückenbindung hydrogen bridge linkage
Wasserstoffdonator hydrogen donor
Wasserstoffion hydrogen ion
wasserstofffrei hydrogen-free
Wasserstoffsuperoxid hydrogen peroxide
Wasserstrahlschneiden waterjet cutting
Wassertemperiergerät water-fed temperature control unit
wassertemperiert temperature-controlled with water: **Die Schmelze ist wassertemperiert** the melt temperature is controlled by means of water
Wasserüberdruck excess water pressure
Wasserumlauftemperiergerät circulating water temperature control unit
Wasserumwälzeinheit water circulating unit
Wasserumwälzung 1. water circulation. 2. water circulating unit
wasserundurchlässig water impermeable
Wasserundurchlässigkeit water impermeability
wasserunempfindlich unaffected by water
Wasserunempfindlichkeit indifference to water, water resistance, resistance to water
wasserunlöslich water insoluble
wasserverdünnbar water-thinnable
wasservergütet water quenched *(steel)*
Wasserversorgung 1. water supply. 2. water supply unit/system
wasserverträglich miscible with water
Wasserwanderung water diffusion
Wasserwiderstand water resistance
Wasser-Zementfaktor water-cement ratio
Wasserzufluß 1. water inlet. 2. water supply
Wasserzufuhr 1. water inlet. 2. water supply
Wasserzugabemenge amount of water added/to be added
Wasserzugabestutzen water inlet port
Wasserzulauf 1. water inlet. 2. water supply
Wasserzwangsführung forced water circulation
wäßrig 1. aqueous. 2. water-borne/-based *(paint)*
Webart weave, weave pattern
Webband woven film tape
Webestreifen woven film tape
Webteppich woven carpet
Webwaren woven fabrics/materials
Wechsel bill (of exchange)
Wechselautomatik automatic changing mechanism
Wechselbeanspruchung cyclic stress
Wechselbelastung, thermische alternating exposure to heat and cold
Wechselbelastungen, thermische changes in temperature, fluctuating temperatures
Wechselbeziehung interrelation
Wechselbiegefestigkeit flexural fatigue strength
Wechselbiegeversuch flexural fatigue test
Wechselfeldspannung a.c. voltage
Wechselfestigkeit fatigue strength, endurance limit
Wechselklima changing climatic conditions
Wechselknicktest folding endurance test
Wechsellasten cyclic stresses/loads
wechselnde Beanspruchung cyclic loading/stress
wechselnde Belastung cyclic stress/loading
Wechselplatte exchangeable disk *(me)*
Wechselspannung a.c. voltage
Wechselspannungserzeuger a.c. voltage source
Wechselstrom alternating current, a.c.
Wechseltorsion alternating torsion
Wechselwirkung interaction
Wechselwirkungskräfte interactive forces
wegabhängig stroke-dependent
Wegänderung change in position
Wegaufnehmer 1. stroke transducer/sensor. 2. position transducer/sensor. 3. position decoder *(me)*
Wegerfassung stroke measurement
Wegeservoventil directional servo-valve
Wegesteuerung stroke control
Wegeventil directional control valve
Weggeber stroke transducer/sensor, position transducer/sensor
Weggebersystem stroke sensing mechanism/system
Wegmeßgeber stroke transducer/sensor, position transducer/sensor
Wegmeßsystem stroke measuring system
wegschwenkbar swing-back, swing-hinged, hinged, swivel-mounted, swivel-type *(machine unit)*
wegunabhängig stroke-independent
Wegwerfgefäß disposable container

Wegwerfgeschirr disposable crockery
Wegwerfhandschuhe disposable gloves
Wegwerfhandtuch disposable towel
Wegwerfwaren disposables
weich 1. soft. 2. flexible
weich eingestellt with a high plasticiser content *(PVC paste or compound)*
Weich PVC Spritzguß injection moulding of plasticised PVC
Weichblockschaumstoff flexible slabstock foam
Weichcompound plastisised (PVC) compound
weiche Einstellung high-plasticiser formulation *(PVC paste or compound)*
weichelastisch flexible, pliable
Weichextrusion extrusion of plasticised PVC
weichflexibel pliable, flexible
weichfließend easy-flow *(e.g. moulding compound)*
Weichfolie flexible film
weichgemacht 1. plasticised *(PVC)*: **weichgemachte Schläuche und Profile** flexible hoses and profiles. 2. flexibilised *(e.g. epoxies)*
weichgemacht, innerlich internally plasticised
Weichgranulat plasticised (PVC) compound
Weichgummi soft rubber
Weichharz soft resin
Weichheitsgrad softness, flexibility
Weichintegralschaumstoff flexible integral/structural foam
Weichkaolin plastic kaolin
weichkörnig soft-particle
weichmachend 1. plasticising *(plasticiser in PVC)*. 2. softening *(solvents etc. acting on certain materials)*
Weichmacher 1. plasticiser. 2. softener *(txt)*
Weichmacherabsorption plasticiser absorption
Weichmacheranteil plasticiser content, amount of plasticiser
weichmacherarm with a low plasticiser content
Weichmacherart type of plasticiser
Weichmacheraufnahme plasticiser absorption
Weichmacherbedarf required amount of plasticiser: **Der Weichmacherbedarf ist für die niedrigviskosen PVC-Typen geringer** low viscosity PVC resins require less plasticiser
Weichmacherdampfabsaugung plasticiser vapour extractor
Weichmacherdämpfe plasticiser vapours
Weichmacherdosierung 1. addition of plasticiser. 2. amount of plasticiser (added): **Erst mit darüberliegenden Weichmacherdosierungen steigt die Weichheit und Flexibilität** only if more plasticiser is added do softness and flexibility increase
Weichmachereffekt plasticiser effect
Weichmachereinfluß effect of plasticiser
Weichmacherextraktion plasticiser extraction
weichmacherfest plasticiser resistant

Weichmacherflüchtigkeit plasticiser volatility
weichmacherfrei unplasticised
Weichmachergehalt plasticiser content
weichmacherhaltig plasticised
Weichmacherkonzentration plasticiser concentration
Weichmachermenge amount of plasticiser
Weichmachermigration plasticiser migration
Weichmachermischung plasticiser blend
Weichmachermolekül plasticiser molecule
Weichmacheröl lubricating plasticiser
weichmacherreich with a high plasticiser content
Weichmacherresistenz plasticiser resistance
Weichmachersortiment range of plasticisers, plasticiser range
weichmachersperrend acting as a barrier against plasticiser migration
Weichmachertyp type of plasticiser
Weichmacherverdampfung plasticiser evaporation
Weichmacherverlust plasticiser loss
Weichmacherverlustrate rate of plasticiser loss
Weichmacherwanderung plasticiser migration
Weichmacherwanderungsbeständigkeit plasticiser migration resistance
Weichmacherwanderungsgeschwindigkeit plasticiser migration rate
Weichmacherwirksamkeit plasticiser efficiency
Weichmacherwirkung plasticiser action
Weichmachung plasticisation
Weichmachung, äußere external plasticisation
Weichmachung, innere internal plasticisation
Weichmachungsgrad degree of plasticisation
Weichmachungsmittel plasticiser
Weichmachungsvermögen plasticising capacity
Weichmischung plasticised (PVC) compound
Weich-PVC plasticised PVC: *(The expression soft PVC, often seen in translations, is not normal usage and should be avoided)*
Weich-PVC-Artikel flexible PVC product/article
Weich-PVC-Folie flexible PVC film/sheeting
Weich-PVC-Masse plasticised PVC compound
Weich-PVC-Schaum flexible PVC foam
Weichschaum flexible foam
Weichschaumblock flexible slabstock foam
Weichschaumkunststoff flexible plastics foam
Weichschaumstoff flexible foam
Weichsegment soft segment
Weichspritzguß injection moulding of plasticised PVC
Weichverarbeitung processing with plasticiser, processing in plasticised form *(PVC)*
Weichverbund flexible composite
Weißanlauf blushing *(of paint film, due to excessive moisture)*
Weißblech tinplate
Weißbruch stress-whitening
Weißbruchbildung stress-whitening
Weißbrucheffekt stress-whitening effect
Weißbruchzone stress-whitened zone

weiße Ware white goods, kitchen equipment
Weißgeräte white goods *(large household appliances such as refrigerators, dishwashers, etc.)*
Weißgrad whiteness
Weißheitsgrad whiteness
Weißlack white paint/enamel
weißlich whitish
Weißpigment white pigment
weißpigmentiert (pigmented) white
Weißpunkt 1. minimum film forming temperature. 2. powder point *(of an emulsion)*
Weite zwischen den Holmen, lichte distance between tie bars
Weiteranstieg further rise/increase
Weiterbrennen continued burning
weiterentwickelt upgraded, improved: *The word also implies progress, development, etc. as shown in the first example*: **Die Spritzgießmaschinen wurden in den letzten Jahren sprunghaft weiterentwickelt** injection moulding machines have, in recent years, progressed by leaps and bounds; **Die Firma XYZ bietet weiterentwickelte Maschinensysteme an** XYZ are offering an improved range of equipment; **Bestimmte Eigenschaften wurden modifiziert und weiterentwickelt** certain properties have been modified and improved **Unsere Maschinen werden ständig weiterentwickelt** our machines are constantly being developed and improved
Weiterentwicklung further devlopment, improved version, improvement: **Die Spritzgießmaschine ABC ist eine Weiterentwicklung der XYZ Maschine** the ABC injection moulding machine is an improved version of the XYZ machine
Weiterkondensation continued condensation
Weiterreißen tear propagation
Weiterreißfestigkeit tear propagation resistance
Weiterreißkraft tear propagation force
Weiterreißversuch tear propagation test
Weiterreißwiderstand tear propagation resistance
Weiterverarbeiter fabricator, converter
Weiterverarbeitung 1. conversion *(usually of flexible products such as film, sheeting or coated fabrics)*. 2. fabrication *(usually of rigid, semi-finished products such as sheet, pipe, etc.). Occasionally - but very rarely - the word can be translated literally, as in this example*: **Ein Strang von 20 mm ⌀ vergrößerte sich infolge von Gasentwicklung auf das zweifache Volumen, sodaß er für eine Weiterverarbeitung unbrauchbar war** an extrudate, 20 mm in diameter, expanded to twice its volume due to the formation of gas, so that it proved unsuitable for further processing; *Sometimes, the word can be omitted altogether, e.g.* **...bei der Weiterverarbeitung des Compounds auf Extrusionsmaschinen** ...when the compound is subsequently extruded
Weiterverarbeitungsmaschine converting equipment: *This word may sometimes have to be interpreted differently, depending on the context. It could, for example, refer to an extruder, an injection moulding machine etc.*
Weiterverarbeitungsmöglichkeiten possibilities for conversion/fabrication: **Die drei erstgenannten Gruppen bieten gute Weiterverarbeitungsmöglichkeiten** the three first named groups are easy to fabricate *(e.g. by welding, bonding etc.)*
weitgefächert wide-ranging, widespread: **ein weitgefächertes Spektrum von Werkstoffen** a wide range of materials
weitgehend löslich largely soluble
Weithalsbehälter wide-neck container
weitmaschig wide-mesh
weitmaschig vernetzt loosely crosslinked
weitreichend far-reaching
Welle 1. shaft. 2. screw *(e)*
Wellenabdichtung shaft seal
Wellenachse screw axis
Wellendichtring shaft seal
Wellendichtung shaft seal
Wellendrehzahl screw speed *(e)*
Wellenlänge wavelength
Wellenlängenverschiebung shift in wavelength
Wellenstrahlung wave radiation
Wellenstummel shaft end
Wellenzahl wave number
Wellplate corrugated sheet
Wellrohranlage corrugated pipe extrusion line
Weltjahreskapazität annual world capacity
Weltjahresproduktion annual world production
Weltumsatz world turnover
Weltverbrauch world consumption
weltweit world-wide
Wendeflügel pivot casement
wendelförmig spiral
Wendelheizer heating spiral
Wendelkanal spiral channel
Wendelnuten spiral grooves
Wendelstrom spiral melt stream
Wendelströmung spiral flow
Wendelverteiler spiral mandrel (melt) distributor *(e,bfe)*
Wendelverteilerkopf 1. spiral mandrel die *(e)*. 2. spiral mandrel blown film die *(bfe)*
Wendelverteilerwerkzeug 1. spiral mandrel die *(e)*. 2. spiral mandrel blown film die *(bfe)*
Wendepunkt point of inflexion *(of a curve)*
Wendestangen reversing rods *(bfe)*
Wendestangensystem reversing rod mechanism/system *(bfe)*
Wendewickler reverse wind-up (unit)
Werbemittel publicity material
Werksanlage plant
werkseitig in-plant
Werkshalle factory shed
Werksleiter works manager

Werksleitung works management
Werkstatt, mechanische engineering workshop
Werkstätte workshop
Werkstattzeichnung working drawing
Werkstoff material: *when the word precedes the actual name of a material, it must be omitted in translation, e.g.* **Die Eigenschaften des Werkstoffes PVC...** the properties of PVC...; **Die Anforderungen an den Werkstoff Polymerbeton...** demands made on polymer concrete...
Werkstoffabtrag material wear/abrasion
Werkstoffanhäufungen material accumulations
Werkstoffanisotropie material anisotropy
Werkstoffauswahl material selection, choice of material
Werkstoffauswahlprogramm material selection program *(me)*
werkstoffbezogen material-related
Werkstoffdaten material constants
Werkstoffeigenschaften material properties
Werkstoffeinsparung material saving: **Dickenmeßgeräte werden heute zur Verbesserung der Dickentoleranzen und damit zur Werkstoffeinsparung vorgesehen** thickness gauges are today being used to improve thickness tolerances and thereby save material
Werkstoffermüdung material fatigue
Werkstofferwärmung heating up of the material
werkstoffgerecht *correct or suitable for a given material* **bei werkstoffgerechter Gestaltung der Formteile...** if the mouldings have been correctly designed...; *As so often with words ending in* **-gerecht**, *this one, too, can sometimes be omitted , e.g.* **Beitrag zur werkstoffgerechten Verarbeitung von PVC auf Zweischneckenextrudern;** *Here, the word is used to underline the obvious - PVC would hardly be processed "unsuitably". Even the word "correct" would be superfluous, so that the sentence should be rendered as* Notes on the extrusion of PVC on twin-screw extruders
Werkstoffgruppe group of materials
Werkstoffkenndaten material constants
Werkstoffkenngröße material constant/property
Werkstoffkennwert material constant/property
Werkstoffklasse group of materials
Werkstoffkosten material costs
werkstoffmechanisch mechanical **werkstoffmechanische Eigenschaften des Klebstoffes** mechanical properties of the adhesive; **werkstoffmechanisches Verhalten des Stahles** mechanical performance/properties of steel.
Werkstoffparameter material constants
Werkstoffprüflabor material testing laboratory
Werkstoffprüfung materials testing

werkstoffspezifisch material-related, specific to the material
werkstoffrelevant material-related
Werkstoffverhalten material behaviour/performance
Werkstoffversagen material failure
Werkstor factory gate
Werkstück workpiece
Werkstück-Handhabungsautomat automatic workpiece handling device
∗**Werkzeug** 1. (moulding) tool *(general term covering all kinds of moulds and dies used for transforming plastics into finished and semi-finished products).* 2. die *(e).* 3. mould *(im): The word die is sometimes encountered in descriptions of injection moulding machines. This usage is incorrect since this word should only be used in an extrusion context (see BS 1755 Part 2, 1974, No. 2549)*
Werkzeug, formgebendes 1. mould *(im).* 2. die *(e)*
Werkzeug, mehrteiliges multi-part mould
Werkzeug, zweiteiliges split mould *(im)*
Werkzeugabkühlung cooling down of the mould: **Automatisches Abschalten der Werkzeugkühlung bei Zyklusstörung, zur Vermeidung starker Werkzeugabkühlung** automatic switching-off of the mould cooling system if there is an interruption of the moulding cycle, to prevent excessive cooling down of the mould
Werkzeugänderungswünsche required mould modifications
Werkzeuganschlußmaße mould mounting dimensions, platen dimensions
Werkzeuganschlußplan mould mounting diagram
Werkzeugatmung mould breathing
Werkzeugaufbau 1. standard mould unit/base *(im).* 2. mould construction: *The plural form* **(Werkzeugaufbauten)** *can only be translated by version 1.*
Werkzeugaufspannfläche platen area *(im):* **Gute Zugänglichkeit zu der Werkzeugaufspannfläche** easy access to the platens
Werkzeugaufspannmaße mould mounting dimensions, platen dimensions *(im)*
Werkzeugaufspannplatte platen *(im)*
Werkzeugaufspannplatte, bewegliche moving platen *(im)*
Werkzeugaufspannplatte, feststehende fixed/stationary platen *(im)*
Werkzeugaufspannung mould attachment
Werkzeugaufspannzeichnung mould fixing diagram, mould fixing details, standard platen details *(diagram showing pattern of holes in platen for mould mounting)(im)*
Werkzeugauftriebskraft 1. mould opening force *(im).* 2. die opening force *(e)*

∗ *Words starting with* **Werkzeug-**, *not found here, may be found under* **Düsen-, Form** *or* **Formen-**

Werkzeugaufwand *the expense or complexity of a mould or die*: **Das Verfahren erfordert einen erhöhten Werkzeugaufwand mit beweglichen Einsätzen** the method requires more elaborate moulds, with movable inserts
Werkzeugausbau 1. removal of the mould *(im)*. 2. removal of the die *(e)*
Werkzeugausführung mould construction
Werkzeugauslegung 1. mould design. 2. mould design department
Werkzeugbau 1. mould making/construction. 2. toolmaking department
Werkzeugbauer toolmaker, mould maker
Werkzeugbauhöhe mould space/height
werkzeugbauseitig *relating to mould construction* **werkzeugbauseitig höhere Anforderungen** greater demands from the mould construction point of view
Werkzeugbauteil mould component
werkzeugbedingt 1. due to the mould *(im)*. 2. due to the die *(e)*
Werkzeugbefestigung mould attachment
Werkzeugbeheizung 1. mould heating. 2. mould heating system
Werkzeugbelag plate-out
Werkzeugbelüftung mould venting *(im)*
Werkzeugbeschädigungen damage to the mould: **...um Werkzeugbeschädigungen zu vermeiden** to prevent the moulds being damaged
Werkzeugbewegung mould movement: **Die Werkzeugbewegung geschieht waagerecht** the mould moves horizontally
werkzeugbewegungsabhängig mould movement-related
Werkzeugbreite die width *(e)*
Werkzeugdatenkatalog record/list of mould data
Werkzeugdorn mandrel, core *(e,bm)*
Werkzeugdruck 1. cavity pressure *(im)*. 2. back pressure *(e)*
Werkzeugdruckverlauf cavity pressure profile *(im)*
Werkzeugdruckverlust drop in cavity pressure
Werkzeugdurchbiegung mould deflection *(im)*
Werkzeugdüse (extrusion) die
Werkzeugdüsenplatte fixed/stationary platen *(im)*
Werkzeugeinarbeitung (mould) cavity *(im)*
Werkzeugeinbau mould mounting: **Der Werkzeugeinbau ist leicht zu bewerkstelligen** the mould is easily mounted
Werkzeugeinbauhöhe mould space/height *(im)*
Werkzeugeinbauhöhenverstellung 1. mould height/space adjustment 2. mould height/space adjusting mechanism
Werkzeugeinbaulänge mould space/height *(im)*
Werkzeugeinbaumaße mould mounting dimensions, platen dimensions
Werkzeugeinbauraum mould space/height *(im)*
Werkzeugeinrichter mould setter *(im)*
Werkzeugeinrichtezeit mould setting time *(im)*

Werkzeugeinsatz mould insert *(im)*
Werkzeugelemente, standardisierte standard mould units
Werkzeugentlüftung mould venting *(im)*
Werkzeugetage daylight *(im)*
werkzeugfallende Teile parts made with the mould used for actual production purposes
Werkzeugfüllgeschwindigkeit mould filling speed
Werkzeugfüllung mould filling: **Da bei der Herstellung verzugsarmer Formteile eine gleiche Werkzeugfüllung wichtig ist...** since, in order to obtain low-warpage mouldings the mould must be filled evenly...
Werkzeugfüllungsgrad amount of material in the mould
Werkzeugfüllvorgang mould filling operation *(for translation example see under* **Formfüllvorgang***)*
Werkzeugfüllzeit mould filling time, injection time *(im)*
werkzeuggebunden mould-dependent *(see explanatory note under* **Maße, werkzeuggebundene***)*
werkzeuggebunden, nicht mould-independent *(see explanatory note under* **Maße, nicht werkzeuggebundene***)*
Werkzeuggegendruck die back pressure *(e)*
Werkzeuggeometrie die geometry *(e)*
Werkzeuggeschwindigkeit mould opening and closing speed
Werkzeuggesenk (mould) cavity *(im)*
Werkzeuggestaltung 1. mould design *(im)*. 2. die design *(e)*: **Bei der Werkzeuggestaltung...** when designing the mould/die
Werkzeuggewicht mould weight: **erhöhtes Werkzeuggewicht** a heavier mould
Werkzeuggriff tool handle
Werkzeuggrundplatte platen *(im)*
Werkzeughälfte mould half
Werkzeughälfte, auswerferseitige ejector (mould) half, moving mould half *(im)*
Werkzeughälfte, bewegliche moving mould half *(im)*
Werkzeughälfte, düsenseitige fixed/stationary mould half *(im)*
Werkzeughälfte, feststehende fixed/stationary mould half *(im)*
Werkzeughälfte, schließeitige moving mould half *(im)*
Werkzeughälfte, spritzseitige fixed/stationary mould half *(im)*
Werkzeughalteplatte die plate *(e)*
Werkzeugharz tooling resin
Werkzeugheizmedium mould heating medium/fluid
Werkzeugherstellungskosten tooling costs
Werkzeughöhe mould space/height *(im)*
Werkzeughöhenverstelleinrichtung mould height adjusting mechanism

Werkzeughöhenverstellung 1. mould height adjustment: **Die Werkzeughöhenverstellung geschieht von Hand** the mould height is adjusted manually. 2. mould height adjusting mechanism
Werkzeughohlraum (mould) cavity
Werkzeughohlraumoberfläche (mould) cavity surface
Werkzeughohlraumtiefe (mould) cavity depth
Werkzeughohlraumvolumen (mould) cavity volume
Werkzeughohlraumwandung (mould) cavity wall
Werkzeughöhlung (mould) cavity
Werkzeuginnendruck cavity pressure *(im)*
werkzeuginnendruckabhängig cavity pressure-dependent, depending on the cavity pressure
Werkzeuginnendruckaufnehmer cavity pressure transducer
Werkzeuginnendruck-Meßwertaufnehmer cavity pressure transducer
Werkzeuginnendruckverlauf cavity pressure profile
Werkzeuginnenraum (mould) cavity
Werkzeuginneren, im inside the mould, in the (mould) cavity
werkzeugintern inside the mould
Werkzeugkasten tool box
Werkzeugkavität (mould) cavity
Werkzeugkennlinie die characteristic
Werkzeugkern (mould) core *(im)*
Werkzeugkonstrukteur mould designer
Werkzeugkonstruktion mould design: **Schon bei der Werkzeugkonstruktion muß die am Spritzling auftretende Schwindung berücksichtigt werden** shrinkage of the moulded part must be taken into account already when the mould is being designed
Werkzeugkontur mould cavity
Werkzeugkontureinsatz (mould) cavity insert
Werkzeugkonzept 1. mould design *(im)*. 2. die design *(e)*. 3. tooling concept
Werkzeugkopf 1. extruder head, die head. 2. die *(see explanatory note under* **Kopf**)
Werkzeugkörper die body *(e)*
Werkzeugkosten tooling costs
Werkzeugkühlkanalanlage mould cooling (channel) system
Werkzeugkühlmedium mould cooling medium
Werkzeugkühlsystem mould cooling system
Werkzeugkühlung 1. mould cooling. 2. mould cooling system
Werkzeuglängsachse longitudinal mould axis
Werkzeuglippen die lips *(e)*
Werkzeugmacher toolmaker, mould maker
Werkzeugmängel mould defects
Werkzeugmaschine machine tool
Werkzeugmaße mould dimensions
Werkzeugmaße, formgebende mould impression dimensions
Werkzeugmittelteil centre part of the mould
Werkzeugnest (mould) cavity

Werkzeugnormalien standard mould units
Werkzeugoberfläche cavity/mould surface
Werkzeugoberflächentemperatur cavity/mould surface temperature
Werkzeugoberteil top mould half
Werkzeugöffnung 1. opening of the mould: **...bis kurz vor der Werkzeugöffnung** ...until shortly before the mould is opened. 2. die gap/orifice *(e)*
Werkzeugöffnungsbewegung mould opening movement
Werkzeugöffnungsgeschwindigkeit mould opening speed
Werkzeugöffnungshub mould opening stroke
Werkzeugöffnungshubbegrenzung mould opening stroke limiting device
Werkzeugöffnungshubeinstellung 1. mould opening stroke setting 2. mould opening stroke setting mechanism/device
Werkzeugöffnungskraft mould opening force
Werkzeugöffnungsvorgang mould opening movement
Werkzeugöffnungsweg mould opening stroke
Werkzeugparallelführung die land *(e)*
Werkzeugpark 1. available moulds *(im)*. 2. available dies *(e)*
Werkzeugplatte mould plate *(im): This word is sometimes wrongly used instead of* **Werkzeugaufspannplatte**, *in which case it must be translated as* platen. **Werkzeugplatte** *is a steel plate or block containing a cavity or a core, whereas* **Werkzeugaufspannplatte** *is that part of the injection moulding machine to which the mould halves are attached*
Werkzeugplattenabstand daylight *(im)*
Werkzeugraum 1. (mould) cavity. 2. mould space
Werkzeugreinigungsvorrichtung mould cleaning device
Werkzeugrohling mould unit/blank
Werkzeugrückdruck die back pressure *(e)*
Werkzeugrückhub mould return stroke *(im)*
Werkzeugsatz set of tools
Werkzeugschließbewegung mould closing movement *(im)*
Werkzeugschließeinheit clamp(ing) unit *(im)*
Werkzeugschließgeschwindigkeit mould closing speed *(im)*
Werkzeugschließkraft 1. clamp(ing) force. 2. locking force *(im) (see explanatory note under* **Schließkraft**)
Werkzeugschließplatte moving platen *(im)*
Werkzeugschließsicherung mould safety mechanism *(im)*
Werkzeugschließsystem (mould) clamping mechanism *(im)*
Werkzeugschließzylinder clamping cylinder *(im)*
Werkzeugschlitten mould carriage *(im)*
Werkzeugschluß 1. closing the mould: **Bei Werkzeugschluß** when the mould closes. 2. (mould) clamping mechanism *(im)*

Werkzeugschluß, sanfter 1. careful/gentle closing of the mould. 2. gentle mould clamping mechanism *(for translation example see under* **Werkzeugschonung**)
Werkzeugschnellspannvorrichtung high speed mould clamping mechanism/device *(im)*
Werkzeugschnellwechselsystem high speed mould changing system
werkzeugschonend without damaging the mould, gently: **werkzeugschonende Schließbewegung der beweglichen Werkzeugaufspannplatte** gentle closing of the moving platen to prevent damage to the mould
Werkzeugschonung careful/gentle treatment of the mould, mould protection: **Werkzeugschonung durch sanften Werkzeugschluß** thanks to the gentle clamping mechanism, the mould is not damaged
Werkzeugschutz mould safety mechanism
werkzeugseitig on the mould: **werkzeugseitig angebracht** attached to the mould
Werkzeugsicherheitsbalken mould safety mechanism
Werkzeugsicherung 1. mould safety/protection: **erhöhte Werkzeugsicherung** increased mould protection. 2. mould safety mechanism
Werkzeugsicherungsdruck reduced mould clamping pressure *(to prevent damage to the mould)*
Werkzeugspannplatte platen *(im)*
Werkzeugstahl tool steel
Werkzeugstandzeit mould life
Werkzeugsteifigkeit mould rigidity
Werkzeugtauchkante vertical flash face
werkzeugtechnisch *relating to the mould:* **werkzeugtechnische Einzelheiten** mould details
Werkzeugteilungsebene (mould) parting surface/line
Werkzeugtemperatur mould temperature
Werkzeugtemperaturfühler mould thermocouple
Werkzeugtemperaturregelung 1. mould temperature control. 2. mould temperature control system
Werkzeugtemperaturverteilung mould temperature distribution
Werkzeugtemperierung 1. mould temperature control. 2. mould temperature control system
Werkzeugtoleranzen mould tolerances
Werkzeugträger 1. mould carrier *(e.g. in a carousel-type blow moulding machine).* 2. platen *(im)*
Werkzeugträgerplatte platen *(im)*
Werkzeugträgerschlitten moving platen *(im)*
Werkzeugträgerseite, bewegliche moving platen *(im)*
Werkzeugträgerseite, feste fixed/stationary platen *(im)*

Werkzeugtrennebene (mould) parting surface/line *(im)*
Werkzeugtrennfläche (mould) parting surface/line *(im)*
Werkzeugtrennkante mould parting line *(im)*
Werkzeugtrennmittel (mould) release agent
Werkzeugtrennverhalten (mould) release properties
Werkzeugumgestaltung re-designing of the mould *(im)*/die *(e)*
Werkzeugumrüstung 1. mould changing 2. mould changing system/mechanism
Werkzeugunterteil bottom mould half
Werkzeugversatz mould misalignment
Werkzeugverschleiß mould wear
Werkzeugwand cavity/mould wall
Werkzeugwandtemperatur cavity/mould wall temperature
Werkzeugwandung cavity/mould wall
Werkzeugwartung und Instandhaltung mould servicing and maintenance
Werkzeugwechsel 1. mould changing: **...gute Zugänglichkeit bei einem Werkzeugwechsel** ...easy access if the mould has to be changed; **Der Werkzeugwechsel ist sehr einfach** moulds are very easily changed. 2. mould changing system
Werkzeugwechselablauf mould changing operation
Werkzeugwechselzeit mould changing time: **Die Werkzeugwechselzeit ist 65 Minuten** the time required to change the mould is 65 minutes
Werkzeugwerkstoff mould material
Werkzeugwiderstand die resistance *(e)*
Werkzeugzeichnung mould drawing
Werkzeugzentrierbohrung mould centring hole
Werkzeugzentrierung mould centring device *(im)*
Werkzeugzuhaltedruck locking force, lock *(im)*
Werkzeugzuhaltekraft locking force, lock *(im)*
Wert 1. figure, value: **Derartig niedrige Werte können mit geschnittenen Fasern nicht erreicht werden** such low figures cannot be achieved with chopped fibres. 2. property
Werte, elektrische electrical properties
Werte, mechanische mechanical properties
Werte, thermische thermal properties
Werteabfall deterioration of properties
Wertebild (general) properties **gutes mechanisches Wertebild** good mechanical properties
Werteniveau (general) properties
Wertigkeit valency
wertmäßig in terms of money
Wertpapiere securities
Wertproduktivität productivity in terms of value
Wertsiegel quality seal
Wertstoff re-usable/recoverable material *e.g. in household rubbish. The comments made under* **Werkstoff** *apply also here, e.g.* **Wie bei den Wertstoffen Glas und Papier kommt**

einer gezielten Sortierung der Kunststoffe erhebliche Bedeutung zu. The proper sorting of plastics is just as important as that of glass and paper.
Wertstoffrecycling recycling of usable materials
wesentlich, verfahrenstechnisch technically important, important from the processing point of view
Wettbewerb competition
wettbewerbsfähig competitive
Wettbewerbsfähigkeit competitiveness
Wettbewerbsprodukt competitive product
wetterbeständig weather resistant
Wetterbeständigkeit weathering resistance
wetterecht weather resistant, weatherproof
Wetterechtheit weathering fastness (of pigments)
wetterfest weather resistant, weatherproof
Wetterfestigkeit weathering resistance
Wetterverhältnisse 1. weather conditions. 2. weathering conditions
Wheatstone-Meßbrücke Wheatstone bridge
∗**Wickel** 1. reel. 2. dolly (c) (see explanatory note under **Puppe**)
Wickelanlage wind-up (unit), winder
Wickelautomat automatic winder
Wickelautomatik automatic winding system
Wickeldorn mandrel (grp) (used in filament winding; the first part of the word need not be translated since its meaning will invariably be clear from the context)
Wickeldurchmesser reel diameter
Wickeleinheit wind-up (unit), winder
Wickelgeschwindigkeit winding/wind-up speed
Wickelgut material being/to be wound up
Wickelharz filament winding resin
Wickelhülse core (c, bfe)
Wickelkern mandrel (grp); for explanatory note see under **Wickeldorn**)
Wickelkonzeption wind-up design (bfe)
Wickelkopf coil end
Wickelkörper winding mandrel
Wickelmaschine 1. filament winding machine. 2. wind-up (unit)
Wickeln filament winding
Wickelrobot filament winding robot
Wickelrohr filament wound pipe
Wickelrovings rovings for filament winding (grp)
Wickelschuß filament wound section
Wickelspannung 1. reel tension. 2. winding tension
Wickelstation wind-up station
Wickelstelle wind-up station
Wickeltechnik filament winding (grp)
Wickelträger coil former
Wickeltrommel wind-up drum
Wickelverfahren filament winding (process)
Wickelvorrichtung wind-up (unit), winder
Wickelwelle wind-up shaft
Wickelwerk 1. filament winding machine. 2. wind-up (unit)
Wickelwinkel winding angle (grp)
Wickelzug 1. reel tension. 2. winding tension
Wickler wind-up (unit), winder
Wicklertyp type of wind-up/winder
Widerlager support
Widerlagerabstand distance between supports
widersprüchlich contradictory
Widerstand resistance
Widerstand, elektrischer electrical resistance
Widerstand, spezifischer resistivity
Widerstand zwischen Stöpseln insulation resistance
widerstandsbeheizt resistance heated
Widerstandsbeiwert drag coefficient/factor
Widerstandsdraht resistance wire
widerstandsfähig resistant, durable
Widerstandsfähigkeit resistance
Widerstandsfühler resistance thermocouple
Widerstandsheizband resistance band heater
Widerstandsheizelement resistance heater
Widerstandsheizkörper resistance heater
Widerstandsheizung resistance heater
Widerstandskraft (power of) resistance
Widerstandsmoment moment of resistance
Widerstandstemperaturfühler resistance thermocouple
Widerstandsthermometer resistance thermometer
Wiederanfahren re-starting (a machine)
wiederaufbereiten to reprocess, to reclaim, to recondition
Wiederaufheizzeit re-heating time, time required for re-heating
wiederaufladbar rechargeable
Wiederaufschmelzen re-melting
Wiederaufspannen re-clamping, re-mounting (e.g. a mould)
wiederaufwärmen to re-heat
Wiederaufwärmstrecke re-heating section
Wiederbeschaffungswert replacement value
Wiedereinrichten re-setting
wiedereinstellen to re-set
Wiedererwärmung re-heating
Wiedergewinnung recovery
wiedergewonnen recovered, reclaimed
wiederholbar repeatable
Wiederholbarkeit repeatability
wiederholgenau reproducible
Wiederholgenauigkeit repeatability, reproducibility
Wiederholungseinheit repeat unit
Wiederholungsmeßreihen repeat test series
Wiederholungsmessung repeat test/determination
Wiederholungsprüfung repeat test
Wiederinbetriebnahme re-starting (a machine)
Wiederverarbeitbarkeit recyclability
Wiederverarbeiten reprocessing, recycling
wiederverschließbar re-sealable

∗ *Words starting with* **Wickel-***, not found here, may be found under* **Aufwickel-**

wiederverwendbar recyclable, re-usable
Wiederverwendbarkeit recyclability, re-usability
Wiederverwendung recycling, re-use
wiederverwertbar re-usable, recyclable
Wiederverwertung recycling
Wiederverwertung, energetische energy recovery/recycling
Wiederverwertung, sortenreine recycling of sorted/segregated/ separated waste
Wiederverwertung, stoffliche material recycling
Wiegebereich weighing range
Wiegegut material to be weighed, material being weighed
willkürlich arbitrary
Winchesterlaufwerk Winchester disk drive *(me)*
Winchesterplatte Winchester disk *(me)*
Winddruck wind pressure
Windgeräusch noise made by wind
Windgeschwindigkeit wind velocity
Windkanal wind tunnel
Windlast wind load/force: **durch Windlast auftretende Spannungen** stresses caused by strong winds
Windleitteil spoiler
Windrichtung wind direction
Windschutzscheibe windscreen
Winkel 1. angle. 2. elbow *(fitting)*
Winkel, tote dead spots
Winkelgeschwindigkeit angular velocity
Winkelkopf crosshead (die) *(e)*
Winkelprobe angle test piece
Winkelspritzkopf crosshead (die) *(e)*
Wirbelbett fluidised bed
Wirbelschicht fluidised bed
Wirbelschichtpyrolyse fluidised bed pyrolysis
Wirbelschichtreaktor fluidised bed reactor
Wirbelschütten fluidised pouring
Wirbelsintergerät fluidised bed coater
Wirbelsintern fluidised bed coating
Wirbelsintern, elektrostatisches electrostatic fluidised bed coating
Wirbelstrom eddy current
Wirbelwirkung vortex effect
Wirkleistung, aufgenommene effective power/ energy
Wirkprinzip mode of action
wirksam effective
Wirksamkeit effectiveness
Wirkstoff active substance/ingredient
Wirkstoffgehalt active substance content
Wirkstoffkonzentration active substance concentration
Wirksubstanz active substance/ingredient
Wirkungsgrad efficiency: **...hat einen höheren Wirkungsgrad als...** ...is more efficient than...
Wirkungsmechanismus mechanism, mode of action
Wirkungsspektrum effective range
wirkungsstark powerful
Wirkungsweise mechanism, mode of action
Wirkwaren knitted fabrics/materials

wirr random: **die Fasern liegen in wirrer Anordnung vor** the fibres are randomly distributed
Wirrfasergehalt random fibre content
Wirrfaserverstärkung random fibre reinforcement
wirtschaftlich low-cost, cost-effective, economic(al), profitable **aus wirtschaftlichen Gründen** for economic reasons
wirtschaftlich machbar economically viable
Wirtschaftlichkeit cost-effectiveness, economic advantages, economic viability, economic aspects, profitability, efficiency: *Since this word covers a wide range of ideas associated with productivity, high outputs, low cost etc., one can often expand the translation, e.g.* **Ein weiterer Beitrag zur Erhöhung der Wirtschaftlichkeit** another contribution towards greater efficiency and increased profits; **der Wirtschaftlichkeit wegen** for economic reasons
Wirtschaftlichkeitsaspekte economic aspects, economic points of view
Wirtschaftlichkeitsbetrachtung profitability study, cost-benefit study
Wirtschaftlichkeitsgründe economic reasons/ considerations
Wirtschaftlichkeitsüberlegungen economic considerations
Wirtschaftsklima economic climate
Wirtschaftspotential economic potential
Wirtschaftsrezession economic recession
Wirtschaftszweig sector of the economy
Wirtsmedium host medium
Wismutoxychlorid bismuth oxychloride
Witterungsbedingungen weathering conditions
witterungsbeständig weather resistant
Witterungsbeständigkeit weathering resistance
Witterungseinflüsse weathering influences
witterungsfest weather resistant
witterungsstabil weather resistant
Witterungsstabilität weathering resistance
Witterungsverhalten weathering behaviour
Witterungsverhältnisse weather conditions
WL 1. *abbr. of* **Wellenlänge**, wavelength. 2. *abbr. of* **Wärmelagerung**, heat ageing
WLD *abbr. of* **Wärmeleitfähigkeitsdetektor**, thermal conductivity detector
WLF-Gleichung WLF equation *(the initials stand for the names of the three men who first proposed the equation - Williams, Landel and Ferry)*
WLF-Kurve WLF curve
WLF-Wert WLF figure/value
Wochenschaltuhr weekly time switch
Wöhlerkurve Wöhler curve
Wöhlerlinie Wöhler curve
Wöhlerversuch Wöhler test
Wohnanlage housing estate
Wollastonit wollastonite
Wolle 1. wool. 2. nitrocellulose
WOM *abbr. of* weatherometer
Wortlänge word length *(me)*

Wortmarke trade mark
Wortprozessor word processor *(me)*
WOT *abbr. of* **Werkzeug-Oberflächentemperatur:** mould surface temperature, cavity surface temperature
Wulst roll/bank of molten plastic, bead of plastic material
Wundbenzin surgical spirit
würfelförmig cube-shaped, diced
Würfelgranulat diced/cubed granules
Würfelschneider dicer, strip pelletiser *(sr)*
Würstchenspritzguß jetting *(im) (see also explanatory note under* **Freistrahl**)
Wursthülle sausage casing
wurzelfest root resistant, resistant to root penetration
Wurzelfestigkeit root resistance, resistance to root penetration.
WVS *abbr. of* **Wärmedämmverbundsystem,** composite thermal insulation system
w.w. *abbr. of* **wahlweise,** optional, if required, as required. *For translation example see under* **wahlweise**
WZ-Faktor water-cement ratio
WZT *abbr. of* **Werkzeugtemperatur,** mould temperature
WZ-Wert water-cement ratio

X

Xenonbogen xenon arc
Xenonbogenstrahler xenon lamp
Xenonbogenstrahlung xenon radiation
X-Naht double butt weld
Xylenol xylenol
Xylol xylene

Y

YI-Wert yellowness index
Yoghurtbecher yogurt/yoghurt cup

Z

zäh 1. tough. 2. viscous
Zähbruch tough fracture/failure, ductile fracture/failure
zähelastisch ductile, tough and resilient
zähfließend viscous
zähflüssig viscous
Zähflüssigkeit high viscosity
zähhart hard and tough
Zähigkeit 1. toughness, strength. 2. impact strength/resistance. 3. viscosity
Zähigkeit, kinematische kinematic viscosity
Zähigkeitserhöhung increase in strength
Zähigkeitsverbesserung improvement in toughness, improvement in impact strength: **Dieser Modifikator zeigte die höchste Zähigkeitsverbesserung.** This modifier proved to be the most effective for improving/increasing impact strength
Zähigkeitsverhalten 1. strength characteristics 2. viscosity characteristics
Zähigkeitsverlust loss of strength
Zähigkeitswert 1. toughness. 2. viscosity
Zählbaugruppe counting module *(me)*
Zahlenbeispiel numerical example
Zahlenmittel number average
Zahlentaste numeric key *(me)*
Zahlenwert 1. numerical value. 2. figure
Zahlenwertgleichung numerical equation
Zahlenzeilen lines of numbers *(me)*
Zählfunktion counting function
Zählimpuls counting impulse *(me)*
Zählmodus counting mode *(me)*
Zählwerk counter, counting mechanism
zähmodifiziert toughened, impact modified
Zähmodifizierung impact modification
Zahnflanke tooth surface
Zahnkette sprocket chain
Zahnkettenrad sprocket wheel
Zahnkettentrieb sprocket chain drive
Zahnkranz gear ring
Zahnrad gear, gear wheel
Zahnradgetriebe gear drive
Zahnradmotor gear motor
Zahnradpumpe gear pump
Zahnradrohling gear wheel blank
Zahnraduntersetzung gear reduction
Zahnraduntersetzungsgetriebe reduction gear
Zahnritzel pinion
Zahnscheibenmühle toothed disc mill
Zahnspachtel spreader/spreading comb
Zahnstange (gear) rack
Zahnstangenantrieb rack-and-pinion drive
Zahntraufel spreader/spreading comb
Zäh-Sprödbruchübergang ductile-brittle failure transition point
Zäh-Sprödübergang rubber-glass transition
zähsteif tough and rigid
Zähverbesserung improvement in impact strength: **Diese Schlagzähmodifikatoren**

bewirken nur eine Zähverbesserung im Raumtemperaturbereich. These impact modifiers only increase impact strength at room temperature
zähviskos high-viscosity, viscous
Zapfendurchbiegung (roll) journal deflection *(c)*
Zarge surround
Zehnerpotenz power of ten
Zeichenbrett drawing board
Zeichenebene two-dimensional plane **in die Zeichenebene abwickeln** to convert into a layflat
Zeichnen, rechnerunterstütztes computer aided drafting, CAD
Zeichnungsvorschrift specification drawing
Zeit, offene open assembly time
zeitabhängig time-dependent
Zeitabhängigkeit time dependence
Zeitabschnitt interval
Zeitabstand interval
Zeitaufwand time required: **geringer Zeitaufwand** little time is needed **Solche Berechnungen sind mit sehr wenig Zeitaufwand durchzuführen** Such calculations can be done very quickly
zeitaufwendig time consuming
Zeitbaustein time module *(me)*
Zeitbedarf (amount of) time required
zeitbezogen time-related
Zeit-Biegewechselfestigkeit flexural fatigue strength
Zeitbruchkurve creep rupture curve
Zeitbruchlinie creep rupture curve
Zeitdauer time, period
Zeitdehngrenze creep stress: *This somewhat misleading word is now obsolete, having been replaced by* **Zeitdehnspannung** *(Stoeckhert, Kunststoff-Lexikon 7th ed. p.568)*
Zeitdehnlinie creep curve
Zeitdehnspannung creep stress
Zeitdehnverhalten creep behaviour
Zeiteinfluß effect of time
Zeiteinheit unit of time
Zeitfaktor time factor
Zeitfestigkeit fatigue endurance
Zeitgeber timer, timing device
Zeitgebersystem timing mechanism
zeitintensiv time consuming
Zeitintervall interval
Zeitkontrolle timer, time control device
zeitlich time-dependent/-related, over a period
zeitlich steuerbar time-controlled
zeitlicher Temperaturverlauf time-temperature relationship
zeitlicher Verlauf variation with time *(e.g. of a parameter) (for translation example see under* **Verlauf, zeitlicher***)*
Zeitmaßstab time scale
Zeitmesser timer, timing device
Zeitmodul time module *(me)*
zeitoptimal 1. time saving. 2. correctly timed
Zeitplanung time planning

Zeitpunkt moment
zeitraffend accelerated *(e.g. tests)*
Zeitraffertest accelerated test
Zeitrafferversuch accelerated test
zeitraubend time consuming
Zeitraum period, interval
Zeitregelung time controller
Zeitrelais time relay
Zeitschalter time switch
Zeitschaltuhr time switch
Zeitschaltwerk time switch
Zeitschritt time interval
Zeitspanne period, interval
Zeitspannung creep stress
Zeitspannungslinie creep curve
zeitsparend time-saving
Zeitstandanlage creep test machine
Zeitstandbeanspruchung creep stress
Zeitstandbiegefestigkeit flexural creep strength
Zeitstandbiegeversuch flexural creep test
Zeitstandbruch creep fracture/failure
Zeitstanddiagramm creep diagram
Zeitstanddruckverhalten compressive creep behaviour
Zeitstanddruckversuch compressive creep test
Zeitstandfestigkeit 1. creep strength. 2. creep rupture strength *(of pipes)*
Zeitstandfestigkeitkurve creep strength curve
Zeitstandfestigkeitslinie creep strength curve
Zeitstandfestigkeitsschaubild creep diagram
Zeitstandinnendruckversuch long-term pressure test, long-term failure test under internal hydrostatic pressure *(test used for plastics pipes)*
Zeitstandkurve creep curve
Zeitstandniveau time-to-failure *(expressed in hours)*
Zeitstandprüfung creep test
Zeitstandschaubild creep diagram
Zeitstandverhalten creep behaviour, long-term behaviour/ performance
Zeitstandversuch creep test
Zeitstandwert time-to-failure *(expressed in hours)*
Zeitstandzugbeanspruchung tensile creep stress, long-term tensile stress
Zeitstandzugbelastung tensile creep stress, long-term tensile stress
Zeitstandzugfestigkeit tensile creep strength
Zeitstandzugprüfung tensile creep test
Zeitstandzugverhalten tensile creep behaviour
Zeitstandzugversuch tensile creep test
Zeitsteuerung time controller
Zeituhr timer, timing device
zeitunabhängig independent of time
zeitunkritisch *this implies that time is not a critical factor.* **Für zeitunkritische Anwendungen** for applications where time is of secondary/minor importance OR where time is unimportant
Zeitverlauf (passage of) time: **Der Zeitverlauf des Einspritzvorganges ist variabel**

Zeitverlust   **zerkleinern**

einstellbar the time taken for injection can be varied
Zeitverlust time wasted/lost
Zeitverzögerungsmethode time-lag method
Zeitvorwahlschalter time pre-selection switch
Zeitwechselfestigkeit endurance/fatigue limit
Zeitwechselfestigkeitsversuch fatigue test
Zelldurchmesser cell diameter
zellenförmig cellular
Zellengefüge cell structure
Zellenstruktur cell structure
Zellgefüge cell structure
Zellglas cellulose film, cellophane: *Although the last named word started life as a trade name, it has now become part of everyday English*
Zellgröße cell size
Zellgrößenverteilung cell size distribution
Zellgummi foam rubber
zellig cellular
Zellorientierung cell orientation
Zellpolyethylen cellular/expanded polyethylene, polyethylene foam
zellregulierend cell-regulating
Zellregulierungsmittel cell regulator
Zellstabilisator foam stabiliser
Zellsteg cell wall
Zellstoff cellulose
Zellstruktur cell structure
zellular cellular
Zellulose- *see* **Cellulose-**
Zellwachstum cell growth
Zellwand cell wall
Zementestrich cement screed
Zementmischung cement mix
Zementmörtel cement mortar
Zementputz cement rendering
Zementschlamm laitance
Zementschlämme laitance
Zementspachtel cement surfacer
Zementstein 1. hardened cement. 2. artificial stone
Zementverflüssiger cement plasticiser
zementverträglich cement-compatible
Zementzumischung addition of cement
zentral central, centralised
zentral angebunden centre-gated, centre-fed *(im)*
zentral angespeist centre-fed *(e, bm, bfe)*
zentral angeströmt centre-fed *(e, bm, bfe)*
Zentralabteilung central department
Zentralanguß central gating, centre feed *(im)*
Zentralausdrückstift central ejector pin *(im)*
Zentralauswerfer central ejector *(im)*
Zentralbereich central sector
Zentralcomputer central processing unit, CPU, main frame (computer)
Zentrale central processor *(me)*
Zentraleinheit 1. central unit *(general term)* 2. central processing unit, CPU, main frame (computer)
Zentraleinspeisung centre feed *(e,im)*
zentralgespeist centre-fed *(e,bm,bfe)*

Zentralhydraulikanlage central hydraulic system
Zentralkühlwasserverteilung central cooling water manifold
Zentralprozessor central processing unit, CPU, main frame (computer)
Zentralrechner central procesing unit, CPU, main frame (computer)
Zentralschmieranlage central lubricating unit
Zentralschmierung central lubricating system
Zentralschnecke main screw
Zentralschneckenextruder multi-screw extruder: *This is a machine made by Krauss-Maffei, described in Kunststoffe 65(1975)12, p.791, operating with a* **Zentralschnecke** *(q.v.) and two* **Nebenschnecken** *(q.v.)*
Zentralspeisung centre feed *(e,im)*
Zentralspindel main screw
zentralumströmt centre-fed *(e,bm,bfe)*
zentralverteilend centre-gated, centre-fed *(im)*
Zentralwelle 1. central shaft. 2. main screw *(e)*
Zentralwickler centre winder *(bfe)*
zentrierbar capable of being centred: **Die Düse ist leicht zentrierbar** the die can be easily centred
Zentrierbohrung centring hole
Zentrierbuchse centring bush
Zentrierbund centring bush
Zentrierelement centring element
Zentriergerät centring device
Zentrierhülse centring bush
Zentrierkonus centring cone
Zentrierring 1. register/locating ring *(im)*. 2. die ring *(e)*
Zentrierschraube centring screw
Zentrierung centring device
Zentriervorrichtung centring device
Zentrifugalabscheider centrifugal separator
Zentrifugalkraft centrifugal force
Zentrifugalpumpe centrifugal pump
Zentrifugaltrockner centrifugal dryer
Zentrifuge centrifuge
zentrifugieren to centrifuge
Zentrumswickler central winder
zerbrechen to break
zerbrechlich fragile
Zerfall decomposition
zerfallen to break up, to dissociate, to disintegrate, to decompose
Zerfallgeschwindigkeit decomposition rate, rate of decomposition
Zerfallsprodukt decomposition product
Zerfallsprozeß decomposition process
Zerfallsreaktion decomposition reaction
Zerfallstemperatur decomposition temperature
Zerkleinerer 1. granulator, shredder *(for reclaiming plastics scrap)*. 2. pelletiser *(for converting extruded strands into pellets)*. 3. size reduction unit *(general term)*
zerkleinern 1. to granulate *(plastics scrap)*. 2. to pelletise *(extruded strands)*. 3. to shred, to cut up, to break down

307

Zerkleinerungsaggregat 1. granulator, shredder. 2. pelletiser. 3. size reduction unit. *(see explanatory note under* **Zerkleinerer***)*
Zerkleinerungsanlage 1. granulator, shredder. 2. pelletiser. 3. size reduction unit. *(see explanatory note under* **Zerkleinerer***)*
Zerkleinerungselmente knives, granulating elements
Zerkleinerungsmaschine 1. granulator, shredder. 2. pelletiser. 3. size reduction unit. *(see explanatory note under* **Zerkleinerer***)*
Zerkleinerungsmühle 1. granulator, shredder. 2. size reduction unit *(see explanatory note under* **Zerkleinerer***)*
Zerkleinerungswucht impact force *(of a granulator rotor)*
zerlegbar capable of being taken apart
zerlegen to take apart
Zerreißfestigkeit tear strength, ultimate tensile strength
Zerreißkraft tensile strength at break
Zerreißmaschine tensile testing machine
Zerreißprobe tensile (test) specimen
Zerreißprüfmaschine tensile testing machine
Zerrüttungsprüfung destructive test
Zerrüttungsuntersuchung destructive test
Zersetzung decomposition
Zersetzungpunkt decomposition temperature
Zersetzungsbereich decomposition range
Zersetzungsbeschleuniger kicker *(in foaming)*
Zersetzungserscheinungen signs of decomposition
Zersetzungsgeschwindigkeit decomposition rate
Zersetzungsmechanisms degradation mechanism
Zersetzungsprodukt decomposition product
Zersetzungsreaktion decomposition reaction
Zersetzungstemperatur decomposition temperature
Zersetzungsverhalten decomposition behaviour
Zerspanbarkeit machinability
Zerspanung machining
zerstäubt atomised
Zerstäubung atomisation
zerstörend destructive *(test)*
Zerstörung destruction
zerstörungsfrei non-destructive *(test)*
zerstörungssicher vandal resistant/proof
Zerteileffekt breaking-down effect *(of solids)*
zerteilen to break down *(e.g. solids)*
Zerteilung breaking down *(e.g solids, agglomerates etc.)*
Zerteilvorgang breaking-down process
Zertifikat certificate
Z-Form four-roll Z-type calender
Ziegelmauerwerk brickwork
Ziegelwand brick wall
Zieglerkatalysator Ziegler catalyst
Ziegler-Natta-Katalysator Ziegler-Natta catalyst
Ziehblenden sizing/draw plates *(e) (used to calibrate very thin tubing)*

Ziehdüse high speed welding nozzle
Ziehen 1. thermoforming. 2. calendering. 3. pultrusion *(grp)*
Ziehklinge trowel, putty knife, scraper
Ziehscheiben sizing/draw plates *(used to calibrate very thin tubing)*
Ziehschleifen honing, superfinishing
Ziehschweißen high speed welding
Ziehtiefe depth of draw *(t)*
Ziehverfahren pultrusion *(grp)*
Zielgröße 1. target quantity/parameter. 2. dependent variable
Zielprodukt target product
Zielsetzung purpose, aim, objective
Ziergitter (radiator) grille
Zierleiste decorative strip/moulding
Ziffernanzeige digital/numeric display *(me)*
Zifferntastatur numeric keypad *(me)*
Zimmertemperatur room temperature
Zinkborat zinc borate
Zinkchlorid zinc chloride
Zinkchromat zinc chromate
Zinkchromatgrundierung zinc chromate primer
Zinkdibutyldithiophosphat zinc dibutyl dithiophosphate
Zinkdithiocarbamat zinc dithiocarbamate
Zinkgelb zinc yellow
Zink-Gießlegierung zinc casting alloy
zinkhaltig containing zinc
Zinkoxid zinc oxide
Zinkoxychlorid zinc oxychloride
Zinkphosphat zinc phosphate
Zinkphosphat-Eisenoxidgrundierung zinc phosphate-iron oxide primer
Zinkpigment zinc dust
zinkreich zinc rich
Zinkseife zinc soap
Zinkstaub zinc dust
Zinkstaubfarbe zinc-rich paint
Zinkstaubformulierung zinc-rich paint
Zinkstaubgrundierung zinc-rich primer
Zinkstaubpigment zinc dust
Zinkstaubprimer zinc-rich primer
Zinkstearat zinc stearate
Zinksulfid zinc sulphide
Zinkweiß zinc white
Zinncarboxylat tin carboxylate
zinnfrei tin-free
Zinnmaleinat tin maleinate
Zinnmerkaptid tin mercaptide
zinnorganisch organo-tin
Zinnstabilisator tin stabiliser
zinnstabilisiert tin stabilised
Zinsertrag interest
Zirkel compass
Zirkonacetylactonat zirconium acetylacetonate
Zirkonoxychlorid zirconium oxychloride
Zirkonpropionat zirconium propionate
Zirkonsilikat zirconium silicate
Zirkonverbindung zirconium compound
zirkulieren to circulate
Zitronensäure citric acid

Zitronensäureester citrate
Z-Kalander four-roll Z-type calender
Zonen, tote dead spots
Zsb. *abbr.* **of Zusammenbau,** assembly
Zubehör ancillary equipment, accessories
Zubehöreinrichtungen ancillary equipment, accessories
Zubehörprogramm range of accessories
Zubehörteile ancillary equipment, accessories
zudosieren to add, to incorporate
Zudosierung addition, incorporation
Zufahren closing *(of mould)*
Zufahrkraft 1. clamping force. 2. locking force *(im)*
Zufahrsicherung mould safety mechanism
Zuführband feed conveyor
Zuführeinrichtung feed equipment
Zuführkanal 1. runner *(im)*. 2. manifold *(e)*. 3. feed channel *(general term)*
Zuführmaschine feed unit, loader
Zuführöffnung feed opening/port/throat *(e)*
Zuführschnecke feed screw *(e)*
Zuführschneckenpaar feed twin screws *(e)*
Zuführungsautomat automatic feeder, automatic feed unit
Zuführungsbohrung feed opening/port/throat
Zuführungskabel power supply cable
Zuführungskanal 1. runner *(im)*. 2. manifold *(e)*. 3. feed channel *(general term)*
Zuführungswalze feed roll
Zuführvorrichtung feeder, feed unit
Zuführwalze feed roll
Zug 1. tension 2. train
Zugabemenge amount added
Zugabeöffnung feed opening/port/throat *(e)*
Zugabetrichter (feed/material) hopper *(e,im)*
Zugang access
Zugang, bequemer easy access
zugänglich accessible
Zugänglichkeit accessibility, access, ease of access
Zuganker tension rod
Zugarbeit tensile energy
zugbeansprucht under tensile stress
Zugbeanspruchung tensile stress
Zugbeanspruchungsrichtung direction of tensile stress
zugbelastet under tensile stress
Zugbelastung tensile stress
Zugbruchlast tensile breaking stres
Zugdeformation tensile deformation
Zugdehnung elongation
Zugdehnung bei Streckgrenze elongation at yield
Zug-Dehnungsdiagramm tensile stress-elongation curve
Zugdehnungsversuch tensile test
Zugdose tensile force transducer
zugeführte Leistung power input
Zugeigenspannung internal tensile stress
zugelassen approved, authorised, permitted, allowed

zugelassen, amtlich officially approved
Zug-Elastizitätsmodul tensile modulus of elasticity
Zug-E-Modul tensile modulus of elasticity
zugeordnet dedicated: **zugeordneter Rechner** dedicated computer
zugeschnitten cut-to-size
Zugfestigkeit tensile strength
Zugfestigkeit bei Bruch tensile strength at break, ultimate tensile strength
Zugfestigkeitsänderung change in tensile strength
zugfrei draught-free
Zuggeschwindigkeit tensile testing rate/speed
zügige Beanspruchung dynamic stress
zügige Belastung dynamic stress
Zugkraft tensile force
Zugkraftaufnehmer tensile force transducer
Zug-Kriechmodul tensile creep modulus
Zug-Kriechversuch tensile creep test
Zuglast tensile stress
Zugmodul tensile modulus
Zugprobe tensile specimen, tensile test piece, tensile bar
Zugprobekörper tensile specimen, tensile test piece, tensile bar
Zugprobestäbchen tensile specimen, tensile test piece
Zugprüfmaschine tensile testing machine
Zugprüfstab tensile specimen/bar, tensile test piece
Zugprüfung tensile test
Zugriff access *(me)*
Zugriffszeit access time *(me)*
Zugscherbelastung tensile shear stress
Zugscherfestigkeit 1. shear strength *(of a bonded joint)*. 2. tensile shear strength
Zugscherversuch tensile shear test
Zugschwellast repeated tensile stress
Zugschwellbelastung repeated tensile stress
Zugschwellbereich region of repeated tensile stress **im Zugschwellbereich** under conditions of repeated tensile stress
Zugschwellfestigkeit tensile fatigue strength
Zugschwellversuch tensile fatigue test
Zugschwingungsversuch tensile fatigue test
Zugspannung tensile stress
Zugspannung bei Streckgrenze tensile stress at yield
Zugspannungsänderungen changes in tensile stress
Zugspannungscraze tensile craze
Zugspannungs-Dehnungskurve tensile stress-elongation curve
Zugspannungsmaximum maximum tensile stress
Zugspannungsregelung 1. web tension control. 2. web tension control unit/mechanism/device
Zugstab tensile specimen, tensile test piece, tensile bar
Zugumformen plug-assist forming using vacuum

Zugverformung tensile deformation, deformation under tensile stress
Zugverformungsrest tension set
Zugverhalten tensile behaviour
Zugversuch tensile test
Zugwalze tension roll
Zugwerk tensioning unit/device
Zug-Zeitstandfestigkeit tensile creep strength
Zugzone tensile zone
Zuhaltedruck locking force, lock *(im)*
Zuhalteeinrichtung locking mechanism *(im)*
Zuhaltekraft locking force, lock *(im)*:
  Spritzgießmaschine mit 250 Mp **Zuhaltekraft** 250 Mp lock injection moulder; *The word can also be translated like* **Schließkraft** *(q.v.)*
Zuhaltekraftreserve reserve locking force *(im)*
Zuhaltemechanismus (mould) locking mechanism *(im)*
Zuhaltezylinder locking cylinder *(im)*
Zukunftsaussichten prospects
zukunftsorientiert with great possibilities for the future, having great potential
Zukunftsperspektiven prospects
zulässig permissible
zulässige Beanspruchung safe working stress *(of pipes)*
zulässige Wandbeanspruchung safe working stress *(of pipes)*
Zulassung, amtliche official authorisation
zulaufend converging
Zulauföffnung feed opening/port/throat *(e)*
Zuleitung feed line
Zuleitungsrohr feed pipe
Zulieferant supplier
Zulieferer supplier
Zuluft incoming air
Zumischung addition *(of a substance to a mix)*
Zunderschicht scale
Zündfähigkeit ignitability
Zündgruppe ignition group
Zündpunkt iginition point
Zündquelle source of ignition
Zündsystem ignition system
Zündtemperatur ignition point
Zündung ignition
Zündversuch ignition test
Zuordnungsdaten allocation data *(me)*
zurückfahrbar retractable
zurückgezogen retracted
zurückziehbar retractable
Zusammenbacken caking *(of powdered materials when damp)*
zusammendrücken to compress
Zusammendrückung compression
zusammenfallen to collapse
Zusammenflußlinie weld/flow line *(see explanatory note under* **Bindenaht***)*
Zusammenflußnaht weld/flow line *(see explanatory note under* **Bindenaht***)*
Zusammenflußstelle weld/flow line *(see explanatory note under* **Bindenaht***)*

zusammenfügen to join
Zusammenhaltskraft cohesive force
Zusammenlaufen coalescence
Zusammenschmelzen fusing (together)
Zusammensetzung composition, constitution
Zusammenspiel interplay
Zusammenstellungszeichnung assembly drawing
Zusatz 1. additive. 2. addition
Zusatzaggregat ancillary/supplementary unit
Zusatzaktien additional shares
Zusatzausgang extra output *(me)*
Zusatzausrüstung ancillary/supplementary equipment
Zusatzbaustein extra/additional module
Zusatzbauteil extra part, ancillary component
Zusatzdaten additional data
Zusatzdosieraggregat additional metering unit
Zusatzdraht welding rod *(w)*
Zusatzeinheit 1. extra/additional module *(e)*. 2. extra unit
Zusatzeinrichtung additional/extra/ancillary equipment
Zusatzextruder ancillary/supplementary extruder
zusatzfrei additive-free, free from additives
Zusatzfunktion additional/extra function
Zusatzgerät additional/extra/ancillary unit
Zusatzheizband extra heater band
Zusatzinformation additional information
Zusatzkosten extra/additional cost
Zusatzlogik additional logic element *(me)*
Zusatzmenge amount added, amount to be added, amount: **...selbst in geringen Zusatzmengen** ...even if added in small amounts
Zusatzmittel additive
Zusatzprüfungen additional tests
Zusatzschnecke ancillary/supplementary screw
Zusatzsteuerungen supplementary/additional controls
Zusatzstoff 1. additive. 2. filler. 3. welding rod *(w)*
Zusatzweichmacher additional plasticiser
Zusatzwerkstoff welding rod *(w)*
Zuschlag additive
Zuschlagkomponente additive
Zuschlagstoff additive
Zuschnitte cut-to-size pieces, blanks *(of sheet stock, prepregs etc.)*
Zustand 1. condition, state. 2. status *(me)*
Zustand, im geformten in the moulded state
Zustand, im ungeformten in the unmoulded state
Zustandsanzeige status display *(me)*
Zustandsbericht status report *(me)*
Zustandsgleichung equation of state
Zustandsinformation status information *(me)*
Zustandsmeldung status signal/report *(me)*
Zustandsprotokoll status record *(me)*
Zuströmkanal feed channel
Zuwachsrate growth rate

Zwangsbeschickung 1. forced feed. 2. forced feed system
Zwangsdosiereinrichtung forced feed unit
Zwangsdurchlüftung forced ventilation
Zwangsentlüftung forced venting
Zwangsförderung 1. forced conveying. 2. forced conveying system
Zwangsförderungseffekt forced conveying effect
Zwangsführung forced circulation *(of cooling air or medium)*
Zwangsfütterung 1. forced feed. 2. forced feed system
zwangsgeschmiert pressure-lubricated
Zwangskühlung intensive cooling
Zwangslauferhitzer forced circulation heater
zwangsläufig *denotes something being done forcibly*: **Kammern in denen der Kunststoff zwangsläufig vorwärts bewegt wird** compartments in which the polymer is forced forwards; **zwangsläufig fördernd** with forced conveying action
Zwangslaufprinzip forced flow/circulation principle
Zwangsluftzirkulation forced air circulation
Zwangsströmung forced circulation
Zwangsumlaufsystem forced circulation system
Zwangsverriegelung automatic locking mechanism
Zwangszuführeinrichtung forced feed mechanism
zweckentsprechend suitable, appropriate
zweiachsig biaxial
zweibasisch dibasic
zweibindig divalent
Zweiblockcopolymer diblock copolymer
zweidimensional two-dimensional
zweidirektional two-directional
Zweietagenspritzen double-daylight moulding
Zweietagenwerkzeug double-daylight mould, three-plate mould, three-part mould *(im); (see explanatory note under* **Etage***)*
Zweifachanspritzung two-point gating *(im)*
Zweifachblaskopf twin-parison die *(bm)*
Zweifachextrusionskopf twin die (extruder) head
Zweifachform two-cavity/-impression mould
Zweifachkopf twin die (extruder) head
Zweifachschlauchkopf twin-parison die *(bm)*
Zweifachspritzblaswerkzeug two-cavity injection blow mould
Zweifachspritzgießwerkzeug two-cavity injection mould
Zweifachwerkzeug 1. two-cavity mould *(im)*. 2. twin die (extruder) head *(e)*
Zweifarbenspritzguß two-colour injection moulding
Zweifarbenspritzgußautomat automatic two-colour injection moulding machine
zweifelsfrei unequivocal
zweifunktionell difunctional
zweigängig two-start, double flighted *(screw)(e)*

Zweihalsrundkolben two-neck round-bottom flask
Zweiholmenausführung two-tie bar design *(im)*
Zweiholmenschließeinheit two-tie bar clamp unit *(im)*
zweiholmig with two tie bars
Zweikammertrichter two-compartment hopper
Zweikanalwerkzeug double-manifold die *(e)*
Zweikavitätenwerkzeug two-cavity/-impression mould *(im)*
Zweikomponentenanstrichfarbe two-pack paint
Zweikomponentenanstrichmittel two-pack paint
Zweikomponentenepoxidharzlack two-pack epoxy paint
Zweikomponentenflüssigsilikonkautschuk two-pack liquid silicone rubber
Zweikomponentengießharzmischung two-part casting compound
Zweikomponentenklebstoff two-pack adhesive
Zweikomponentenlack two-pack paint
Zweikomponentenlacksystem two-pack paint
Zweikomponentenmasse two-pack/-component compound
Zweikomponenten-Niederdruckspritzgießmaschine two-component low-pressure injection moulding machine
Zweikomponentenpolyurethanlack two-pack polyurethane paint
Zweikomponentenspritzgießen two-component injection moulding, sandwich moulding
Zweikomponentenspritzpistole two-component spraygun
Zweikomponentensystem two-pack/-component system
Zweikomponentenverfahren two-component process
zweikomponentig two-component/-pack
Zweikreisgerät dual-circuit unit
Zweikreishydrauliksystem dual-circuit hydraulic system
Zweikreissystem dual-circuit system
Zweiphasenmorphologie two-phase morphology
Zweiphasenstruktur two-phase structure
Zweiphasensystem two-phase system
zweiphasig two-phase
Zweiplattenwerkzeug two-plate mould, two-part mould, single-daylight mould *(im); (see explanatory note under* **Etage***)*
Zweipunktregler two-point controller
Zweischalenbauweise double-shell construction
zweischalig double-shell
Zweischichtbreitschlitzwerkzeug two-layer slit die
Zweischichtdüse two-layer coextrusion die *(e)*
Zweischichtfolienblasanlage two-layer film blowing line

Zweischichtfolienblaskopf two-layer blown film die
Zweischichthohlkörper two-layer blow moulding
Zweischichtkoextrusion two-layer coextrusion
Zweischichttafelherstellung two-layer sheet extrusion
Zweischichttiefziehfolienanlage two-layer thermoforming film extrusion line
Zweischichtverbund two-layer composite
Zweischnecke twin screw
Zweischneckenaustragszone twin screw metering section *(e)*
Zweischneckeneinzugszone twin screw feed section *(e)*
Zweischneckenextruder twin screw extruder
Zweischneckenmaschine 1. twin screw extruder. 2. twin screw compounder
Zweischneckensystem twin-screw system
Zweistationenblasmaschine two-station blow moulding machine
Zweistationenmaschine two-station machine
Zweistellenaufwicklung twin-station wind-up (unit)
Zweistufenauswerfer two-stage ejector *(im)*
Zweistufenblasverfahren two-stage blow moulding (process)
Zweistufendoppelschneckenextruder two-stage twin screw extruder
Zweistufeneinspritzaggregat two-stage injection unit
Zweistufenextruder two-stage extruder
Zweistufenextrusionsblasformen two-stage extrusion blow moulding
Zweistufenprozeß two-stage process
Zweistufenreckprozeß two-stage stretching (process) *(for film)*
Zweistufenspritzblasen two-stage injection blow moulding
Zweistufenspritzstreckblasverfahren two-stage injection stretch blow moulding (process)
Zweistufenverfahren two-stage process
zweistufig two-stage
zweiteilig two-part, split *(mould)*
zweitourig two-speed
Zweitträger secondary backing web *(of carpet)*
Zweitweichmacher secondary plasticiser
Zweiverteilerwerkzeug twin-manifold die *(e)*
Zweiwalze two-roll mill
Zweiwalzenglättwerk twin-roll polishing stack
Zweiwalzen-I-Kalander two-roll vertical/superimposed calender
Zweiwalzenkalander two-roll calender
Zweiwalzenlabormischer laboratory two-roll mill
Zweiwalzenmaschine two-roll mill
Zweiwalzenstuhl two-roll mill
Zweiwegestromregelung 1. two-way flow control. 2. two-way flow control unit
Zweiwegventil two-way valve
Zweiwellenextruder twin screw extruder
Zweiwellengranuliermaschine twin screw pelletiser
Zweiwellenkneter twin screw compounder

Zweiwellen-Knetscheibenschneckenpresse twin screw compounding extruder
Zweiwellenmaschine 1. twin screw extruder. 2. twin screw compounder
Zweiwellenschnecke twin screw
Zweiwellensystem twin screw system/assembly
zweiwellig twin screw
zweiwertig 1. bivalent, divalent. 2. dihydric *(if an alcohol)*
Zweizonenschnecke two-section screw
Zwickel 1. intermeshing zone *(between twin screws)*. 2. wedge. 3. gap, interstice
Zwickelbereich intermeshing zone *(between twin screws)*
Zwischenabkühlung intermediate cooling
Zwischenanstrich undercoat
Zwischenbereich intermediate range
Zwischenform intermediate form
Zwischenkornluft air between the particles *(e.g. of resin)*
Zwischenlagenhaftung inter-layer adhesion
zwischenlagern 1. to store temporarily. 2. to mature, to age: **Deshalb müssen diese Blöcke vor der Weiterverarbeitung mindestens 12 Stunden zwischengelagert werden.** These blocks must therefore be allowed to age/mature for at least 12 hours before they are processed further.
zwischenmolekular intermolecular
Zwischenphase intermediate phase
Zwischenprodukt intermediate (product)
Zwischenraum interstice
Zwischenreaktion intermediate reaction
Zwischenring adaptor/spacer ring
Zwischenschichthaftung inter-layer adhesion
Zwischenshorehärte intermediate Shore hardness
zwischenspeichern to store temporarily *(e.g. data)(me)*
Zwischenstadium intermediate stage
Zwischenstück adaptor
Zwischenstufe intermediate stage
Zwischentermin intermediate target date
Zwischenwert intermediate figure
zwitterionisch zwitterionic
Zwölffach-Spritzwerkzeug twelve-cavity/-impression injection mould
zyklisch cyclic
Zyklisierung cyclisation
Zyklon cyclone
Zyklonabscheider cyclone separator
Zykloneffekt cyclone effect
Zyklus cycle
Zyklusablauf 1. cycle sequence. 2. end of cycle: **nach Zyklusablauf** after completion of the cycle
Zyklusdauer cycle time
Zyklusende end of the cycle
Zyklusgedächtnis cycle memory
Zyklusgesamtzeit total cycle time
Zykluskontrolle (moulding) cycle control

Zyklusstörung interruption of the moulding cycle *(for translation example see under* **Zyklusunterbrechung***)*
Zyklusüberwachung 1. moulding cycle monitoring/control. 2. moulding cycle monitoring/control system
zyklusunabhängig cycle-independent, independent of the moulding cycle
Zyklusunterbrechung interruption of the moulding cycle: **Abschalten der Werkzeugkühlung bei Zyklusunterbrechung** switching off of mould cooling if the moulding cycle is interrupted
Zykluszähler cycle counter
Zykluszeit cycle time
Zykluszeitfaktor factor influencing cycle time: **Einer der Zykluszeitfaktoren ist die Werkzeugwanddicke** one of the factors which influences cycle time is the mould wall thickness
Zykluszeitreduzierung cycle time reduction: **zum Zweck der Zykluszeitreduzierung** to reduce cycle time
Zykluszeitverkürzung cycle time reduction *(for translation example see under* **Zykluszeitreduzierung***)*
Zylinder 1. barrel *(e)*. 2. cylinder *(im)*
Zylinderabmessungen 1. barrel dimensions *(e)*. 2. cylinder dimensions *(im)*
Zylinderauskleidung 1. barrel liner *(e)*. 2. cylinder liner *(im)*
Zylinderaußenfläche 1. outer barrel surface *(e)*. 2. outer cylinder surface *(im)*
Zylinderbeheizung 1. barrel heaters *(e)*. 2. cylinder heaters *(im)*
Zylinderbohrung 1. barrel bore *(e)*. 2. cylinder bore *(im)*. 3. hole *(in a barrel or cylinder, to accomodate a bush, gauge etc.)*: It is normally advisable to elaborate on this in a translation, e.g.: **Es genügt Sensoren in den Zylinderbohrungen anzubringen um den Druck zu messen** it is sufficient to fit pressure sensors into the holes drilled into the barrel wall for this purpose
Zylinderbuchse barrel bushing *(e)*
Zylindereinzug feed section/zone *(e)*
Zylindereinzugsteil feed section/zone *(e)*
Zylindereinzugszone feed section/zone *(e)*
Zylinderelement barrel section *(e)*
zylinderentgast with a vented barrel *(e)*
Zylinderentgasung 1. barrel venting. 2. barrel venting system: **Der Extruder arbeitet mit Zylinderentgasung** the extruder has a vented barrel; **In diesem Zusammenhang sei noch ein Nachteil der Zylinderentgasung beim Vorliegen oxidationsempfindlicher Materialien erwähnt**; There are two ways of translating this, using venting *or* deaeration. The difference is explained under **Entgasen** (q.v.), e.g.: In this connection we should mention a disadvantage of using a vented barrel when processing oxidation-sensitive materials OR: In this connection we should mention a disadvantage of deaerating oxidation-sensitive materials in the barrel

Zylinderentgasungsextruder vented barrel extruder
Zylinderentgasungszone vent zone, devolatilising section *(e)*
zylinderförmig cylindrical
Zylindergarnitur 1. cylinder (assembly) *(im)*. 2. barrel (assembly) *(e)*
Zylindergehäuse (extruder) barrel
Zylindergranulat cylindrical pellets
Zylinderheizelement 1. barrel heater *(e)*. 2. cylinder heater *(im)*
Zylinderheizkreis 1. barrel heating circuit *(e)*. 2. cylinder heating circuit *(im)*
Zylinderheizleistung 1. barrel heating capacity *(e)*. 2. cylinder heating capacity *(im)*
Zylinderheizung 1. barrel heaters *(e)*. 2. cylinder heaters *(im)*
Zylinderheizzonen barrel heating zones *(e)*
Zylinderinnenfläche 1. barrel liner *(e)*. 2. cylinder liner *(im)*
Zylinderinnenwand 1. barrel liner *(e)*. 2. cylinder liner *(im)*
Zylinderkopf 1. barrel head *(e)*. 2. cylinder head *(im)*
Zylinderkopfhaube cylinder head cover
Zylinderkühlaggregat barrel cooling unit *(e)*
Zylinderkühlung 1. barrel cooling. 2. barrel cooling system: **Zum Verarbeiten von PE geeignete Extruder können daher auf eine Zylinderkühlung verzichten** barrels of extruders designed for processing polyethylene need not, therefore, be cooled
Zylinderkühlzonen barrel cooling sections *(e)*
Zylinderlänge 1. barrel length *(e)*. 2. cylinder length *(im)*
Zylinderreinigungsmittel cylinder purging compound *(im)*
Zylinderrohr (extruder) barrel
Zylinderrollenlager roller bearing
Zylinderschneckensystem barrel-screw combination *(e)*
Zylinderschubspeicher reciprocating barrel accumulator *(bm)*
Zylinderschuß barrel section *(e)*
Zylindersegment barrel segment/section *(e)*
Zylinderspeicher melt accumulator *(bm)*
Zylinderstift guide pin *(m)*
Zylindertemperatur 1. barrel temperature *(e)*. 2. cylinder temperature *(im)*
Zylindertemperierung 1. barrel *(e)*/cylinder *(im)* temperature control. 2. barrel *(e)*/cylinder *(im)* temperature control system
Zylinderverschleiß 1. barrel wear *(e)*. 2. cylinder wear *(im)*
Zylinderwand 1. barrel wall *(e)*. 2. cylinder wall *(im)*
Zylinderwandtemperatur 1. barrel wall temperature *(e)*. 2. cylinder wall temperature *(im)*
Zylinderzone barrel section/zone *(e)*
zylindrisch cylindrical

APPENDIX

Technical translators should never forget that the rendering of words from one language into another is not just a simple matter of exchanging literal meanings, but the need to find appropriate idioms and contexts which will make sense to the speaker of the target language. Translation, in other words, is - or should be - an act of creative writing, a distillation of essentials. Good technical translations can only be achieved by departing from the original, by careful editing, by remoulding sentences. It is ideas which must be transferred into the other language, not words.

The translator's job is to spot and eliminate passages which repeat, in slightly different form, what has just been said. Padding must be removed, as well as irrelevancies which serve no useful purpose.

The passages that follow illustrate the principal shortcomings of present-day written technical German and ways of dealing with them in translation. There are, no doubt, other ways in which these passages can be translated, possibly even better ways. My main concern has been to show how obscurely worded, muddled texts can be converted into clear, intelligible English.

1. **Damit eine Übersicht über die Werte der Wasseraufnahmeprüfung der Produktionsüberwachung erreicht und Abweichungen leichter erkannt werden, wird das Anlegen von Kontrollkarten empfohlen.**
   A rather involved way of giving simple advice. The word **Produktüberwachung** *contributes nothing to the meaning and should be omitted. This is the sort of thing to aim for:* The results of water absorption tests should be recorded on a chart, to enable any deviations to be more easily recognised.

2. **Daher stehen auch verschiedene spezielle Silicon-Trenmittel zur Verfügung, um dem Verarbeiter eine für den jeweiligen Anwendungsfall optimale Lösung der Trennprobleme zu ermöglichen.**
   Here, the stilted final part of the sentence needs careful editing if the result is not to sound equally stilted. This is why there are various special silicone release agents on the market, to enable the user to select the one most suited for his particular purpose.

3. **Für unsere Mitarbeiter in diesem Bereich haben sich die Arbeitsplätze wesentlich verbessert.**
   If one replaces **Arbeitsplätze** *by* **Arbeitsbedingungen** *the meaning of the sentence immediately becomes clear.* Working conditions in this plant have improved considerably.

4. **In der Verseifungsbeständigkeit liegen terpolymere XYZ- Dispersionen im oberen Bereich der Wertskala.**
   A rather pompous way of putting it. Simplify, and you get this: Terpolymer XYZ dispersions have very good resistance to saponification.

5. **Diese Gleichgewichtsfeuchte kann durch die Anwesenheit von löslichen Salzen verändert werden, daß derartige Salzverbindungen in der Lage sind, die Gleichgewichtsfeuchte des Baustoffs zu erhöhen.**

The two parts of this long and involved sentence mean more or less the same thing. Reduce by about half, and you get this: The presence of soluble salts in the masonry will increase its equilibrium moisture content.

6. **Eine besondere Bedeutung im Sinne der chemischen Korrosion kommt dem Baustoff Stahlbeton zu.**
   One can just about guess what the writer is trying to say - and then produce something like this: Reinforced concrete is particularly vulnerable to chemical corrosion.

7. **Dadurch erfolgt eine Griffbeinflussung nach der harten Seite.**
   An unnecesssarily complicated sentence - which can be reduced to this: This imparts a harder feel.

8. **Da bei den Kunststoffen und insbesondere bei den thermoplastischen Kunststoffen die Gebrauchseigenschaften bereits unter den klimatischen Bedingungen im normalen praktischen Einsatz durch Einwirkung von Wärme und Kälte erheblich verändert werden können, ist das Verhalten der Eigenschaften bei Temperatureinwirkung in Form von Kennfunktionen zu ermitteln.**
   A rambling sentence full of padding. The only way of dealing with it is to rewrite it completely. This is the result: Since the properties of plastics - especially thermoplastics - can be considerably affected by heat and cold, the effect of temperature on properties must be ascertained.

9. **Die Profile können in Profilarten eingeteilt werden. Für die Praxis erscheint im Profilsektor eine Klassifizierung, wie dargestellt, am zweckmässigsten.**
   This is what the author meant to say - but didn't. Profiles can be divided into the following groups.

10. **Eine Ergänzung zur Kalanderfertigung ist die Extrusion von Folien und Platten. Bei dem dazu benötigten Breitschlitzwerkzeug erfolgt eine bevorzugte Verformung in die Breite.**
    Another case for drastic editing to remove the irrelevant words, especially from the second sentence. The result should look something like this: Sheets and films can be produced not only by calendering but also by extrusion, using slit dies.

11. *This was the opening sentence of a chapter headed* **Extrusion von Rohren: Seine Form erhält das plastische Material im Schlauchspritzkopf der an den Extruder angeschlossen ist.**
    This is how the author could have expressed his thoughts: Pipes are made by extruding the polymer melt through a pipe die coupled to the extruder.

12. **Die Umformung von Granulaten thermoplastischer Formmassen zu Spritzgußteilen erfolgt heute fast ausschließlich auf Spritzgießmaschinen mit Schubschnecken. Diese sorgen sowohl für die homogene Aufbereitung der Schmelze als auch für deren Transport in den Werkzeughohlraum des Spritzgießwerkzeuges.**
    Here the translator must do some ruthless editing, cutting away the

Appendix

jungle of unnecesssary words to get at the simple fact the author has tried to convey. The result will read as follows: Nowadays, thermoplastics are injection moulded almost exclusively on reciprocating-screw injection moulding machines, which convert the compound into an homogeneous melt and force it into the mould cavity.

13. **Zur Erzielung von optimalen Eigenschaften im Fertigteil sollten Massetemperaturen von 240 bis 260°C gewählt werden.**
    The tortuous introduction must be edited and condensed to give this: Melt temperatures of between 240 and 260°C are ideal for achieving perfect quality mouldings.

14. **Aufgrund der kunststoffspezifischen Freiheit der Formgestaltung im Spritzgießverfahren...**
    A rather involved way of expressing a simple fact. Some rearranging will produce something far more readily understood: Since plastics injection moulding allows great design freedom...

15. **Während man also mit einer hydrophobierenden Behandlung im wesentlichen die in Wasser gelösten Schadstoffe am Eindringen in den Beton hindern kann, das heißt im wesentlichen die Aufnahme von Säuren und Salzen reduzieren kann, können fach- und sachgerecht ausgewählte Anstrichsysteme auch die gasförmige Schadstoffaufnahme oder die Schadgasdiffusion entscheidend beeinflussen.**
    An ill-constructed, undisciplined sentence. The words immediately following **das heißt** repeat what has just been said. The text up to the second comma must therefore be condensed and made into a separate sentence. The padding is then removed from the final part to give the following: Application of water repellents to concrete prevents the penetration of salts, acids and other harmful substances dissolved in water into the material. The application of suitable paints, on the other hand, also helps to reduce the absorption and diffusion of harmful gases.

16. **Zur Gewährleistung einer einwandfreien Verarbeitung ist es zweckmäßig, die Topfzeit nicht voll auszunützen, sondern die Masse in möglichst niederviskosem (dünnflüssigem) Zustand zu verarbeiten.**
    Wrong words used in the wrong places have almost totally obscured what the writer intended to say. Extensive editing is needed to restore meaning - and produce this: The material should be used well before the pot life has expired, i.e. whilst it is still in a free-flowing state, which makes application that much easier.

17. **Beim Tiefdruck handelt es sich um ein Druckverfahren, mit dem qualitativ hochwertige Drucke erzeugt werden.**
    Some very necessary tightening up and you get this: Rotogravure printing produces high quality results.

18. **Diese Emulsionen werden in der dem Fachmann geläufigen Weise, also durch Verwendung von Pigmenten und den sonst üblichen Hilfsmitteln, zu lagerstabilen Farbsystemen rezeptiert.**
    Translation of everything that has been written would result in a very

Appendix

strange sounding sentence. The offending words should therefore be removed. These emulsions are converted into stable paints by incorporating pigments and the usual additives.

19. **Druck and Werkzeugtemperatur im Formnest...**
    Woolly writing even in so short a part sentence. *This is what the author intended to say:* Cavity pressure and temperature...

20. **Siliconölemulsion ABC ist eine hochkonzentrierte Siliconölemulsion eines mittelviskosen, nichtreaktiven Siliconöls der Reihe XYZ.**
    Repetition at its very worst. Silicone fluid emulsion ABC is a highly concentrated product based on a medium-viscosity, inert silicone fluid in the XYZ range.

21. **...bietet sich an wenn auf einen möglichst geringen Emulgatorgehalt Wert gelegt wird, weil der Tensidgehalt aus anwendungstechnischen Gründen möglichst gering gehalten werden soll.** *The second part of this sentence says exactly the same as the first, with slight variations. The passage must therefore be condensed to give something like this:* ...is recommended if, for technical reasons, the emulsifier content has to be kept to a minimum.

22. **Der Ablauf einer Steinkonservierungsmaßnahme muß spezifisch für jedes Objekt geplant werden. Von größter Bedeutung ist dabei die sogenannte Voruntersuchung am Objekt. Bei dieser Voruntersuchung werden die für das jeweilige Bauwerk oder Bauteil typischen Materialschäden festgestellt. Außerdem wird die Verwitterungstiefe ermittelt bezw. abgeschätzt.**
    A long-winded passage, full of unnecessary words. Best to start again from scratch, to produce something like this: The method used to preserve stone depends on the nature of the object to be treated. Preliminary tests should be carried out to ascertain the nature and extent of the damage that has occurred and to find out how deep down the effects of weathering are evident.

23. **Es kann festgestellt werden, daß bei silicium-organischen Verbindungen die UV-Beständigkeit in hohem Maße gegeben ist.**
    A pompous sounding sentence which can be expressed very simply like this: Organo-silicon compounds have excellent UV resistance.

24. **Von den anwendungstechnischen Eigenschaften konfektionierter Polysiloxane ist wohl die exzellente Hydrophobie auf Baustoffen, die auf die Einführung der organischen Radikale zurückzuführen ist, besonders vorzuheben.**
    This is a good example of authors using **konfektioniert** *to mean whatever they want it to mean, usually in place of quite an ordinary - but less grand-sounding - word. In this example, the context will have told the translator that polysiloxane paints are being discussed. This, then, would be an acceptable translation:* The most notable property of polysiloxane paints is the excellent water repellency they impart to building materials, thanks to the presence of organic radicals.

Appendix

25. **Wenn man nun den bekannten Schadensbildern an Fassaden wie Abplatzungen, Kreidung, Ausblühungen, Farbtonveränderungen auf den Grund geht, wird sehr schnell offensichtlich, daß in diesen komplexen Zerstörungsmechanismen das Medium Wasser eine tragende Rolle spielt.**
    Another sentence full of padding which must be removed in the interests of clarity, to produce something like this: The root cause of all damage to exterior walls such as spalling, chalking, efflorescence and changes in colour, is water.

26. **Die Ergebnisse mechanischer Prüfungen kennzeichnen die Standard-Acrylformmassen als Werkstoffe hoher Steifigkeit und Festigkeit über einen weiten Temperaturbereich.**
    The introductory four words contribute nothing to the meaning the author is trying to convey and should therefore be ignored. Slight editing of what remains will produce this: Standard acrylic moulding compounds retain their great rigidity and strength over a wide range of temperatures.

27. **Das Eigenschaftsprofil der ABC-Produkte basiert auf einer einzigartigen Kombination von Eigenschaften.**
    Unnecessary repetition which must be removed, to give this: ABC products exhibit a unique combination of properties.

28. **Recht universell einsetzbar sind die auf anorganischer Basis konzepierten Produkte.**
    Remove the padding and condense. The translation will then read as follows: Inorganic products are suitable for general use.

29. **Die werkseitig auf ihre optimalen Eigenschaften eingestellten pulverförmigen Klebemörtel bieten aus heutiger Sicht ein Optimum in Hinblick auf Verarbeitungs- und Gebrauchseigenschaften, sowie in punkto Gebrauchssicherheit.**
    Cut out the irrelevancies and you get a nice, crisp sentence: Correctly formulated powdered adhesive mortars are easy to use and give dependable results.

30. **Die technische Entwicklung hat immer bessere Dämmstoffe gefordert, sowohl hinsichtlich der Herstellung seitens der Produzenten als auch bezüglich der immer höher geschraubten Anforderungen der Anwender bezw. Verbraucher.**
    This long sentence contains several words which have no business to be there. Remove them, to produce this: Technological progress needs insulating materials whose qualities are constantly upgraded to meet end users' increasingly exacting demands.

31. **Die vielfältig auf Silicon-Basis formulierten Bautenschutzmittel dienen mehr oder weniger einem einheitlichen Zweck. Sie rufen eine Imprägnierung der z.T. ziemlich porösen mineralischen Baustoffe hervor und diese Wirkung wird vor allem bei den derartige Merkmale in besonderem Maße zeigenden historischen Bauten geschätzt.**
    This long-winded sentence contains much that is irrelevant and misleading. **Vielfältig, mehr oder weniger einem einheitlichen**

Appendix

Zweck *and* **z.T. ziemlich** *can be omitted without altering the meaning the author is trying to convey. The whole thing is far too stilted, so that the translator must use his own words to express its true meaning, to give something like this:* Silicone-based masonry water repellents used to impregnate porous mineral building materials are specially suited for conserving historic buildings.

32. **Zur Bestimmung des Ausmaßes, in dem ein Lack sich thixotrop verhält, eignet sich unter den verschiedenen rheologischen Meßgeräten nur das Rotations-Viskosimeter.**
    Some editing - and you get this: The rotational viscometer is the only instrument suitable for determining the thixotropic behaviour of a paint.

33. **Die Kontrolle der hergestellten Produkte findet zu 100% statt. Das bedeutet, es wird Charge für Charge an einer repräsentativen Probe geprüft.**
    The author evidently thought the first sentence was not clear, which is why he felt compelled to explain its meaning in the second. He might as well have omitted the first sentence altogether - which is what the translator must do. The English version should therefore read like this: Representative samples of every batch produced are tested.

34. **Jeder Ansatz enthält eine Chargen- oder Ansatznummer. Die Kontrolle der produzierten Produkte wird durch Überprüfung jeder einzelnen Charge (100% Prüfung) in den Betriebslabors durch speziell ausgebildetes Personal durchgeführt.**
    In the first sentence, **Chargen** *and* **Ansatz** *mean the same thing - as do* **Kontrolle** *and* **Überprüfung.** *The bracketed words are an unclear explanation of what has already been stated quite clearly. And is it really necessary, in this particular context, to stress that tests are carried out by specially trained staff? Here's the suggested English version:* Each batch produced is given a number and checked in the works laboratory.

35. **Für die Prüfung der Eigenschaften müssen Meßgrößen wie die Viskosität - die eine sehr komplexe Aussage über die Eigenschaften der Dispersion macht - als Meßparameter für die Herstellung benutzt werden.**
    The real meaning of this sentence is almost imposssible to discover. Fortunately, the preceding sentence provided a hidden clue. The first five words can be omitted for they are irrelevant. The translator must then discover what the author is trying to say - and use his own words to say it. Viscosity provides important information about dispersion properties and is therefore used to monitor the polymerisation process.

36. **Die Prüfungen der Gebrauchseigenschaften eines Produktes eines Grundstoffherstellers - die ja darauf abzielen, die Qualität des Produktes zu charakterisieren - können allenfalls Modellversuche für die Qualität des Produktes in der Gebrauchsrezeptur, wie z.B. bei einem formulierten Klebstoff oder Mörtel, sein.**
    This long-winded sentence needs ruthless editing to reduce it by half -

Appendix

like this: The results of tests on raw materials can only serve as a guide to their performance when incorporated in end products such as adhesives and mortars.

37. **An der Oberfläche einer in den festen Zustand übergehenden flüssigen Lackschicht tritt ja gewöhnlich eine Verarmung an Lösemittel ein.**
Here, needlessly complicated language has been used to express a simple, well-known fact. The word **gewöhnlich,** *incidentally, is completely out of place. One short sentence says it all:* As a paint film dries, solvent is lost through evaporation.

38. **Der Reaktionsharzbeton ist unter der Bezeichnung "Polymerbeton" bekannt geworden. Häufige Anwendungen finden sich in...**
This is a trap for the unwary. The first sentence must be omitted for two reasons. The obvious one is that the English equivalents of both words are the same. The other is that the author has altered the facts to suit his purpose, which seems to have been to coin a grander sounding word than the old established one and then pretend that this (i.e. **Reaktionsharzbeton**) *is the standard expression. The translation should thus read:* Polymer concrete is widely used in... (OR: for...)

39. **Blends müssen als eigenständige Werkstoffe verstanden werden; die Methoden zu deren Herstellung ergänzen die Synthesewege der Polymerchemiker um eine Variante, die mehr ist als nur ein einfacher, mit mehr oder weniger know-how behafteter Compoundierschritt. Sie bedeutet auch die intelligente Verknüpfung von Polymerisations-, Polykondensations- und Polyadditionschemie untereinander.**
Overloaded with irrelevant words and phrases, this needs ruthless editing to produce something like this: Polymer blends should be regarded as materials in their own right. They are produced by polymerisation, polycondensation and polyaddition rather than by mere mixing and compounding.

40. **Silikatfarben schneiden in Hinblick auf ihre Verwendungsmöglichkeiten auf den verschiedenen mineralischen Untergründen sehr günstig ab.**
A rather stilted sentence which needs loosening up to produce this: Silicate paints are ideal for application to a variety of mineral substrates.

41. **Rückblickend auf die Entwicklung der verschiedenen Fassadenbeschichtungssysteme kann zusammenfassend festgehalten werden, daß durch die Einführung der Dispersionsfarben im Hinblick auf wasserabweisende Wirksamkeit gegenüber den bis dahin verwendeten Mineralfarben ein großer Fortschritt gelungen war.**
The essential meaning is contained in the second part of this long sentence, after the word **daß.** *The rest contains only one relevant word -* **Fassadenbeschichtungssysteme** *- the others can be discarded. The result will look something like this:* The introduction of emulsion

Appendix

paints for exterior use marked a major advance in paint technology, for these proved to be far more water repellent than the silicate paints used up to then.

42. **Silicone sind polymer. Sie besitzen damit ein für den Bau organischer Makromoleküle typisches Merkmal.**
A confusing second sentence, due to the unnecessary introduction of the words **den Bau.** *Take these away, and the sentence becomes clear immediately.* Silicones are polymers and thus have the characteristics of organic macromolecules.

43. **Silikonharzfarben stellen aber in jedem Fall ein völlig selbständiges System dar, das vor dem Hintergrund der wichtigen bauphysikalischen Parameter Gasdurchlässigkeit und Wasserdichtigkeit als qualitativ hochwertige Alternative zu den konventionellen Anstrichsystemen, sei es nun auf Basis Kunstharz oder Wasserglas, anzusehen ist.**
This muddled, longwinded sentence is best expressed in the translator's own words, like this: Silicone resin paints are, however, paints in their own right. In view of their excellent permeability to gases, and their impermeability to water, they can be regarded as the ideal alternative to conventional synthetic resin and waterglass based paints.

44. **Die Freisetzung umweltrelevanter Stoffe, z.B. Phenol, Formaldehyd, Lösemittel und dergl. kann bei verschiedenen Arbeitsgängen wie Mischen, Imprägnieren, Trocknen u.ä. Vorgängen erfolgen.**
A good example of an obscure buzz word being used in place of one that would have been immediately understandable: **toxisch.** *This is the suggested translation:* Toxic substances such as phenol, formaldehyde, solvents etc. are produced in certain operations such as mixing, impregnating, drying etc.

45. **Eine Verbesserung der natürlichen Eigenschaften des Aluminiums, insbesondere die Gewährleistung eines hohen Korrosionsschutzes und die Optimierung dekorativer Eigenschaften, wird hier bekanntlich durch Aufbringen einer organischen Beschichtung erreicht.**
Verbosity at its worst. All the author intended to say was this: Painting aluminium increases its corrosion resistance and enhances its appearance.

46. **Diese Materialeigenschaft begünstigt in Verbindung mit rationellen Verfahren zur Herstellung der Formteile ganz wesentlich die Wirtschaftlichkeit der Produktionsabläufe - sowohl für Verpackungen in selbsttragender Konstruktion als auch für deren Einlagen.**
Another example of needless and pointless repetition. Tighten it up and you get this: Moulded packaging units - including self-supporting packs as well as inserts - can therefore be economically produced.

47. **Man geht aber einen kleinen Kompromiß in Richtung der Transparenz ein.**

Appendix

All this sentence means is: There will, however, be a slight loss of transparency.

48. **Doch ist nicht zu verkennen, man hat sowohl den Grad physiologischer Bedenklichkeit als teilweise auch das Umweltverunreinigungspotential bei Aromaten grundsätzlich meist höher anzusetzen als bei den für Lösezwecke verwendeten aliphatischen Spezies.**

    The only way of dealing with this tortuous sentence is to express it entirely in one's own words, since any attempt to even approximately follow the author's words would end in disaster. The result could be something like this: It must, however, be remembered that aromatic solvents are usually more toxic than aliphatic ones, and that they are a greater danger to the environment.

49. **Bei laufender Produktion werden darüber hinaus die Maschinenüberwachung bezüglich der Einstellgrößen Durchsatz, Abzugsgeschwindigkeit und mittlere Foliendicke vereinfacht und eine werkstoffsparende Fahrweise möglich.**

    Cut out all the unnecessary words and you get this: Furthermore, it is easier to monitor throughput, haul-off speed and mean film thickness during production as well as saving material.

50. **Zusatzstoffe, die den Harzen vor dem Aushärten zugefügt werden, ermöglichen eine optimale Anpassung des Werkstoffes an die Verarbeitungstechnik und an die Anforderungen des fertigen Formteils.**

    A needlessly involved way of expressing a simple and straightforward idea. The translator must use his own words and produce something like this: Additives can be incorporated in the resins prior to curing, to make them easier to process and to achieve specific end product properties.

51. *The following comes from an article on the use of GRP in chemical plant:* **Im Schadensfall sind Notlaufeigenschaften gefordert, die durch Verzögerung beispielsweise eines Stoffaustritts ein Eingreifen ermöglichen, bevor es zu einer Belastung der Umwelt kommt.**

    The meaning of this sentence is almost totally obscured by one word having been wrongly used. Ruthless editing is the only way of knocking it into shape to produce this: In case of damage, it must be possible to carry out emergency repairs before chemicals start to leak and pollute the surroundings.